INGENIOUS MECHANISMS
FOR DESIGNERS AND INVENTORS

VOLUME II

INGENIOUS MECHANISMS

FOR DESIGNERS AND INVENTORS

VOLUME II

Mechanisms and Mechanical Movements Selected from Automatic Machines and Various Other Forms of Mechanical Apparatus as Outstanding Examples of Ingenious Design Embodying Ideas or Principles Applicable in Designing Machines or Devices Requiring Automatic Features or Mechanical Control

Edited by

FRANKLIN D. JONES

INDUSTRIAL PRESS INC.

200 MADISON AVENUE, NEW YORK 10016

Industrial Press Inc.
200 Madison Avenue
New York, New York 10016-4078

INGENIOUS MECHANISMS
FOR DESIGNERS AND INVENTORS—VOLUME II

34 36 38 40 37 35 33

CONTENTS

SECOND VOLUME OF INGENIOUS MECHANISMS

THIS additional volume of INGENIOUS MECHANISMS FOR DESIGNERS AND INVENTORS has been published as a companion book to Volume I in order to present illustrated descriptions of a large variety of mechanisms and mechanical movements not at hand when Volume I was produced. The continual demand for Volume I from engineers and machine designers, both here and abroad, is not only a tribute to the value of this treatise, but an indication of the need for information on outstanding mechanical movements. The publication of this second volume makes it possible to present many additional mechanisms of great practical value to designers of automatic machines or other devices, as well as to students of the general subject of mechanism.

Many of the main sections or chapters in Volume II have titles similar to those found in the first volume to assist the user of both books in locating all the information on a given subject, but the mechanisms illustrated and described in the two volumes are entirely different in design. While the second volume is a continuation of the first one, each book is an independent treatise; taken together, they constitute an unusually complete work of reference on the very important subject of mechanism. The numerous mechanical movements featured in Volume II, like those in Volume I, have been applied successfully to automatic machines and many other forms of mechanical apparatus. While it is not feasible in any work of this kind to include mechanisms that are directly applicable to every type of machine and operating condition, it is believed that the numerous designs found in Volumes I and II embody mechanical principles which may be utilized in the solution of practically any mechanism designing problem likely to be encountered.

CHAPTER I

CAM APPLICATIONS AND SPECIAL CAM DESIGNS

Cams in their various forms doubtless are more useful to designers of automatic and semi-automatic machines than any other type of mechanical device. Mechanical movements which would be difficult or impracticable to obtain by other means may, in numerous instances, be derived readily either from a single cam or from two or more cams used in combination. The cams which follow illustrate a variety of interesting applications taken from different classes of mechanical equipment. Other applications of cams and cam-operated mechanisms will be found in Chapter I, Volume I, of INGENIOUS MECHANISMS FOR DESIGNERS AND INVENTORS.

Indexing Cam for Varying Stroke of Follower.—For a given number of strokes of a slide, almost any variation in the length of each successive stroke may be produced by means of an indexing cam mechanism like that shown in Fig. 1. The construction of this cam is economical and the design is unusually simple, when the movements involved are considered. The cam member consists of a core *A* in which are secured eight cam inserts *B*. Each insert is tapered at a different angle and has a throw corresponding with the required movement of the follower roll *C*.

The core is keyed to a shaft turning in bearings on the slide *D*, which is reciprocated through a rack and gear by a member of the machine in which the cam is used. To one

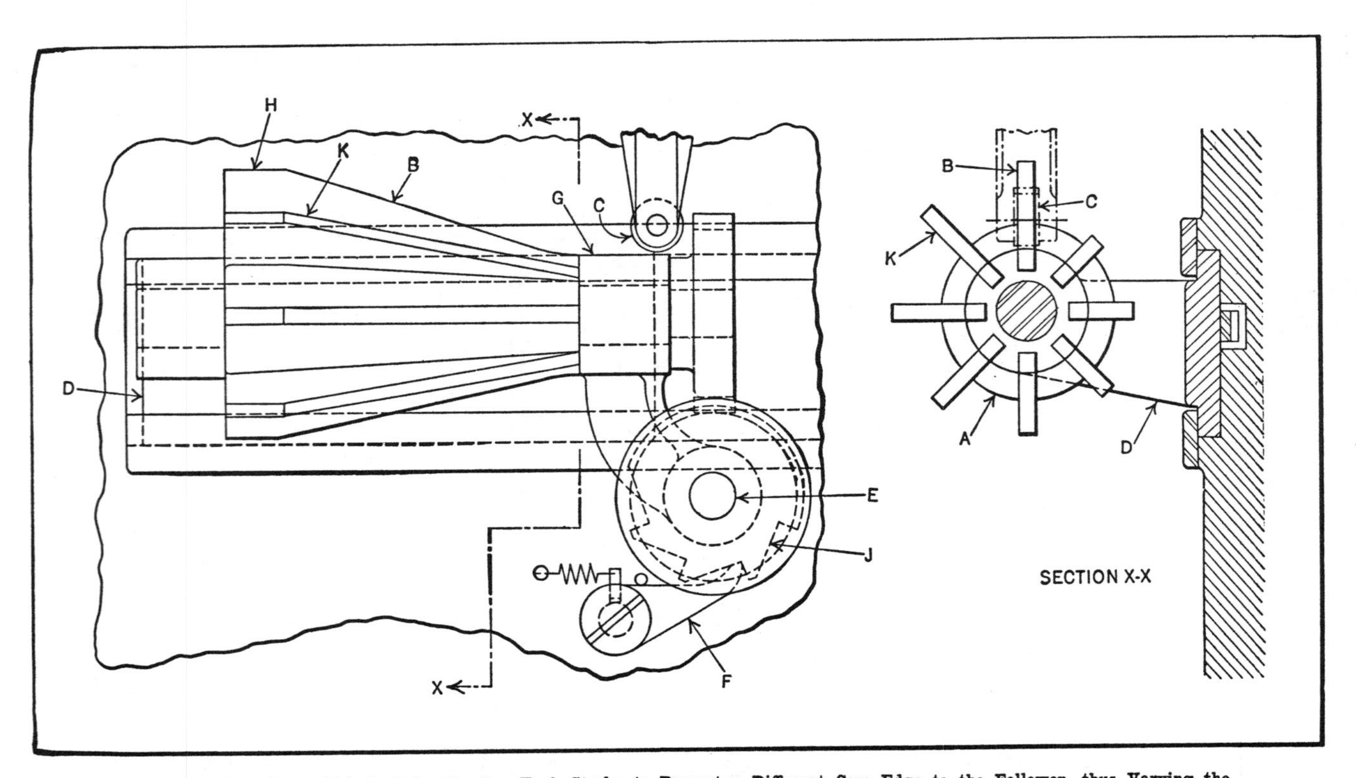

Fig. 1. Sliding Cam which is Indexed after Each Stroke to Present a Different Cam Edge to the Follower, thus Varying the Follower Stroke

end of the core shaft is keyed a helical gear, meshing with a similar gear on the vertical shaft *E*. This shaft, running in two bearings cast integral with the slide, carries a ratchet wheel *J*, which is operated by the pawl *F*, pivoted to the machine base. There are as many teeth in the ratchet wheel as there are inserts.

The various movements are obtained in the following manner: From the position indicated, the slide moves toward the right, causing the follower roll to ride along the bearing *G* and on the insert *B* to point *H*. The slide now returns, during which time the follower roll is also returned by means of a coil spring (not shown). Toward the end of the return stroke, as the roll dwells on bearing *G*, a tooth in ratchet J engages the pawl *F*. Upon the continued movement of the slide, the pawl forces the ratchet wheel around one tooth, causing the core to rotate until insert *K* is in line with the follower roll. Thus, on the return stroke, the roll rides on insert *K*, which imparts a shorter movement to the follower than the preceding one. In this way, each succeeding movement of the follower is varied until the core has been indexed one revolution. At this time, the roll will again be in line with the insert *B* and the cycle of movements will be repeated.

Although not shown, a friction brake should be applied to either the core or the ratchet-wheel shaft to prevent overrun of the cam due to the momentum imparted by the pawl. Other combinations than that shown here may be obtained by using different inserts to vary the throw or a different number of inserts to increase or decrease the number of follower movements per cycle. In the latter case, the number of teeth in the ratchet wheel must be changed to correspond with the number of inserts.

Cam-Plate with Four Adjustable Lobes.—In developing an automatic machine, it was necessary to provide means for transmitting an oscillating motion to an arm or lever from a rotating shaft. The arm was attached to a slide

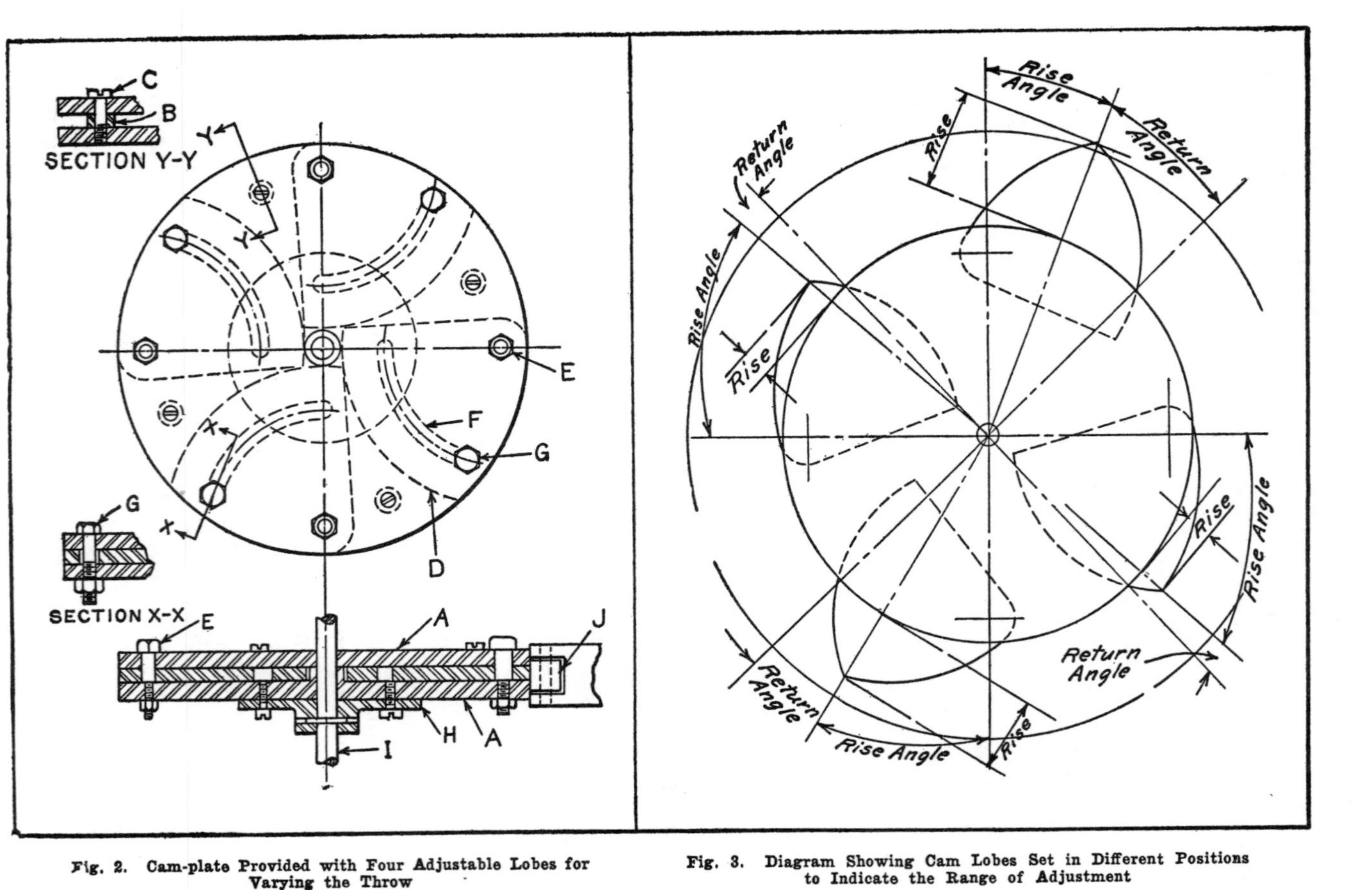

Fig. 2. Cam-plate Provided with Four Adjustable Lobes for Varying the Throw

Fig. 3. Diagram Showing Cam Lobes Set in Different Positions to Indicate the Range of Adjustment

which was returned to the zero position by means of a coil spring after having reached its maximum position.

The variation in the sizes of the product made necessary an occasional change in the length of travel or movement of the slide. To obtain the desired adjustability with the least number of actuating parts, the adjustable cam-plate shown in Fig. 2 was designed. The two side plates *A* are spaced a given distance apart by spacers *B* and are clamped together by screws *C*. Four cam-plates *D*, spaced 90 degrees apart, are held on pivot studs *E*. Circular slots *F* are milled in these plates through which clamp bolts *G* are inserted.

These plates are a sliding fit between the plates *A*, so that when bolts *G* are tightened, cam-plates *D* are held securely in place. To the outer side of one end plate *A* is fastened the flanged bearing *H*, which is held by a cross-pin to the driving shaft *I*. The lever arm roller *J* was made wide enough to allow it to ride on both the central cam lobe and on the periphery of side plates *A*. The various distances to which the cam lobes can be projected and the angles of rise and fall are shown in Fig. 3. This particular cam has four adjustable lobes, but a larger or smaller number of lobes can be used.

Cam for Guiding a Follower Along a Square Path.—A mechanism for guiding a pointer along a square path is illustrated in Fig. 4. The rotating shaft *A*, through the action of cam *B* and dovetailed slides *C* and *D*, causes the pointer *E* to follow the square contour indicated by the dot-and-dash outline. The horizontal slide *D* is mounted in the stationary member *F*, which also serves as a bearing for the shaft *A;* and the vertical slide *C* is mounted in the slide *D*. Elongated holes are provided in both slides so that the slides will clear the shaft in operation.

Alternate vertical and horizontal movements of slide *C* are obtained through the action of the positive cam *B*, the lay-out of which is shown in Fig. 5. Here it will be seen that sections *PQ* and *RS* are concentric with the shaft *A*, and sections *RP* and *SQ* are drawn by scribing arcs having centers at *S* and *R*, respectively, to form the rises of the cam.

Referring again to Fig. 4, pointer *E*, attached to slide *C*,

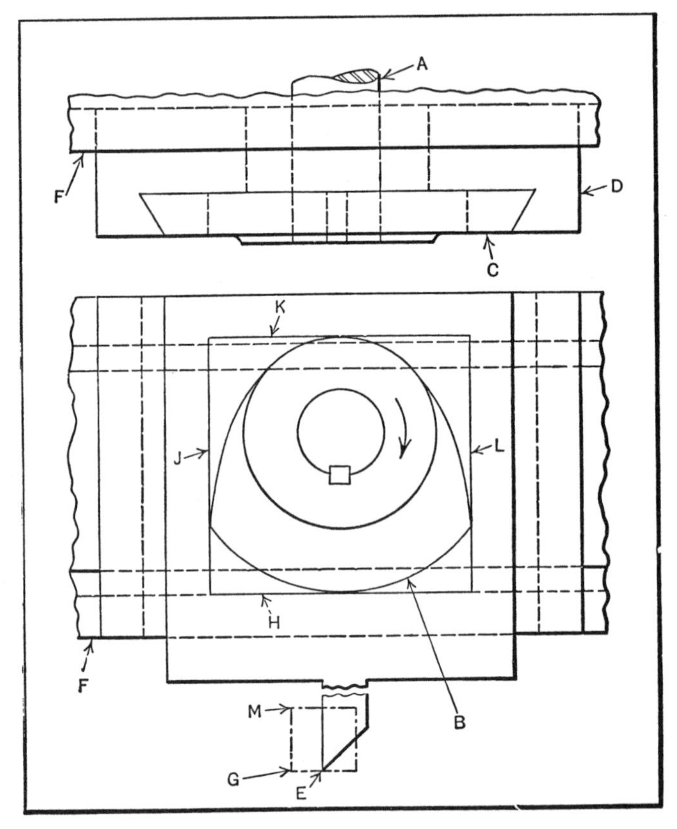

Fig. 4. Positive Cam Movement for Guiding Pointer E along Square Path Indicated by Dot-and-dash Lines

is in its lowest position and has just completed one half of the horizontal movement imparted by the cam *B*. As the cam continues to rotate in the direction of the arrow, slide *C* will move toward the left, carrying pointer *E* to position *G*. During the entire horizontal movement the concentric por-

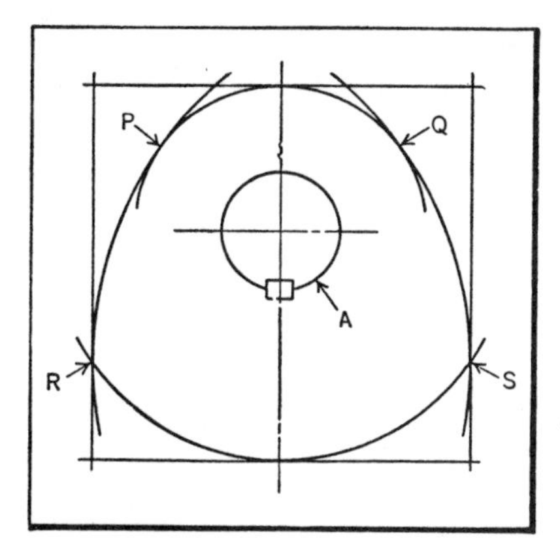

Fig. 5. Lay-out of Cam Used on Mechanism Shown in Fig. 4

tions of the cam are in contact with the surfaces *K* and *H*, preventing slide *C* from moving vertically. However, when the pointer *E* reaches position *G* these concentric surfaces contact with surfaces *J* and *L* and prevent horizontal movement of slide *C*. In the meantime, the cam rises come in contact with the surfaces *K* and *H* and raise the slide *C* until the pointer reaches position *M*. These alternate vertical and horizontal movements guide the pointer *E* in its required path.

Combination Cam and Parallel Motion for Guiding Spindle in Square Path.—The mechanism shown in Fig. 6 was designed to guide the center of the spindle *A* along a square pathway indicated by lines *M* and *N*. It is used in conjunction with a woodworking machine for gouging out an endless grooved recess of square contour into which a decorative insert is fitted. The movement involves two separate motions—a cam motion and a parallel motion. The former is the actuating member which imparts the movement to the follower, while the latter serves merely to maintain the direction of motion of the follower. By the use of interchangeable cams and follower plates, as explained later, the follower can be made to follow paths of various dimensions.

The mechanism is mounted on the machine frame *B*, and consists chiefly of cam *C*, follower *D* which carries the cutter-spindle *A*, and the parallel motion links *E*, *F*, and *G*. The follower is connected to the stationary bracket *H* through these links. With this arrangement, the angular position of the follower will remain unchanged, regardless of its location relative to the cam. The cam is of the triangular type, imparting movement in the four directions required.

From the position shown the cam is rotated in a clock-

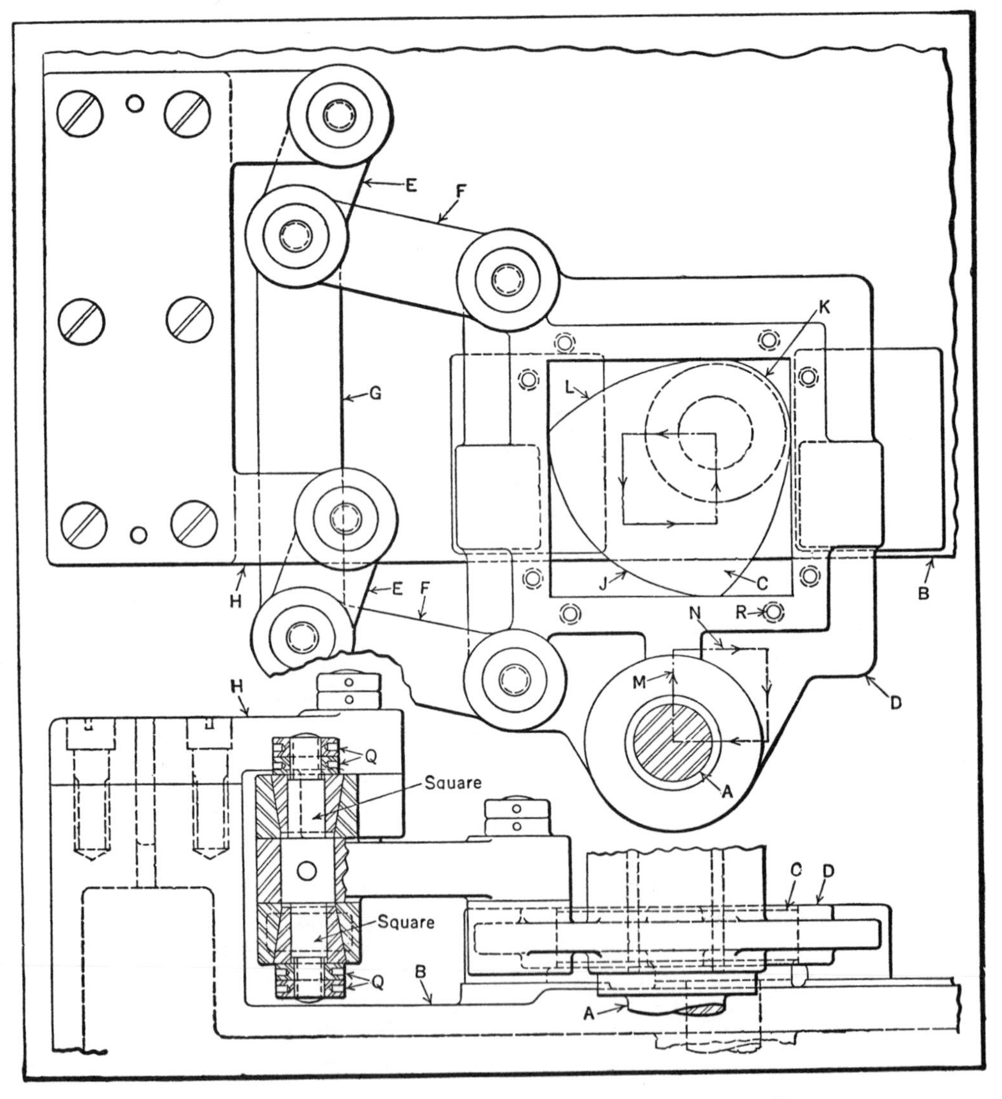

Fig. 6. Mechanism for Guiding Tool along a Path of Square Outline

wise direction. Since edges *J* and *K* are concentric with the camshaft, continued rotation of the cam will not impart a horizontal movement to the follower; but the upper edge *L* of the cam will raise the follower so that the center of the spindle will move along a path coinciding with the line *M*. When the curved edge *J* becomes tangent to the top cam surface of the follower, the center of the spindle will coincide with the top end of line *M* and the vertical movement of the follower will cease. Edge *L* will now force the follower toward the right so that the spindle center will follow a path coinciding with line *N*.

The action of the cam and follower is the same for each side of the square over which the spindle center passes. This cam is of the positive type, since the distance between the two points at which the edges of the cam intersect a line passing through the center of the camshaft is the same, regardless of the angularity of the line.

If the spindle is required to follow a square path of smaller dimensions, the cam surfaces of the follower are lined by means of four flanged plates, and a cam giving the required throw is substituted for the one shown. The camplate can be quickly attached by means of screws which pass through the plate flanges into tapped holes *R*.

Owing to the movement of the spindle in a plane normal to its axis, the upper end of the spindle is provided with two universal joints and a sliding sleeve. This provides a flexible connection with the upper driving shaft of the machine. Since, however, this arrangement is a common one, it is not shown. The follower is supported in its overhanging position from the bracket *H* by two pads integral with the follower, which rest on finished pads cast on the machine frame. In order to compensate for wear in the lever and link connections, the connection pins were designed as shown in the cross-section. With this arrangement, any wear can be taken up by tightening the check-nuts *Q*.

Changing Cam Speed by Transforming Uniform Circular Motion into Periodic Variable Motion.—Fig. 7 shows a mechanism by means of which a uniform circular motion is transformed into a periodic variable circular motion. The driven shaft *A* and the driving shaft *B* rotate in bearings located on the same axis. Disks *C* and *D* are securely

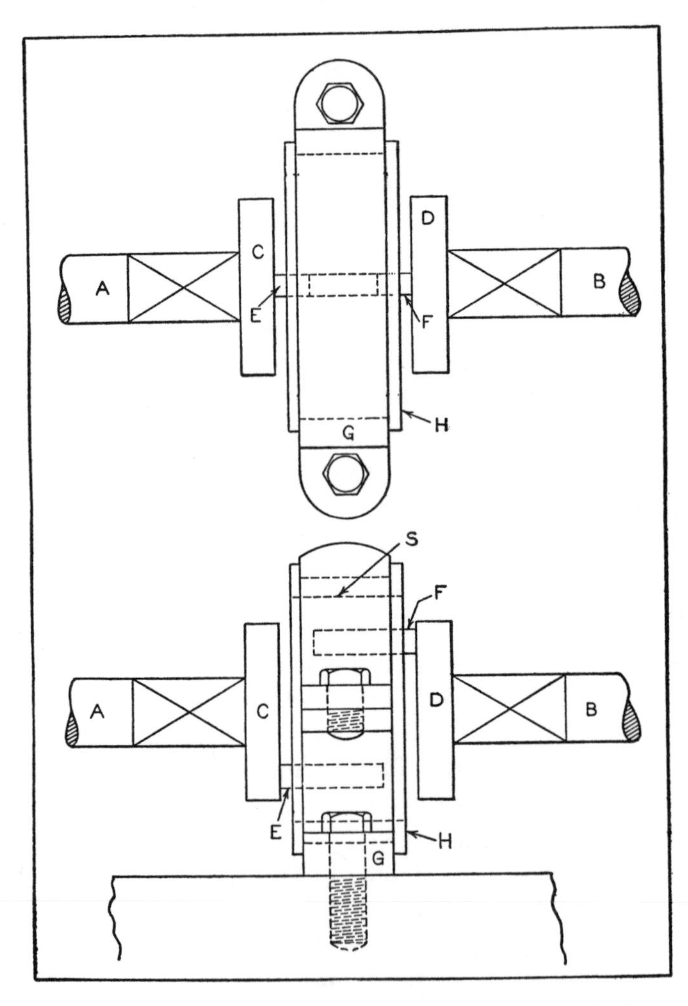

Fig. 7. Mechanism for Accelerating Speed of Driven Shaft During a Portion of Each Revolution

mounted on shafts *A* and *B*. The bearing *G* carries the flanged disk *H*, which is slotted at *S* to receive the pins *E* and *F* in the disks *C* and *D*. The feet of bearing *G* are slotted so that the position of the bearing may be changed in relation to shafts *A* and *B*. The motion of shaft *B* is transmitted to shaft *A* by the pins *E* and *F*, which act in the slot *S* in disk *H*.

In the position shown in the illustration, bearing *G* is so located that the axis of disk *H* coincides with that of shafts *A* and *B*, in which case the motion of shaft *B* is transmitted uniformly to shaft *A*. If the bearing *G* is moved to one side, the axis of disk *H* is thrown out of alignment with those of shafts *A* and *B*. As disk *H* then revolves in the same plane but on a different axis from shafts *A* and *B*, the pins *E* and *F* will alternately approach and recede from the center of disk *H*, thus imparting a periodically fast and slow motion to disk *C*. The amount of variation in the motion given shaft *A* is controlled by the amount of movement given bearing *G*.

An interesting application of this mechanism was made on a machine on which shaft *A* carried a cam. The speed with which the operating point of the cam passed under the follower was varied by shifting the bearing *G*.

Right- and Left-Hand Threaded Cam for Converting Rotary into Oscillating Motion.— A simple mechanism for converting rotary into oscillating motion consists of a cylinder having a right- and left-hand thread and a half-nut made as shown in Fig. 8. This mechanism was incorporated in a specially constructed printing press for the purpose of imparting a reciprocating motion to the rollers which assists in distributing the ink. A similar arrangement can be used in numerous other applications, when the speed of rotation is not too high and the load is not too great.

In the application referred to, three rollers were used for distributing the ink. The two outside rollers were operated by a double rocker arm actuated by the crank-arm *A*, which is fitted with a half-nut *B*. The right- and left-hand threaded cylinder *C* at one end of a rotating shaft serves to oscillate or move the end of arm *A* forward and back. The center ink-distributing roller is moved by a single rocker arm driven by another threaded cylinder similar to the one shown at *C*. The rocker arms are pivoted

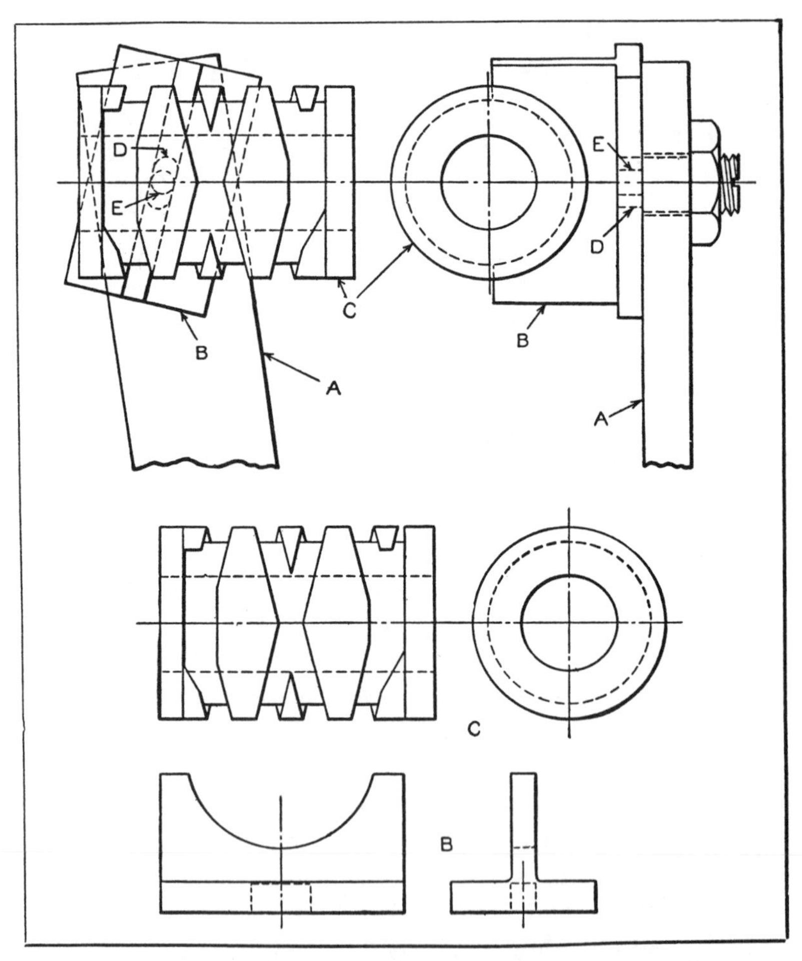

Fig. 8. Cylinder Cam C with Right- and Left-hand Threads Designed to Reverse Direction of Travel of Half-nut B at Each End of Stroke, thus Imparting an Oscillating Motion to Lever A

and carry ball-bearing pins that work against the flanges of spools on the ink-distributing rollers. Thus, as the rocker arms move back and forth, they transmit the required motions to the ink-distributing rollers.

The half-nut *B* is made from a T-shape, the thickness of the stem being equal to the width of the thread groove. The stem is formed to a concave shape to fit the contour of the root diameter of the thread, while its over-all length is made somewhat greater than the outside diameter of the thread. Its minimum length must be such as to more than span the gap made by the crossing of the right- and left-hand threads. At the center of the T-shaped bar is an elongated hole *D*, which slides over a pin *E* attached to the crank-arm. Thus, pin *E* causes the crank to rock back and forth with the longitudinal travel of the nut. An elongated hole is necessary for pin *E*, since the arm swings in an arc while the nut travels in a straight line.

When the half-nut approaches the end of its travel in one direction, its axis is on an angle with the center line of the shaft. This angle is equal to the pitch angle of the screw. In order to reverse the travel, the axis of the half-nut must pivot about pin *E* until it is in the proper angular position for the reverse traverse motion imparted by the thread of the opposite hand lead.

The last thread on the cylinder *C* is cut back a sufficient distance to allow the half-nut to pivot, and the "following" edge where the thread runs out at the end is filed back sufficiently to allow the nut to clear this surface and the end flange. The nut is also beveled at the edge where it enters the thread. The threaded cylinder *C* and the half-nut *B* are shown separately in the views to the left.

This mechanism operates smoothly, having a short dwell at each end of the stroke while the nut reverses and picks up the opposite thread. In the printing press application, the two outside rollers are operated by a double rocker arm which causes them to move an equal amount in opposite

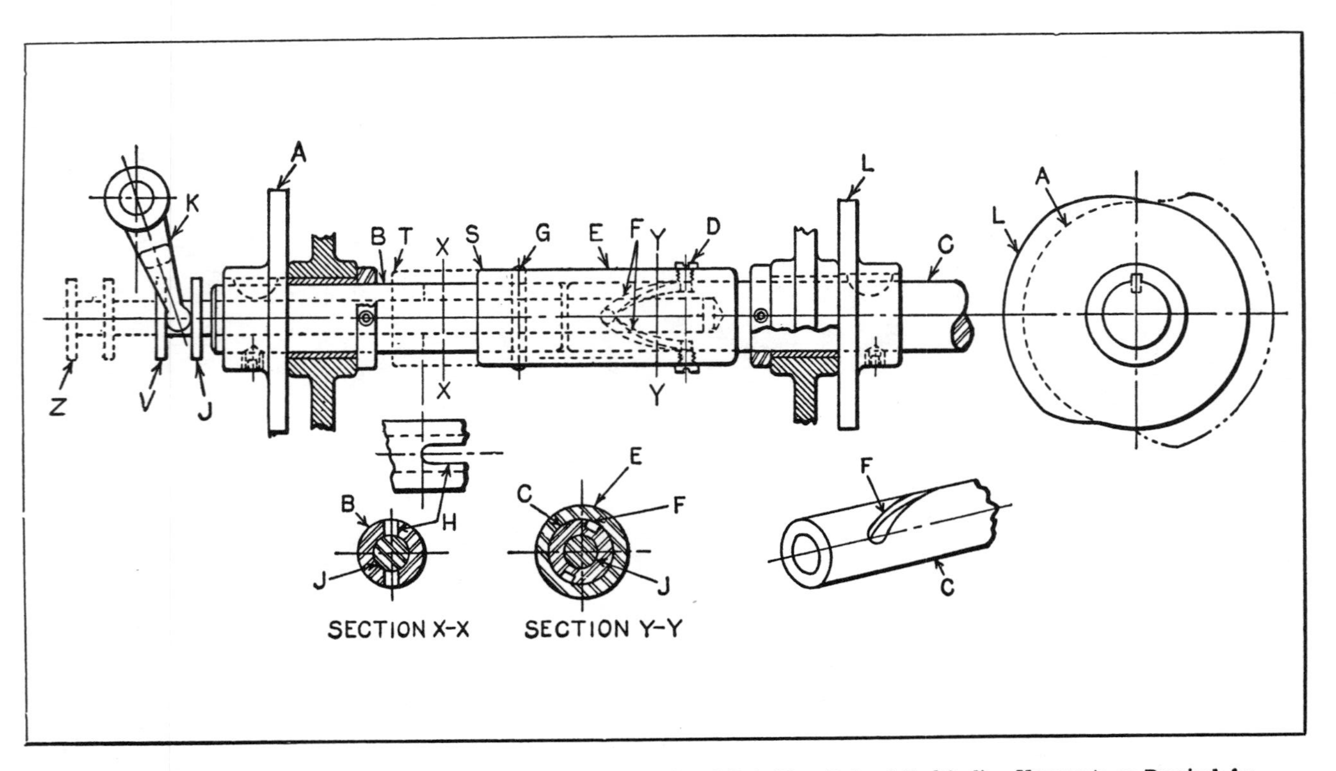

Fig. 9. Mechanism for Changing Angular Positions of Feed-cams A and L to Vary Rate of Tool-feeding Movements as Required for Different Machining Operations

directions. It is desirable to introduce as much variety as possible into the motion of the three rollers in order to smooth out the ink more effectively. For this reason, the leverage for the crank-arm of the center roller is made somewhat different from that for the outside rollers. In this case, the length of the thread on the cam for actuating the crank-arm of the center roller is longer than that of the cam for the outside rollers. With this arrangement, the center roller continuously varies its position in relation to the outer rollers.

Mechanism for Making Quick Change in Angular Positions of Feed-Cams.—The staggered production requirements and the available tool equipment for rough-turning several parts of similar design necessitated changing the angular relationship of the two principal feed-cams on one shaft for each tool set-up. The arrangement provided to permit the positions of the cams to be changed quickly to suit the machining requirements of the different parts is shown in Fig. 9.

Cam *A* on shaft *B* is driven by shaft *C* through keys *D* in sleeve *E*. Keys *D* operate in spiral slots *F*. Pin *G* fits in sleeve *E* and extends through slots *H* of shaft *B*. Pin *G* also extends through the shifter shaft *J* in shaft *B*. Axial movement of shifter shaft *J*, by means of lever *K*, from position *V* to *Z* causes cam *A* to advance clockwise in relation to cam *L*. This movement of sleeve *E* from position *S* to *T* causes keys *D* to operate in slots *F* of shaft *C*. Shaft *J* is piloted in shaft *C* to maintain the alignment of shafts *B* and *C*. The follower on cam *A* is released when changing cam positions. Shifter lever *K* is provided with conventional means (not shown) for locking in any of the required positions.

Cam and Eccentric Combinations.—The vertical ram *B*, Fig. 10, is given the required motion by combining a cam and a crank or eccentric motion. In this mechanism, gear *C* supports cam *D* and gear *E* supports the eccentric *A*. The sliding member *F* which carries the fulcrum of the lever *G* is driven by means of a connecting-rod *H*. The movements of both the cam and the crank serve to give the lever *G* a long stroke. A small slideway at the outer end of lever *G* provides the connection with the ram *B*.

The sliding member *F* may be omitted in some cases and the end of lever *G* connected directly to an eccentric on a

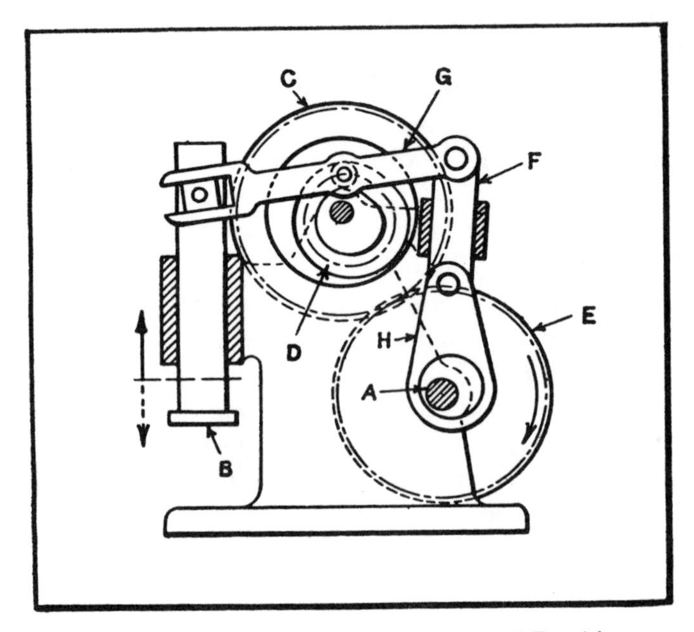

Fig. 10. Combination of Gears, Crank, Cam and Eccentric for Operating Ram

crankpin, as shown in the upper view, Fig. 11. For special cases, the cam profile may also take the form of an eccentric disk. In that event, the form of the mechanism can be changed, so that the cam is replaced by an eccentric which actuates a rod. The crank-arm can also be replaced by an eccentric *A*, as shown in the lower view. This gives a simple mechanism having two eccentric rods coupled together by means of a link *G*.

Compound Cam Drive to Reduce Cam Rise.—The possibilities of mechanisms consisting of links, gears, and cams for imparting oscillating movements to bellcranks are indicated by the illustrations Fig. 12. These mechanisms are incorporated in a shoe-sewing machine. The first design, shown in the view to the left, consists of a simple cam drive. The motion of the swinging member *A* is trans-

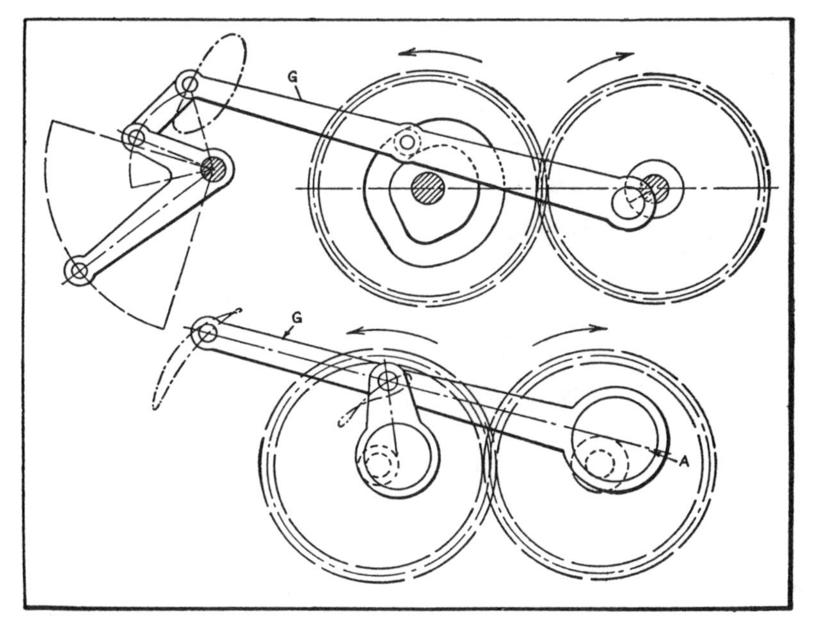

Fig. 11. Levers Operated by Cam and Eccentric Combinations

mitted by means of a rod *B* to the bellcrank *C*, which is required to oscillate or swing back and forth through an angle of approximately 80 degrees about the center of shaft *D*.

To obtain this movement, the driving cam *E* must have a rise of about 1 21/32 inches. The quick rise in the cam groove required to meet this condition, however, prevented the mechanism from being satisfactory for this particular

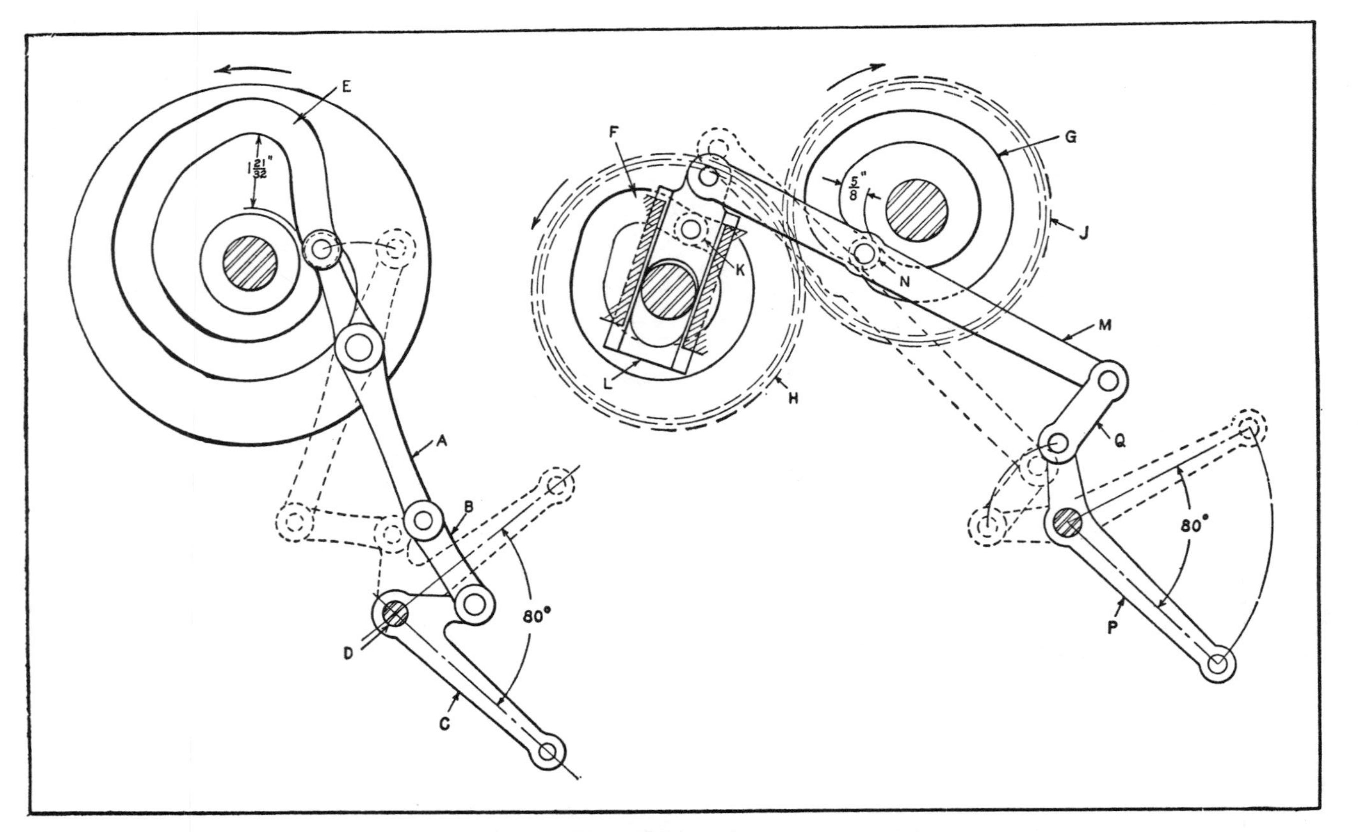

Fig. 12. Combination Link and Cam Mechanisms Used to Impart Oscillating Movement to Bellcrank

application. An enlargement of the base diameter of the cam disk or a change in the distance between the fulcrum of the lever and the cam axis to overcome this difficulty was impractical.

For this reason, the improved mechanism shown in the view to the right was designed. In the latter mechanism, an additional motion is imparted to the swinging lever by the use of another cam. The two cams *F* and *G* are driven in opposite directions by spur gears *H* and *J*, which are of equal size. These gears revolve on the same axis as the cams that they drive. Cam *F* actuates roller *K* attached to the slide *L*, which is mounted between guides on the cam-plate. The lever *M* is attached to the slide *L* and receives an additional motion from the cam disk *G* through the roller *N*.

The motion of bellcrank lever *P* is derived from lever *M* through rod *Q*. As shown in the illustration, the stroke of lever *M* is greatly increased by the two cams *F* and *G* and the slide *L*. The rise of each cam in the new mechanism is reduced to about 5/8 inch. The mechanism described works satisfactorily at speeds ranging from 400 to 500 revolutions of the driving gears per minute.

Long-Stroke Cam of Small Diameter with Rapid Return.— Cylindrical cams of the usual type for imparting a relatively long and powerful stroke to the follower must necessarily be large. Frequently this is undesirable, especially in a machine of light construction. In designing a certain machine for inserting the packing in stuffing-boxes, a rather long stroke of a slide was required to press the packing into place. The return or idle stroke was to be rapid. Because of the light construction of the machine and the elevated position of the cam, it was desirable to have the cam of small diameter, as well as light.

To meet these requirements, the cam mechanism illustrated in Fig. 13 was developed. It consists of the cylin-

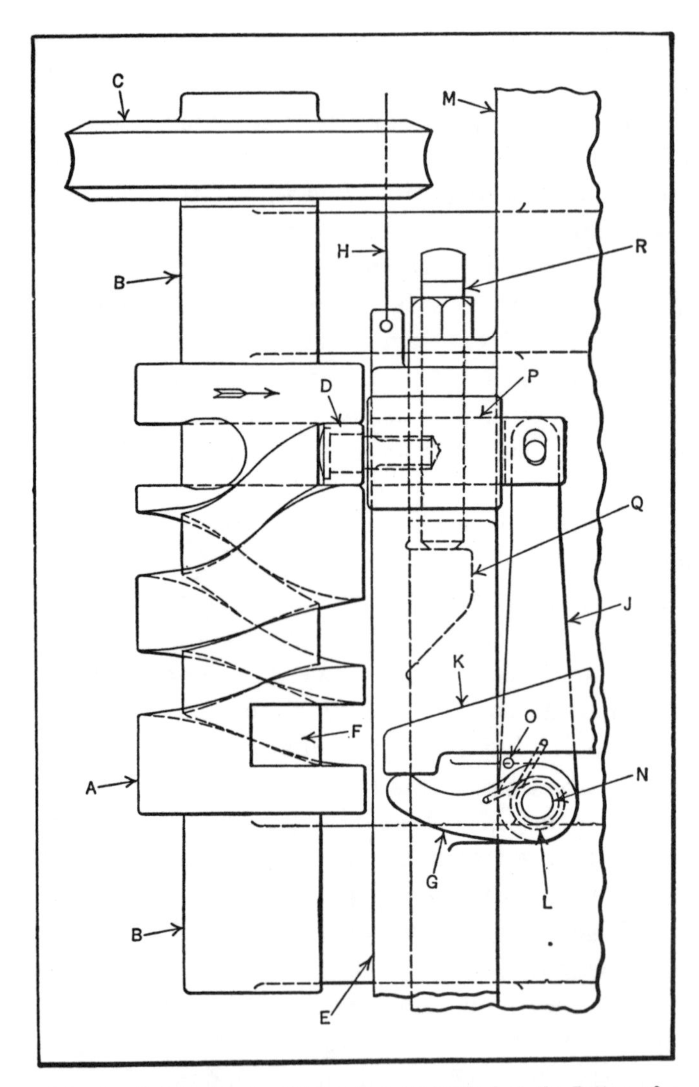

Fig. 13. Rotary Cam in which Roll is Disengaged at the Bottom of the Stroke to Allow the Slide to be Returned Rapidly by a Counterweight. Re-engagement Takes Place at the Top of the Stroke

drical cam *A*, secured to a shaft running in the stationary bearings *B*. The cam is rotated by means of a worm (not shown) and worm-gear *C*, and is engaged by the follower roll *D* on the slide *E*. This slide is mounted on the machine column *M* and carries the levers *G* and *J* on the shaft *N*.

Although both levers are free to rotate on the shaft, the lower lever *G* is normally held against the pin *O* in lever *J* by means of the coil spring *L*. The upper end of lever *J* is connected to a plunger *P*. This plunger slides in a bearing cast integral with the slide and carries the follower roll. A counterweight on cable *H* returns the slide to its upper position, the upward movement being limited by stop *Q* on the slide and adjusting screw *R* on the machine column. Stop *K* is fastened to the machine column and serves to operate levers *G* and *J* for engaging the roll with the cam groove at the top of the stroke as explained later.

In the position shown, the slide is about to begin its downward stroke. As the cam is rotated in the direction of the arrow, the slide moves downward until the roll has reached the part of the groove at *F*. Here the bottom of the groove is sloped gradually toward the outside of the cam; thus when the cam continues to rotate, the roll is forced out of the groove and the slide is returned to the upper position by the counterweight. Just before the slide reaches the upper position, the lever *G* comes into contact with the stop *K* and swings lever *J* with plunger *P* toward the left carrying the end of the roll stud against the cam. At the top of the stroke the roll is forced into the groove through the action of coil spring *L*. As the cam continues to rotate, the slide is once more carried downward.

Axial Movement from Mating Cam Sections Rotating at Different Speeds.— Certain copper tubes used in connection with steam-heating apparatus are covered with strips of copper, the strip being wound around the tube and soldered. The strip and the solder must be removed from the ends of the tubes to provide a bare length of 1 inch

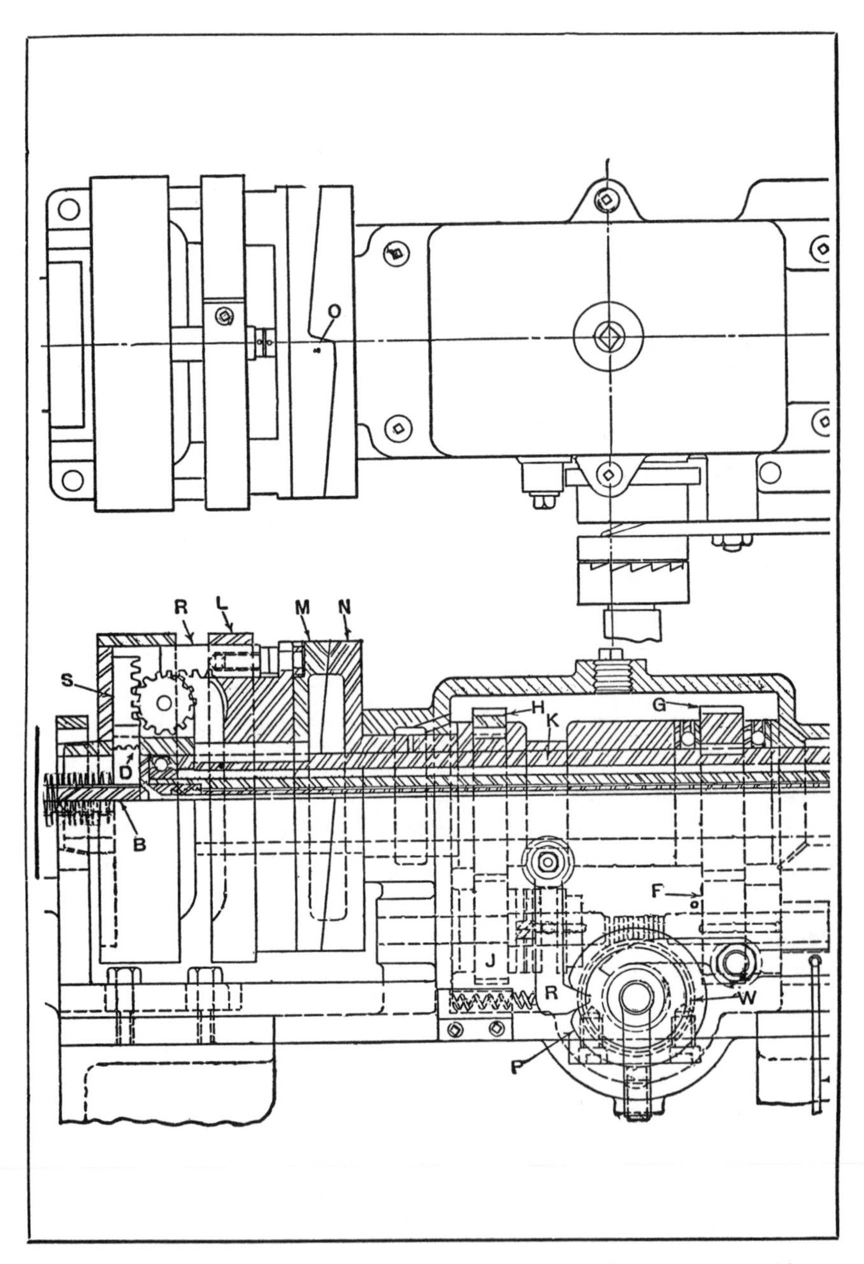

Fig. 14. Sectional and Plan View Showing Tool-feeding Mechanism Equipped with Mating Cam Sections which Rotate at Different Rates to Provide the Motion Required

for connection to a tank or header. This "stripping" of the tube ends is done by using a machine having three cutters, which are held radially and feed inward as the cutter-head rotates about the tube. The machine used for this work is shown by the sectional and plan views.

The end of a wound tube (represented by the zigzag lines) is pushed over a stationary pilot *B*, Fig. 14, which fits snugly inside the tube. An air-operated clamp is next tightened and the tube is ready for the stripping operation. The head of the machine, which contains the three cutters (one of which is shown at *D*), revolves continually at the rate of 600 revolutions per minute, and when a clutch is tripped by a foot-pedal, the three tools feed inward a distance of 3/4 inch at the rate of about 0.018 inch per revolution. The mechanism for obtaining and controlling this feeding movement is the interesting feature of the machine.

The drive from the motor to the cutter-head is through gears *F* and *G*. Gear *G* is attached to the main spindle *K*, which connects with the cutter-head. A head *L*, which is rotated by the cutter-head proper, is free to slide for a limited distance along spindle *K*. Attached to sliding head *L* there is a cam *M* which fits a mating cam *N*. Cam *N* is free to revolve on spindle *K*, and it has attached to it a gear *H* which meshes with the gear *J*.

Before the tool feeding movement begins, cam *M* drives the mating section *N* through the step or shoulder *O* (see plan view), and gears *H* and *J* revolve idly. When the tools are to be fed inward, cam *N* is rotated 40 1/2 revolutions to 40 revolutions of cam *M*. The result is that cam *N* exerts a wedging effect against *M*, causing the latter, with head *L*, to slide along the spindle. When this sliding movement occurs, racks *R*, attached to sliding head *L*, transmit this movement through pinions to racks *S*, attached to the cutter-holders. The method of obtaining this differential movement between cam sections *M* and *N* will now be described.

In order to start the tool-feeding movement, a clutch trip lever is raised by depressing a foot-pedal. This releases a clutch dog or plunger connecting plate P through a clutch with the shaft of worm-wheel W, which is rotated continually from the driving shaft. As soon as plate P begins to revolve, the dog or clutch lever R is forced out of the notch in plate P, thus connecting, through a clutch, the driving shaft with gear J; consequently, cam section N is now driven from shaft E through gears J and H, and since it rotates 40 1/2 revolutions to 40 revolutions of cam M, the wedging action and traversing movement previously referred to occurs. This difference in the speeds of cams M and N is due, of course, to the ratios of gears F and G as compared with gears J and H. Gear F has 25 teeth and G 40 teeth; hence, for each turn of gear G, F makes 40/25 turn. Therefore, 40 turns of G require $\frac{40}{25} \times 40 = 64$ turns of shaft E and gear F. For each turn of gear J, H makes 31/49 turn, as J has 31 teeth and H 49 teeth; hence, if J makes 64 turns then H will make $\frac{31}{49} \times 64 = 40\ 1/2$ turns.

While the driving shaft is turning sixty-four times in order to complete one cycle in the movement of the feeding mechanism, plate P is turned 64/65 revolution, as the worm-wheel W has sixty-five teeth. At the end of the cycle, clutch lever R is again opposite the notch in plate P and gear J is disconnected from the driving shaft, thus stopping the feeding movement automatically. Shoulder O on cam N is also around to the point where section M can slide back into engagement, which it is forced to do by means of springs concealed in the cutter-head. The difference in the speeds of the two cam sections is so slight that this re-engagement occurs easily and without objectionable shock.

Duplex Cam Action for Turning Cam Contour and Maintaining Proper Cutting Angle of Tool.—A special movement, embodied in a camshaft lathe, controls the turning tool by two sets of cams, so that the cutting angle in relation to the cam outline will always be the same. A partial cross-section of the lathe showing the carriage slide can be seen in Fig. 15. Both the cam at the top and the one

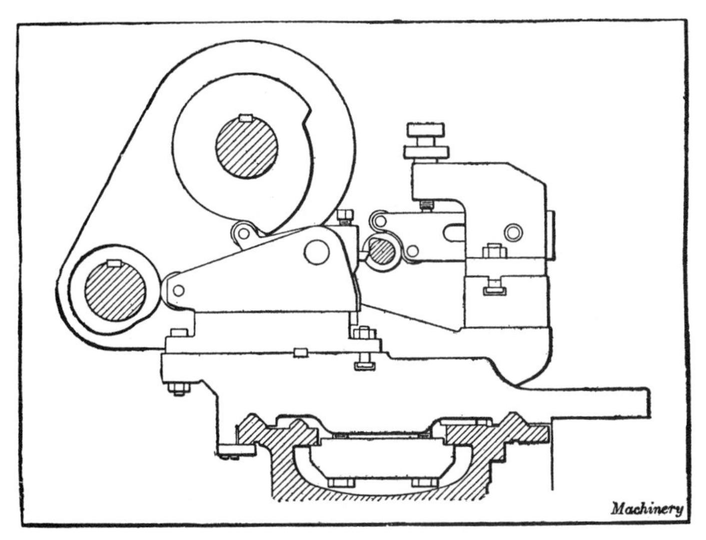

Fig. 15. Cross-section of Lathe, Showing Cams that Maintain Constant Cutting Angle of Tool

at the left revolve at the same number of revolutions per minute as the camshaft to be machined. The cam at the left is used as a master while the top cam controls the swinging motion of the tool about the horizontal axis in such a way that the cutting angle remains constant.

With the combined movements of both cams, the desired result is obtained. The master cam at the left is ground accurately in the lathe by using a grinding wheel in place of the cam roller and of the same size, while a suitable member engages a revolving cam of correct form on a shaft between the lathe centers.

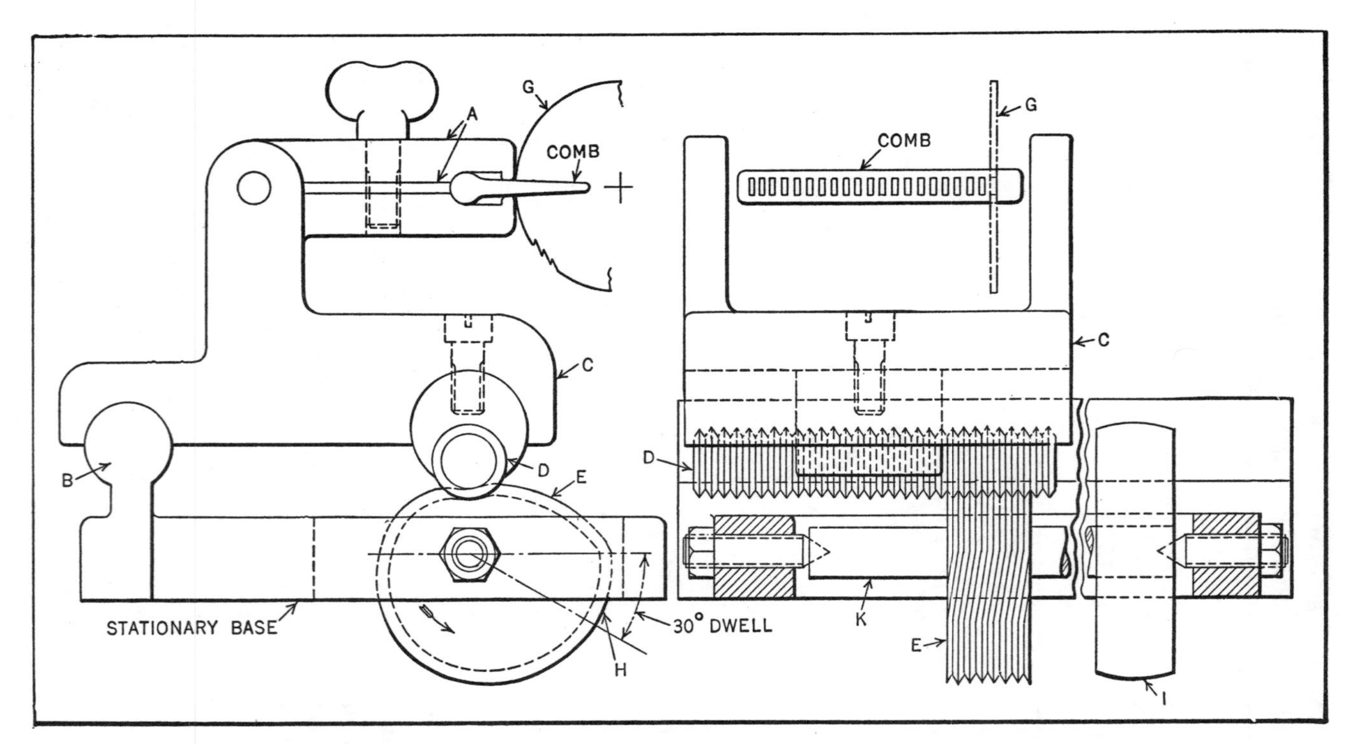

Fig. 16. Mechanism in which a Cam Imparts an Oscillating Movement Followed by a Horizontal Indexing Movement

Double-Acting Cam which Oscillates Follower and also Indexes it Horizontally.—In the mechanism shown in Fig. 16, a cam is used to impart an oscillating movement, as well as a horizontal indexing movement, to the table of a machine for sawing teeth in combs. A comb is clamped rigidly between the straps *A* on the table *C*. This table oscillates about the bearing *B* and receives its motion from the cam *E* acting against the follower *D*. The cam is driven by a belt passing over the pulley *I*. The circular saw *G* revolves in stationary bearings, the comb being fed to it by the oscillating action of the table.

The principal feature of this cam motion is the manner in which the horizontal movement is imparted to the comb for cutting the successive teeth. On the outside of the cam is cut a continuous V-groove which engages corresponding grooves in the follower *D*. The several turns of the groove on the cam follow a parallel plane perpendicular to the center line of the shaft *K* until they approach the dwelling portion *H*, where they are deflected to one side a distance equal to the pitch of the groove. This pitch is also equal to that of the slots being cut in the comb. The grooves in the follower, however, are not continuous but are a series of separate grooves.

In the position shown, the table is at its lowest point and the saw has just completed cutting a tooth in the comb. As the cam continues to revolve in the direction of the arrow an upward movement is imparted to the table. When the dwelling surface of the cam has come in contact with the follower, the comb is clear of the saw, and the horizontal or indexing movement of the table begins, continuing until the follower has passed over the angular portion of the cam groove. Further movement of the cam carries the table downward, causing the comb to be fed against the saw for cutting the next tooth. This completes the cycle of operations. The follower *D* does not revolve in actual operation, but can be adjusted to present new wearing surfaces. The design of this mechanism permits the use of interchangeable cams and followers for slotting combs having teeth of different pitches.

Straight-Line Movement Applied to a Cam Follower.—A practical application of a straight-line movement obtained by means of a link and a lever is shown in Fig. 17.

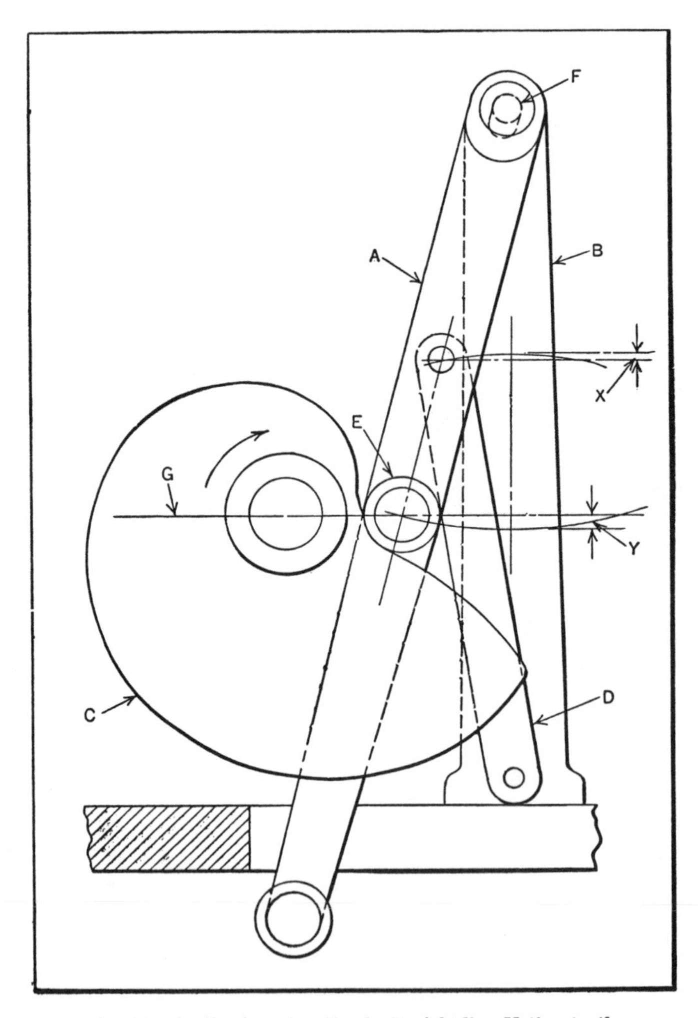

Fig. 17. Application of a Simple Straight-line Motion to the Follower Roll E

This movement is applied to the follower roll of a cam on an automatic machine intended for sawing slots in latch needles.

The roll *E* moves 3 inches forward and backward, and the return movement is effected during one-twelfth of a revolution of cam *C*. Originally, the roll lever *A* was pivoted to the stationary bracket *B*, and was not equipped with the auxiliary link *D*. Consequently, the center of the roll *E* followed a curved path, and on the return of the lever *A*, the roll had a tendency to leave the cam surface, especially when the machine was operated at high speed. This action caused the roll to strike the low point of the cam with an appreciable impact during each cycle.

To overcome this condition, the pivot hole in the upper end of the lever *A* was elongated and the link *D* added to force the roll *E* to travel in a straight instead of a curved path. The center distance between the pivot *F* and the roll *E* is equal to that of the holes in the ends of the link. The lower end of the link is pivoted to the bracket, while its upper end is pivoted to the lever. This pivot is located in such a position that the distance *X* equals one-half of the distance *Y*.

As the cam rotates from the position shown, the upper end of lever *A* is gradually lifted and lowered through the action of the link, so that the center of the roll *E* follows very closely the center line *G*. Thus, when the steep incline of the cam is reached, the roll is returned along the same straight line and remains in contact with the incline instead of leaving the cam surface, as when the roll followed a curved path.

Varying the Cam Dwell with Two Adjustable Follower Rolls.—An increase in the variety of products manufactured in one plant made it necessary to alter some of the wire-forming machines so that the dwelling periods of their slides could be varied. To do this, instead of employing one follower roll for each slide, two adjustable rolls were used, as shown in Fig. 18. The two dwelling periods of the slides can thus be varied to suit requirements. The cam, indicated at *A*, is secured to the driving shaft and engages both rolls *B* and *C*. The rolls are mounted on flanged bushings and secured to slide *D* by studs. They can be adjusted to any position along the curved T-slot *E*.

The amount of dwell and the timing of the rise and fall of the slide depend upon the distance between the two rolls

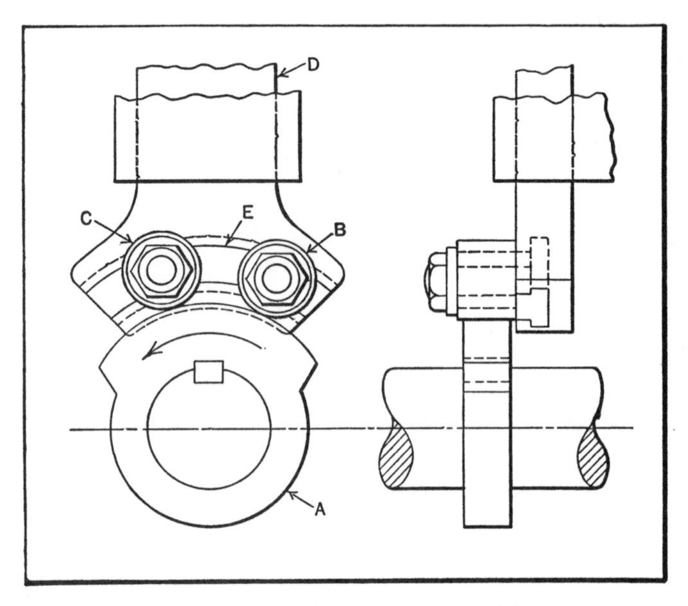

Fig. 18. The Dwelling Period and Timing of This Cam can be Varied by Simply Changing the Positions of the Two Follower Rolls

and their location along the T-slot. For instance, if the slide were required to dwell longer in its upper position, the distance between the rolls would be increased. On the other hand, if the dwelling time in the upper position was to be decreased, the rolls would be brought closer together. The time at which the rise and fall of the slide occurs may be varied by adjusting the rolls along the T-slot without changing their center distance.

Double-Faced Cam for Rapid Rise without Excessive Side Thrust.— The cam for operating the slide of a certain machine required a rapid rise without excessive side thrust. To meet this requirement, a double-faced cam was used (see Fig. 19). Each face or edge of this cam *C* has a rise equal to one-half the total rise required. The cam has a sliding fit on shaft *A*, and it is revolved by the driving gear *G* which meshes with gear teeth extending around the center of the cam.

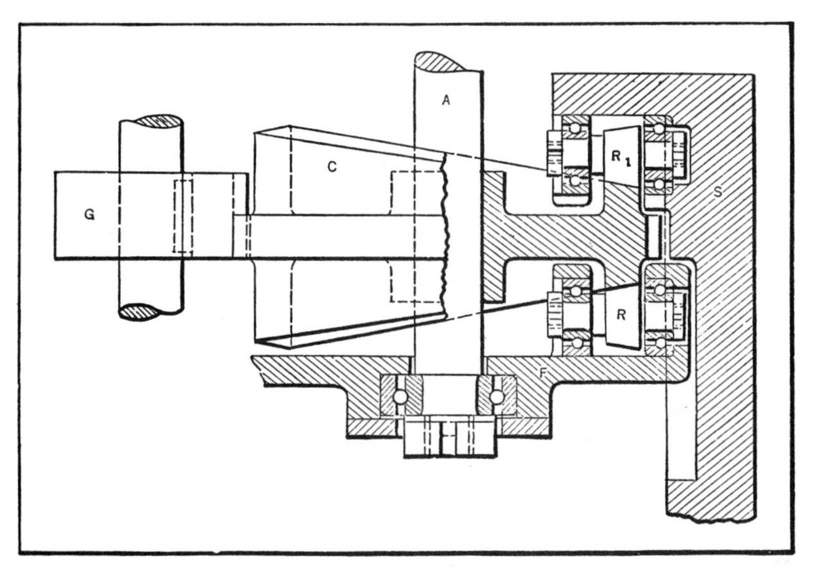

Fig. 19. Double-faced Cam which Moves Driven Slide a Distance Equal to Sum of Leads of Both Faces

As the cam rotates it rises, owing to the fact that it rests on a roller *R*, which is supported by the machine frame *F* and remains stationary except for rotation about its own axis. Bearing against the top face of the cam is another roller R_1, which is supported by slide *S;* this slide is the one that is operated by the cam. It will be evident that when the cam makes one revolution, slide *S* moves a distance equal to the sum of the leads of both cam faces, but roller R_1 and the slide take the thrust of only one cam face.

Reciprocating Motion to Square Bar from Cam Made of Helical Gear Segments.—A novel and what proved to be a very practical application of helical gear segments and pinions is shown by Fig. 20. Shaft *A* has an intermittent rocking movement, which is alternately clockwise and counter-clockwise. The range of these movements is through an angle of about 5 degrees. This rocking lever is required to impart an endwise movement to the square bar or shaft *B*. For this purpose, a segment of a single helical

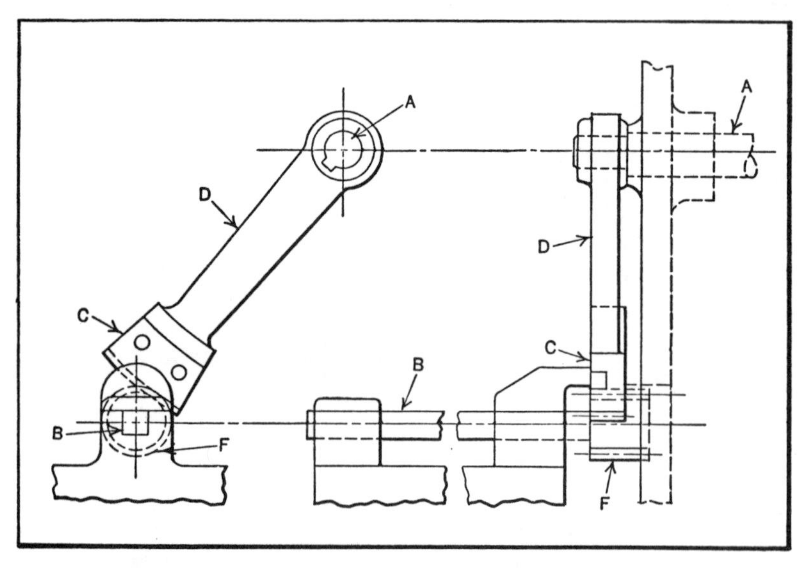

Fig. 20. Helical Gear Segment and Pinion Used as Cams to Produce Longitudinal Reciprocation from Rocking Movement

gear *C* is attached to lever *D*, and a helical pinion *F* of equal angle, but of opposite hand, is fitted to the shaft *B*. Shaft *B*, being square, cannot rotate, and is therefore forced to move endwise.

The helical segments and the helical pinions used in this construction were much less expensive than cams. A complete ring gear furnishes enough segments for several machines. By making the number of teeth in the pinion a multiple of 4 and cutting four keyways in the shaft hole, it is possible to bring new teeth of the pinion into the working position when wear takes place by changing the position of the gear on the shaft. When the square shaft *B* can be made to serve equally well in any position, only one keyway is necessary, as the shaft and gear can be keyed together as a solid unit and relocated in one of four positions to bring unworn teeth into contact with the segment *C*. The segment *C* is supported on each side, a roller

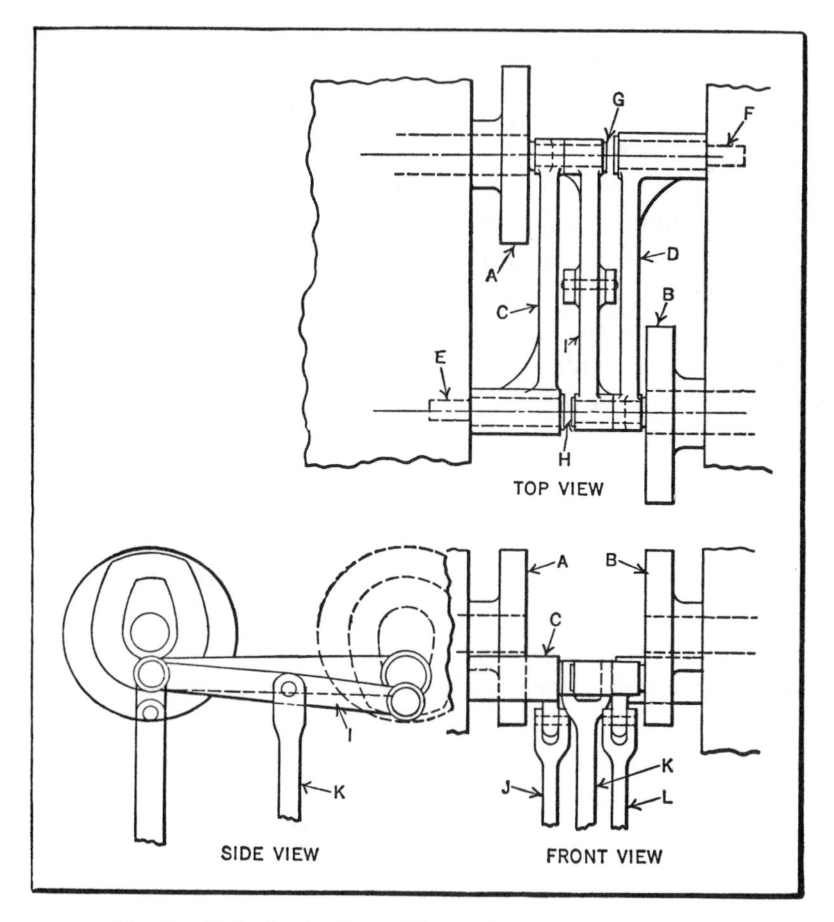

Fig. 21. Mechanism for Transmitting Motion to Three Levers of a Wire-bending Machine

support being used when necessary to reduce the friction load. Gears with teeth having a helix angle of 45 degrees or more give satisfactory performance in this kind of service.

Double Cam Drive for Three Reciprocating Rods.—The lever-motion mechanism shown in Fig. 21 is used on a wire-forming machine to obtain the motions described in the following: Two rods *J* and *L* are given a reciprocating motion, the timing relationship of which must be adjustable. Each of these rods must pass through a complete cycle of motions for each revolution of the drive shaft, although they never operate simultaneously. A third rod *K* is given a similar reciprocating motion of lesser magnitude. The latter rod, however, must pass through two complete cycles for each revolution of the drive shaft, each cycle being performed simultaneously with the cycle of the other two rods. Any change in the timing relationship between the first two rods must be automatically transmitted to the third rod.

The cams *A* and *B* operate at the same speed, and impart the required oscillating movements to the levers *C* and *D*. Cams *A* and *B*, although similar in outline, are set with their lobes approximately 180 degrees apart, and each cam can be adjusted slightly in its timing relationship with the other cam. Lever *C* fulcrums on stud *E*, while lever *D* fulcrums on stud *F*. Studs *E* and *F* are so located that levers *C* and *D* are in a horizontal position when their oscillating ends are held at their lowest points by the cams.

Lever *I* is supported on studs *G* and *H*, carried on the oscillating ends of levers *C* and *D*, respectively. Rods *J*, *K*, and *L* are attached to levers *C*, *I*, and *D*, respectively, and serve to transmit the motion to the required points. As lever *C* is oscillated by cam *A*, lever *I* is given a similar motion, being pivoted on stud *H*, which is held in a fixed position by cam *B*.

After lever *C* has passed through its cycle and come to rest, lever *I* is given a similar movement at the opposite end by cam *B* through lever *D*. As the movement of lever *I* is produced entirely by levers *C* and *D*, it must always operate in exact synchronism with these levers, regardless of the adjustment of cams *A* and *B*. The front view shows the levers at rest. The side view shows rod *K* moved to its lowest point by the action of lever *I*.

Obtaining Instantaneous Movement of Cam-Operated Lever.— One of the best known means of imparting a very quick movement in one direction to a reciprocating part of an automatic machine is by a cam and spring mechanism, such as shown in Fig. 22. The member to be actuated (not shown) is attached to the upper end of link *A*. The other end of this link is connected to the rocker lever *B*, pivoted on stud *C*. Lever *B* acts in conjunction with cam *J* through roller *E* and spring *F*.

The left-hand diagram shows the mechanism just at the end of a dwell period of the lever *B*. Further rotation of the cam in the direction indicated by the arrow will result in roller *E* dropping into the recess of the cam and thus producing a quick downward movement of lever *B* and link *A*. It is clear that no matter how heavily spring *F* is loaded, there is a relatively slow accelerating movement of lever *B* while point *G* of the cam moves from the position shown to point *K* on roller *E*, or along the arc *GK*. Only when the cam has made an angular movement equivalent to angle *GDH* does lever *B* completely lose the restraint imposed upon it by the cam and roller and allow spring *F* to pull lever *B* downward with a quick motion. The point *H* is found at the intersection of the cam outline with an arc *KH* swung about stud *C* as a center and tangent to roller *E*.

The angle *GDH*, through which the cam rotates during the delayed action, depends primarily on the length of the radius of the roller *E* and to a much smaller extent upon the lengths *CK* and *GD*. This angle represents, in terms of angular velocity of the cam, the delay in the time of the

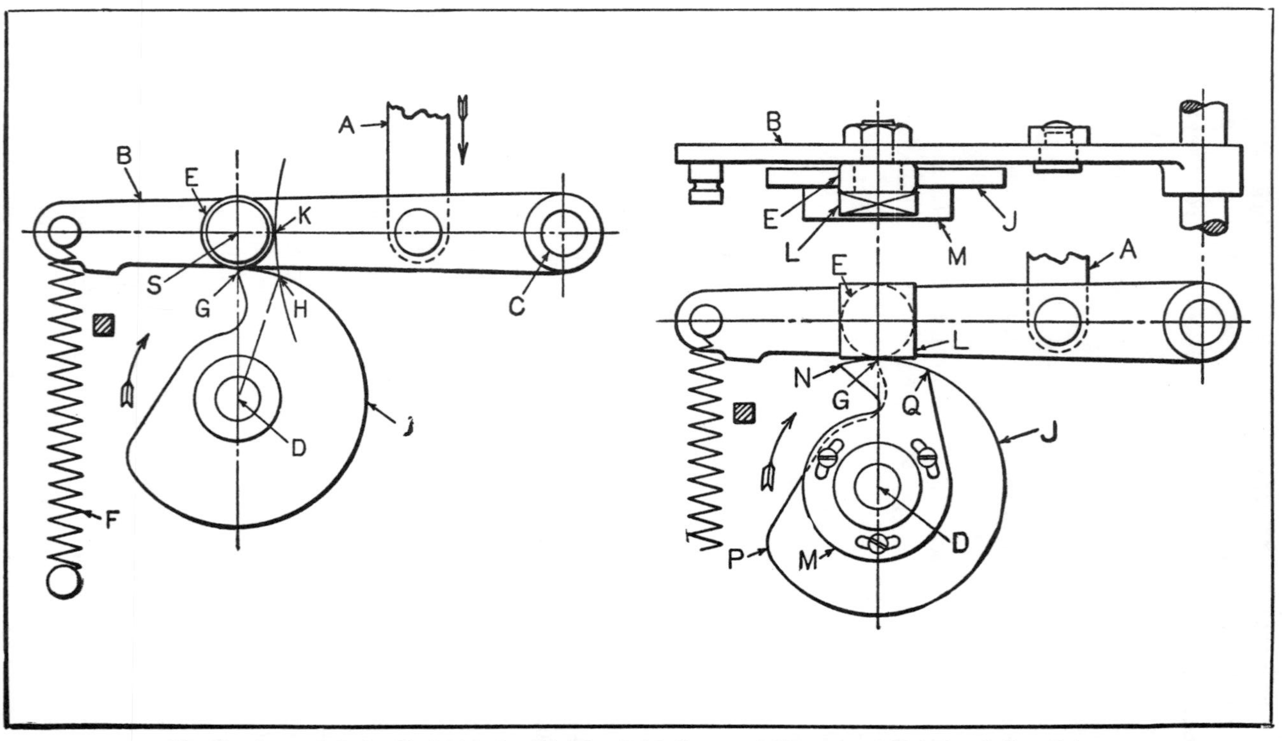

Fig. 22. Cam and Spring Mechanism for Obtaining Quick Downward Movement of Link A and Slow Return

snappy spring action, the delay being greater the larger the roller size, the shorter the roller arm *CS*, and the slower the rotation of the cam.

In most cases, this delay is not objectionable. It may even be welcome in some cases, as it results in much less shock to the mechanism. However, there are occasions when this delay must be eliminated, as for instance, when a hot fluid which sets very quickly must be pumped into a mold. Under such conditions, a very sudden action on the fluid-forcing pistons is desired. The right-hand diagram shows how this action can be effected by the addition of a few parts.

At the side of cam *J* is mounted an auxiliary shoe *M*, which is rotated about shaft *D*. This shoe engages the square block *L*, which is held rigidly to the lever *B*. Lobe *NQ* of the shoe *M* remains in contact with block *L* for some time after cam *J* has lost contact with the roller *E*. During the time cam *J* and shoe *M* are rotating from the position shown to the point of release, lever *B* remains nearly stationary, as the lobe *NQ* of shoe *M* slides underneath the flat face of block *L*. Further movement of the cam and shoe in the same direction results in an instantaneous drop of lever *B*.

Following the sudden drop of lever *B*, shoe *M* and block *L* are inoperative. After the desired dwell, the follower is restored to its initial position by the lobe *P* of the cam, which acts upon the roller *E* alone. Just before the end of the cycle of shaft *D*, both the roller and the block engage the cam and the shoe simultaneously. A little care in the design of the details insures smooth operation.

The angular margin between points *N* and *G* can be materially reduced without danger of the roller interfering with the cam during the sudden drop. If desired, this angular margin can be increased, provided lobe *NQ* of the shoe is made of sufficient size. This is a very desirable feature, as the exact moment of the drop can be adjusted within fairly wide limits, independent of the exact moment of the withdrawal movement.

Cam Mechanism that Returns Lever to its Starting Position when Machine is Stopped.—In a certain type of machine, an oscillating lever is required to return to its starting position, or very near it, regardless of the part of the cycle in which the machine is stopped. This lever, which is indicated at *A* in Fig. 23, controls the movement of an independent feeding device on the machine.

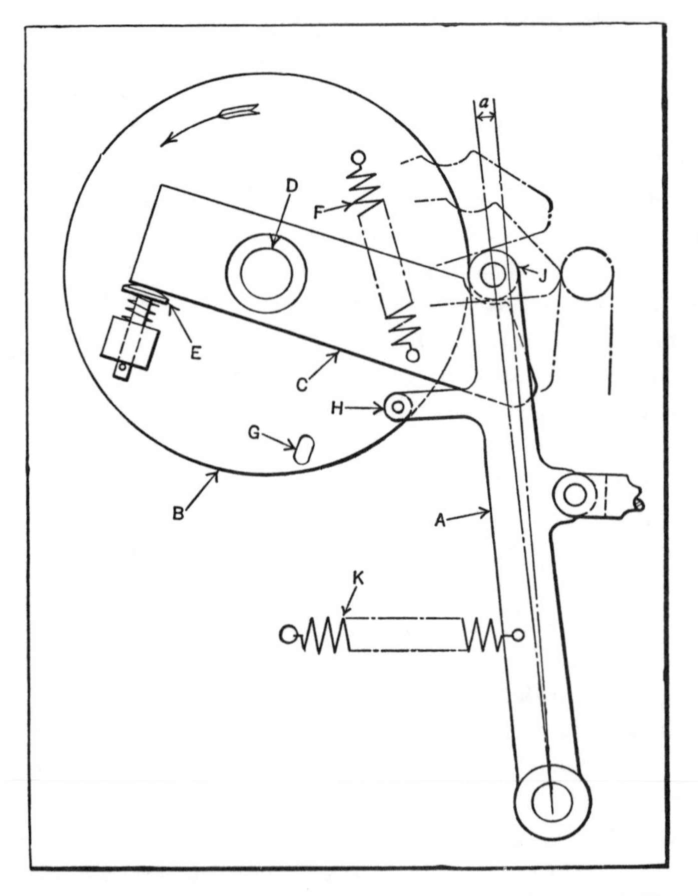

Fig. 23. Cam-actuated Lever that Always Returns to its Starting Position, Regardless of the Part of the Cycle in which the Machine is Stopped

The mechanism consists chiefly of a disk *B* which rotates continuously in the direction of the arrow and a cam *C* which is pivoted on the disk shaft *D* and held normally against the spring bumper *E* by spring *F*.

As the disk rotates from the position shown, cam *C* and lever *A* become locked and remain stationary until the lug *G*, secured to the disk, comes in contact with the roll *H* on the lever. Further movement of the disk then forces the lever toward the right, so that the upper roll *J* on the lever will move out of the hooked part of the cam. The remaining part of the stroke of lever *A* is imparted by the cam through the action of spring *F*. As the end of the cam passes roll *J*, the lever is immediately returned to its starting position by the spring *K*.

With this arrangement, it is obvious that the movement of the lever is obtained through a trigger action between cam *C* and roll *J* and regardless of the position in which disk *B* may stop, the lever will always return to its starting position when roll *J* is released.

If, however, the disk is stopped during the angular movement *a* of lever *A*, that is, before the roll *J* is released, the lever also will stop and will not return to its starting position. In the present application, however, the lever does not begin to function until it has moved through this angle, and hence is sufficiently near its starting position to fulfill the conditions required.

Single Cam Action Performs Four Different Functions.— An excellent example of a multiple cam action in which four movements are obtained essentially by one simple edge-cam is shown in Fig. 24. It is applied to a device used for capping bottles, and although two cams are used here, they are identical and impart the same movements simultaneously. The cam arrangement is such that by swinging the forked lever *G* toward the right, a split collar or "table" grips the neck of the bottle, the table being automatically locked in this position while continued movement

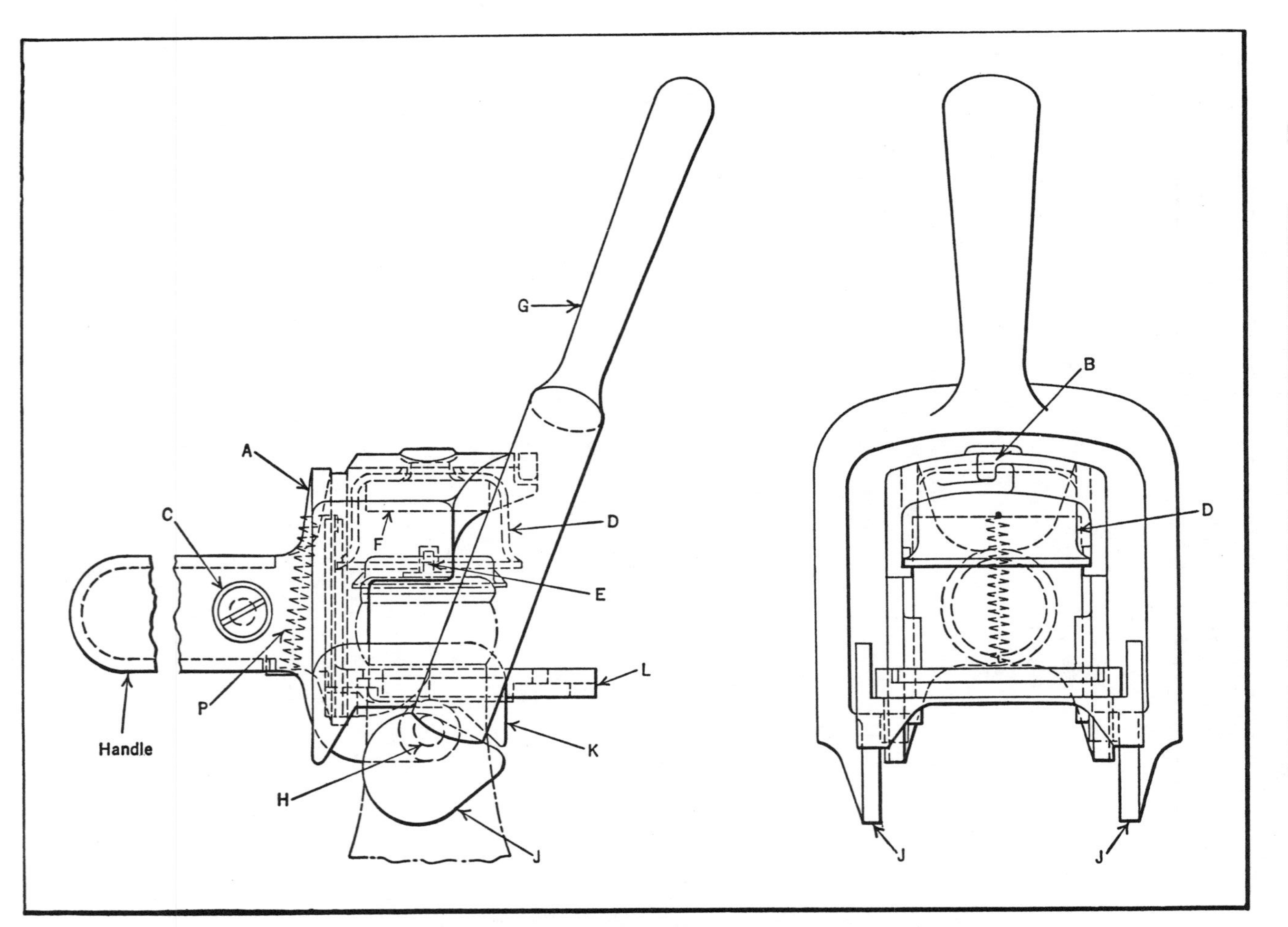

Fig. 24. Arrangement in which Only One Cam-lever is Required for the Operation of a Bottle-capping Device

of the lever causes the cap to be forced in place. To remove the device from the bottle after the capping operation, the lever is merely returned to its original position.

One of the outstanding features in the design of this device is that only one screw is required in its assembly. No machining is done on any of the parts, as sufficient clearance has been allowed to permit the use of unfinished castings. This arrangement resulted in an inexpensive product which in no way affects its utility. The body *A* is cast in two parts, which are held together by the interlocking hooks at *B* and the screw *C*. The cup-shaped capping hood *D* is held in place between the two halves of the body and prevented from rotating by the two lugs *E*. Inside the capping hood is a rubber pad *F* which is forced into place and held by a stem projecting through a hole in both the hood and the body.

The forked capping lever *G* has two pins *H* cast integral with it. These pins serve as a pivot for the lever and engage holes in the lower part of the body. On each side of the forked lever is a cam *J*. The most important part of the device is the split collar or table which consists of two parts—the lifting cam-plate *K* and the guide plate *L*, the latter having a sliding fit in the body. Both parts of this table are interlocked, as shown in Fig. 25.

The cams *J* on the capping lever impart four different movements. When lever *G* is in its farthest position toward the left, the table halves are separated in order to permit the open end of the bottle to pass through. Separation of the table halves, as indicated in both sectional views, is accomplished as the point of the cam engages the projection *M* on the cam-plate. Referring to the extreme left-hand view, it will be noted that the table halves are together, in position to grip the neck of the bottle. This is done with the portion *R* of the cam as it engages the projection *N* on the cam-plate when lever *G* is swung toward the right. Continuation of this lever movement (see ex-

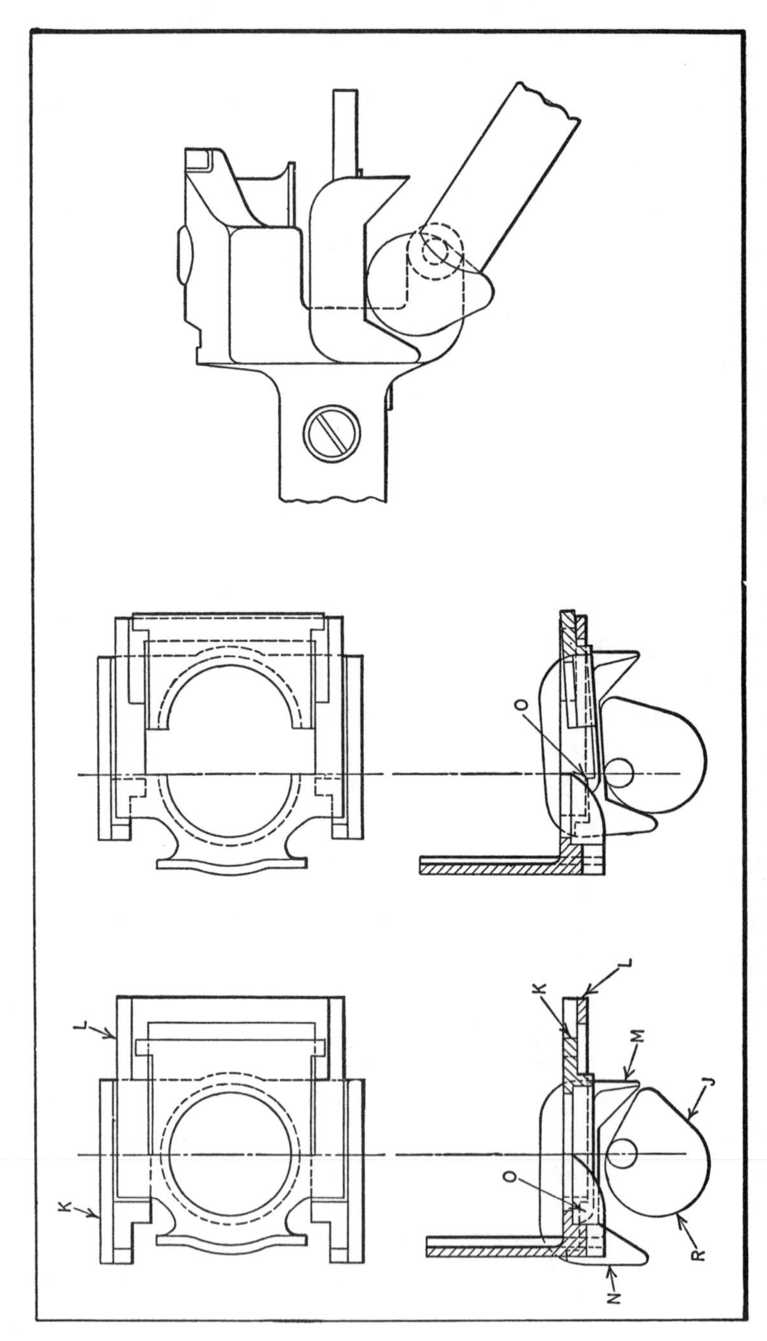

Fig. 25. Views Showing the Closing, Locking, Capping, and Opening Actions Obtained by the Twin Cams

treme right-hand view) causes the cam to rotate into a position where it has forced the cap down over the bottle top, thus completing the operation.

During the capping operation, there is a side thrust on the cam-plate *K* which tends to separate the halves of the table. This would, of course, permit the bottle neck to pass through the table. To prevent this, latches *O* are provided on the cam-plate. These latches hook over the end of the guide plate and lock the two halves after they have been closed around the neck of the bottle.

When the lever *G* is swung toward the left to open the table halves, the cams on this lever tilt the cam-plate *K* enough to disengage the latch and permit it to pass under the guide plate *L*. This is shown clearly in the central view. A spring *P*, Fig. 24, keeps the cam in contact with the cam-plate. One end of this spring is fastened to the body and the other end to the guide plate.

Switching Arrangement for Cylindrical Cam with Intersecting Grooves.—Cylindrical cams having intersecting roll grooves are sometimes used when a cam of small diameter is desired, or when two revolutions of the camshaft are required to one cycle of the follower. These cams have also found application in sewing machines, gas engines, etc. In the ordinary cam of this type, the break in the grooves at their intersection necessitates the use of a follower of special design, because a roll would become wedged at this point. The roll is usually replaced by an oblong shoe, the sides of which curve inward at the ends so that the shoe will be a sliding fit in any part of the groove.

This arrangement is not always satisfactory when a smooth action of the follower is required, owing to the increased clearance around the shoe at the intersection of the groove. Moreover, at this point, the pressure of the sides of the shoe against the corners of the groove causes a great deal of wear on both members. These objections

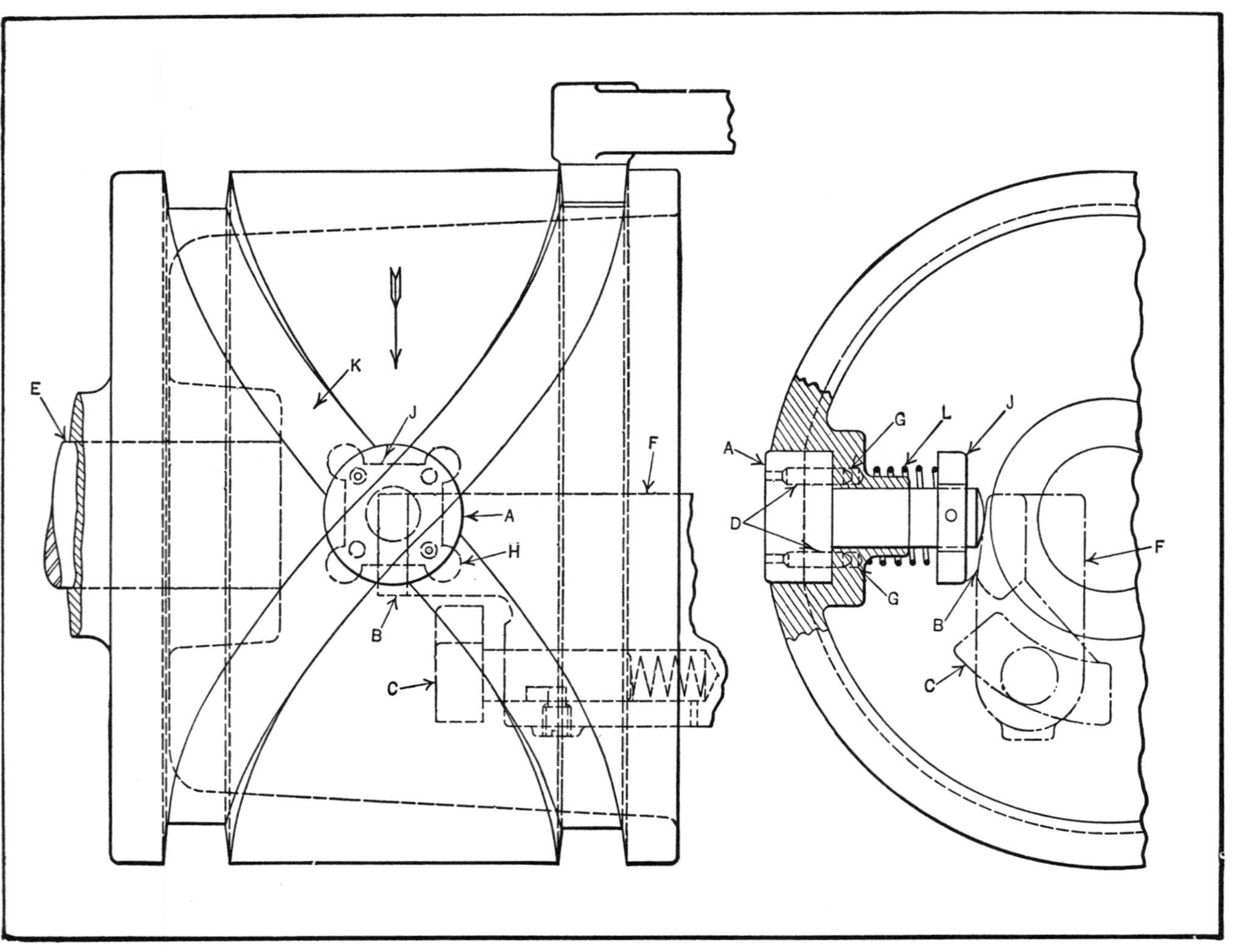

Fig. 26. Cam Having Two Grooves and an Automatic Switch at their Intersection

are overcome, however, by the rather ingenious switching arrangement on the cam illustrated (Fig. 26). It is entirely automatic and provides a continuous groove for the roll, regardless of which groove the roll is in.

The arrangement consists of the grooved plunger *A*, member *J* secured to the plunger shank, and the stationary cams *B* and *C*. These cams are mounted on the arm *F* extending within the cored portion of the cylindrical cam and serve to rotate the plunger 90 degrees for every revolution of shaft *E*. Cam *C* has a shank which is a sliding fit in a hole bored in the arm *F*. The shank is backed up by a coil spring to compensate for the interference of cam *C* when engaging with member *J*. Pins *D* lock the plunger in position after each indexing movement.

In the position shown, the lower end of plunger *A* has engaged cam *B*. Further rotation of the cylindrical cam in the direction of the arrow will cause cam *B* to force the plunger outward until pins *D* have been withdrawn from holes *G*. The plunger is now free to rotate. As the cylindrical cam continues its rotation, the end of cam *C* comes in contact with lobe *H* on member *J* and rotates the plunger 90 degrees. In this position, the pins *D* are directly over another set of holes like those at *G*, and the plunger is seated through the action of the coil spring *L* and locked in position by the pins as they enter these holes.

The groove in the plunger is now aligned with cam groove *K* in which the roll is guided as the cylindrical cam continues to rotate. For each succeeding revolution of this cam, the indexing action of plunger *A* is repeated, so that the plunger groove is always in line with the proper cam groove. In designing a cam of this type, it should be remembered that the cam grooves must cross at an angle of exactly 90 degrees; otherwise inaccurate alignment of the cam and plunger grooves will result. Hardened bushings in the cylindrical cam may also be provided for the indexing pins to reduce wear at these points.

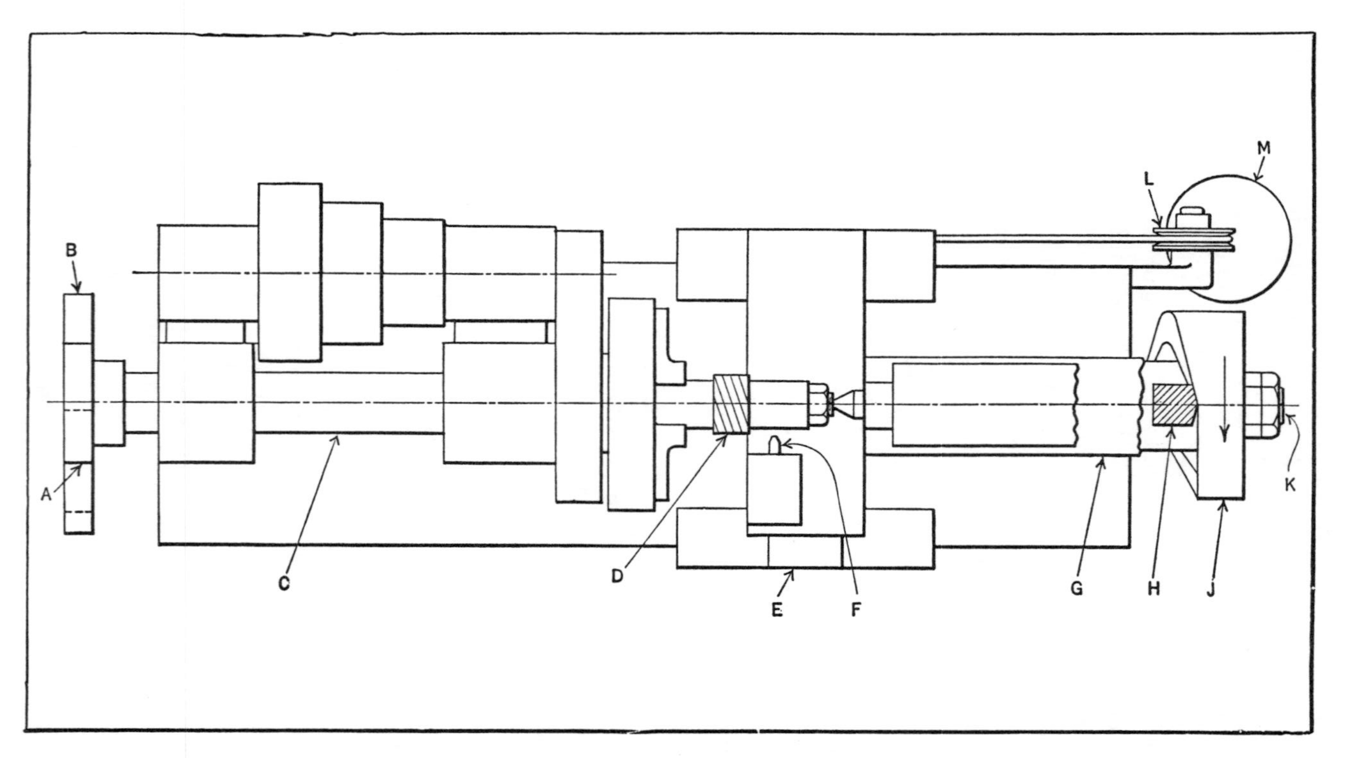

Fig. 27. Redesigned Automatic Threading Lathe in which a Simple Cam and Change-gears Serve to Index the Work when Chasing Quadruple Threads by Rotating the Work 1 3/4 Turns to One Cam Revolution

Cam-Operated Threading Tool and Automatic Indexing for Multiple Thread Cutting.—To meet the demand for an economical method of chasing quadruple threads on the short sleeves, an automatic threading lathe was redesigned. The usual lead-screw was replaced by a cam, and the cross-feed was arranged to feed once in every fourth pass of the tool. By employing a cam and proportioning the change-gears correctly, it was possible to index the work automatically from thread to thread with each longitudinal cycle of the tool.

A diagrammatic plan view of the lathe is shown in Fig. 27. The work, indicated at *D*, is mounted on an arbor and supported in the lathe in the usual manner. The gear *A* on the spindle and gear *B* on the camshaft *K* are connected by means of an idler. On the right-hand end of the camshaft is keyed the cam *J* which imparts an intermittent reciprocating movement to the carriage *E* through the bronze follower *H* attached to the slide *G*. In order to maintain contact between the follower and cam, a weight *M* was provided, which is connected to the carriage by a cable passing over pulley *L*.

The ratio of gears *A* and *B* is such that, for every revolution of the cam, the work rotates 1 3/4 revolutions. In other words, the work assumes a new angular position at the beginning of each cut. This change in position is equivalent to 90 degrees. Thus the work is indexed smoothly from thread to thread without employing a complicated indexing mechanism.

Calculating Gear Ratio and Developing Cam.—The method of calculating the gear ratio and developing the cam for the multiple-threading operation will now be described. In Fig. 28, the line *ON* was drawn equal to the cam circumference, and on this line the cam was developed. Line *ON* was divided into seven equal parts by vertical lines numbered as indicated. Since the thread to be cut was quadruple, each one of these equal parts was assumed to repre-

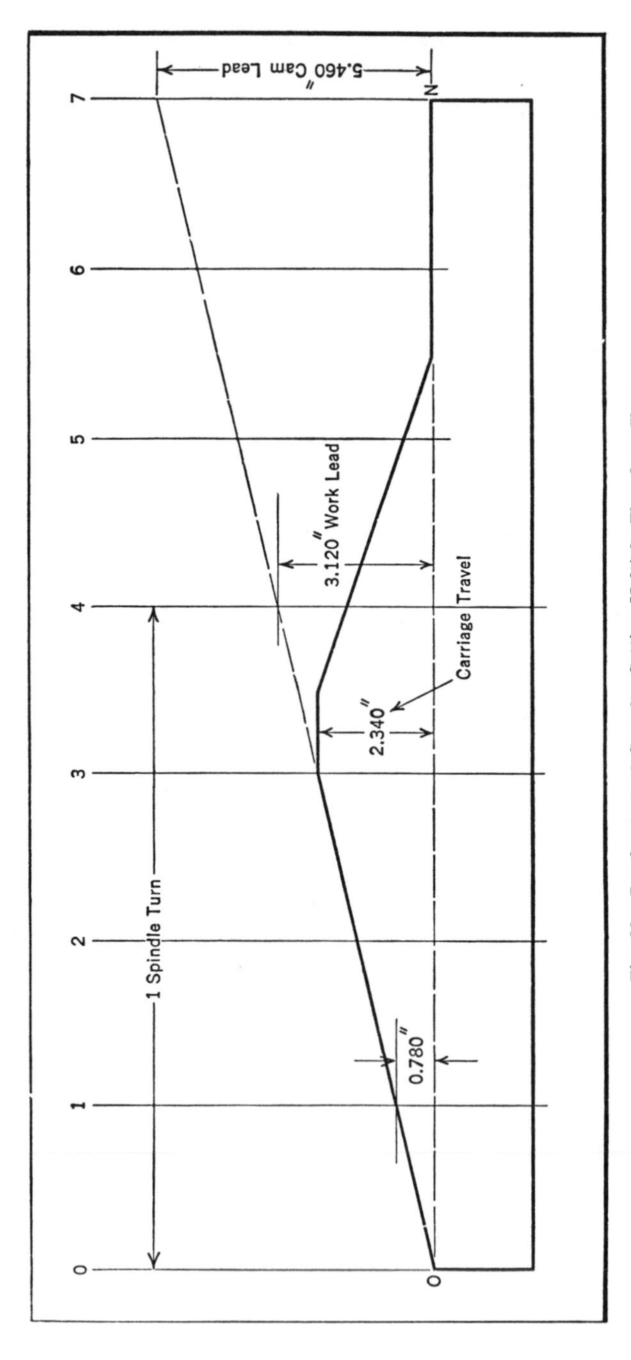

Fig. 28. Development of Cam for Cutting Multiple Thread on Sleeve

sent one-fourth of a spindle turn. Thus the ratio of the spindle turn to the cam turn is 7 to 4; that is, when the cam completes one turn, the spindle makes 7/4 or 1 3/4 turns. Gears corresponding to this ratio have 40 and 70 teeth. The larger gear or the one having 70 teeth is mounted on the camshaft.

Now it is obvious that the part of the cam that forms the thread must be an accurate helix, the development of which is a straight line. To develop this line, a point was located on vertical No. 4, 3.120 inches (the thread lead) above line *ON*. Through this point a straight line was drawn from point *O* to vertical No. 7. It was found that stopping the thread-forming portion of the cam on vertical No. 3 provided ample carriage travel for cutting the thread.

The exact dimensions for the cam rise were found by first dividing the thread lead (3.120) by 4 to obtain the rise for one-quarter revolution of the spindle, or 0.780 inch. This rise was then multiplied by the number of spaces from *O* corresponding to three-quarters of a revolution of the spindle, and the total cam rise, 2.340 inches, was obtained. The lead (5.460 inches) for the working portion of the cam was found by multiplying the entire number of spaces by the rise during one-fourth revolution of the spindle.

The dwell at the top of the cam allowed time for backing the tool out of the thread before the carriage started on its return movement. The longer dwell at the bottom allowed time for moving the tool forward and for the functioning of the cross-feed. With this arrangement, the work rotates continually in the same direction. To enable multiple threads of different leads to be cut on this machine, the sizes of suitable cams and gears can be computed by the method described.

This lathe can also be used for chasing internal threads in short bores. In this case, the action of the cross-feed is reversed, so that the cutting tool will be moved toward the center of the bore at the end of each cut.

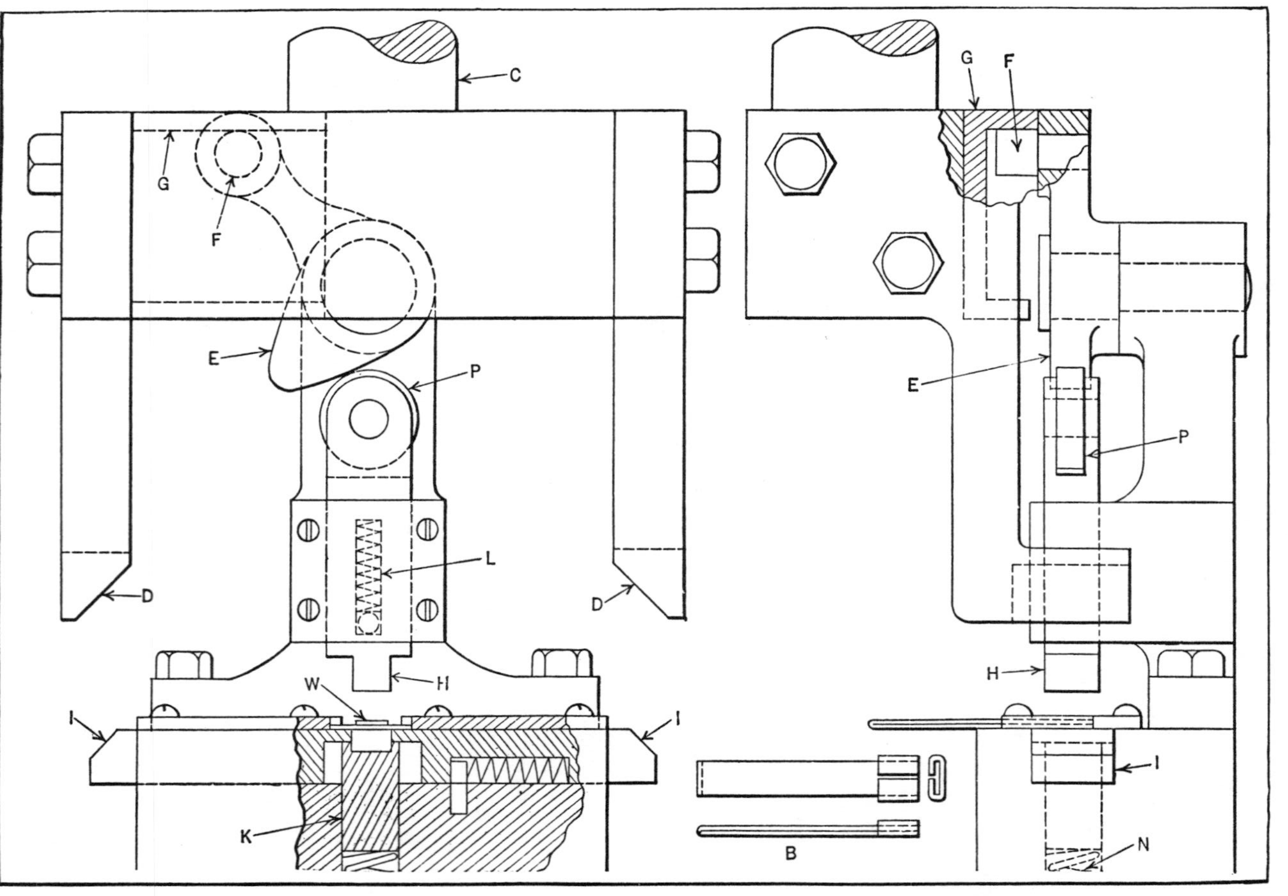

Fig. 29. Folding Die with Pivoted Cam for Rapidly Advancing and Withdrawing Plunger H in Advance of Cams D

Double-Acting Pivoted Cam Mechanism for Folding Die.— The forming plunger *H* and slides *I* of the die shown in Fig. 29 are so actuated by an ingenious mechanism that the two tabs of a flat blank are folded and tightly clenched over the central portion, as shown at *B*, in one stroke of the punch-holder *C*. After placing the piece in the position shown at *W*, the press is tripped.

The upper surface of part *G*, coming in contact with stud *F*, causes cam *E* to act on plunger *H*. Plunger *H*, acting on the work, forces it between the ends of slides *I*, causing the tabs of the work to be bent upward. As the lobe of the cam *E* passes the roller *P*, the spring *L* in plunger *H* reacts on cam *E*, causing it to swing quickly to the right until it is restrained from further movement by stud *F* coming in contact with the lower flange of part *G*, as shown in Fig. 30. This has the effect of causing plunger *H* to rise rapidly and thus avoid interference with the inward movement of the slides *I*. As the ram reaches its bottom position, the slides *I* are operated by the cams *D*, causing the tabs in the work to be folded over.

Fig. 30 shows the die with the ram at the bottom position. As the ram ascends, the slides *I* return to the positions shown in Fig. 29, while the cam *E*, being returned to its original position, again acts on plunger *H*, causing it to press tightly on the folded tabs of the work. On this stroke of plunger *H*, there are three thicknesses of metal under it, whereas on the first stroke there were only two thicknesses. This causes plunger *K* to recede a distance equal to one thickness of the stock. Thus the pressure on the work will always be equal to that exerted by the spring *N*, and can never be great enough to crush the work.

Operating Two Slides in Opposite Directions with One Single-Groove Cam.—In redesigning a tapping machine to be used for tapping opposite sides of a part simultaneously, two tapping heads were required to travel in opposite directions. This movement, as first suggested, was to be obtained by means of individual cams. A simpler method was devised for transmitting movement to both heads from a single cylindrical cam being used, as indicated by the diagram, Fig. 31.

The diagram is so clear that it hardly requires a descrip-

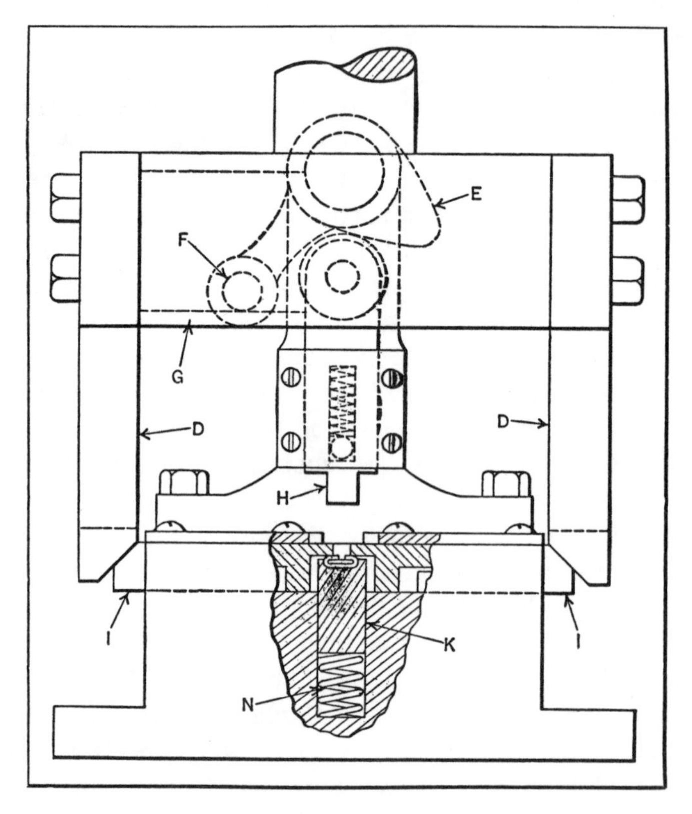

Fig. 30. Die Shown in Fig. 29 in Final Tab-clinching Position

tion. There are numerous cases to which this idea may be applied, with a great reduction in construction and upkeep cost. As indicated, one cam groove serves both heads (not shown). In stationary guides *A* and *B*, on opposite sides of the cam, are slides *C* and *D*. These slides carry the rolls *E* and *F*, both of which engage the same cam groove. The slides are connected to their respective heads by the tie-rods *G* and *H* through which the required movement is transmitted.

Sliding Triangular Cam for Reducing Cam Size and Stroke.— In many automatic machines, sliding cams are employed for transmitting a straight-line movement to the tool or the work-holder, followed by a dwell to permit load-

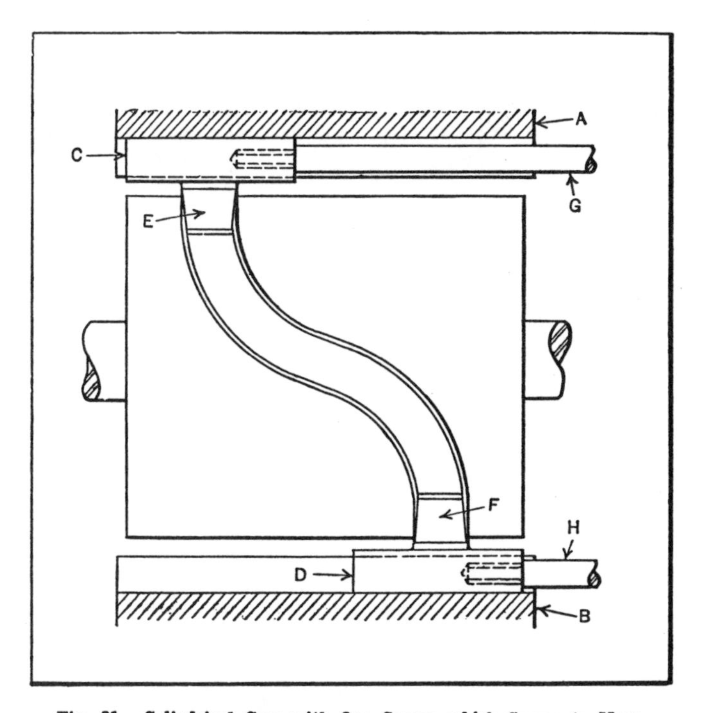

Fig. 31. Cylindrical Cam with One Groove which Serves to Move Two Slides in Opposite Directions

ing and unloading of the work. Frequently this dwell is unusually long, and the movement of the cam follower considerable; hence, a cam of the usual design would not only be large in proportion to other parts of the machine, but also would require a comparatively long stroke. These objections are overcome with the cam shown in Fig. 32. This cam is positive and compact.

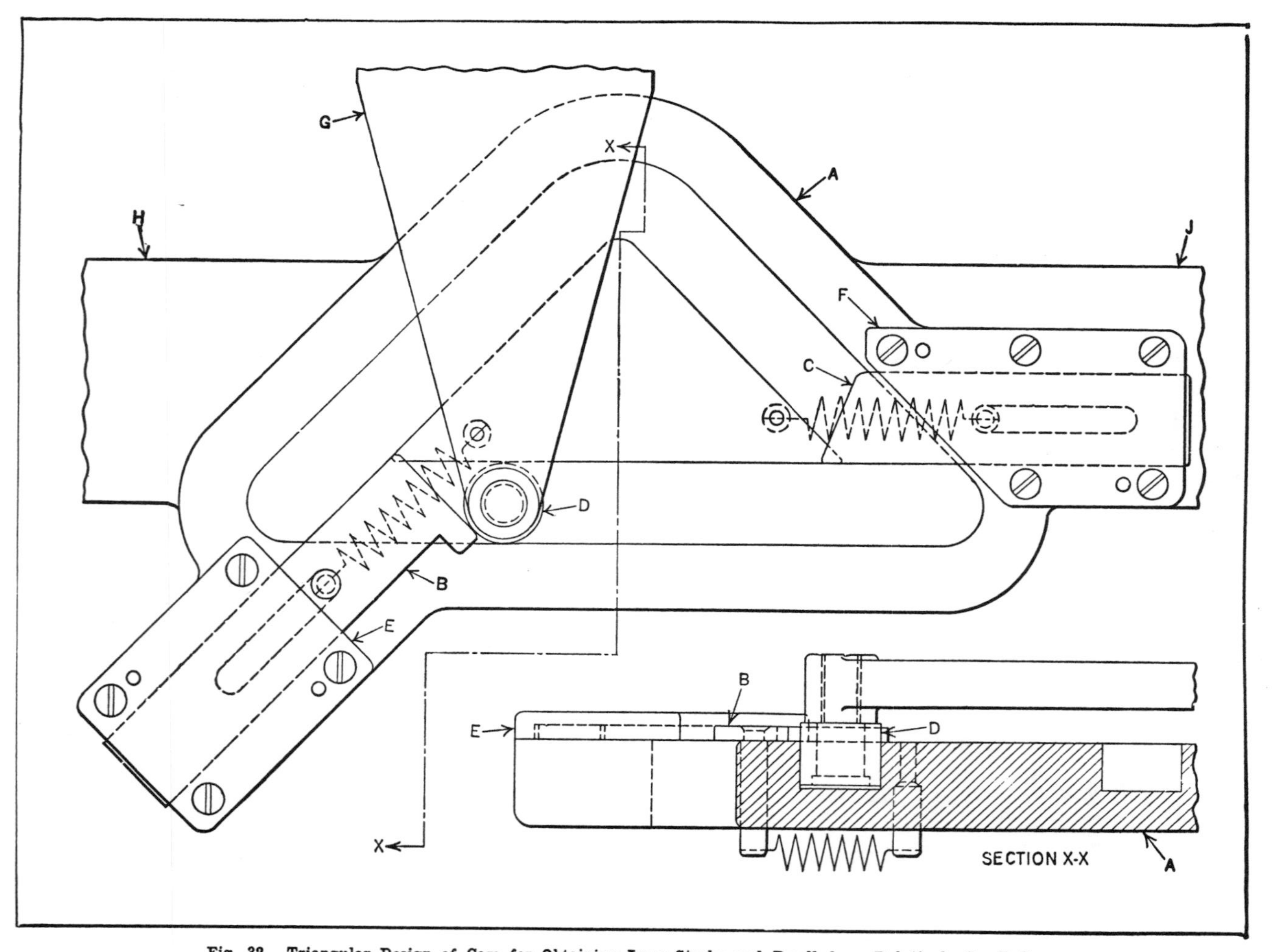

Fig. 32. Triangular Design of Cam for Obtaining Long Stroke and Dwell from Relatively Small Cam

The cam *A* is supported at the ends *H* and *J* by suitable bearings, and is given a reciprocating horizontal movement by some member of the machine. It has a continuous roll groove following a triangular path, and is equipped with locking plates *B* and *C* for retaining the roll *D* in the proper section of the groove. These locking plates are a sliding fit in the caps *E* and *F*, respectively, and are normally held in the position shown by coil springs.

In the position indicated, the cam has nearly completed its stroke toward the right. Further movement of the cam will cause the roll to depress the locking plate *B;* and at the end of the stroke, when the roll has reached the end of the horizontal section of the groove, the plate will once more return to the position shown.

As the movement of the cam is reversed, the roll is forced upward along the edge of the plate and finally into the groove, imparting a vertical upward movement to the follower arm *G*. This movement of the follower arm continues during the first half of the cam stroke. During the remainder of the stroke, however, the follower arm is returned to its starting point, after having passed the locking plate *C*, which is similar to plate *B*.

At this point, the movement of the cam is reversed and the roll simply rides in the horizontal section of the groove for the entire return stroke of the cam. During the latter stroke no vertical movement is imparted to the roll, and hence the follower arm *G* dwells at this time. This completes the cam cycle. The distance that the cam follower moves, as well as the timing, may be varied by changing the angle of the angular groove sections.

Double-Action Cam that Rotates Follower and Moves it Axially.—The interesting mechanism Figs. 33 and 34 is used in a four-slide spring-winding machine. Springs 1/2 inch in diameter and 1 1/4 inches long are made in this machine. At one station, the spring is coiled and cut off. It is then carried, by means of a transfer arm, to another

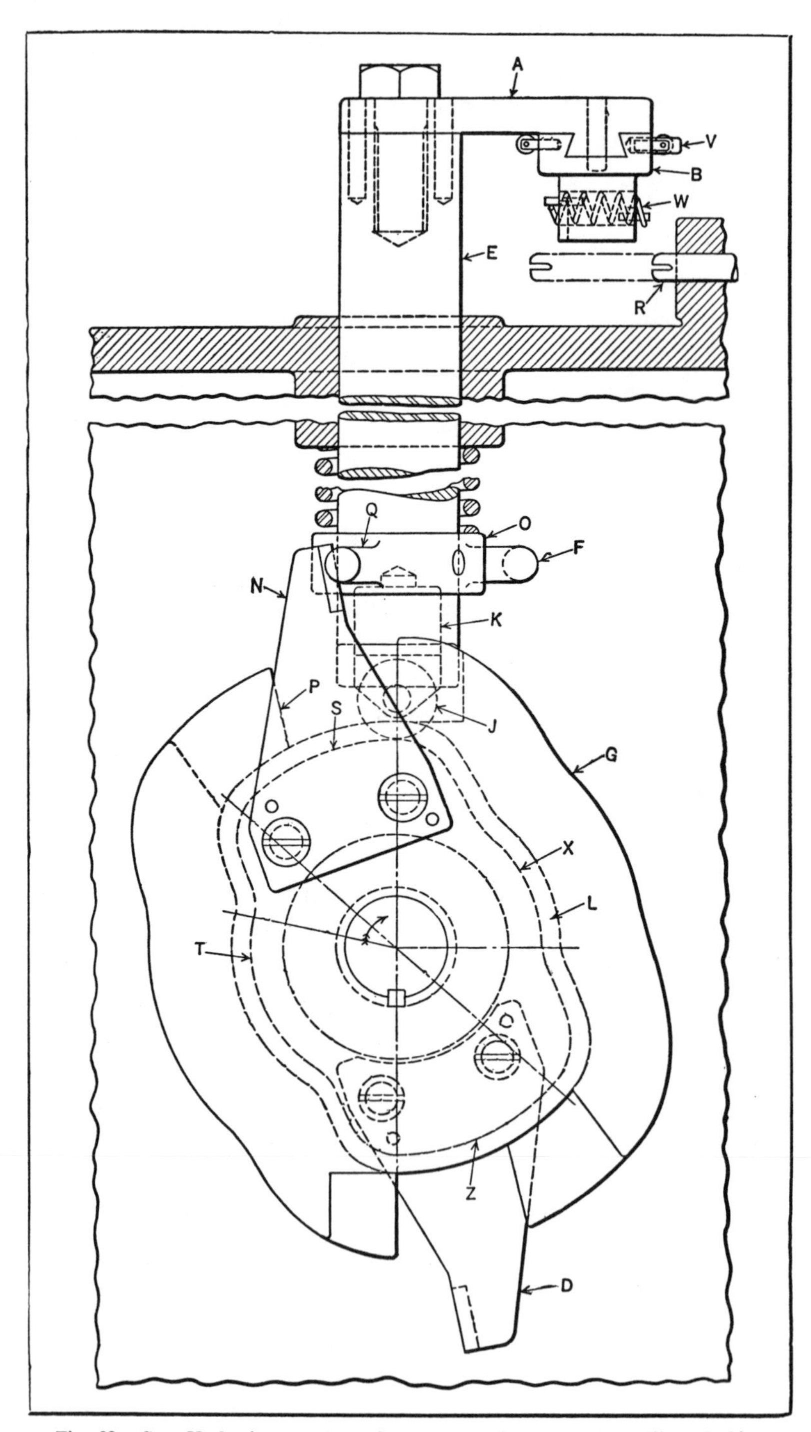

Fig. 33. Cam Mechanism for Operating Transfer Arm of Spring-coiling Machine

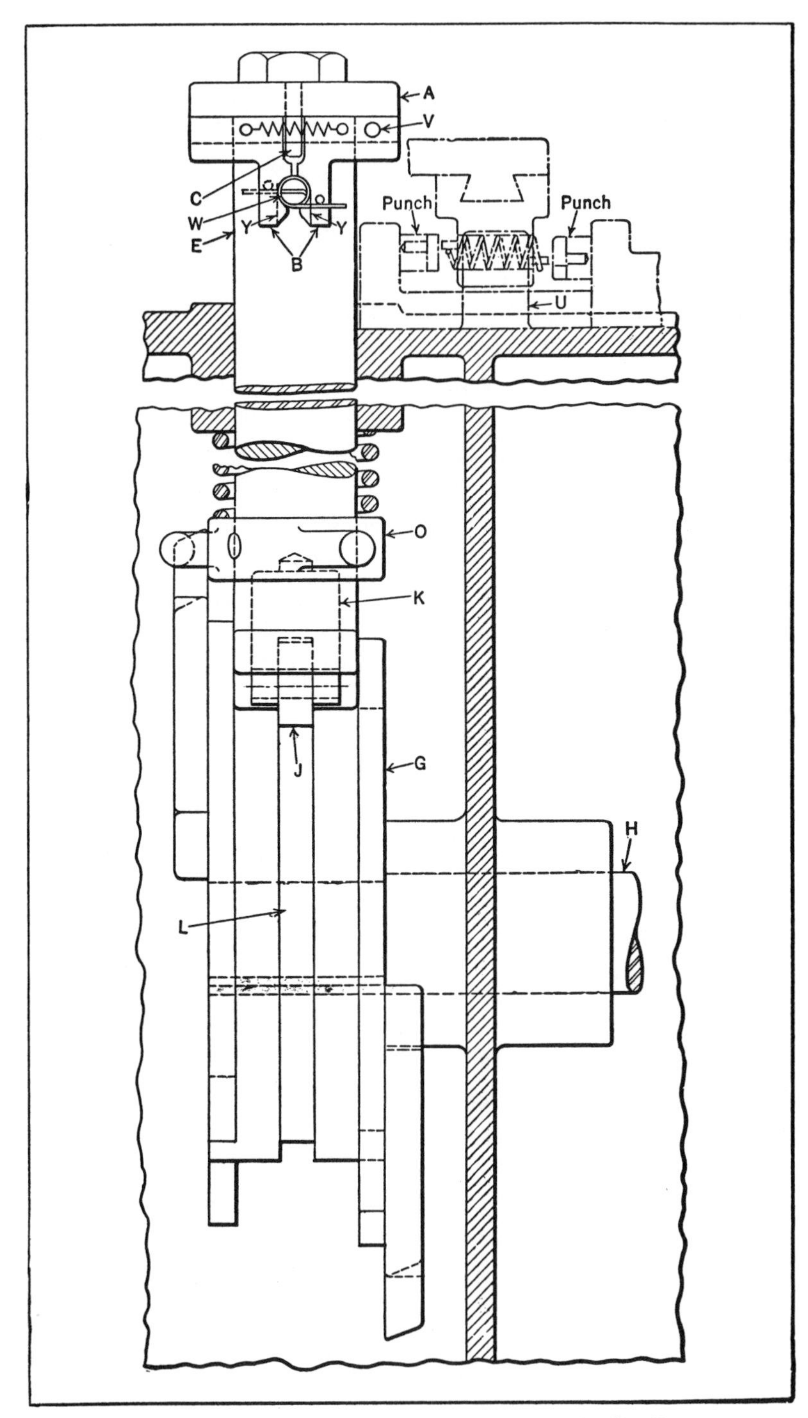

Fig. 34. Side View of Cam Mechanism Shown in Fig. 33

station, where the ends are bent parallel with its axis, after which the completed spring is ejected from the machine. The transfer arm moves through three different planes during each cycle, yet all its actions are controlled by a single cam. The reason for forming the spring ends at a separate station is that another spring is being wound while the preceding one is being formed. With this arrangement, the production is practically double that obtained when the forming was done on the coiling mandrel. As a matter of fact, it would have been extremely difficult to perform all the operations at one station.

The transfer arm *A* has two jaws *B* mounted on its overhanging end. Between the jaws is gripped the coil spring *W*, on which the coiling and cutting operations have been performed. The jaws are centralized by the pin *C* against which they are held normally by the coil spring shown. The arm is bolted and doweled to the vertical plunger *E*, which is a free fit in a long bearing cast in the machine frame. The plunger is given a combined vertical and rotary movement by means of the cam *G* mounted on the drive shaft *H*.

The connection between the cam and the plunger is made by the roll *J* pivoted in the plug *K*. The plug, in turn, is a free fit in a hole bored in the lower end of the plunger. Thus the plunger can be rotated to any position, yet the roll will remain in the same plane, being constrained by the continuous cam groove *L*. As indicated, jaws *B* have grasped spring *W* and elevated it vertically to the position shown, through the action of cam *G*. Incidentally, the coiling arbor *R* has automatically receded to permit the spring to pass upward. The lower end of plunger *E* is square and is a sliding fit between the flanges of the cam. Thus, when the square end is confined between the flanges, the plunger cannot rotate. However, at certain points in the flanges, gaps are provided to permit rotation of the plunger for swinging the transfer arm 90 degrees to the forming station indicated in dot-and-dash outline at the upper part of Fig. 34.

The finger *N* is fastened to the cam for the purpose of rotating the plunger at this time. This finger engages a lug on the collar *O*, pinned to the plunger, and starts the rotation of the plunger. The rotary movement is then picked up and continued as the end *P* of the flange comes in contact with the squared end of the plunger. This action is more clearly illustrated in Fig. 35. Here the cam is rotating in a clockwise direction and the finger *N* is about to swing the plunger in the direction indicated by the arrow.

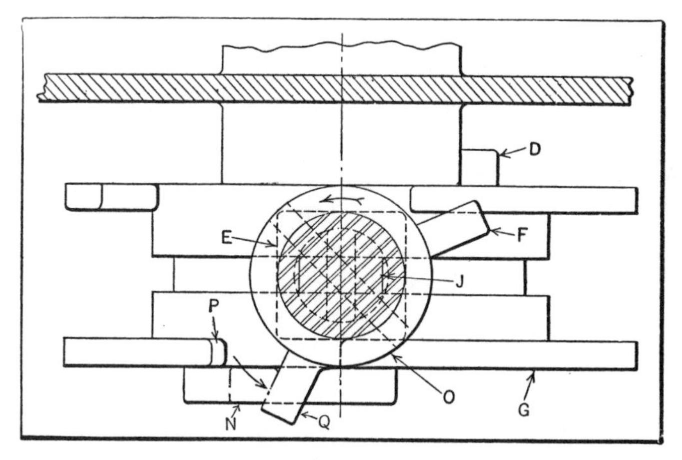

Fig. 35. Plan View of Collar O, Fig. 33, Showing Finger N About to Rotate Plunger E

As the cam rotates the finger *N* engages lug *Q* and rotates the plunger until the flange end *P* comes in contact with the squared end of the plunger and continues the rotation of the latter until it has completed its 90-degree movement. At this position, the squared end of the plunger enters between the flanges, thus preventing further rotation of the plunger, and in addition, the finger and lug absorb the entire starting torque. The rotary movement of the plunger occurs while the roll is passing over the concentric portion *S* of the cam; hence, the height of the arm remains constant at this time. However, as the cam continues to rotate, the roll descends to the low concentric part of the cam lobe at *T* and dwells, causing the arm also to descend and dwell. In descending, jaws *B* enter between two stationary guides *U* which prevent the jaws from opening during the forming operation. The operation at this station consists of bending the projecting ends of the spring outward so that they will be parallel with the axis of the spring. This is done by the automatically controlled punches which advance, with their slides, and bend the ends over the corners *Y* of the jaws. The position of the spring ends relative to the jaws is maintained by the two pins indicated.

When the ends have been formed, the cam raises the arm vertically to its former height. At this time, the roll engages the cam surface at *Z* and, as on the opposite side of the cam, the plunger and arm are brought back to their original position. In this case, however, finger *D*, engaging lug *F*, starts the rotation of the plunger, after which the corresponding flange end completes the 90-degree movement. When the arm is swung back, a latch (not shown) engages the pin *V* and opens the right-hand jaw, allowing the completed spring to drop into a chute. The cam then allows the arm to dwell until the succeeding spring has been coiled and cut off. Next, the roll passes to the cam surface *X*, causing the plunger and arm to descend until the jaws snap over and grip the spring. This completes the cycle.

The heavy coil spring on the plunger insures constant engagement of the roll with the cam. Although this mechanism was designed primarily for a two-station machine, the same principle can be used for three or more stations by merely modifying the cam throws and adding the required fingers and lugs. In order to facilitate the machining of the cam, the cam is made in two sections and fastened together with screws, the parting line coinciding with the side of the roll groove. For the purpose of simplification, this sectional construction is not shown.

CHAPTER II

INTERMITTENT MOTIONS FROM GEARS AND CAMS

The term "intermittent motion" is applied to mechanisms for obtaining a "dwell" or possibly a series of dwells or moving and stationary periods of equal or unequal lengths. Many different designs of intermittent motions are in use because they are required on so many different types of automatic and semi-automatic machines. The intermittent motions illustrated and described in this and the two following chapters, supplement the two chapters on this general subject found in Volume I of INGENIOUS MECHANISMS FOR DESIGNERS AND INVENTORS.

Advancing Reciprocating Motion with Dwell at Each Point of Reversal.—In coating certain parts of household appliances with enamel by means of a combination dipping and baking machine, the parts are hooked on an endless conveyor chain and passed through a bath of enamel and then through an adjacent heating oven for drying the coated surfaces quickly. In order to facilitate the spreading of the enamel while the parts are passing through the bath, the chain is given an advancing reciprocating movement. The chain advances to deliver the parts to the oven.

This movement of the chain is obtained by the mechanism shown in Fig. 1. The mechanism is mounted on the base *A* of the machine. It consists essentially of a combination of planetary gearing and a double intermittent gear arrangement. The intermittent gearing provides the reciprocating movement, while the planetary gearing is necessary to transmit this movement to the chain sprocket.

Beginning with the intermittent gearing, ring gear *B*

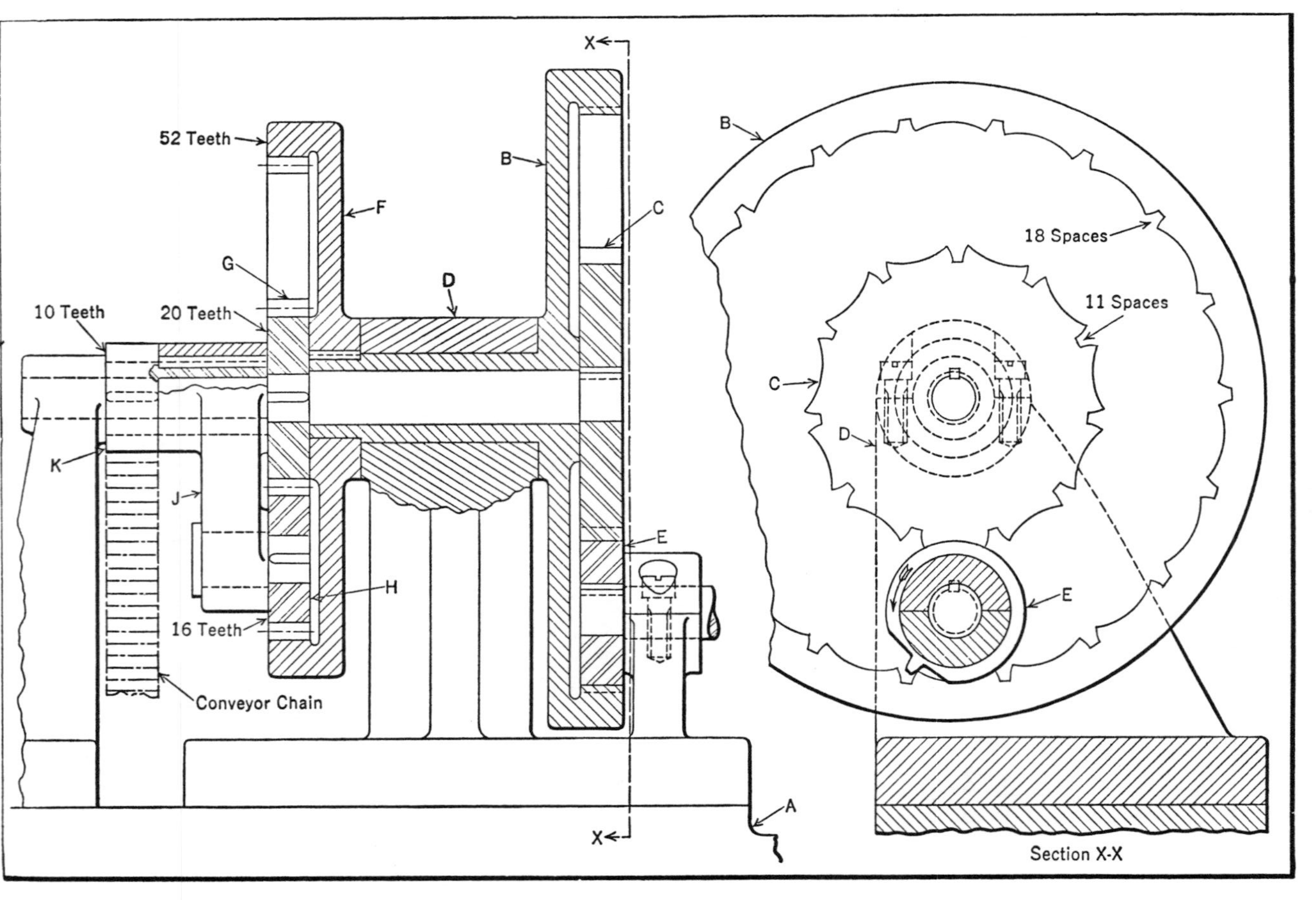

Fig. 1. Mechanism for Imparting Advancing Reciprocating Motion to Conveyor Chain of Enamel Dipping and Baking Machine

and center gear *C* are supported in the stationary bearing *D* and mesh with the driving pinion *E*, which rotates in a stationary bearing. At the left-hand end of the sleeve that forms the journal for gear *B* is keyed an ordinary internal ring gear *F*, and on the shaft to which gear *C* is secured is keyed the pinion *G*. Gear *H* is free to turn with the stud in arm *J* and meshes with internal gear *F* and pinion *G*. The arm *J* is keyed to an extension sleeve integral with the conveyor chain sprocket *K*, the sleeve being free to rotate on the center shaft.

When driving gear *E* rotates in the direction indicated by the arrow, the single tooth will engage the adjacent tooth space in gear *B* and rotate the latter 1/18 revolution. During this movement, gear *C* will be locked in a stationary position by gear *E*. Hence, the partial rotation of gear *B* will rotate gear *F* and cause gear *H* to roll around the stationary pinion *G* and swing arm *J*, with the sprocket *K*, in the same direction.

As the gear *E* continues to rotate, its cylindrical portion locks gear *B* and the single tooth engages a tooth space in the center gear *C*, rotating the latter 1/11 revolution, after which the cylindrical portion of gear *E* locks it in a stationary position. Rotating gear *C* in this way causes gear *G* to rotate and roll gear *H* on the now stationary gear *F*. In this manner, gear *H* carries arm *J* and sprocket *K* around the center shaft in a direction opposite to that of the driving gear *E*. This completes one cycle of movements.

The required angular movements of the sprocket are as follows: 14 1/2 degrees, or approximately 0.04 revolution, in a clockwise direction, as observed from the right-hand end of the mechanism. The sprocket then dwells and reverses its movement, rotating 9 degrees, or 0.025 revolution. The angular advance of the sprocket for each cycle is $0.04 - 0.025 = 0.015$ revolution, or about 5 1/2 degrees. In calculating the ratios and the number of teeth and tooth

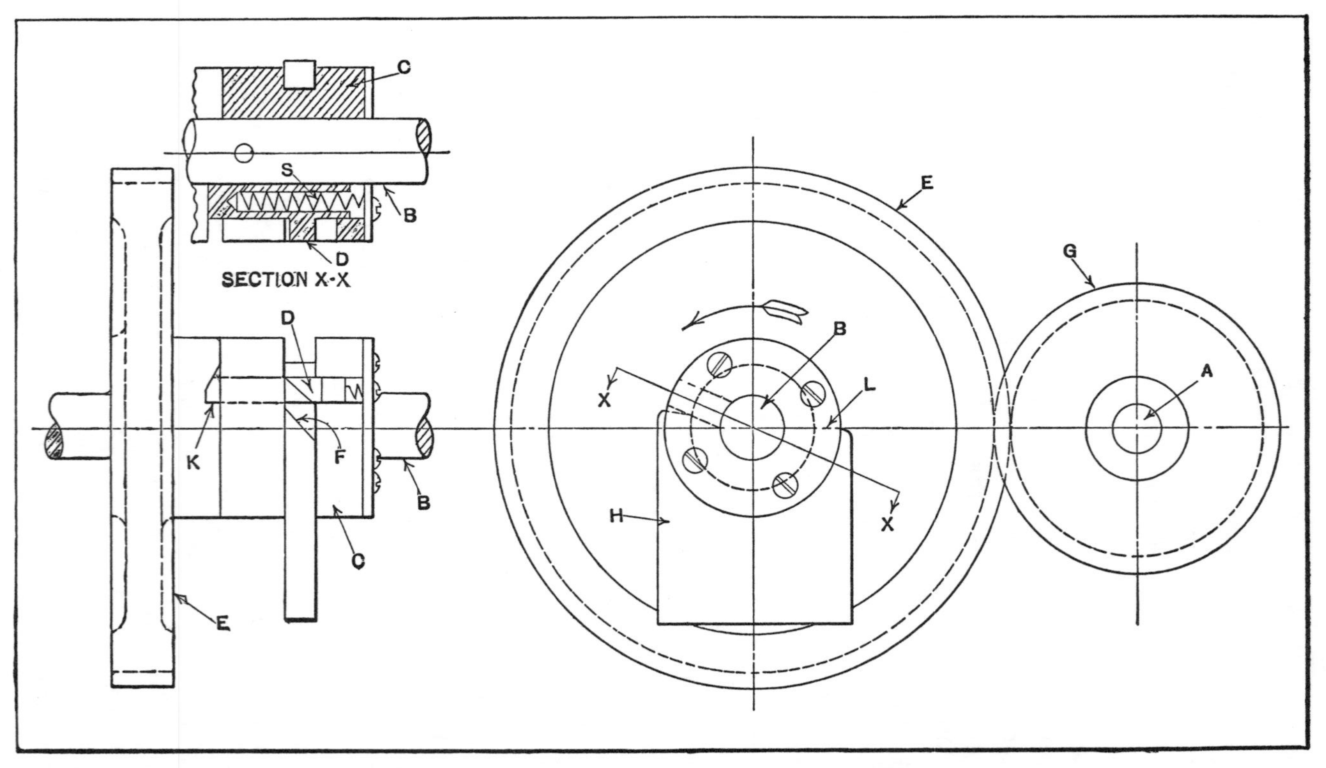

Fig. 2. Intermittent Gear Drive Actuated by Cam-operated Clutch Dog

spaces in the gears, two separate conditions are involved: First, the sprocket movement when gear *E* rotates gear *B* while gears *C* and *G* are locked; and second, the sprocket movement in the opposite direction when gear *E* rotates gear *C* while gears *B* and *F* are locked.

One Revolution of Shaft is Followed by Dwell Equivalent to One Revolution.—The shaft *A* of the drive shown diagrammatically in Fig. 2 is required to make a revolution and then stop or dwell for a period equivalent to one revolution. Shaft *A* is driven by shaft *B*, which rotates continuously. The gear *G* is keyed to shaft *A* and meshes with the gear *E*, which is a running fit on shaft *B*. The sleeve *C* is pinned to the driving shaft *B*. When the drive is in operation, the clutch dog *D*, which is a sliding fit in a slot in sleeve *C*, drives gear *E* one-half revolution; then as the angular face on the dog comes in contact with the angular or cam face *F* of the stationary piece *H*, the dog is withdrawn from contact with gear *E* at point *K*, allowing gear *E* to remain stationary while shaft *B* makes one-half revolution.

After shaft *B* has made one-half revolution, the dog *D* passes out of contact with the piece *H* at point *L*, allowing the dog to re-engage gear *E* through the action of spring *S*. Gear *E* then makes one-half revolution, following which the cycle of movements described is repeated. Thus gear *E* rotates one-half revolution and then remains stationary while shaft *B* rotates one-half revolution.

As gears *E* and *G* have a driving ratio of 2 to 1, gear *G* is given the required intermittent motion. A wide range in the timing of the intermittent motion may be obtained by varying the ratio of the gears and the length of the actuating or cam surface of the piece *H*.

Positive High Speed Intermittent Rotary Motion.—The mechanism shown in Fig. 3 provides the intermittent rotary motion required for operating the conveyor of a wire stitcher. The member *A* receives its intermittent rotary motion from the continuously rotating shaft *B* through the positive indexing action of an eccentric strap *C* operated by the eccentric *L* keyed to shaft *B*. A sprocket or gear—not shown—attached to the hub or face of member *A* transmits the intermittent motion to the conveyor.

At each revolution of shaft *B* the member *A* is indexed 1/11 revolution by pin *D* which engages one of the eleven evenly spaced slots *S*. During the idle or return movement of pin *D*, the member *A* is locked in position by pin *E*. The ratio of the idle time between the indexing movements, to the indexing time, is 73 to 107 in the design illustrated.

At the beginning of the indexing movement, member *A*

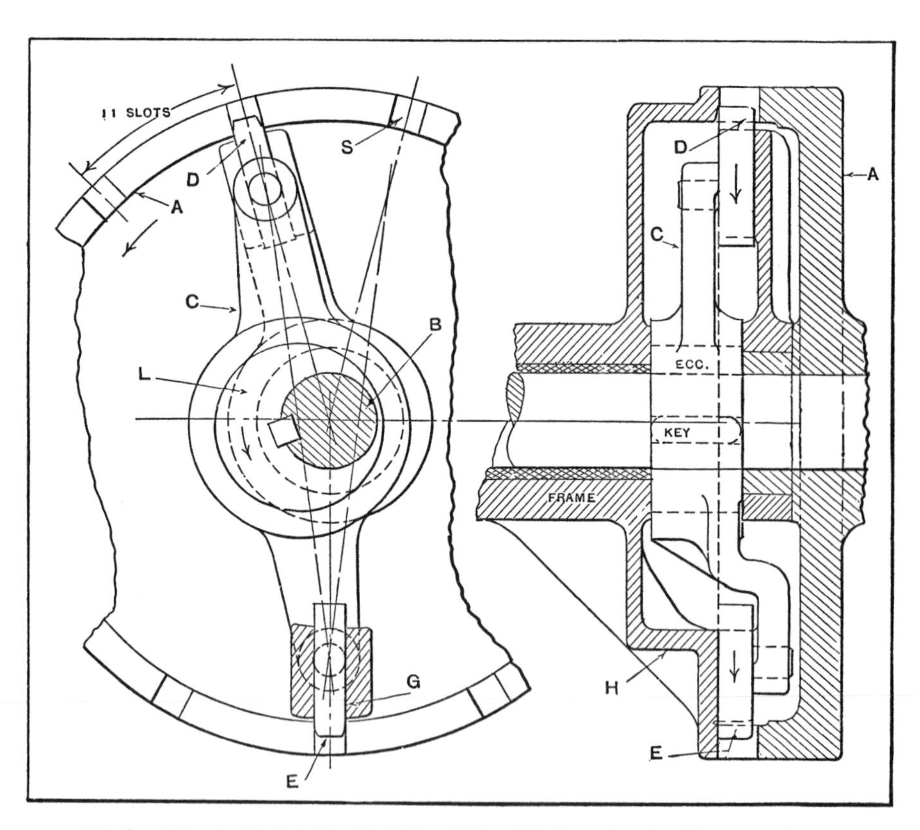

Fig. 3. A Constantly Rotating Shaft B and Eccentric Imparts a Positive Intermittent Motion to Member A which is Locked During the Idle Period

moves slowly, but the speed increases rapidly to the maximum and then slows down as the end of the movement is reached. As the mechanism stops the load slowly and without shock, it can be operated at high speeds, as compared with the usual ratchet wheel and pawl mechanism, which is difficult to balance and has a tendency to "overthrow" under appreciable loads.

The member *A* is always engaged by one or both of the actuating teeth *D* and *E*. These teeth are pivoted to the extreme opposite ends of the eccentric strap *C*. Overthrow is prevented by the locking tooth *E*, which is a free sliding fit in the groove *G* machined in the frame *H*. This slot restricts the movement of the pivoted tooth *E* so that it is forced to travel in a vertical direction.

The view at the left shows the parts of the mechanism in the positions they occupy at the completion of the indexing movement. It will be noticed that the tooth *E* has entered one of the slots in member *A* before tooth *D* has become disengaged from another slot of the member.

The peculiar motion imparted to the eccentric strap *C* by the eccentric *L*, due to the vertical path which its lower end is forced to follow, causes the top end of the strap to move in an elliptical path, carrying with it the actuating tooth *D*. Tooth *D* is always held in a radial position by a slot in the guiding arm, which is a free fit on the hub of member *A*. The elliptical motion and the radial guide force the actuating tooth *D* to engage and disengage successively the slots in the edge of member *A*, thus converting constant rotary motion into a positive intermittent motion.

The mechanism is equally efficient when operated in either direction. In adapting it for other purposes, the following characteristics should be considered: The slot spacing in member *A* controls the amount of intermittent motion and also the idle time. There must be an odd number of slots if the best action is to be obtained, but as a gear or sprocket drive of the proper ratio can be used to suit individual requirements, this characteristic is not a vital fault.

If too few slots are used, the eccentric throw will be excessive and the action of the mechanism will not be so smooth as with a greater number of slots. It is well to bear in mind that fewer slots decrease and more slots increase the idle time.

The interior of the mechanism shown is filled with grease, but the flanged parts of the frame could be fitted with a cover and a packing ring could be provided on the hub of member *A* so that the mechanism could be filled with oil. With this form of lubrication, the carrying power would be increased to handle greater loads at higher speeds.

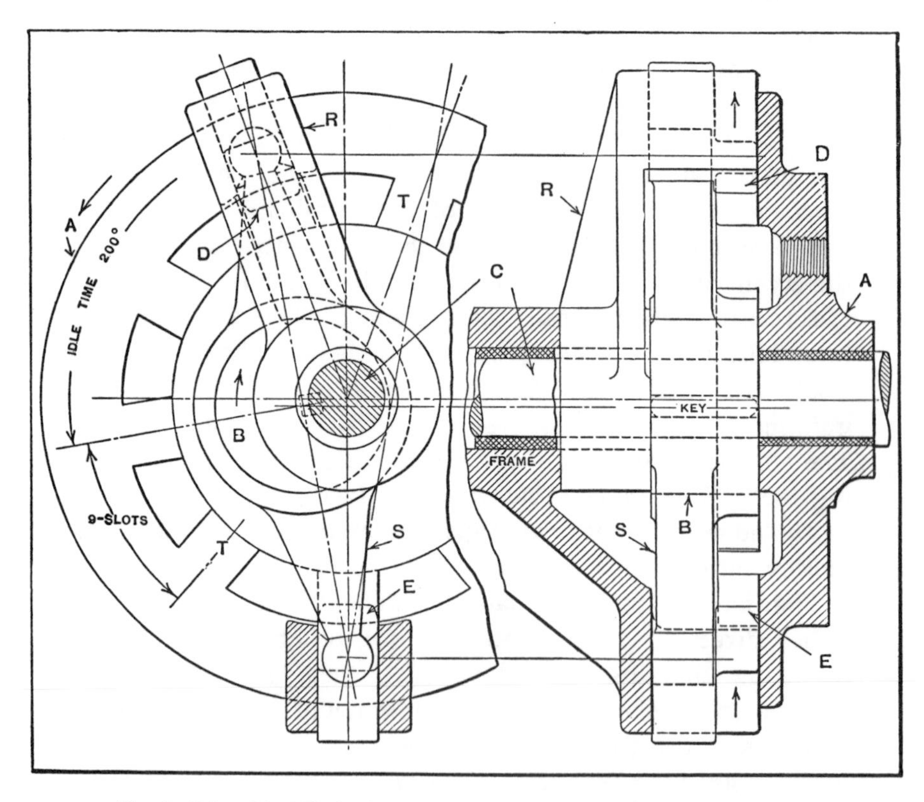

Fig. 4. Intermittent Mechanism which Provides a Longer Idle Period than the Design Shown in Fig. 3

Another Design of Positive High-Speed Intermittent Motion.— The intermittent rotary motion just described is suited for use where the idle time is short, as compared with the feeding time. The total idle time between the rotary or feeding movements in the case of the mechanism about to be described is equivalent to more than 180 degrees per revolution, while in the case of the previously described mechanism it was less than 180 degrees.

The principal difference in the designs is found in the location of the actuating teeth *D* and *E* (see Fig. 4) in relation to the member *A* to which the intermittent or indexing motion is imparted. The actuating teeth operate on the outside of the slotted ring of member *A*, so that the idle time occurs while the eccentric throw travels above the center line during the return stroke of strap *S*. The forward or feeding movement, therefore, occurs while the eccentric throw travels below the horizontal center line. With this arrangement, the longer throw of the eccentric is utilized for the idle or return movement of arm *S*, while the shorter throw is employed for the feeding movement. It will be noted that the shaft and its eccentric driver *B* rotate in a direction opposite to that of the driven member *A*, whereas in the design previously described, these members rotate in the same direction.

When the mechanism is in operation, the central shaft to which the eccentric driver *B* is keyed rotates at a constant speed. The lower end of the strap *S* is restricted to a vertical motion by the tooth *E*. Tooth *E* is pivotally mounted on strap *S*, having a wide bearing on the strap, and slides freely in a groove in the rigid frame. The tooth *D* at the opposite end of strap *S* has a similar pivoted connection to the strap and slides freely in a groove cut in the radial rocker guide *R*, which is a free turning fit on the central shaft bushing. The guide rocker *R* serves to maintain the tooth *D* in a radial position with respect to the central shaft. When the mechanism is in operation, the eccentric *B* causes the teeth *D* and *E* alternately to engage and disengage the equally spaced slots *T* cut in the annular ring and produces the intermittent motion of member *A*.

The vertical motion of the bottom end of the strap *S*, combined with the rotary motion at the center imparted by the eccentric *B*, gives the top end of the strap carrying tooth *D* a peculiar elliptical motion. This motion is such that tooth *D* alternately engages and disengages successive slots *T*, thereby imparting the required intermittent motion to the driven member *A*. The length and shape of the teeth are such that either or both teeth are always in engagement with slots in member *A*, thus giving a positive control over the motion. Tooth *E* locks the part *A* in position while tooth *D* is on its return or idle stroke. Tooth *D* engages a slot preparatory to the forward movement before the locking tooth *E* is disengaged. A hub at the side of member *A* is provided so that a sprocket or gear can be attached to it, through which the motion is transmitted to other parts.

This mechanism has several desirable characteristics, such as its slow starting and stopping action, absence of over-throw, positive locking between movements, compactness, and ability to operate in either direction. It can also be operated at relatively high speeds. The idle time is determined by the number of slots, and is always equivalent to more than 180 degrees per revolution. With nine slots, as in the design illustrated, the idle time is 200 degrees per revolution. With fewer slots, the idle time would be greater, and with a greater number, the idle time would be less. For the best action, there should be an odd number of indexing slots.

Planetary Intermittent Gearing.— Intermittent gearing of the planetary type may be used to advantage in cases where the driving and driven shafts must be in line with each other, and where a large number of dwells per revolution of the driven shaft is required. A drive of this type is shown in Fig. 5.

The ring gear *A* is stationary, and the single-tooth gear *B* is driven by means of the shaft *C* through the gears *D* and *E*. Gear *D* is keyed to shaft *C*, while gear *E* is integral with gear *B*. Both gears *E* and *B* are free to turn on the shaft *J*, mounted in the arms *F* and *G*. A hub on the lower end of arm *F* turns freely in the stationary bearing *H*, the intermittent movement being taken from this hub. Gears

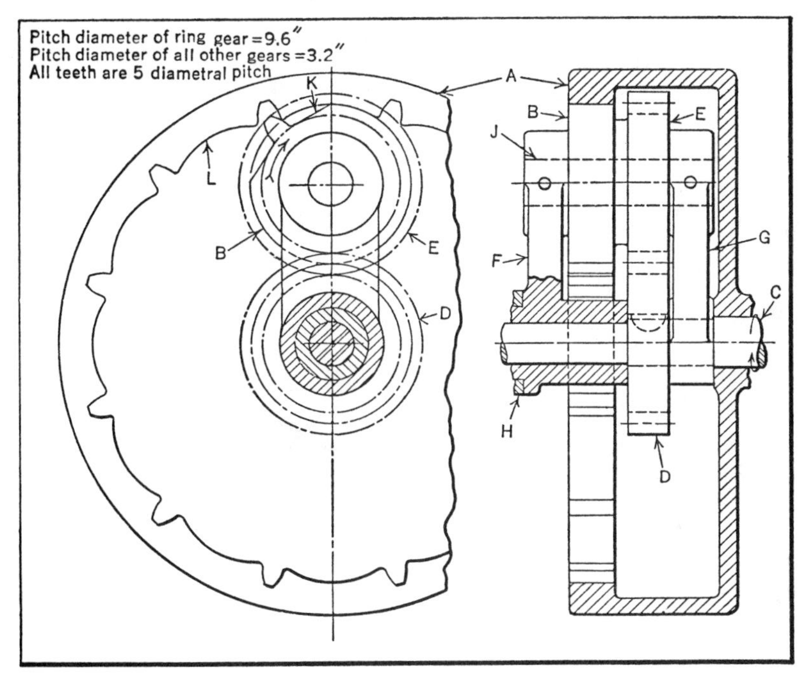

Fig. 5. Intermittent Gearing of Planetary Type

D, E, and *B,* in this case, have the same pitch and pitch diameters; hence, according to the principles of epicyclic gearing, one-third of a revolution of shaft *C* is required to index the arm *F* one division.

In the position shown, the single tooth in gear *B* is about to engage a tooth space in the ring gear. As soon as this engagement occurs, arms *F* and *G* will start to rotate on shaft *C* in a counter-clockwise direction. Rotation of the

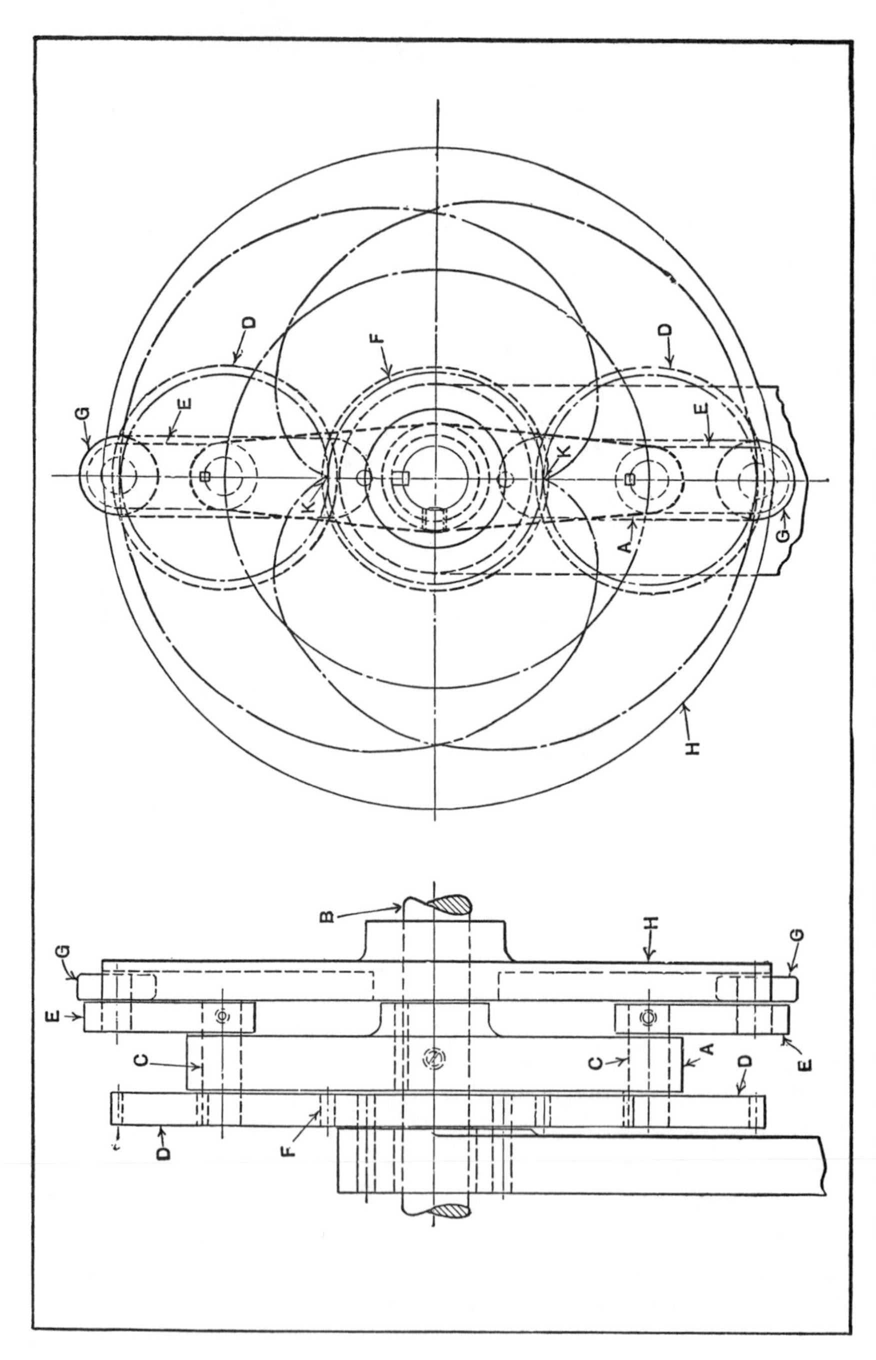

Fig. 6. Planetary Type of Crank Motion for Gradually Varying Speed of Driven Member from Zero to Maximum Velocity and Back to Zero

arms continues until the single tooth has left the tooth space, at which time the concentric portion of gear *B* engages the corresponding cylindrical surface *L* in the ring gear, locking gear *B* and causing arms *F* and *G* to dwell. The arms continue to dwell until the tooth in gear *B* has engaged the next tooth space in the ring gear, which causes the arms to move toward their next dwelling position. In designing the single-tooth gear *B*, sufficient clearance should be provided, as indicated, at *K;* otherwise, interference with the ring gear will result.

Rotary Motion which Varies from Zero to Maximum and Vice Versa.—The purpose of the mechanism illustrated in Fig. 6 is to produce an intermittent rotary motion which will start and stop a driven member without shock and yet keep it under positive control throughout the cycle. This motion is obtained by the practical application of the mathematical curve known as the epicycloid, which is the curve traced by a point on a circle as the latter revolves on the outside of another fixed circle.

The arm *A* is keyed to the driving shaft *B* which revolves continuously at a constant rate. At each end of arm *A* is a revolving shaft *C*, which has a revolving gear *D* keyed to one end and a short arm or crank *E* keyed to the other. The two revolving gears *D* mesh with a fixed gear *F* which is concentric with the driving shaft. All three gears are of equal diameter. At the end of each crank *E* is a roller *G*, the center of which lies on the pitch circle of the corresponding gear *D*. The circular plate *H* revolves freely on the driving shaft *B* as it is driven by rollers *G*, each of which engages a radial slot in the plate. The drive is taken from plate *H* in any desired manner.

It will be seen from the illustration that the centers of the rollers *G* trace the epicycloids shown by the broken curves in the end view. As the arm *A* rotates, the angular velocity of the centers of the rollers and of the driven plate gradually increases from the zero point at *K* until, in the position shown, the angular velocity is greater than that of the driving shaft; from this point and during the following half revolution of the driving arm, the angular velocity of the roller center and of the driven plate is gradually reduced again to zero at point K. Since the gears are of equal

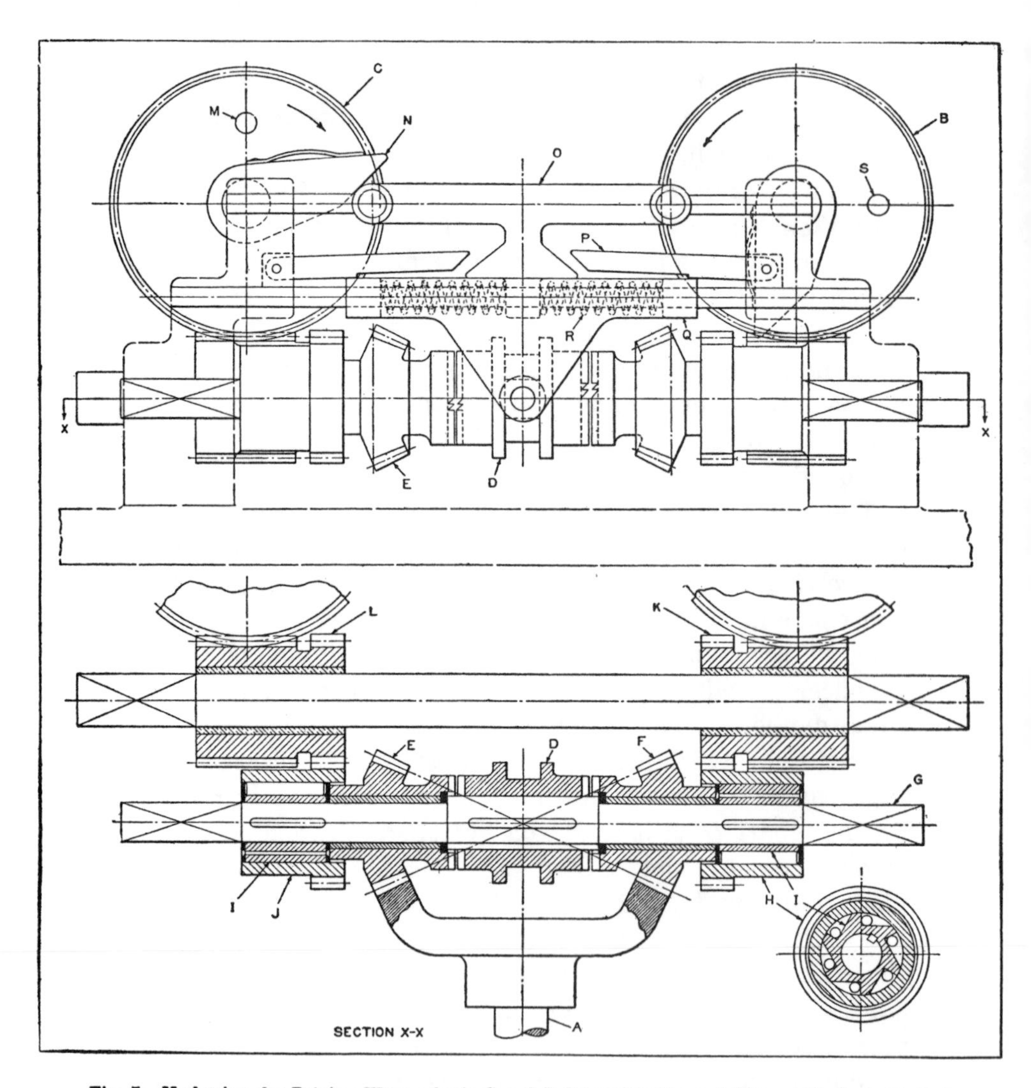

Fig. 7. Mechanism for Driving Worm-wheels C and B Intermittently and Alternately from Shaft A

diameter, the roller centers and the driven plate make a complete revolution in the same time as the driving shaft.

The angular velocity of the driving shaft is constant. If θ equals the angular position of the driving arm from the zero point K, and ω equals the angular velocity of the driven plate H, then $\omega = \frac{6a\,(1 - \cos\theta)}{5 - 4\cos\theta}$. The maximum angular velocity of the driven plate is 1.3 times that of the driving arm.

Mechanism for Driving Two Shafts Intermittently and Alternately.—The mechanism shown diagrammatically in Fig. 7 was designed for use on a special machine. In operation, the constantly rotating shaft A, through a gear train, drives worm-wheel B one revolution in the direction indicated by the arrow, after which the gear-shifting mechanism functions automatically, causing worm-wheel B to dwell and the driving motion to be transmitted to worm-wheel C through another gear train. After worm-wheel C has been driven one revolution, the gear-shifting mechanism again functions, causing worm-wheel C to dwell while the driving action is again transmitted to worm-wheel B, thus completing the cycle, which is continuous as long as the driving shaft A rotates.

The clutch member D, which is slidably keyed to shaft G, is shown in the neutral position, but when the mechanism is in operation, this clutch is in engagement with either pinion E or F, causing shaft G to rotate in one direction or the other, depending upon which pinion is engaged. The driving of shaft G in either direction from the crown gear on shaft A is made possible by the "free-wheeling" type friction clutches, consisting of two members H and I, and the friction rollers arranged as shown in the section view in the lower right-hand corner of the illustration. Two members I of the proper hand are keyed to the shaft and the two friction members H and J are slipped over them. Thus, when the clutch member *D* is in mesh with pinion *F*, the wedging action of the friction rollers serves to lock members *H* and *I* together as one piece, while member *J* runs freely over its mating member *I*. When the clutch engages pinion *E*, the direction of rotation of shaft *G* is reversed, the drive being through friction clutch *J*, while member *H* runs free. Thus, the direction of rotation of shaft *G* is controlled by the movement of clutch member *D*.

Spur gear teeth on members *H* and *J* mesh with the spur gears *K* and *L*, which have worms cut on their hubs that mesh with the worm-wheels *B* and *C*, respectively. The manner in which the clutch is automatically controlled to give the worm-wheels *B* and *C* their respective intermittent movements is shown by the upper view.

Assume that the mechanism is in operation and that clutch *D* is in engagement with gear *E*, so that worm-wheel *C* is being driven in the direction indicated by the arrow. When pin *M* on the worm-wheel comes in contact with the flat spring on the swinging arm *N*, which is a free turning fit on the worm-wheel shaft, it causes the swinging arm to come in contact with the roller on shifter lever *O*. The latch *P* would be down at this time instead of in the position shown. In the down position, a step on the latch engages a collar *Q* on the shifter slide, preventing the slide from moving to the right. Thus, continued rotation of the arm *N* serves to compress the spring *R* until a cam surface on the shifter lever lifts latch *P*, releasing the spring, which forces clutch member *D* to the right into mesh with gear *F*, engaging the drive to worm-wheel *B*, and allowing worm-wheel *C* to dwell. When this takes place, the shifter lever *O*, being released from the pressure exerted by spring *R*, also moves to the right and the arm *N* is rotated past the roller on lever *O* by the flat spring, previously compressed by pin *M*.

When worm-wheel *B* has rotated through the required angle, the pin *S* comes into contact with the flat spring on a swinging arm similar to the arm N previously described. The movement of the clutch member D into engagement with gear E is accomplished automatically, the same as the movement in the opposite direction. This cycle of operations is repeated automatically.

Adjustment of Intermittently Driven Sprockets while Drive is Operating.—Motion picture projectors are designed to give the film a rapid intermittent movement, stopping it sixteen times every second. These dwelling periods in the movement of the film are so timed that the light is projected through the film only when it is stationary. Moving pictures, therefore, are in reality a series of sixteen stationary pictures projected on the screen every second at normal operating speed, and owing to the "persistence of vision" this rapid succession of still pictures causes the successive views to blend into one another and produce the effect of continuous motion.

The edges of the film have accurately spaced perforations which are engaged by teeth on a double sprocket E (see Fig. 8). The drive to this sprocket is through the internal driving gear A, the driven gear B, driving disk C of the intermittent motion, and driven member D on the sprocket shaft. This driven member turns 1/4 revolution for every complete turn of driver C, and this quarter turn of D occurs during one-fifth of the revolution of C. During the remaining four-fifths of a turn of C, member D is locked in the stationary position.

This mechanism has an original feature which makes it possible to shift the film sprocket E from, say, its lowest position, which is the one illustrated, to a higher position, without interfering with the intermittent drive and while this drive is in operation. It was discovered accidentally that four points in a plane may be so located relative to one another that two of these points, if moved along straight lines perpendicular to each other, will cause a third point to describe an arc about the fourth; thus, if points a, b,

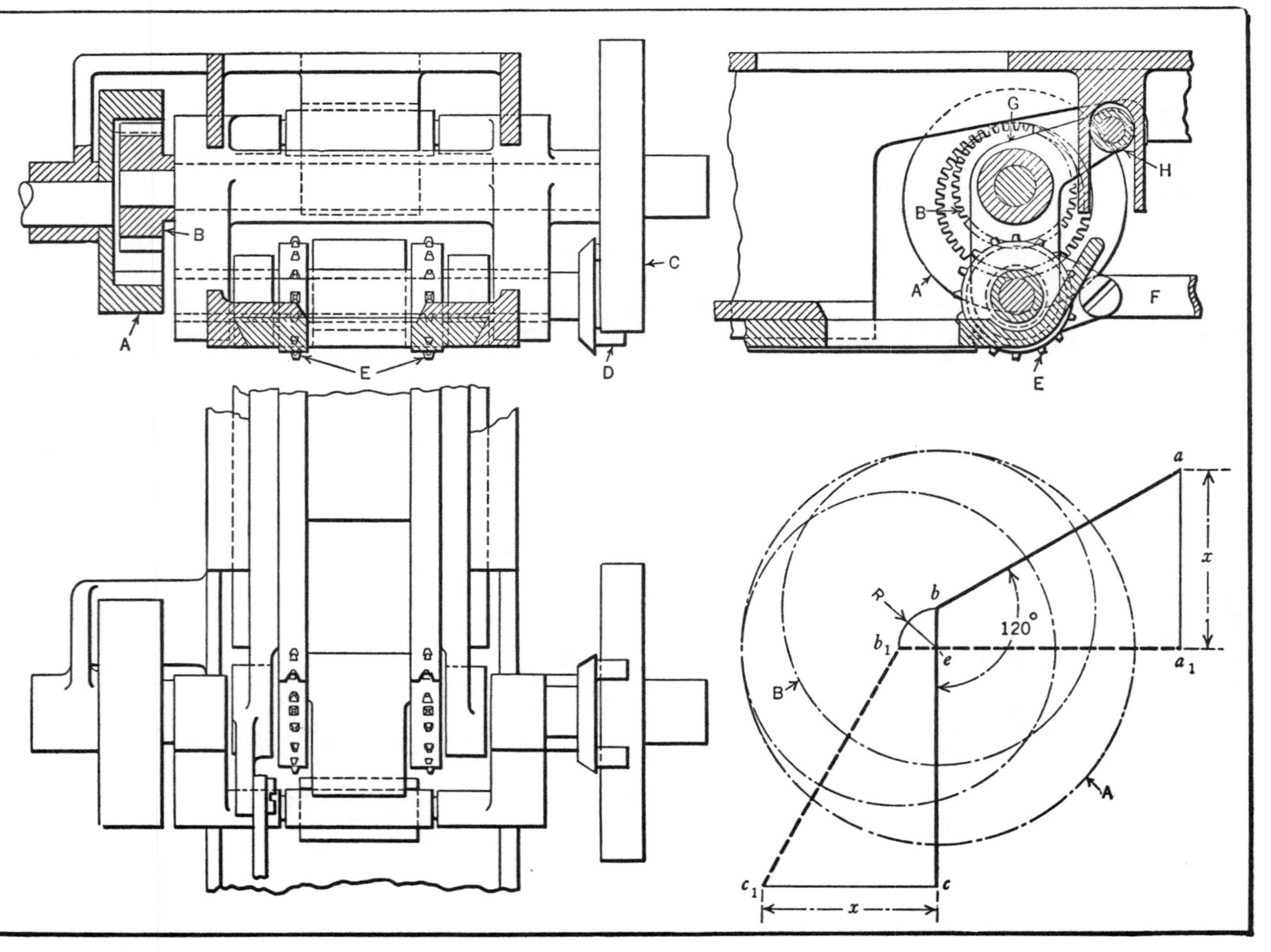

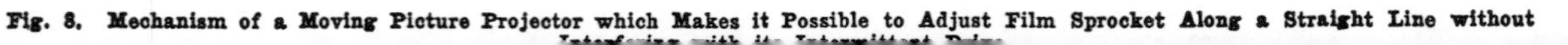
Fig. 8. Mechanism of a Moving Picture Projector which Makes it Possible to Adjust Film Sprocket Along a Straight Line without

and c on the diagram are located properly, the movement of point a to a_1 and of c to c_1 will cause point b to describe an arc of radius R about point e.

Before describing the essential requirements in this design, its practical application will be explained. This application is illustrated by the diagram just referred to, in conjunction with the sectional view just above it, which represents the actual mechanism. The line abc of the diagram represents the center line of the arm G. When the axis of the sprocket E is shifted along a straight line, as indicated on the diagram at cc_1 the axis of roller H moves along a perpendicular straight line aa_1. In conjunction with these two straight-line movements, the axis of driven gear B (represented at b on the diagram) describes an arc bb_1 of 90 degrees. As this arc is concentric with the axis of driving gear A, driven gear B continues to mesh properly with A during the straight-line movement of sprocket E, which is the requirement.

From what has preceded it will be evident that, in designing this mechanism, the problem is to so proportion the angular arm abc that when c and a move along straight lines at right angles, point b will follow a circular arc having a radius R equal to one-half the difference between the pitch diameters of gears A and B. To obtain this circular movement of point b, the design must be according to the following requirements:

Points a and c must be equidistant from point b.

The angle between arms ab and bc must be 120 degrees. If x equals the length of the straight-line movement, and y equals the dimension ab or bc, then,

$$y = \frac{x}{\cos 30^\circ\ (2 \cos 30^\circ - 1)} = 1.57x$$

$$\text{Radius } R = \frac{x}{2 \cos 30^\circ - 1} \times \left(\frac{1}{\cos 30^\circ} - 1 \right) = 0.21x$$

The movement of sprocket E is effected by a hand-lever

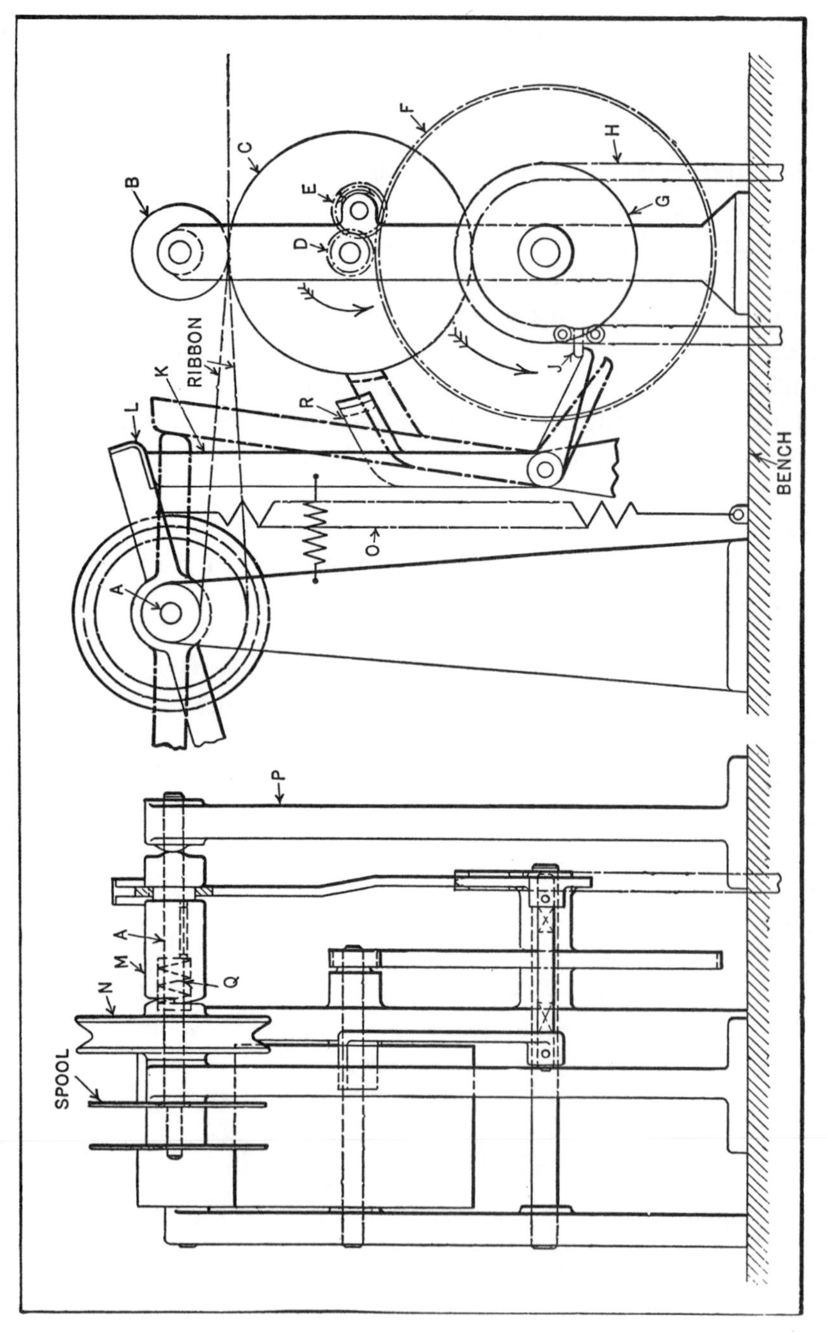

Fig. 9. Mechanism for Winding a Predetermined Length of Typewriter Ribbon on a Spool

connected with link *F*. This adjustment in the position of the sprocket is only used when an improperly made splice in the film requires what is known as "framing."

Intermittent Rotation for Measuring Typewriter Ribbon as it is Wound on Spool.—In winding typewriter ribbon on spools, a device like the one shown in Fig. 9 is used for stopping the rotation of the spool when a predetermined length of ribbon is wound up on it, at which time the ribbon is cut off. The spool is slipped on the end of the power-driven shaft *A*, and the ribbon is drawn from between the two rolls *B* and *C*. A coil spring (not shown), acting upon the upper roll, serves to keep a constant pressure of the rolls on the ribbon, so that as the ribbon is wound on the spool, both rolls are rotated.

Roll *C*, through the medium of the gears *D*, *E*, and *F*, rotates the chain sprocket *G*, over which the chain *H* is hung. Protruding from one of the links in this chain is the pin *J* which, through the levers *K* and *L*, disengages the clutch *M* from the driving pulley *N* and thus discontinues the rotation of the spool. The length of the ribbon wound on this spool depends upon the circumference of the roll *C*, the ratio of the gears, the number of teeth in the sprocket, and the number of links in the chain.

For every cycle of this chain, the pin depresses the lower end of the lever *K*, and in doing so, forces the upper end of the lever toward the right, allowing the hand-lever *L* to swing in a clockwise direction under the pull of the spring *O*. This hand-lever is secured to the clutch member *M*, at the right-hand end of which is a cam-shaped projection engaging a similar projection on the stationary hub of the bracket *P*. As the end of lever *L* moves downward, the clutch member is oscillated on the shaft *A* and the cam projections are disengaged, allowing the coil spring *Q* to force the clutch member to the right and disengage its teeth from those of the driving pulley, thus discontinuing the rotation of the shaft and the spool.

To prevent further rotation of the rolls (due to inertia) after the clutch has been disengaged, the brake arm *R* is provided. This arm is attached to the pivot shaft of the lever *K*, so that just as soon as this lever has been tripped, or immediately after the clutch has been disengaged, the end of the hand-lever *L* swings downward and wedges against the edge of lever *K*, forcing the brake-shoe against the roll *C*.

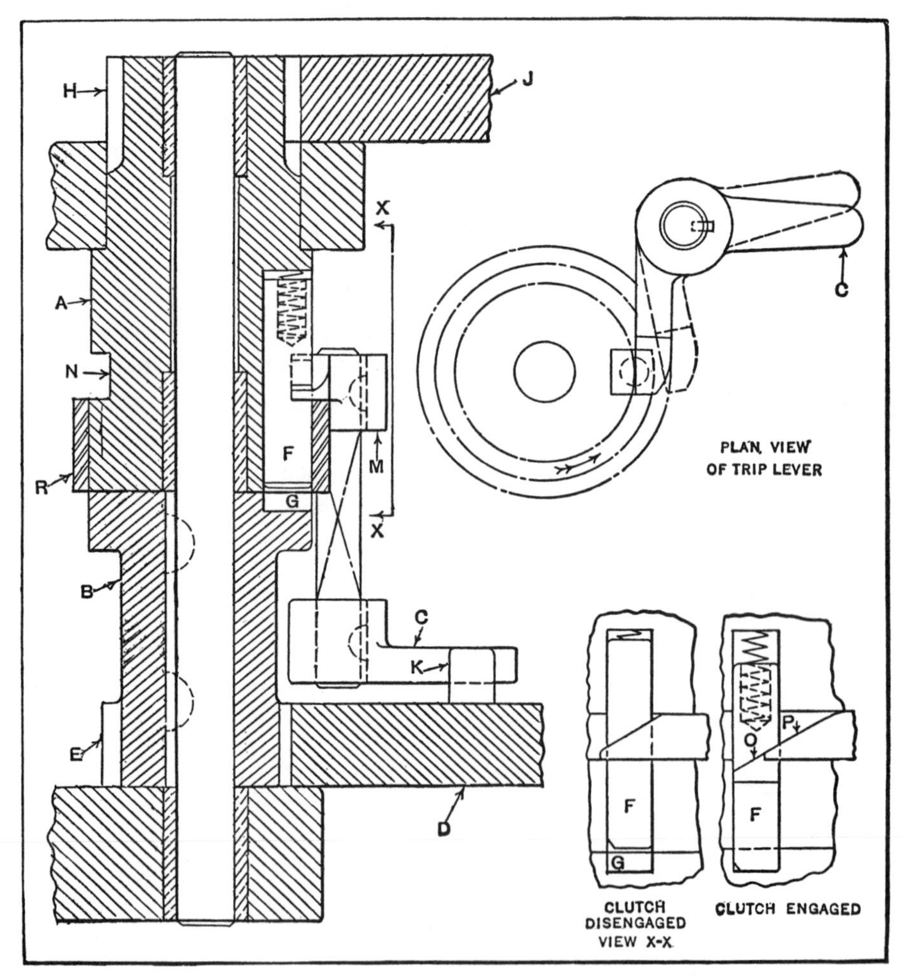

Fig. 10. This Clutch is Operated Intermittently through the Action of a Trip-lever

In the design shown, roll *C* has a circumference of 12 inches, and the ratio of the gears and sprockets is such that when the roll rotates once, the linear movement of the chain is equal to the pitch of the chain; hence, the number of links in the chain corresponds to the number of feet of ribbon upon the spool. Although designed primarily for winding typewriter ribbon, this device could doubtless be used successfully for other applications.

Intermittent Movement from Continuous Rotary Motion.—A mechanism for transforming continuous rotary motion into intermittent rotary motion is shown in Fig. 10. A movement such as this is often applied to indexing plates or tables of multi-stage drilling or chucking machines. The mechanism consists chiefly of two clutch members *A* and *B*, which are automatically disengaged at uniform intervals by means of a key actuated by the trip-lever *C*.

The driving gear *D* is rotated uniformly, receiving its motion from some other member of the machine. This gear meshes with the pinion *E* on the lower clutch member *B*. In the upper clutch member *A* is a sliding key *F*, which is backed up by a coil spring. This key is forced by the spring into the slot *G* when the lower clutch member is rotated into a position where the key and slot are in alignment. When this engagement occurs, both members of the clutch are locked together. The ring *R*, which is shrunk on the lower part of the clutch member *A*, serves to retain the key in its slot.

Pinion *H* is integral with the upper part of the clutch member *A* and meshes with the gear *J*. This gear, in turn, serves to drive an indexing plate (not shown). The clutch is engaged through the action of the pin *K* in the driving gear *D;* and at a certain point in the rotation of this gear, the pin trips the lever *C* so that the lever and the dog *M* assume the position indicated by the dotted lines in the detail plan view at the right. The shaft on which the lever and dog are keyed rotates in a stationary bearing secured to the machine. As soon as the point of the dog is lifted out of the wedge-shaped slot in the key, the latter is free to drop down into the slot *G*, provided the two members of the clutch are located in the proper position radially. When the key enters the slot *G*, the gear *J* rotates. In the meantime, the pin *K* has passed the lever *C*, and the dog *M* is returned to its original position in the annular groove *N* by a spring (not shown).

When the clutch has rotated nearly a complete revolution, a bevel face *O* on the key *F* (see detail in lower right-hand corner) comes into contact with a bevel *P* on the dog; and as the clutch continues to rotate, the key is forced upward and out of the slot *G*, as shown in the detail view.

This mechanism will operate satisfactorily at speeds up to 50 revolutions per minute, but at higher speeds, the key is not given sufficient time to drop into the slot *G*. If accurate indexing is required, the usual plunger arrangement for the indexing plate is used in conjunction with the mechanism described. A mechanism of this type lends itself very well to jobs where it is necessary to vary the indexing ratios. To obtain the various ratios, the gear *H* may be made demountable with respect to clutch member *A*. In this way, gears *H* and *J* can be changed to suit the required indexing ratio.

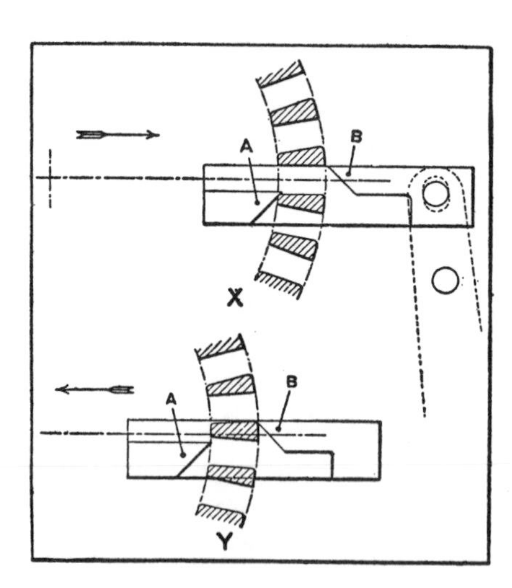

Fig. 11. Simple Indexing Mechanism which is Operated Rapidly by One Lever

Escapement Type of Indexing Mechanism.—A simple indexing mechanism consisting of a rotating ring having a number of radially milled slots and sliding indexing fingers or cams is shown in Fig. 11. The device is operated by means of a hand-lever, which is indicated by dotted lines. At *X*, the finger *B* has just left a slot, and the inclined face

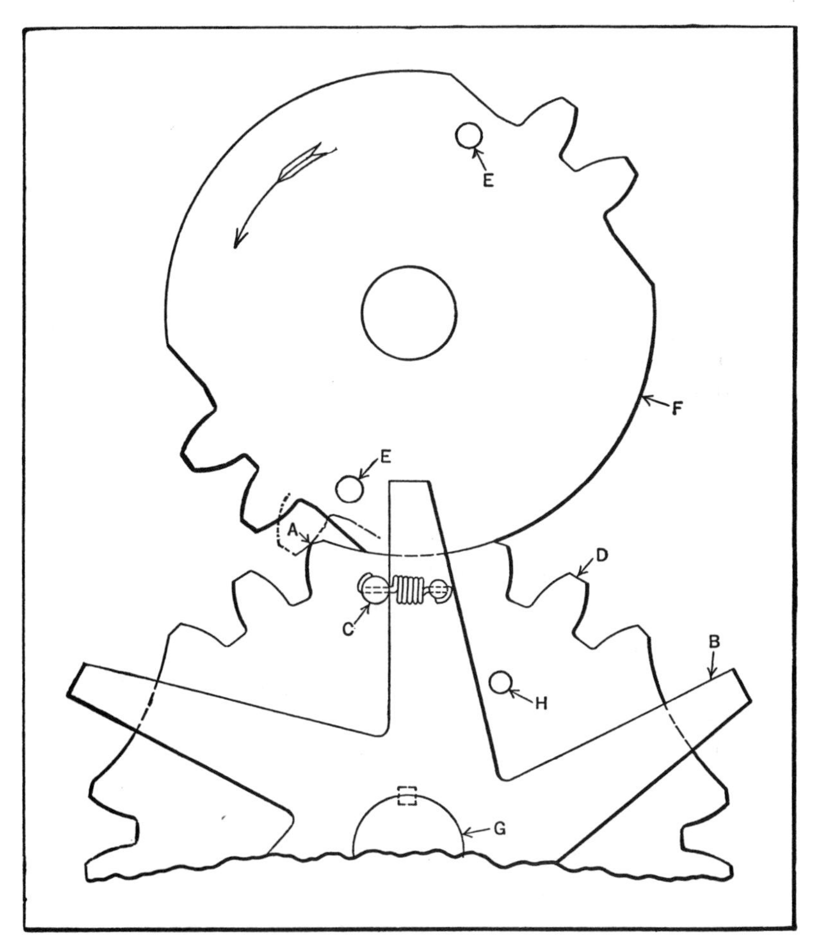

Fig. 12. High-speed Intermittent Gearing with Arrangement for Reducing Tooth Impact

of the finger *A* is engaging a corner of the same slot. As the finger continues toward the right and enters the slot, the ring is moved through a little more than half a division. The movement of the finger is now reversed and the finger *B* engages the corner of the next slot, as seen at *Y*. As the movement continues toward the left, the finger *B* enters the slot and pushes the ring around to its correct indexing position.

Shock Absorber for High-Speed Intermittent Gearing.—One of the greatest objections to the intermittent type of gearing when used for transmitting high-speed movements is the impact of the mating gear teeth at the beginning of each intermittent movement. This action is due, of course, to the inertia of the driven gear and the offset position at which tooth contact takes place, as indicated at *A* in the illustration.

The greater part of the wear, tooth breakage, and noisy operation resulting from the tooth impact is prevented by means of the arrangement shown in Fig. 12. Here a steel spider *B* having as many arms as there are dwelling positions in the driven gear *D* is mounted on the shaft *G*. This spider, although free to rotate on the gear-shaft, is held normally by a coil spring against pin *C* in gear *D*. The movement of the spider on the shaft is limited by pin *H*.

Just before the tooth contact at *A* occurs, one of the pins *E* in gear *F* forces the top arm of the spider toward the right, causing the spring to exert a pull on pin *C* and start gear *D* gradually. Thus the inertia of gear *D* is overcome before the contact at *A* occurs; hence the force of the impact at the point of engagement of the teeth is greatly reduced.

Auxiliary Friction-Driven Gear to Reduce Starting Shock of Intermittent Gearing.— In the operation of intermittent gear trains the impact of the teeth at the beginning of each movement may not be serious at lower speeds, but for higher speeds, the operation of the mechanism is likely to be noisy and the leading teeth are either soon battered out of shape or broken. To overcome this condition in an intermittent gear train operating an automatic hopper, a second set of gears was incorporated, as shown in

Fig. 13. These gears *A* and *B* serve to start the driven shaft *G* rotating with very little shock just before the leading teeth in the intermittent gears come into contact. Another advantage is that the starting torque is borne by a number of teeth in gears *A* and *B* instead of by two teeth only, as in the usual type of intermittent gear train, thus reducing tooth wear.

The intermittent gears, which are keyed to their respective shafts, are indicated at *D* and *C*. The second set of gears is also mounted on these shafts. Gear *A*, however, is free to turn on its shaft, while gear *B* is keyed to the lower shaft (not shown). Both the gears *A* and *B* have teeth all around their circumference, the tooth pitch and pitch diameters being the same as in the corresponding gears *D* and *C*. It will be noted that gear *A* is confined between friction washers, which tend to transmit a turning movement to the driven shaft. The pressure of the washers against the gear can be varied by adjusting the lock-nuts, which changes the tension of the coil spring. With this arrangement, the pitch-line speed of gear *A* and of gear *D* (when in motion) are the same.

In operation, the driving gears *C* and *B* rotate in the direction of the arrow. With the gears in the position shown, it is obvious that unless special provision is made, the entire force of impact in starting the indexing movement will be at point *H*. In the present design, however, part of the force is divided between several teeth in gears *A* and *B*. As soon as point *F* has passed point *E*, gear *D* begins to rotate, through the action of the friction drive, before the leading teeth in gears *D* and *C* come into contact. This rotation is started with practically no shock, and continues until the teeth of both intermittent gears are properly meshed.

Some experimenting may be required before the check-nuts are adjusted so that the tension on the coil spring is sufficient to balance the normal load imposed on the mech-

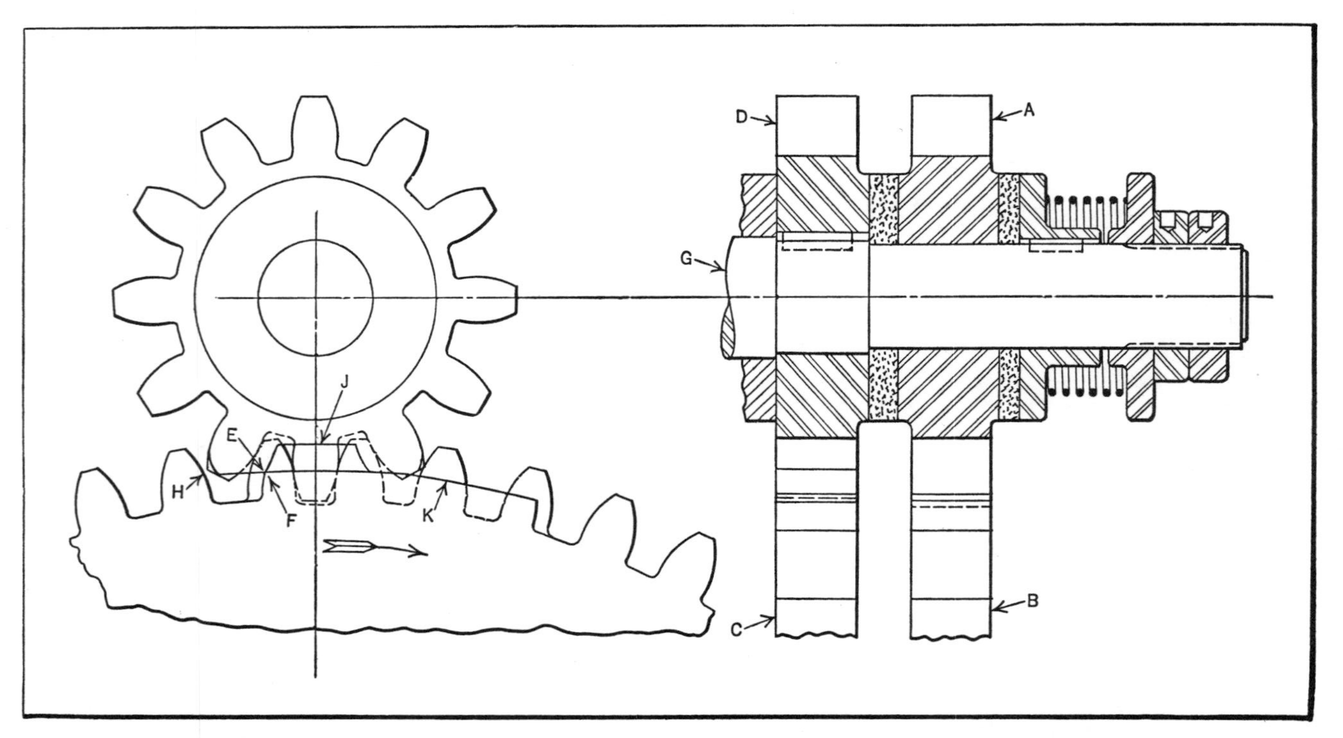

Fig. 13. Intermittent Gear Train in which Impact Shocks are Reduced by a Friction Gear Drive that Starts Each Indexing Movement

anism. The clearance J in gear D also deserves some mention. By removing the metal at this point, a longer dwelling surface K is obtained, thus reducing the time, at the beginning and end of each dwell, in which the gears are in their unlocked positions. It should be understood that this mechanism is suitable for light loads only. If the load is too great, the wear on the dwelling surfaces of the intermittent gears will be excessive due to the torque produced by the friction drive during each dwell. Rapid wear, however, can be prevented by the use of hardened inserts in the dwelling surfaces.

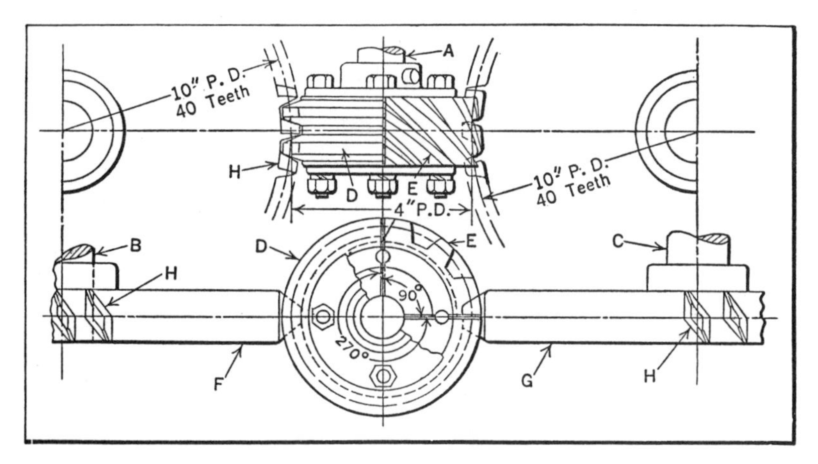

Fig. 14. Gear Drive with Special Gears Designed to Have Shaft A Drive Shafts B and C Intermittently

Gear Drive for Imparting Intermittent Motion Alternately to Parallel Shafts.—In designing a transformer tap changer, provision had to be made for alternately moving the arms of two tap adjusters with a dwell between each movement. Also the arms were required to be locked between movements. It was desirable to have the driving shaft at right angles to the shafts that operated the tap adjuster arms. The speed reduction was required to be approximately 1 to 12.

These conditions were fulfilled by the mechanism shown

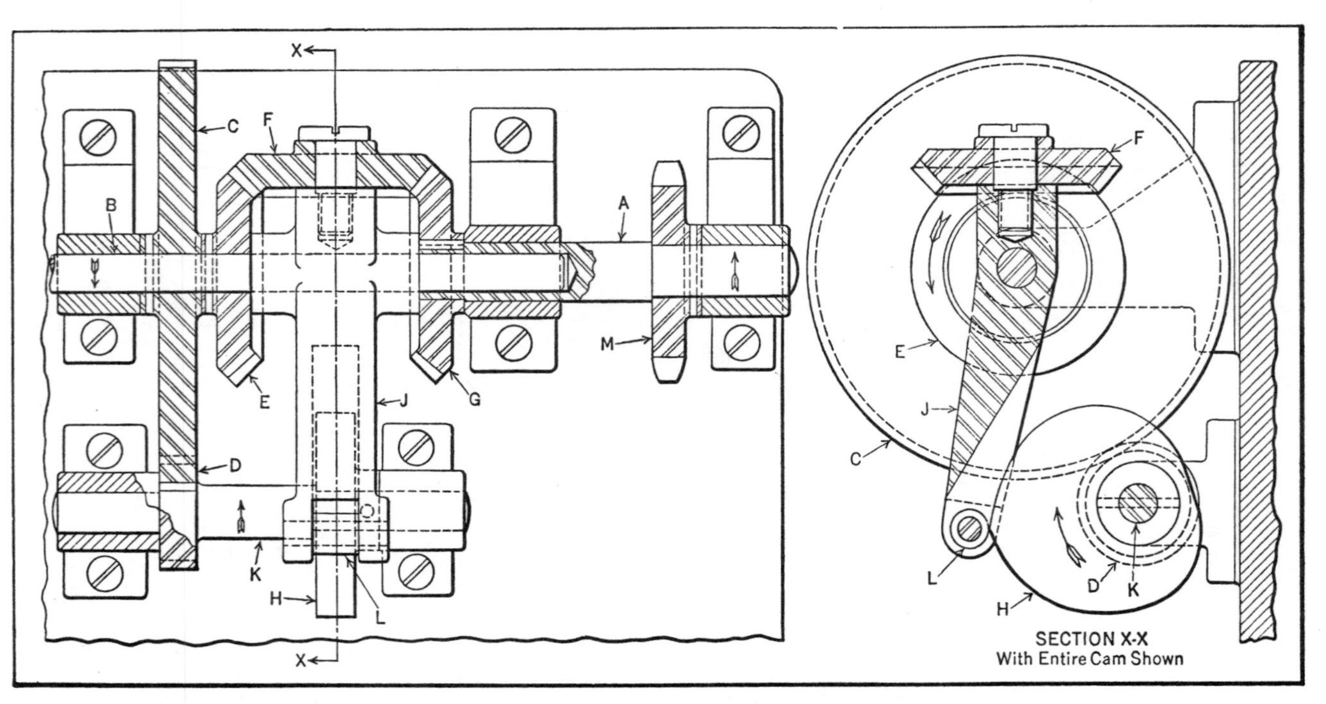

Fig. 15. Sprocket Drive for Conveyor with Mechanism that Causes Conveyor to Dwell at Operation Stations without Starting or Stopping Shock

in Fig. 14. The gear drive consists of a combination worm on the driving shaft *A* which meshes with two gears on the shafts *B* and *C* connected to the tap adjusters. The worm is built of two parts *D* and *E*. Part *D* has tooth spaces that appear simply like annular grooves. This part comprises a segment of 270 degrees. The other part *E* has a helix angle of 53.1 degrees. These two parts have grooves in their sides into which annular ribs on the side plates fit when the four members are bolted together as shown.

The two gears *F* and *G* are alternately in mesh with both parts *D* and *E* of the worm. This is accomplished by making gears *F* and *G* with teeth, as indicated at *H*, which will mesh with the teeth in both section *D* and section *E* of the driving worm. When the teeth in *F* and *G* are in mesh with the teeth in section *D*, no motion is transmitted from the driving to the driven shaft. One rotation of segment *E* past *F* or *G* serves to rotate either one of these gears through an angle of 27 degrees. The gears *F* and *G* each have 40 teeth. The section *E* is a 90-degree segment of a 12-thread worm. With this arrangement, gear *F* and then gear *G* will be turned through an angle of 27 degrees. There is a stop or dwell between each movement corresponding to three-fourths revolution of the driving shaft. Both pinions are locked between their respective rotational movements.

Combination Cam and Differential Gear Movement for Chain Conveyor.—Sprocket chain conveyors are used extensively for conveying containers through filling machines, and frequently the drive is arranged so that the chain dwells at regular intervals to permit the filling of the containers. One rather interesting drive for obtaining this intermittent conveyor movement is shown in Fig. 15. Its design embodies a cam which transmits a rocking movement to a differential planet gear for controlling the rotation of the driving sprocket of the conveyor chain. This mechanism has its application in a machine for filling glass vials with liquid. To prevent the spilling of the liquid from the vials as they pass along on the conveyor, provision is made to eliminate shock in stopping and starting the chain.

The intermittent movement is transmitted to the sprocket shaft *A* from the constantly rotating drive shaft *B* through the spur gears *C* and *D*, miter gears *E*, *F*, and *G*, and also through the cam *H*. Gears *C* and *E* are pinned to the drive shaft, the end of which turns freely in the end of the driven shaft *A*. On this shaft is keyed gear *G* which meshes with gear *F*. Gear *F* is mounted on the arm *J*, which is free to turn on shaft *B*. The outer end of arm *J* carries a follower roll *L* which engages the cam *H*, the latter being pinned to the pinion shaft *K*. In order to synchronize the conveyor movement with that of the rest of the machine, each intermittent cycle of the conveyor chain must occur during one-quarter revolution of the drive shaft *B*.

There are four vial stations to each length of conveyor chain equivalent to the pitch circumference of the sprocket; hence, in order to cause the chain to dwell as each station passes the filling valve, the sprocket *M* must dwell after each quarter revolution. It was found by experiment that a vial could be filled in the same time that it takes shaft *B* to rotate one-eighth revolution. Thus, having determined the angular movement of this shaft during the dwell period, it remains to proportion the gears and cam to impart the required rocking motion to arm *J* for obtaining this dwell; that is, to cause gear *F* to roll on gear *G* without rotating the latter and the sprocket.

Assuming that arm *J* is stationary, one-eighth revolution of gear *E* in the direction of the arrow would rotate gear *G* the same amount in the opposite direction. Now suppose that during this one-eighth revolution of gear *E*, arm *J* were rotated one-sixteenth revolution in the same direction. Then gear *F* would merely roll on gear *G* and the latter would remain stationary. Since we know the movement of arm *J* required to cause gear *G* and sprocket *M* to dwell during one-eighth revolution of the drive shaft, the contour of the cam can be developed. The throw of the cam will, of course, correspond to the angular movement of the arm. One complete cycle of the cam is required for each one-quarter revolution of the drive shaft. Therefore, the ratio of gears *C* and *D* must be 4 to 1.

Thus, while the drive shaft *B* rotates one-eighth revolution from the position shown, the cam will rotate one-half revolution and gear *F* will roll on gear *G*, causing the latter and the sprocket to dwell. During the next one-eighth revolution of shaft *B*, however, the cam will complete its revolution, swinging the arm in the opposite direction and causing gear *F* to rotate gear *G* one-quarter revolution, or twice the amount it would rotate if arm *J* were stationary. In this way, it will be seen that shafts *B* and *A* have the same angular movement for each station movement, although shaft *A* rotates at a higher velocity, owing to lost motion resulting from its dwell.

By observing the contour of the cam, it will be noted that it is developed to impart a constant rise for the first half revolution. This constant rise is important if a steady dwell is to be obtained. For the remaining half of the cam, the contour is such that the beginning of the upward movement of the arm is accelerated and then retarded at the top. This accelerating and retarding of the arm, when transmitted through the gears, results in a corresponding movement being imparted to the conveyor chain, the shock to the chain being so slight that spilling of the liquid in the vials does not occur. The working torque transmitted through the gears is sufficient to maintain engagement of the follower roll on the cam.

Parallel Slides with Latch and Cams for Operating One Slide Intermittently.—In connection with a certain extrusion process, it was found necessary to withdraw two sliding members of a stripping mechanism up to a predetermined point, after which one slide had to remain stationary while the other completed its full travel. On the return stroke, the stationary slide had to be "picked up" and carried along with the other slide. Fig. 16 shows how this is accomplished by the use of a swinging latch, the principle of which might well be applied to other devices where one of a pair of slides must have a temporary dwell. This latch operates between two flat profile cams with oppositely disposed notches shaped to receive the rollers on the latch.

The upper cam *A* is secured to the moving platen or slide

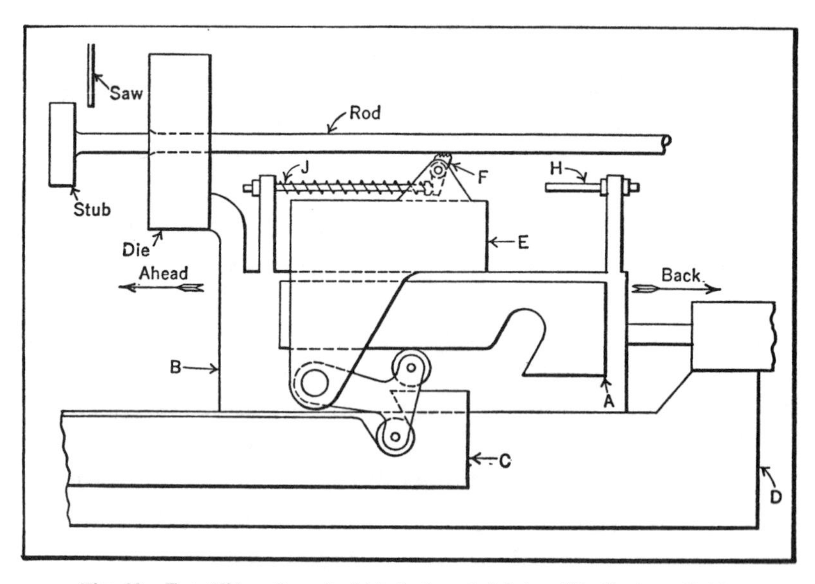

Fig. 16. Two Slides, One of which is Operated Intermittently by a Latch Actuated by Opposing Cams

B, while the lower cam *C* is fixed to the bedplate *D*. The carriage *E* has a limited range of sliding movement on the slide *B* equal to the longitudinal distance between the two cam notches when the platen is in its "back" position. At both ends of the carriage travel, adjustable trip-rods *J* and *H* engage and disengage pawls *F*, respectively, these pawls gripping the extruded rod.

As soon as the "stub" has been severed from the rod by the saw, the slide *B* is started ahead. At this time the rod is held stationary by the pawls *F* on the carriage, which is now anchored to the bedplate; as the die is attached to the moving slide, the effect is to strip the die from the rod. When the two notches in the cams come opposite each other, the latch swings out of the lower notch and into the upper one, so that the carriage is picked up and carried along with slide *B*. Just before the latch swings upward, at which time the rod is clear of the die, the rear trip-rod *H* releases the pawls, leaving the rod free to be removed.

Intermittent Movement of Reciprocating Slide.—Many ingenious slide movements are to be found in the various types of wire-forming machines. One intermittent movement, applicable to these machines, is shown in Figs. 17 and 18. In this design, the two adjacent slides *A* and *B* are actuated by the connecting-rod *C*. Slide *A* is connected directly to this rod and is given a continuous reciprocating movement. Slide *B* operates intermittently. For each cycle of the mechanism, slide *B* moves with slide *A* for one working and one return stroke, dwelling for three succeeding working and return strokes.

Both slides operate in the stationary guideway *D*. On slide *A* is mounted a locking device consisting of housing *E*, locking plunger *F* (Fig. 18) which engages bushing *G*, and cam *H* with its indexing pins *J*. This device is actuated by the spring pawl *K*, which slides in a boss on the guideway.

In the position shown, the slides are locked together by the plunger *F;* consequently, both slides are moving together. They have just completed their working stroke and are about to return in the direction indicated by the arrow (Fig. 17). On the return stroke, pawl *K* engages one of the pins *J* and rotates the cam 90 degrees, causing the projection *L* (Fig. 18) to slide upward along the deep notch in the housing *E* and drop into one of the three shallow notches *M*. This results in the plunger being withdrawn from bushing *G* in the lower slide just before the return stroke is completed. Hence, some means must be provided

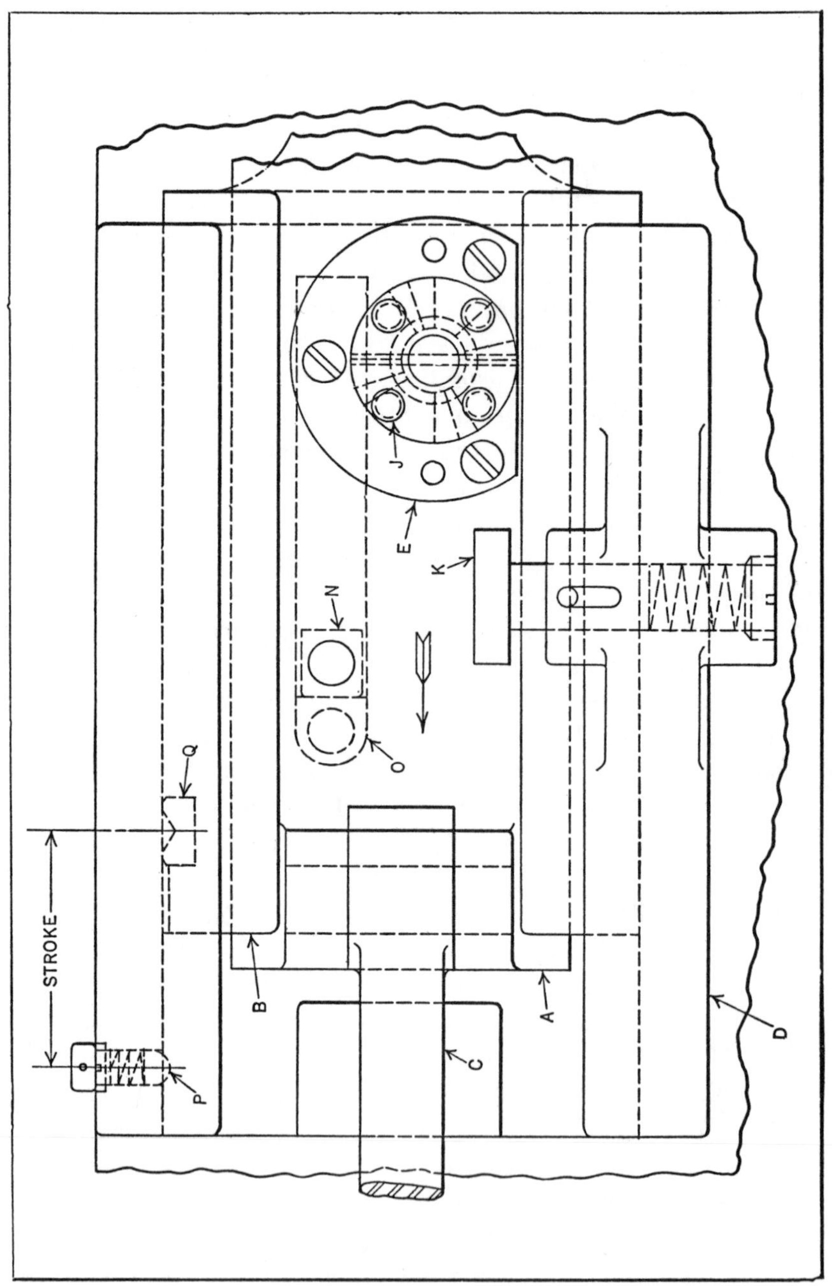

Fig. 17. Plan View of Double-slide Movement Shown in Fig. 18

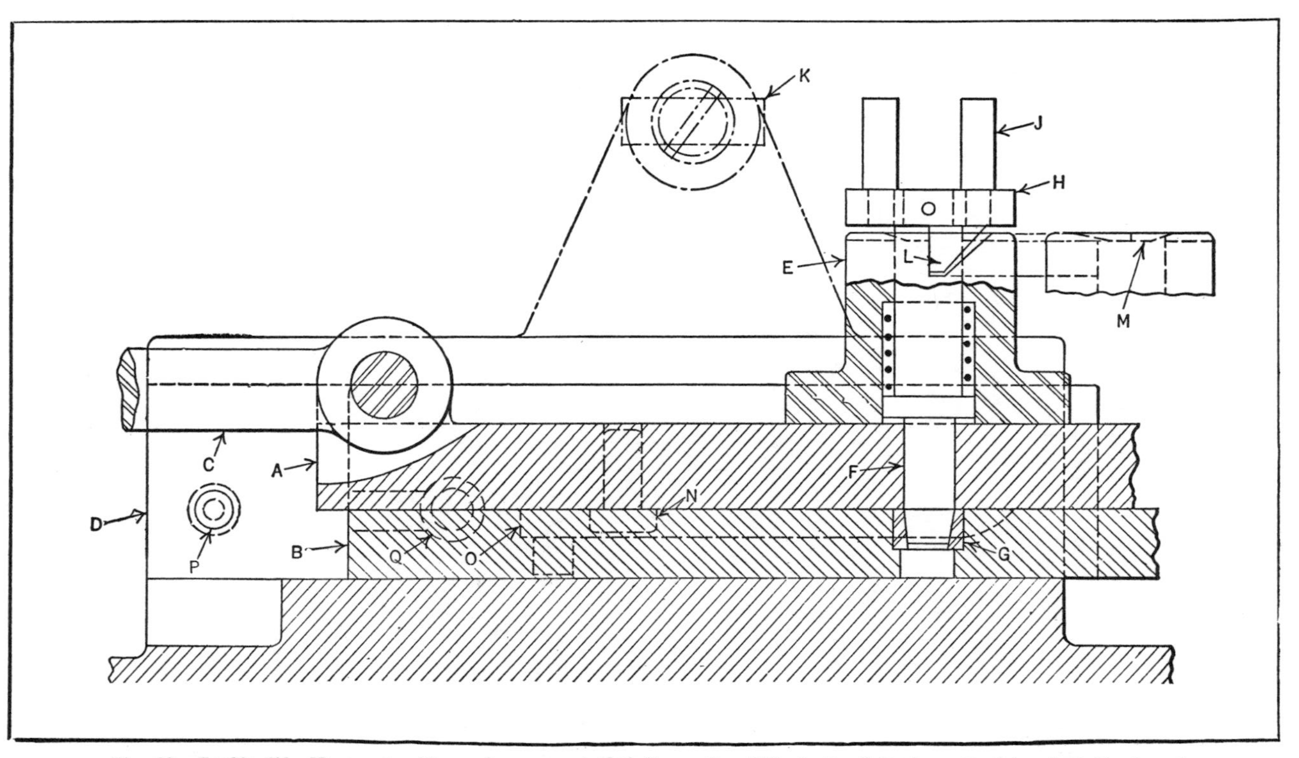

Fig. 18. Double-slide Movement with an Arrangement that Causes One Slide to Dwell During a Predetermined Number of Strokes of the Other Slide

for completing the return movement of slide *B*. Stops *N* and *O* serve this purpose. As these stops are in contact with each other at this time, the upper stop resumes pushing the lower slide to the end of its return stroke. At this point, spring button *P* engages the depression in the pad *Q* and prevents slide *B* from moving toward the right (during its dwell), due to frictional contact with slide *A*.

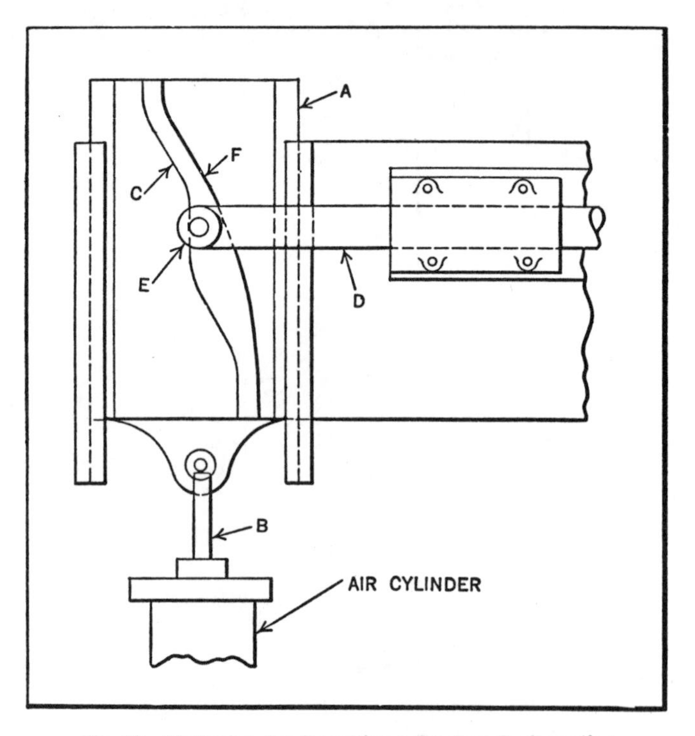

Fig. 19. Mechanism for Converting a Constant Reciprocating Movement into an Intermittent Movement

For each of the two succeeding working and return strokes of slide *A*, cam *H* is indexed 90 degrees, as previously described; but as the projection *L* enters a shallow notch for all three indexing movements, slide *B* remains stationary for three working and three return strokes of slide *A*. On the last return stroke, however, cam *H* is again indexed. This time the projection enters the deep notch, allowing plunger *F* to drop down and enter bushing *G*. At the end of this return stroke the cycle is completed, and on the succeeding working and return strokes both slides travel together. The object of the shallow notches in housing *E* is to prevent the cam from reversing its movement after being indexed due to the back drag on the pins *J* when they leave the pawl.

Intermittent Movements from a Constant Reciprocating Movement.—A feeding mechanism operated by an air cylinder was required to convert the constant reciprocating movement of the air piston into an intermittent movement on the outward stroke. This movement was to be at right angles to that of the piston. On the return stroke, the motion was to be continuous and at a constant speed. The mechanism for obtaining these movements is shown diagrammatically in Fig. 19. The reciprocating piston *B* is attached to the slide *A*. Slide *A* has a cam groove with the side *C* formed with a dwell to impart the required intermittent movement to the feeding plunger *D* on the outward movement of the piston.

On the return movement, the side *F* of the cam groove returns the plunger *D* to the starting position without the intermittent motion required on the outward movement. There is sufficient friction in the mechanism to keep the roller *E* of the plunger *D* in contact with the sides *C* and *F* on the outward and inward movements, respectively. Automatically operated air valves control the dwell at the end of each stroke. The speed of the feeding and return movements of plunger *D* is governed by the rate at which air is admitted to the cylinders by the air valves.

Adjustable Clock-Controlled Intermittent Mechanism.—The mechanism shown in Fig. 20 is used in a bottle-cap counting machine. It is the function of this mechanism to swing a pivoted chute alternately from one position to another, allowing the chute to remain in each position long

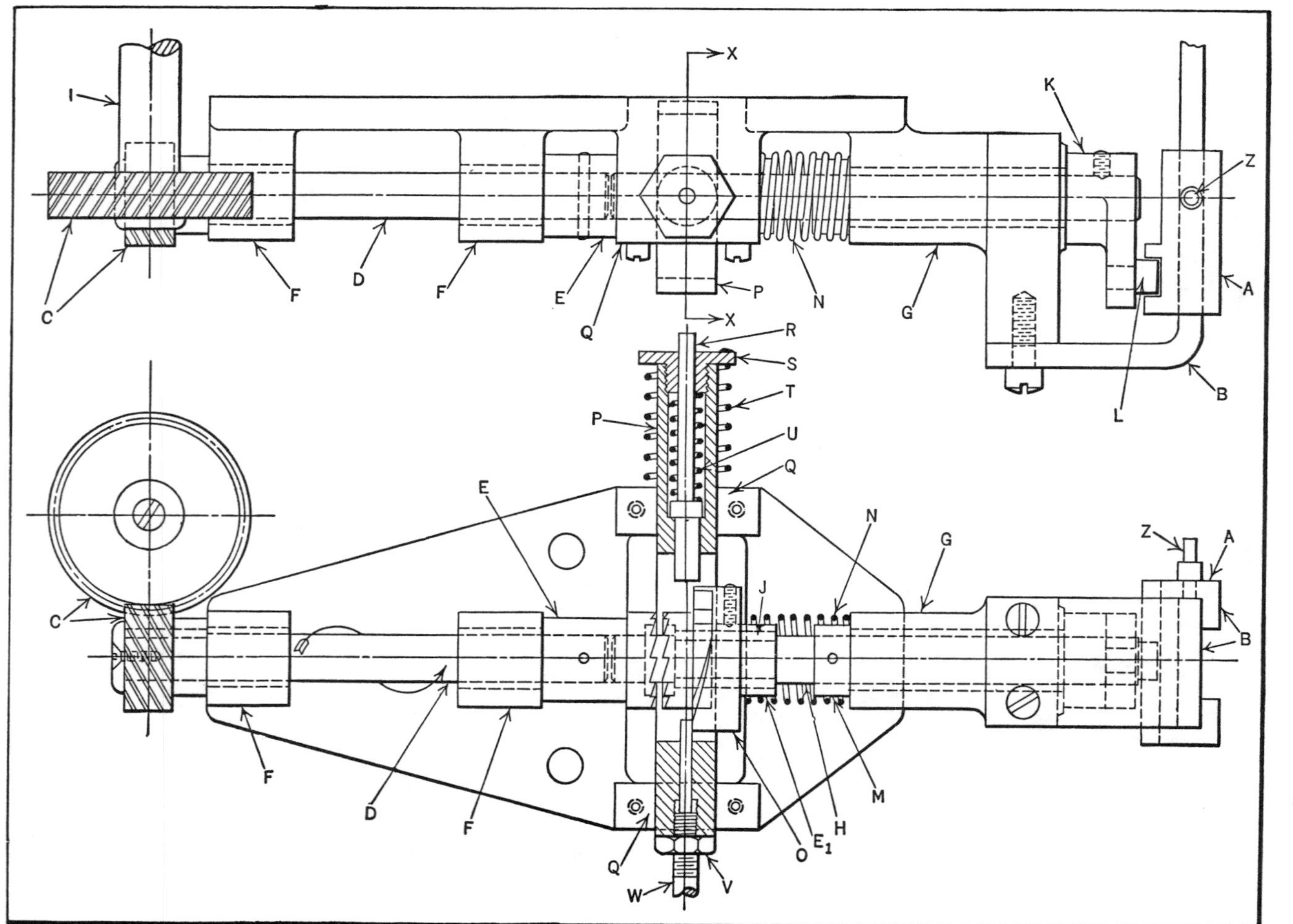

Fig. 20. Clock-controlled Mechanism for Imparting Intermittent Reciprocating Motion to Slide A

enough to permit a packing case to be filled with bottle-cap crowns which are delivered by the chute. Although the movements obtained with the mechanism here illustrated could be duplicated by other mechanical arrangements, none of the available types met the particular requirements of the counting machine. The mechanism here described has proved successful.

The required movements are transmitted to the chute by the slide *A*. The pin *Z* in slide *A* engages a slot on the under side of the pivoted chute. When slide *A* is in the position shown in the upper view, Fig. 20, the chute discharges into one of the packing cases. As soon as the packing case is filled, a clock, having its actuating lever connected to rod *W* of the yoke *P*, releases the latter member, which causes the clutch to engage the driving and driven shafts and then disengage them after the driven shaft has carried the crank-arm *K* around one-half revolution.

The pin *L* of the crank-arm engages a slot in slide *A* and carries the slide to the opposite position, where it remains while the packing case under the chute is being filled. The clock then acts again, and the chute is automatically transferred to the other filling position. This cycle of operations is repeated continuously, the mechanism being driven by the constantly rotating shaft *I* through the helical gears *C* and shaft *D*. With this arrangement, the transfer movement of the chute is accomplished very quickly and smoothly. The timing of the movements and the duration of the rest periods are controlled by the clock, which can be adjusted to meet any operating requirements.

Construction and Operation of the Clock-Controlled Intermittent Mechanism.—Referring now to the construction of the mechanism, the toothed clutch member *E* is fastened to the continuously rotating shaft *D*, mounted in the bearings *F*. Another part of the mechanism is supported in the bearing *G*, and consists of shaft *H*, on which the toothed clutch member E_1 is a sliding fit. Rotation of E_1 on shaft H is prevented, however, by the two small feather keys J, which are fixed in the shaft and are a close sliding fit in the keyways in E_1.

The crank-arm K, previously referred to, is fastened to

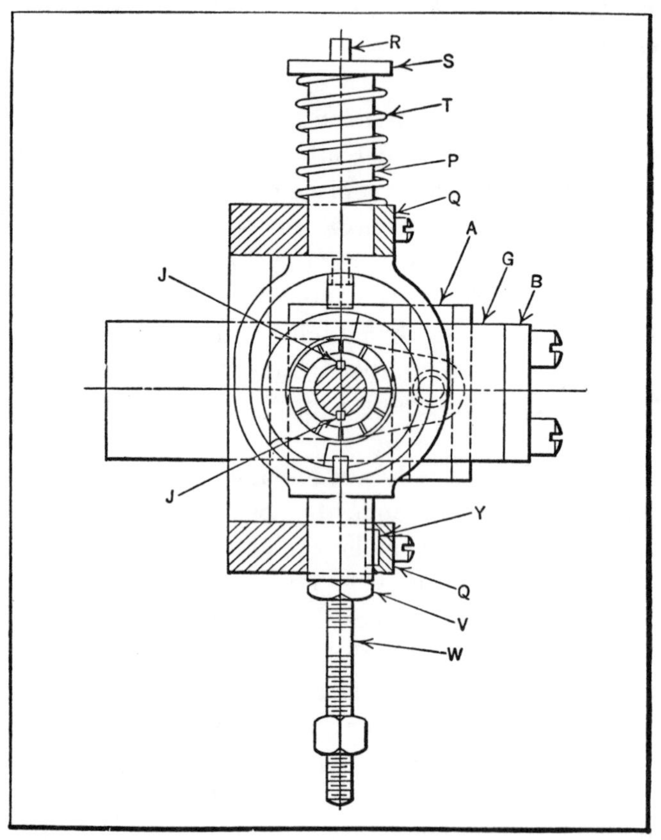

Fig. 21. Section X-X, Fig. 20

the outer end of shaft H. The collar M prevents any lateral movement of the shaft and at the same time serves as a guide for the spring N. When the mechanism is released, the spring N forces the clutch element E_1 into engagement with element E, thus providing for the positive rotation of the crank K through one-half revolution.

The cam *O*, which causes the slide *A* to pause at the end of each stroke, or one-half revolution of shaft *H*, is fastened to the clutch element E_1. The contour of cam *O*, when rolled out in a flat position, is shown in the views to the right, Fig. 22. It will be noted that there are two gradual

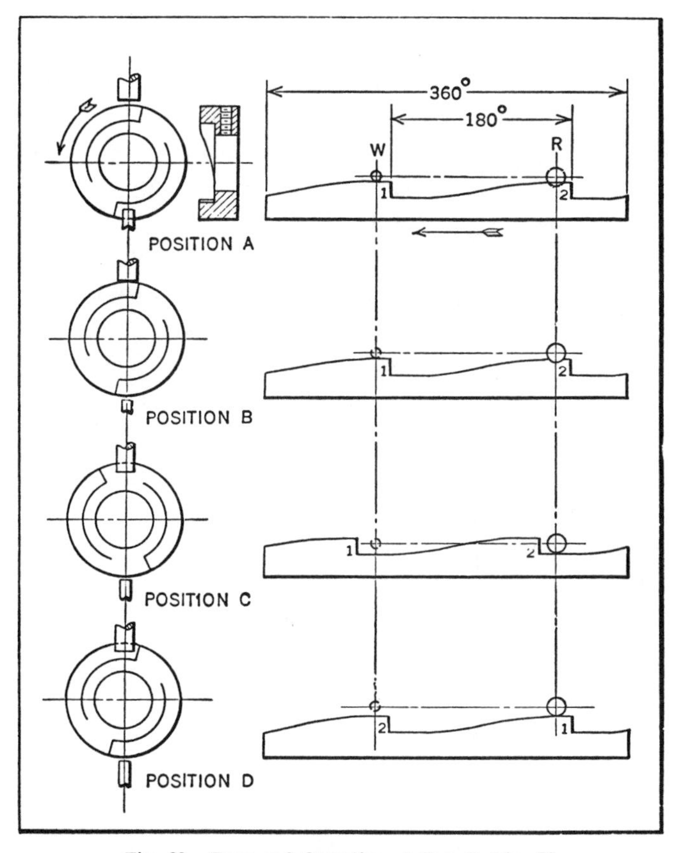

Fig. 22. Form and Operation of Cam O, Fig. 20

rises in the cam surface, after which a sudden drop follows. These drops are just 180 degrees apart.

To disengage the clutch elements, it is only necessary to move the clutch element E_1 a certain distance, depending upon the depth of the clutch teeth plus a reasonable amount of clearance between the teeth. In this case, the depth of the teeth was 1/16 inch. The rise of the cam contour was 7/32 inch, giving a clearance between the teeth of 5/32 inch. This large clearance is necessary, as will be made clear from the following description of the cam followers. The cam, which is machined in the form of a ring, is fastened to the clutch element E_1 by a set-screw, so that these two members are free to move laterally along the shaft *H*.

The cam follower arrangement is somewhat more elaborate than the average type of follower, and is the most interesting and unique feature of the mechanism. Referring to Figs. 20 and 21, the yoke *P* is guided in the bearings *Q*. These bearings are split for the purpose of facilitating the assembling of the mechanism. To prevent the yoke from rotating in its bearing, a small key *Y* is provided and a keyway is cut in the lower stem of the yoke. This restricts the yoke to a vertical movement.

The lower stem of the yoke is provided with the adjustable pin follower *W*, which may be locked in place by nut *V*. The adjustment of this pin is very important, as it determines the amount of movement necessary in the lever mechanism (not shown here), which is attached to the pin follower *W*. This lever mechanism is connected to the clock that times the movements. The clock-operated mechanism will remove the lower pin and permit the clutch elements to make contact under the action of the spring *N*. The upper stem of the yoke is counterbored and provided with the plunger pin follower *R* and the light spring *U*, which is held in place by the nut *S*. The spring *T* keeps the yoke in the upper position with its shoulder against the lower side of the bearing *Q*.

The function of the two pin followers will be more easily understood by referring to the four diagrams in Fig. 22, which show the main positions of the pin followers with relation to the cam contour. At *A* is shown the normal position of the pin followers at the starting position. The lower pin is in contact with the cam contour, while the upper pin is free. As soon as yoke *P*, Fig. 20, is pulled downward, the lower pin is removed and the upper pin strikes the outer edge of the cam ring, as shown in position *B*, Fig. 22.

It will be noted that the upper pin is larger in diameter than the lower one. The purpose of this feature will become obvious on further consideration of the mechanism. When the upper pin strikes the outer edge of the cam ring, the spring *U*, Fig. 20, is compressed. However, this condition only exists momentarily, inasmuch as the spring *N* forces the clutch elements into contact as soon as the lower pin is removed from the cam contour, resulting in the rotation of the cam and all its attached parts.

As soon as rotation begins, the upper pin is freed and drops down under the action of the spring *U*, so that it makes contact with the cam contour as illustrated in position *C*, Fig. 22. When the cam has rotated 180 degrees, the clutch elements are separated and the rotation ceases. This last step is illustrated by position *D*, where the point marked 1 has been replaced by the point marked 2 under the upper pin follower *R*. In the meantime, the crank has traversed from one end of its stroke to the other and stopped. The lower pin is still out of contact with the cam contour, the upper pin having performed the action of separating the clutch elements. As soon as the yoke is released, it is raised by the spring *T*, Fig. 20, and at the same moment, the upper pin is removed from the cam contour and replaced by the lower pin. This explains the necessity for having the upper pin slightly larger in diameter than the lower one.

The upper pin causes the cam contour to move, or be set back slightly from the edge of the lower pin follower. This permits the lower pin to rise freely into position opposite the cam contour. The cam and the clutch member E_1 move toward the clutch member *E* as the upper pin leaves the contour of the cam, but this movement is stopped by the lower pin.

Two very important details should be noted. First, the distance between the ends of the upper and lower pins must be such as to bring the lower pin opposite the cam contour before the upper pin is entirely removed from contact with the cam. If this condition does not exist, the spring *N* will force the clutch elements into contact before the lower pin is in place to hold it back when the upper pin is removed. The second important detail is to note that the rise of the cam contour is determined by the size of the upper pin. The upper pin must obviously be able to fall in with the lower part of the cam contour before it can perform its function.

If the yoke is released before the cam has rotated through 180 degrees, the lower pin itself will perform the function of separating the clutch elements and leave the upper pin inactive as far as contact with the cam is concerned. In this particular case, there was no absolutely definite time release for the yoke, so that a positive operating arrangement had to be provided which would allow a rotary motion of only 180 degrees at each releasing movement of the yoke *P*, regardless of how long the yoke was held in the lower position. This feature accounts for the use of two cam followers instead of one.

It might be of interest to mention here that this mechanism is operated at a speed of about 100 revolutions per minute with no difficulty. However, it might be necessary to provide small depressions in the surface of the cam contour at the points where the followers rest if the speed is much above 200 revolutions per minute. This will prevent overrunning of the cam due to the inertia developed in the rotating parts.

Intermittent Reciprocating Motion Derived from Cam Operated by a Chain.— At times it is necessary to obtain a positive reciprocating motion, followed by a period of dwell, from a moving chain. Such a motion can be imparted by each link of the chain with the device shown in Fig. 23; or by omitting certain cam-rolls, the device can be made to operate only as sections of the chain pass it.

Referring to the illustration, the chain *J* is constructed of flat steel links which are joined together by two lengths of tubing, one within the other, in a way to permit a free turning action at the link joints. Each link is equipped with a spindle *G*, the top end carrying the work-holder (not shown) while at the lower end is mounted a set of three rolls. The smallest roll is a slip fit on the hub of one of the larger ones and acts against the flat cam *A* when the chain is in motion. The two larger rolls come in contact with the bar *C*, thus providing the necessary support for the chain while under the action of the cam. The cam is fastened by screws and dowels to the swinging arm *M*, which is pivoted in the bracket *B* by the pin *N*, held in the bracket by the set-screw *O*. The bracket is secured to the machine table by screws, the supporting bar *C* being mounted on its upper part. There are also two other plain brackets (not shown) to support the extreme ends of this bar. Connected to a projection on arm *M* is the link *P* which carries the reciprocating motion to the required part of the machine.

In operation, the roll *D*, as shown in the side view, is about to force the point *K* of the cam away from the chain, and as the cam is pivoted on pin *N*, the end *L* will move toward the chain between the two rolls *F* and *E*. Upon further movement of the chain, edge *X* will come in contact with the roll *E*. At this time, the center of roll *D* has passed the point *K*, so that as roll *E* forces edge *X* away, point *K* swings toward the chain and between rolls *D* and *F*. The projection *Y* on the cam prevents the point *K* from swinging further than is shown toward the center of the chain. The cam is in action only during a movement of the chain approximately equal to the diameter of the cam-rolls, and as the projection to which link *P* is connected is integral

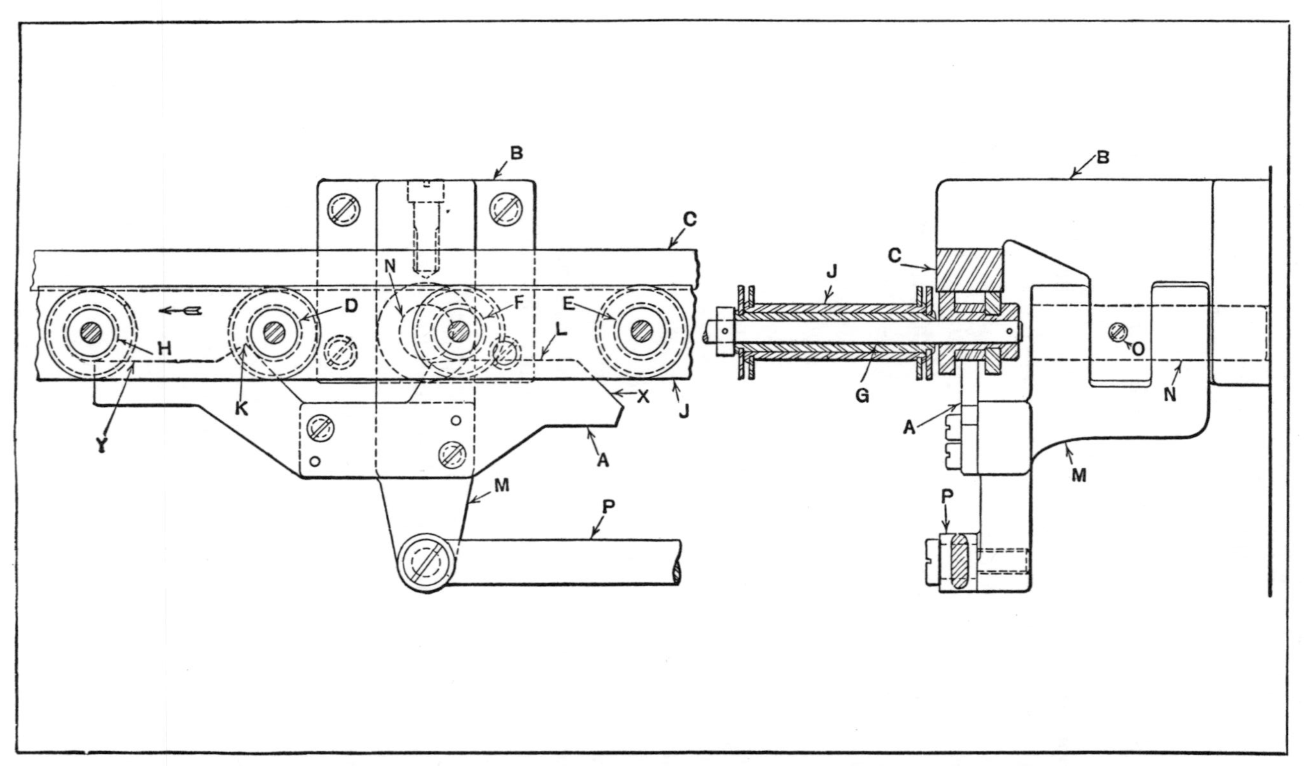

Fig. 23. Intermittent Reciprocating Motion, Produced by the Movement of a Chain Past a Cam

with the cam, the motion of the latter, as described, will produce the required reciprocating movement of the link *P*.

Intermittent Motion for High Rotary Speeds.—Various forms of intermittent motions have been designed for driving machine parts that must alternately turn through part of a revolution and then dwell or remain stationary between each fractional turning movement. Some mechanisms of this class, however, are not adapted to high speeds owing to excessive shocks each time the driven member is started. The design here illustrated (see Fig. 24), which is similar in principle to those used on motion picture machines, although much larger, operates quietly and smoothly at high speeds.

This particular mechanism is used on a milk-bottle cap-making machine. The driver, which is 24 inches in diameter, makes four revolutions to one of the driven member, which has four equally spaced arms each equipped with a roller, as the illustration shows. The speed of the driver is 960 revolutions per minute, and it requires a minimum movement of 90 degrees to operate the driven member smoothly and quietly at this speed. The action may be extended over a larger angle, thus permitting higher speeds and shortening the idle time.

General Design of High-Speed Intermittent Motion.—The rollers on the driven member engage a large annular track (see end view), and after each quarter turn, the driven shaft is securely locked during the idle period. As the driver turns in the direction of the arrow, it rotates the driven shaft intermittently in the same direction. The surface at *G* (Fig. 24) of the outer track acts against roller *B* until roller *A* enters the groove *H*. When roller *A* has fully entered groove *H*, roller *C* begins to enter groove *J*. When point *P* on the driver passes roller *B*, roller *A* is about half way through groove *H*, and as roller *B* begins to engage surface *K*, roller *A* emerges from track *H*.

When roller *A* has reached the position marked A_1,

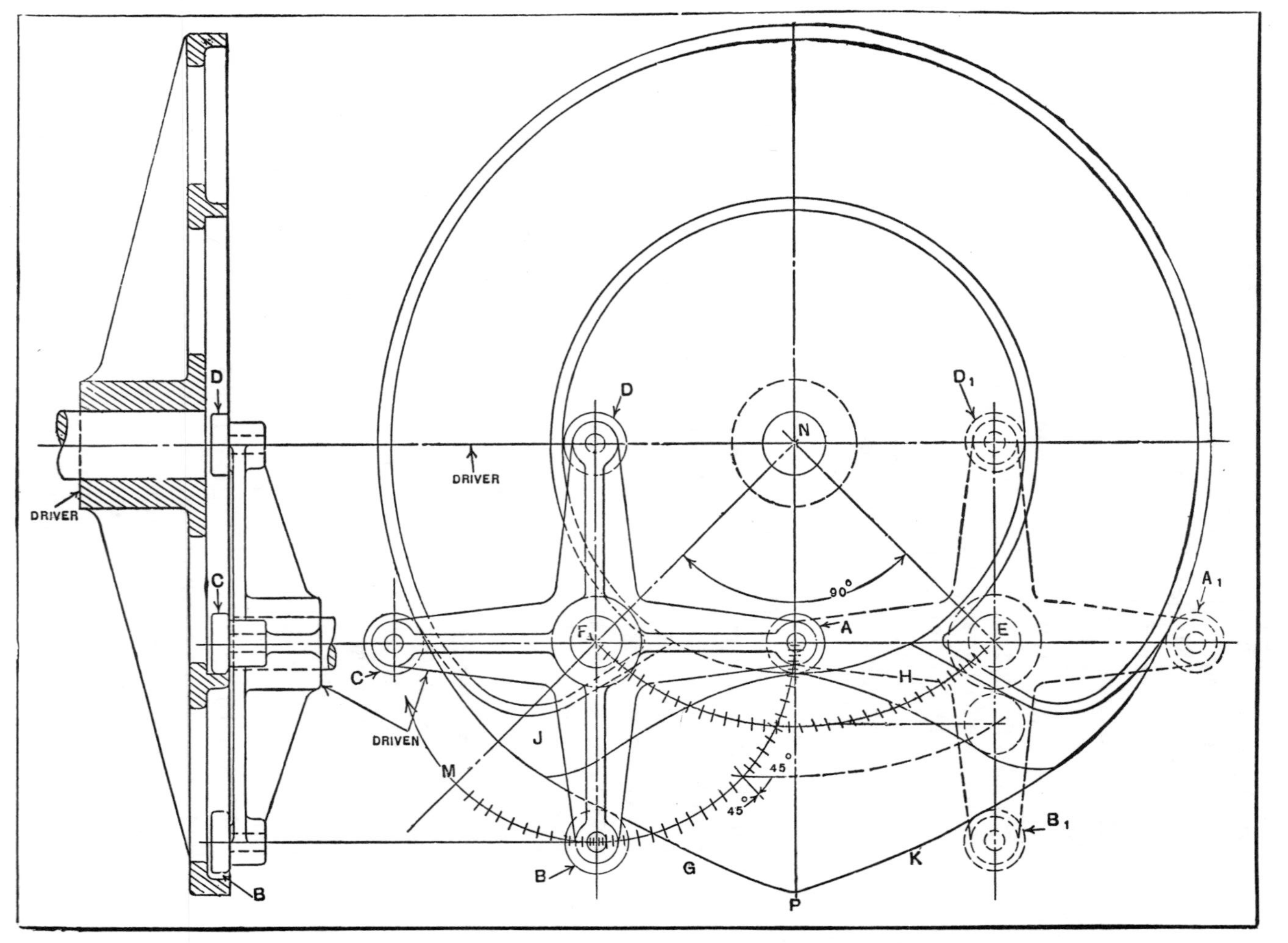

Fig. 24. High-speed Intermittent Motion with Dwell of 270 Degrees or Less

roller *D* has swung around so that it is again in contact with the inner track as at D_1, roller *B* is at B_1, and the driven member has turned one-fourth of a revolution. The entrance to groove *H* now passes roller *C*, which follows roller D_1 around the inner track. At no time is the driven member free to turn in either direction, except as it is revolved by the tracks or cam grooves, and two or more rollers are always in engagement with the driver.

Laying Out the Cam Curves.—In order to avoid shocks, especially at high speeds, it is necessary to gradually accelerate and then gradually retard the movement of the driven member. The method of developing or laying out the cam curves to obtain this result will now be explained. With *N* as a center, draw an arc *FE* through the axis of the driven shaft and divide this arc into thirty-six equal spaces. Next, with *F* as a center, draw an arc through the axes of rollers *A* and *B* and extend this arc 45 degrees to point *M*. Beginning at the center of roller *A*, lay off a division of 1 degree, then a division of 2 degrees, followed by one of 3 degrees, 4 degrees, and so on, up to and including 9 degrees. The total number of degrees thus laid off equals 45, since the nine divisions progressively increase from 1 degree up to 9 degrees by increments of 1 degree.

This procedure is now reversed; that is, the divisions begin at the 45-degree point and progressively decrease from 9 degrees down to 1 degree. Beginning at the center of roller *B*, the order is again reversed, the divisions beginning with 1 degree and increasing up to 9 degrees, ending at *M*.

Each division from *A* to *M* is now bisected; consequently, between the centers of rollers *A* and *B* there are now thirty-six divisions, the same as between the centers *F* and *E*. Assume that the divisions from *F* to *E* are numbered from 1 to 36, and that the divisions from *A* to *B* are also numbered from 1 to 36.

From these divisions we shall now proceed to locate various points on the center lines of the cam grooves *J* and *H*. With *N* as a center, draw an arc through division number 1 on arc *AB*, extending it to the right and left of the vertical center line *NP* a short distance. Draw another arc through division number 2 and continue up to division number 36 next to the center of roller *B*. The arcs through these various division points need not be continuous, but they should be located to the right and left of the vertical line *NP* far enough, as near as can be judged, to intersect the center line of the cam grooves.

Now set the compass to the radius of the driven member, or from the center of shaft *F* to the center of one of the rollers. With division number 1 (adjacent to *F*) as a center, draw an arc intersecting arc number 1 struck from center *N* and to the *right* of the vertical line *NP*. Continue until arcs of the radius of the driven member have been struck from each of the thirty-six divisions on *FE*, thus intersecting all of the thirty-six arcs (to the right of *NP*) struck from center *N*.

The thirty-six centers thus located lie on the center line of the cam groove *H*, and various points along the sides of this groove are located by setting a bow pencil or bow pen to the radius of the driven rollers and drawing a series of arcs. The sides of groove *H*, which are tangent to these arcs, can then be drawn.

To lay out the cam groove *J*, again use the compass set to the radius of the driven member, and in striking arcs to intersect those extended to the left of *NP* from *N*, work from *E* to *F*; for example, with division number 36 as a center (adjacent to *E*) intersect arc number 1 struck from center *N* and to the *left* of the line *NP*; continue until finally division number 1 (adjacent to *F*) is used in striking an arc intersecting arc number 36 struck from *N*. In this way, the series of roller centers for groove *J* can be located.

In laying out the curve *GPK* of the outer track of the driver, proceed as follows: With *N* as a center, and from each of the eighteen divisions on the arc from the center of *B* to *M*, strike eighteen arcs to the right of center *P* and eighteen to the left, all adjacent to the surfaces *K* and *G*; then with the compass set to the radius of the driven member, intersect the thirty-six arcs just struck by another series of thirty-six arcs from the divisions on arc *FE*.

If we assume that the thirty-six arcs adjacent to *G* and *K* are numbered from 1 to 36 from left to right, then division number 1 on *FE* will be used to intersect arc number 1, and so on, until the thirty-sixth division, adjacent to *E*, is used to intersect the thirty-sixth arc at the extreme right of surface *K*. Arcs equal to the roller radius are now struck from these thirty-six centers to locate points along the profile *GPK*.

Although the ratio of this particular mechanism is 4 to 1, other ratios are possible by adding more arms and rollers to the driven member and extending the cam action over a longer arc on the driver. This mechanism has one objectionable feature—the driven shaft must end at the driver and cannot be supported on each side, as will be evident by examining the end view. However, the continuous positive relation between the driving and driven members, the good distribution of wearing surfaces, and the adaptability to high speeds with smooth action compensate for the objectionable feature mentioned.

CHAPTER III

INTERMITTENT MOTIONS FROM RATCHET GEARING

Intermittent motions so designed that a ratchet mechanism constitutes an important element will be found in this chapter. These motions of the ratchet type have been segregated to facilitate finding an intermittent motion likely to meet the requirements of any given design, as, for example, when the application of some form of ratchet gearing appears to offer the best solution.

Ratchet Mechanism for Uniform Intermittent Movement and Heavy Duty.—The ratchet and pawl mechanism illustrated in Fig. 1 was designed for moving, intermittently and accurately, a heavy load at medium speed. The ratchet wheel is positively locked during the idle period, and a positive stop prevents over-travel and insures uniform intermittent movements.

The ratchet wheel *W* is free to turn on the driving shaft, which is shown in section. Behind the ratchet wheel and attached to the driving shaft there is an eccentric *E* connecting with the short arm of bellcrank *B*, which is pivoted at *P*. The operating pawl *O* is pivoted to the long arm of the bellcrank. Pawl *R*, which locks the ratchet wheel during the idle period, is pivoted at *U* and is shown in the locking position. Both pawls *O* and *R* are normally held in engagement with the ratchet wheel by coil spring *C*, which is attached to each pawl. At the upper end of the long arm of the bellcrank there is a steel plate *S* which engages a flat spring *F*, thus lifting the locking pawl *R* to which spring *F* is attached.

This ratchet mechanism will operate with the constant-speed driving shaft turning in either direction in relation to the ratchet wheel. The eccentric E attached to the shaft starts and stops the load gradually like a crank; the full and dotted lines show the extreme positions of bellcrank B and pawl O. Before pawl O comes into engagement with

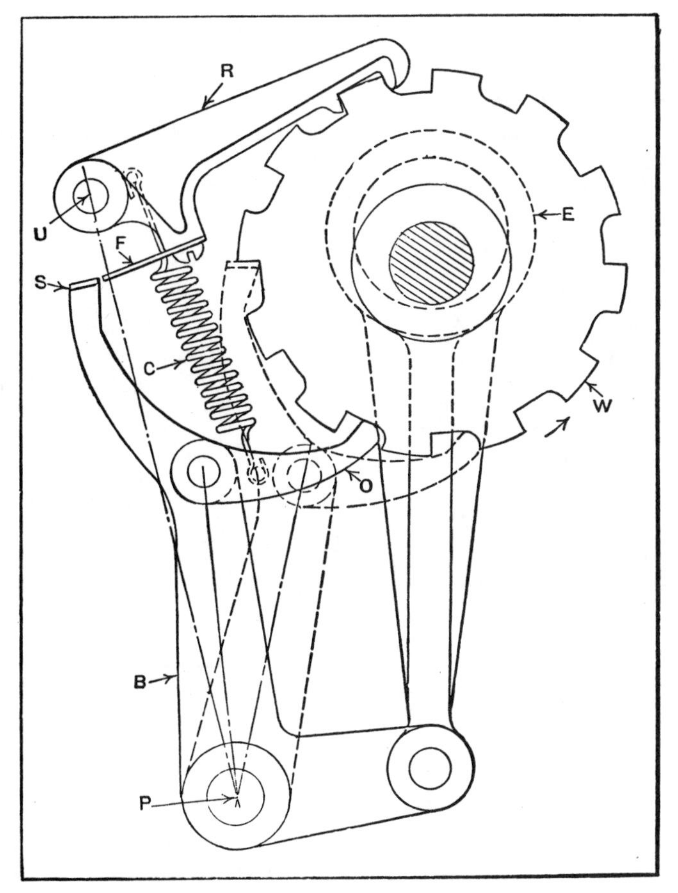

Fig. 1. Ratchet Mechanism Designed to Move a Heavy Load Intermittently and Accurately

a tooth on wheel W, plate S, by engagement with spring F, lifts pawl R, thus unlocking the ratchet wheel.

A short movement of plate S causes it to pass the center line between pivots P and U; consequently, it is soon disengaged from spring *F*, but not until pawl *O* has moved wheel *W* about half a tooth, so that when plate *S* passes spring *F*, the hook end of pawl *R* falls on top of the next approaching tooth. Before pawl *O* reaches the end of its forward movement, plate *S* enters a space ahead of the radial face of an approaching tooth, so that this tooth face comes into contact with plate *S* at the end of the stroke, and any over-travel is thus prevented. During the backward movement of plate *S* to the starting position, it strikes

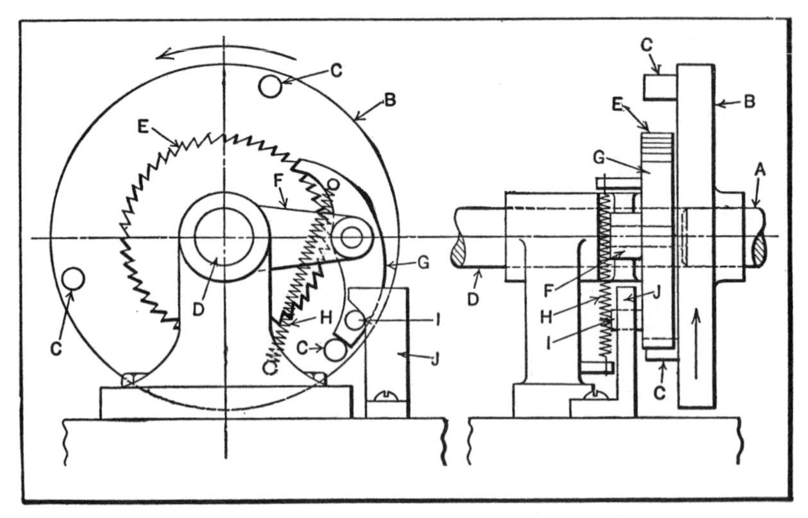

Fig. 2. Intermittent Motion Drive Mechanism Used on Wire-forming Machine

spring *F* and bends it upward slightly, which insures seating the locking pawl firmly. This ratchet mechanism has a ratio of 12 to 1, there being twelve turns of the driving shaft to one complete turn of wheel *W*. The connection between the ratchet wheel and the driven member which it operates is through gearing not shown.

Intermittent Rotary Motion from Constantly Rotating Shaft.—The mechanism shown in Fig. 2 is designed to transmit an intermittent rotary motion to the shaft *D* from the constantly rotating shaft *A*. This intermittent movement operates the feeding device on a wire-forming machine which requires three partial revolutions of the driven member for each rotation of the driving member. Drive shaft *A* with the attached disk *B* rotates continuously in the direction indicated by the arrow. The disk *B* has three equally spaced pins *C* on one side. The ratchet wheel *E* is keyed to shaft *D*. Lever *F* carries the pawl *G* and is free on shaft *D*. Spring *H* serves to keep pawl *G* in contact with the ratchet wheel *E* and also tends to rotate lever *F* in a direction opposite to that in which member *B* is driven.

Pawl *G* is so shaped that when the actuating end is in contact with ratchet wheel *E*, the opposite end lies in the path of the pins *C*. As one of the pins *C* makes contact with pawl *G*, the pawl is carried with it, causing ratchet wheel *E* to rotate. When the pin *I* on pawl *G* comes in contact with the cam *J*, pawl *G* is disengaged from ratchet wheel *E*, which then stops moving.

Continued movement of disk *B* causes the end of pawl *G* to be further depressed by the action of cam *J* until the pawl slips under pin *C*. The two views of the mechanism show pawl *G* about to be disengaged from ratchet wheel *E*. As soon as the pawl has discontinued positive contact with pin *C*, the action of spring *H* causes pawl *G* and lever *F* to rotate in a direction opposite to that of the driving member *B*, until the upper end of lever *F* strikes the end of cam *J* which limits its movement and controls the angular movement of ratchet wheel *E*. The driving movement is repeated as each pin *C* comes into contact with the pawl.

High-Speed Ratchet-Feeding Mechanism with Positive Lock.—The positive intermittent indexing or ratchet-feeding mechanism shown in Fig. 3 is designed for operating a paper feed-roll *R* used in connection with a printing unit of a machine. One end of the roll *R* and the ratchet wheel *Z*, with its actuating and locking pawls, are mounted

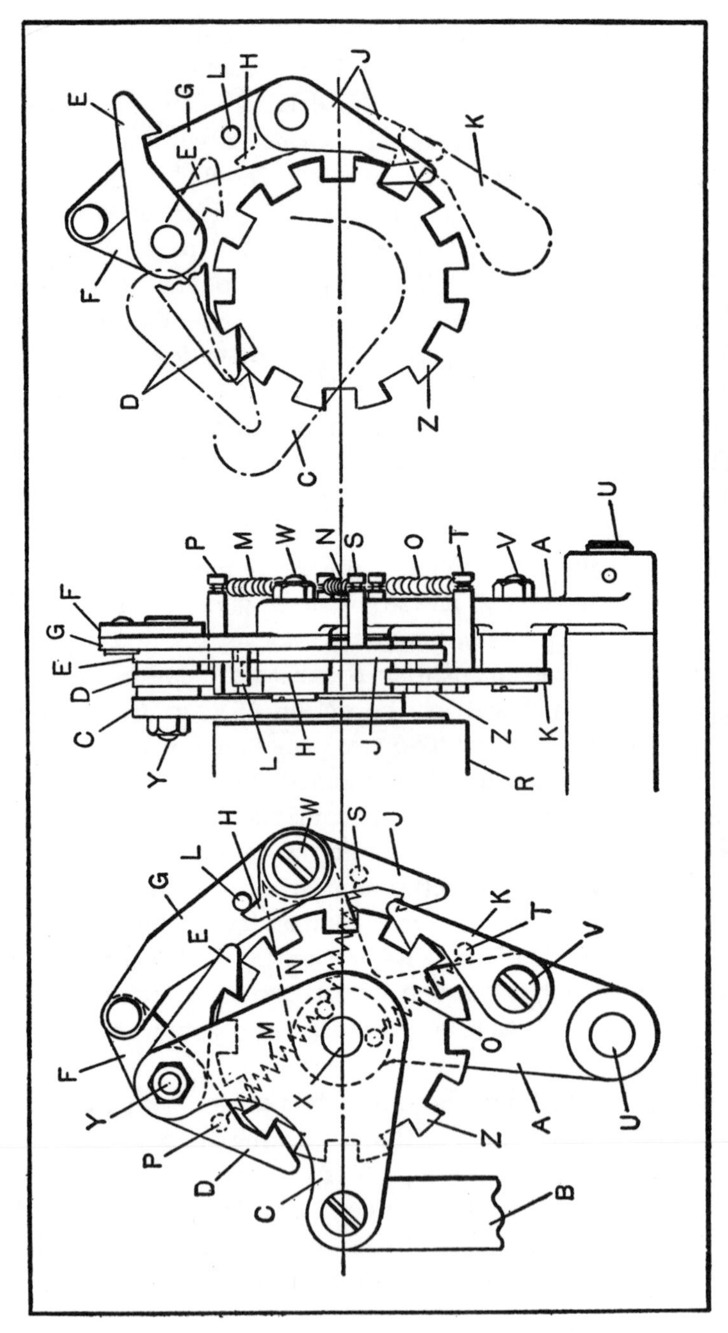

Fig. 3. Positive Ratchet-feeding Mechanism Designed to Operate at High Speed

on arm *A*. Because of the rocking motion of arm *A* about shaft *U*, it is necessary to have roller *R* positively locked against rotation during the return movement of the indexing pawls *D* and *E*. For this reason, the use of a friction feeding device was impracticable.

The indexing mechanism is not affected by the rocking motion, as the ratchet-operating link *B* has a movement corresponding to or parallel with that of arm *A* when it is not being used to actuate the indexing pawls. The indexing pawl *E* and lever *F* are pinned together, and locking pawl *J* is pinned to pawl latch *H*, but this is not shown in the illustration.

To index the roller *R*, link *B* is given a reciprocal motion which rotates the bellcrank *C* clockwise. The indexing pawls *D* and *E* are connected to bellcrank *C* by stud *Y*. Pawl *D*, being held in position by spring *M*, engages ratchet wheel *Z* which is connected to shaft *X*. Pawl *E* is held in position by the connections to levers *F* and *G*.

The locking pawl *J* is spring-connected to lever *G*, and the latching pawl *K* is held in engagement with the ratchet wheel *Z* by the spring *O*. The pawls *J* and *K* are attached to rocker arm *A* by studs *W* and *V*. As the bellcrank *C* rotates, the indexing pawl *D* rotates ratchet wheel *Z*. The indexing pawl *E* is disengaged by the action of levers *F* and *G*. As lever *G* moves outward, pin *L* allows locking pawl *J* to strike the top of the ratchet wheel *Z* through the action of spring *N*.

As the ratchet tooth passes, the spring *N* finally pulls locking pawl *J* into engagement with the tooth. Latching pawl *K* is disengaged by the tooth of ratchet wheel *Z* and is snapped into engagement at the end of the stroke by spring *O*. The positions of the pawls *D*, *E*, and *J* at the end of the forward or indexing stroke are shown by the full lines in the view at the right. The dot-and-dash lines in this view show the positions of the latches and pawl *K* when the bellcrank has almost completed its return stroke.

The dot-and-dash lines of pawl latch *H* and locking pawl *J* show the positions of these pawls when *J* rides against the top of ratchet wheel *Z*.

On the return stroke, the ratchet wheel *Z* is held stationary by pawl *K* which has returned to its original position. Pawl *D* is disengaged by the tooth of the ratchet wheel, and pawl *E* is returned by the action of levers *F* and *G*. Pawl *J* is not disengaged until the last quarter of the stroke of bellcrank *C*, when it is disengaged by pin *L* which forces the pawl latch over against the pull of spring *N*. Referring to the dot-and-dash outlines of the pawls *D*, *E*, and *J* in the view at the right, pawl *J* is shown in the position it occupies just after disengagement with the ratchet wheel *Z*; pawl *E* is shown about to engage the ratchet wheel; and pawl *D* is shown riding against the top of the ratchet wheel.

It will be noted in the central view and the view at the left that all spring connections are made by bringing studs *P*, *S*, and *T* out from the pawls. This was done to simplify the assembling and disassembling of the unit. The delayed action of locking pawl *J* is the main feature of this unit. On the return stroke of bellcrank *C*, lever *F*, when rotated about one-third of its total movement, causes lever *G* to move only a small fraction of its total movement. After the second third of the movement of lever *F*, lever *G* will have moved one-half of its stroke. Thus pawl *J* is not disengaged until after three-quarters of the return stroke of bellcrank *C* is completed.

Ratchet Motion which Varies Movement of Driven Shaft Twice per Revolution.—A ratchet mechanism that automatically increases and decreases the movement of the driven shaft twice in each revolution is shown in Fig. 4. This motion is applied to a wire-forming machine to produce a constantly varying rate of feed. The gear *B* and the ratchet wheel *C* are mounted on shaft *A* and revolve with it. The oscillating lever *D* is free on shaft *A* and carries the pinion *E*, which meshes with gear *B*. Lever *D* also carries the pawl *J*, which engages ratchet *C*.

Lever *F* is attached to the side of pinion *E*, and is connected to the link *G* at its lower end. The upper end of link *G* is carried on stud *K*, which also carries one end of the rod *H*. Stud *K* fits into a hole in the block *M*, which slides in a dovetail groove on the upper end of lever *D*.

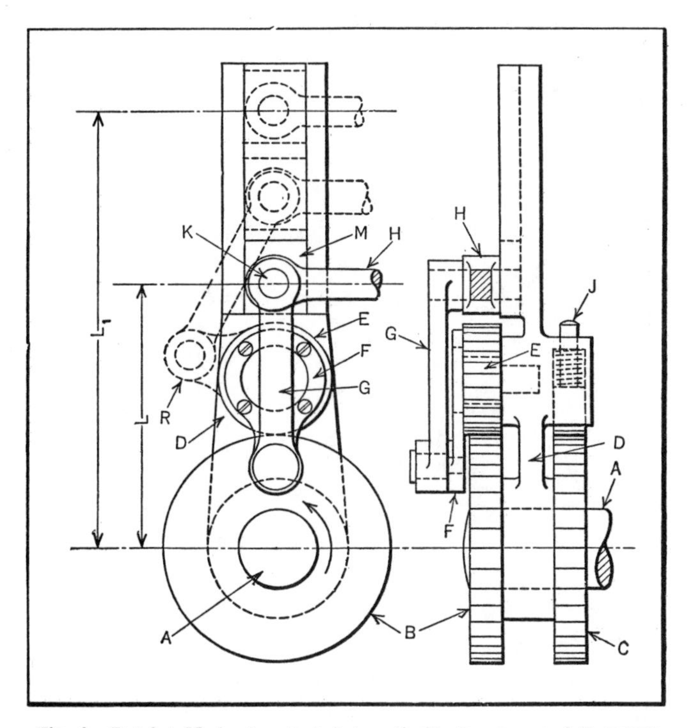

Fig. 4. Ratchet Mechanism that Automatically Increases and Decreases Effective Length of Ratchet Arm Twice During Each Revolution

In operation, rod *H* is given a reciprocating motion by a cam. The movement of rod *H* produces an oscillating motion of lever *D* on shaft *A*. On the forward stroke, pawl *J* engages with ratchet *C*, causing the entire assembly to make a partial revolution in the direction indicated by the arrow on gear *B*. On the return stroke, pawl *J* rides over the teeth of ratchet *C*, while shaft *A*, with gear *B*,

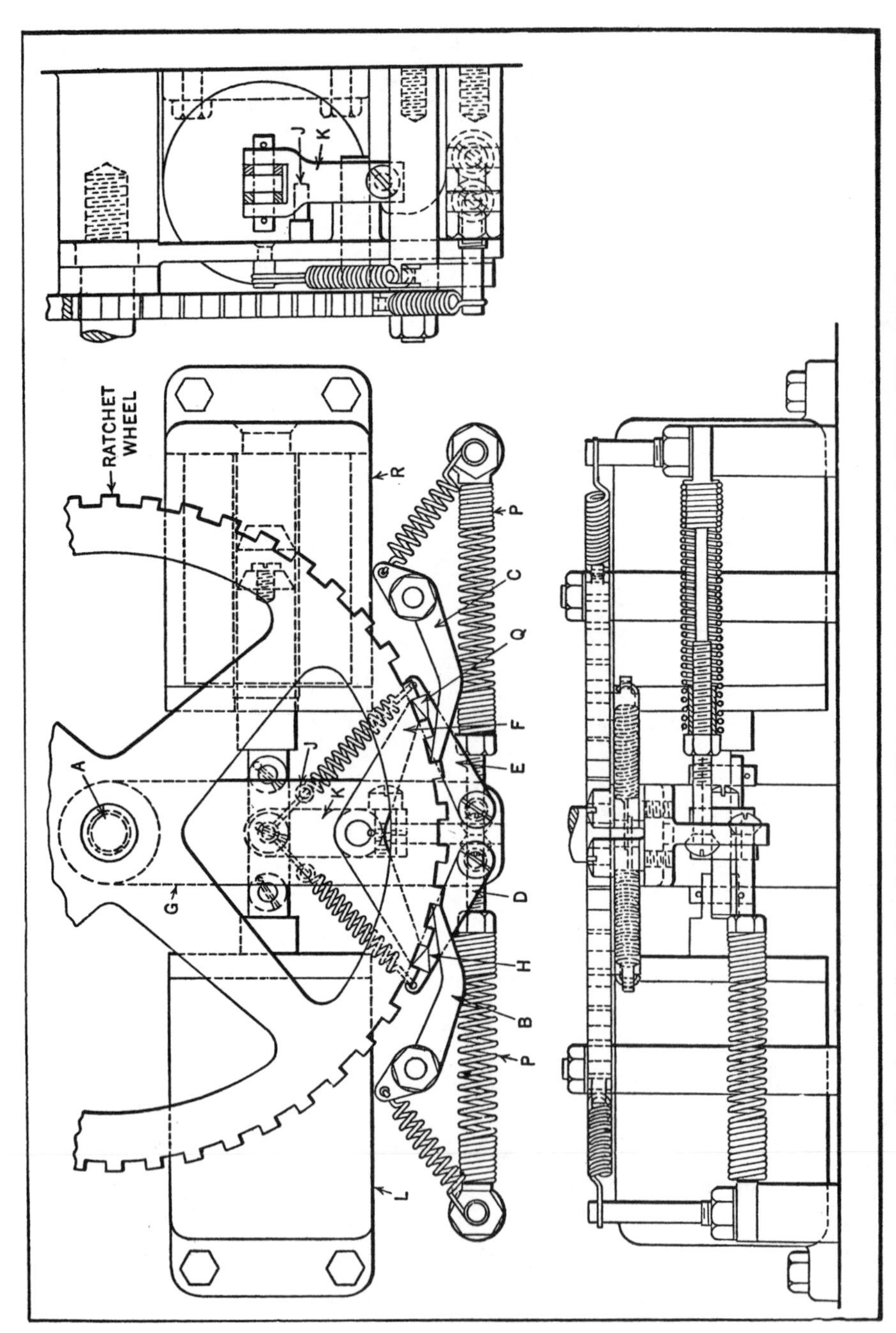

Fig. 5. Electrically Operated Ratchet Mechanism that can be Indexed Rapidly in Either Direction

remains stationary. This causes pinion *E*, which meshes with gear *B*, to make a partial revolution.

The movement of pinion *E* carries lever *F* to the position indicated by the dotted lines *R*, causing the link *G*, connected to rod *H* by stud *K*, to move upward, increasing the center distance between stud *K* and shaft *A*. In this manner, the length of the lever arm is varied from *L* as a minimum to L_1 as a maximum. The number of cycles performed per revolution of shaft *A* is determined by the ratio between pinion *E* and gear *B*, which, in this case is 2 to 1.

Solenoid-Operated Reversible Ratchet Mechanism Adapted to Remote Control.—The mechanism shown in Fig. 5 has interesting possibilities as a means for controlling machines or equipment from a distance. By sending current through one of the two electromagnets or solenoids *L* and *R* one of the two centering springs *P* will be stretched so that it will index the ratchet wheel one tooth in one direction when the current to the solenoid is cut off. Sending electric current through the other solenoid indexes the ratchet wheel one tooth in the opposite direction. The solenoids act very quickly and the spring centers the mechanism without shock. Thus the ratchet wheel can be indexed rapidly in either direction at the will of the operator. No switching arrangement for sending the current through either of the solenoids is shown, but copper contacts set in the periphery of a revolving fiber or Bakelite disk can be arranged to furnish the intermittent electrical impulses necessary to energize the magnets so that each impulse will move the ratchet wheel one tooth in the desired direction.

The reversible ratchet mechanism shown could be used to actuate an elevator position indicator, for example. By using some of the parts and eliminating others, a self-locking device that will index in one direction only could be obtained. Such a device would be suitable for an automatic feed for a notching press. Numerous other applications are possible for this device when used as a reversible ratchet with solenoid control or when controlled by mechanical means. In some cases, it has been used as a single-direction ratchet with either magnetic or mechanical control.

The electromagnets are used to set the mechanism and to stretch or extend one of the adjustable tension springs *P*. These springs are adjusted to overcome the friction of the mechanism and of the machine part actuated by the mechanism. When these springs contract, they gradually exert less force on the parts actuated, so that there is less shock to the mechanism when the stopping pawl drops into its proper notch in the ratchet wheel.

A solenoid should not be used to move the ratchet wheel, because the pull of a solenoid increases very rapidly as the length of the air gap in the solenoid is decreased. For instance, with an air gap 1 inch long, we might obtain a pull of, say, 5 pounds, but when the gap of the same solenoid has been decreased to 1/32 inch, the pull may be as high as 2000 pounds. The hammer blows delivered by the application of such force would flatten the end of the stopping pawl and the sides of the ratchet wheel. It would also cause a rebound of the parts, which would not give sufficient time for the stopping pawl to become properly seated, and the shock of the sudden stopping action might upset the parts actuated. Hence, an adjustable initial tension spring of sufficient strength to overcome the working friction of the mechanism and the parts actuated by it is preferable.

Operation of the Reversible Ratchet.—In considering the operation of the mechanism, let us first assume that an electric current is sent through the solenoid *R*. This causes arm *K* to be pulled to the right until it strikes the right-hand pin *J* on the arm *G*. This, in turn, causes the claw *F* on arm *K* to turn and move the pawl *E* so that the triangular projection *Q* is released from the slot in the ratchet wheel. Next, arm *G* moves to the right about the pivoting point *A*, pulling the triangular boss *H* of pawl *D* up the side of the tooth of the ratchet wheel. This action lifts the stationary locking pawl *B* from its notch, as shown in Fig. 6, and drops boss *H* into the notch formerly occupied by pawl *B*. At this point of the operation only pawl *C* is engaged. This pawl prevents the ratchet wheel from being moved in a counter-clockwise direction by the friction developed by the moving parts.

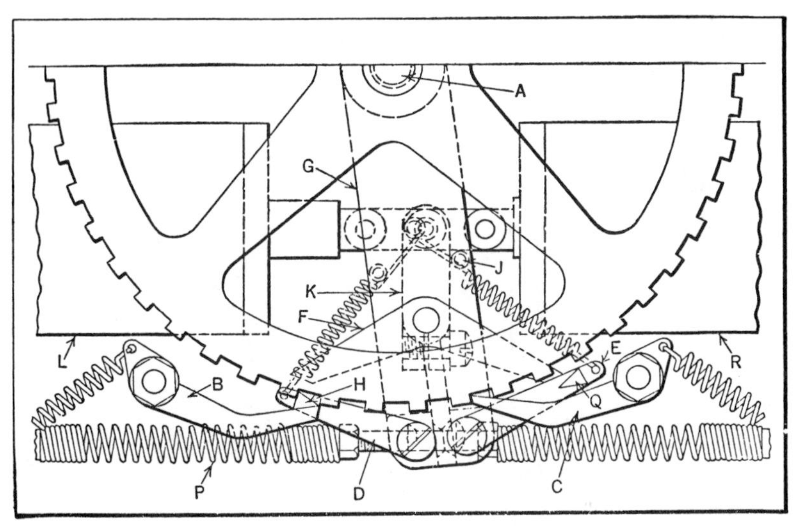

Fig. 6. Mechanism Shown in Fig. 5, with Working Parts in Positions Occupied on Application of Electric Current

The tension spring *P* has now been extended, and when the current through the solenoid is broken, this spring returns arm *G* to the central position. The triangular projection *H* on pawl *D* pushes the ratchet wheel one tooth in a clockwise direction. Pawl *B* rides down the side of the triangular projection *H* and stops the movement of the ratchet wheel. Thus the ratchet wheel is rotated an amount equivalent to one tooth space and positively stopped each time the electrical circuit is completed and opened. The direction of rotation is controlled at the will of the operator, rotation in the opposite direction being obtained

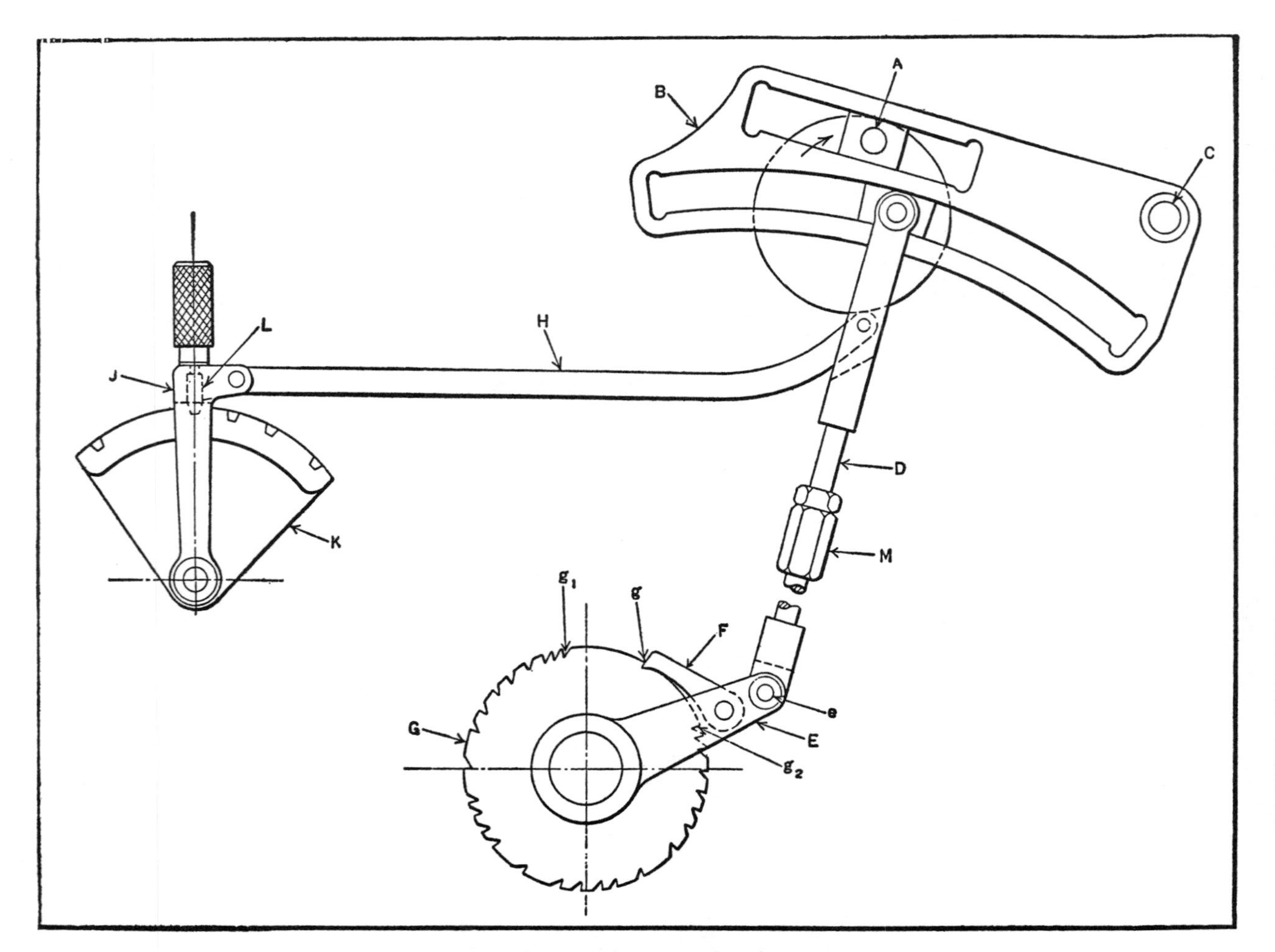

Fig. 7. Design of Ratchet Feed Mechanism with Link Motion Adjustment

by making and breaking the circuit through solenoid *L* instead of solenoid *R*.

The springs for pawls *B, C, D,* and *E* are designed to give just sufficient tension to operate their respective pawls satisfactorily, while the initial tension centering springs *P* are adjusted to give just sufficient tension to move the work, the mechanism, and the plunger cores of the solenoids.

Fig. 5 shows the mechanism in a neutral position—that is, with no electric current applied to the magnets. The arm *K* is shown in a central position for clearness in illustrating the details, although this arm would probably never remain in this position when the mechanism was in use. It may be noted here that there is no need for the spring to be so adjusted that the arm *K* will be exactly centered when no current is passing through either of the solenoids.

Ratchet Feed with Link-Motion Adjustment.—The ratchet and pawl feed shown in Fig. 7 was designed to fulfill the following conditions: (1) The rate of feed must be changeable without stopping the machine; (2) the pawl must always terminate the feeding stroke in the same angular position; (3) after making four complete revolutions, the feeding movement must cease for an interval and the pawl must always engage the same notch on the final movement.

The feeding movement is derived from the crankpin *A*, which imparts a non-varying angular reciprocating movement to plate *B* about the fulcrum stud *C*. By having the crankpin *A* rotate in the direction shown by the arrow, a quick-return motion is obtained for the pawl. The link *D* transmits the movement to pawl lever *E*, and pawl *F* transmits the feed to ratchet wheel *G*.

The swinging link *H* is connected to the link *D* and the anchor lever *J*, which can be swiveled to various positions along the segment *K* by withdrawing the spring plunger *L*. The operator can easily adjust this member on the return stroke of the pawl *F* when the parts are not under a load.

Adjustment of lever *J* causes the upper end of link *D* to slide along the lower slot in plate *B*. Thus by locating the upper end of lever *D* either nearer or farther from the fulcrum *C*, a shorter or longer movement of the pawl *F* may be obtained, as desired.

The lower slot in plate *B* is formed to a radius of the same length as link *D*. Thus, when plate *B* is in the highest position, as shown in the illustration, the center *e* of the arc-shaped slot will always be in the same place. It will be seen that when the motion is arrested in the position shown, the upper end of link *D* can be traversed the whole length of the lower slot without imparting any movement to the pawl *F*. The turnbuckle *M* provides the adjustment required for locating the point *e* accurately.

The ratchet wheel *G* is given a rather unusual form in order to meet the third requirement. The number of indexing movements per cycle ranges from 20 to 36, and as four revolutions of the ratchet wheel are completed per cycle, we have $20 \div 4 = 5$ notches and $36 \div 4 = 9$ notches. The number of notches required for the different numbers of indexing movements within this range are obtained in the same manner. The essential feature is that the first notch for all feeding movements shall be located at *g*.

With this ratchet feed, it is obvious that the coarsest feed will require a minimum angular movement of pawl *F*, equivalent to $360 \div 5 = 72$ degrees, and that the finest feed will require an angular movement of $360 \div 9 = 40$ degrees. The notches nearest notch *g*, that is, g_1 and g_2, are located 40 degrees each side of point *g*, or 80 degrees apart.

Now, if the anchor lever *J* is moved during the running period to increase the feed to the maximum amount, the first one or two movements following the change may be erratic and the pawl may fall short of the required movement; but even if it happens to engage notch g_1, which rightly belongs to the 40-degree feed or the 36-movement indexing cycle, a swing of 72 degrees plus, say, 3 degrees for clearance, will be insufficient to engage the tooth at g_2, which, as previously stated, is 80 degrees from g_1. Thus no movement of ratchet *G* will occur during the next feeding stroke, and there will be no indexing movement until pawl *F* advances to and engages the correct notch *g*. Ratchet *G* will then be rotated until notch *g* reaches the correct finishing point.

From the preceding description it will be obvious that the coarsest feed must be slightly less than twice the finest feed in order to insure proper functioning. Thus, if we let *S* equal the smallest number of divisions and *G* the greatest, then *G* must equal $2S - 1$. In the case described, $S = 5$; therefore, $G = 9$. If $S = 8$, $G = 15$.

Theoretically, we could then use all the numbers from 9 to 14, inclusive. In practice, however, some of these divisions would interfere, but this trouble could be avoided by having two ratchets mounted side by side with the divisions split up between them and the zero notches on both ratchets in alignment. Individual pawls, would, of course, be necessary. If a slight irregularity in feed is not objectionable, one ratchet could be used, employing the maximum number of equally spaced notches consistent with strength, but with the spaces from g_1 to *g* and *g* to g_2 left blank. Each of these spaces would equal $360 \div 15 = 24$ degrees.

Intermittent Rotation of Ratchet Wheel During Forward and Reverse Movements of Pawl Lever.—The mechanism shown in Fig. 8 was designed to provide an automatic infeed for the reciprocating table of a grinding machine. The mechanism is so designed that the table is fed inward at the moment of reversal at each end of the stroke. Ratchet wheel *G* turns counter-clockwise when left-hand pawl *E* moves downward, and *G* also turns in the same direction when right-hand pawl *E* moves downward. Pawls *E* are attached to and operated by lever *D*. The cross-slide or unit *B*, which is actuated by the feeding mechanism, is

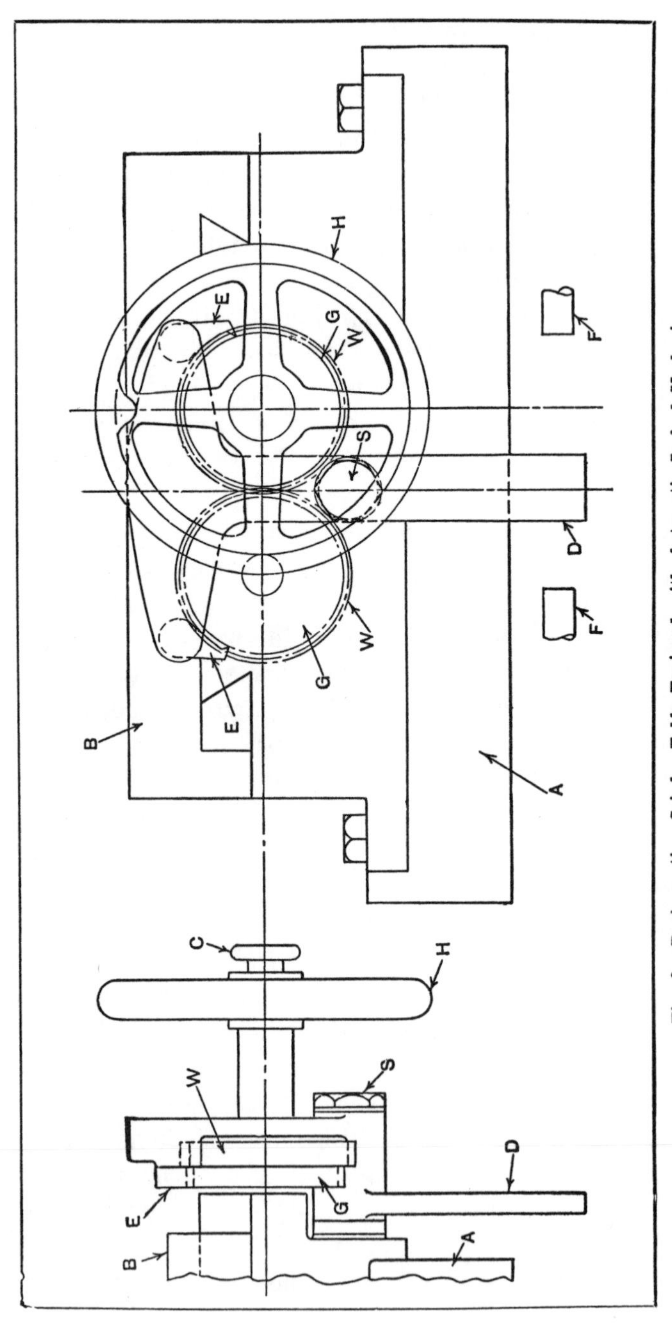

Fig. 8. Reciprocating Grinder Table Equipped with Automatic In-feed Mechanism

mounted on the reciprocating table *A* of the grinding machine. The member *D* is mounted on a fulcrum stud *S*. When the table *A* is about to reverse at either end of its travel, the lower end of the lever strikes one of the adjustable stops *F*. This action causes member *D* to pivot about stud *S*, thus transmitting a rotary motion to one of the ratchet wheels *G* through the pawl *E*. The ratchet wheels are secured to meshing gears *W*, one of which is mounted on the cross-slide feed-screw. Thus the pivoting or angular movement of member *D* causes the feed-screw to be rotated a certain amount in the same direction when the table reverses at each end of the stroke.

The amount of angular movement of member *D* is determined by the position of the stops *F* with respect to the position of the table at the moment of reversal. After member *D* has been actuated by coming in contact with one of the stops *F*, it is inclined in the proper direction for operation by the stop *F* at the opposite end of the table.

The handwheel *H* can be connected with the cross-slide feed-screw by setting the knob *C* to engage an internal clutch. When the handwheel is thus connected, it provides a means for feeding the cross-slide in or out. With the automatic feeding mechanism applied as shown in the illustration, however, the pawls *E* must be thrown up out of contact with the ratchet feeding wheels in order to permit the slide to be fed inward by the handwheel.

Intermittent One-Revolution Drive.—The mechanism to be described is designed to revolve a disk intermittently, so that it will make one revolution at the conclusion of the feeding movement of a certain machine. The ratchet wheel *A*, Fig. 9, which is keyed to shaft *B*, revolves continuously at a constant speed. Shaft *B* rotates freely within the boss of disk *C*, but when pawl *D*, attached to disk *C*, is allowed to engage ratchet *A*, the parts *A* and *C* revolve as one unit. Gear *E*, which is keyed to shaft *B*, then transmits motion to a slide (not shown).

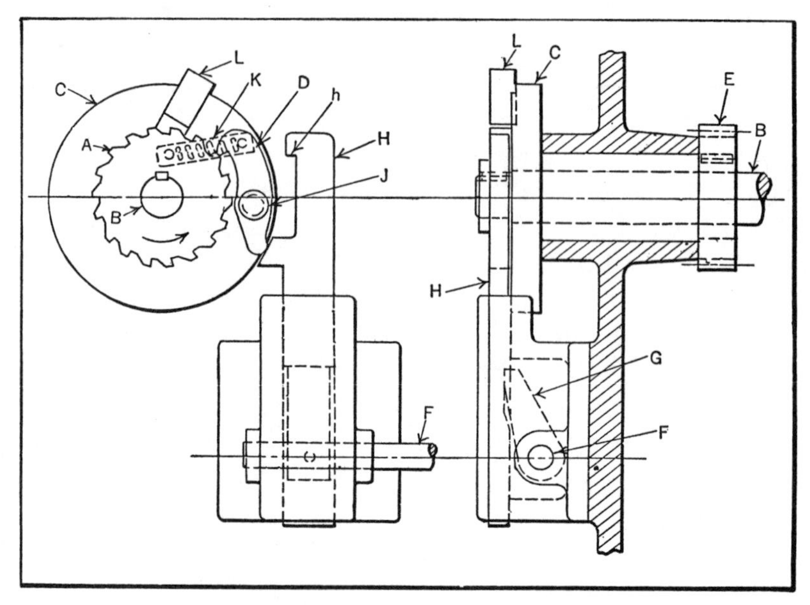

Fig. 9. Mechanism by which Rotating Shaft B Imparts One Revolution to Gear E when Shaft F Releases Latch G

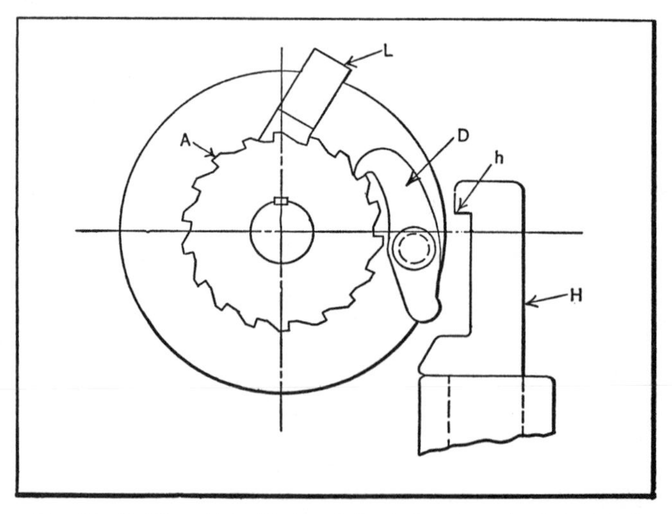

Fig. 10. Position of Bar H after Being Released by Latch G, Fig. 9

At the completion of the feeding movement, shaft *F* is rocked in a clockwise direction until latch *G*, keyed to shaft *F*, moves clear of the drop-bar *H*. This allows bar *H* to drop and clear the end of pawl *D*, which then pivots on stud *J*, under the action of the tension spring *K*, and engages the revolving ratchet wheel in the manner illustrated in Fig. 10.

In order to limit the rotation of disk *C* to one revolution, the projecting lug *L*, affixed to and rotating with disk *C*, engages the upper hook *h* on bar *H*, as shown in Fig. 11. The lug *L* then lifts bar *H* high enough to allow latch *G*, Fig. 9, to re-enter the notch in bar *H*, and hold the latter member in the upper position. This repositioning of bar *H* takes place before one revolution is completed. The continued movement of disk *C* causes the end of pawl *D* to come in contact with the lower projection on bar *H*, thus forcing the pawl out of engagement with the revolving ratchet *A* and arresting the movement of disk *C* after it has completed one revolution. When bar *H* is down, the upper hook *h* must clear the end of pawl *D*, and the lower hook on bar *H* must clear projection *L*, in order to have the mechanism function correctly.

With proper attention to lubrication, bar *H* will function satisfactorily, but dirt or gummy oil will cause it to become stuck in the upper position, as only the weight of the bar itself is relied upon to cause the return movement. It was therefore decided to change this detail. Although a spring could have been used to exert a downward pull on bar *H*, a lever *M*, as shown in Fig. 12, was substituted for bar *H*. This construction was less expensive than the first one, and the rotating movement about fulcrum stud *N* produced less frictional resistance than the sliding movement of bar *H*.

The only other difficulty experienced in the design illustrated in Fig. 9 was with the time interval required to free latch *G* after the drop-bar had been allowed to fall. The rate of the rotation of ratchet wheel *A* is comparatively

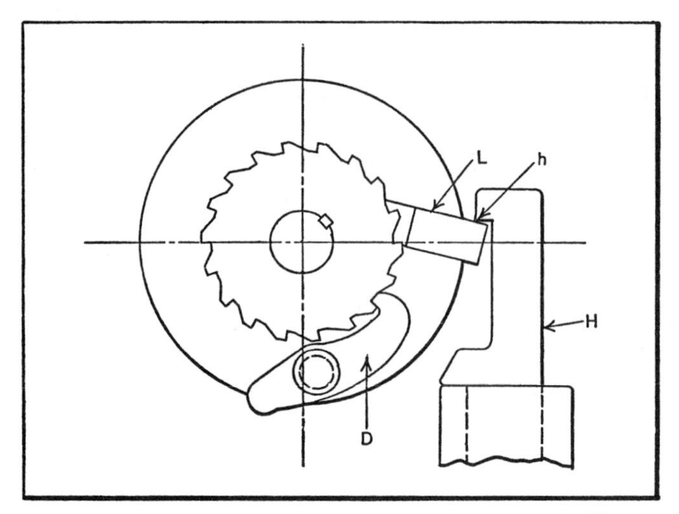

Fig. 11. Lug L Engaging Hook h by which it Lifts Bar H

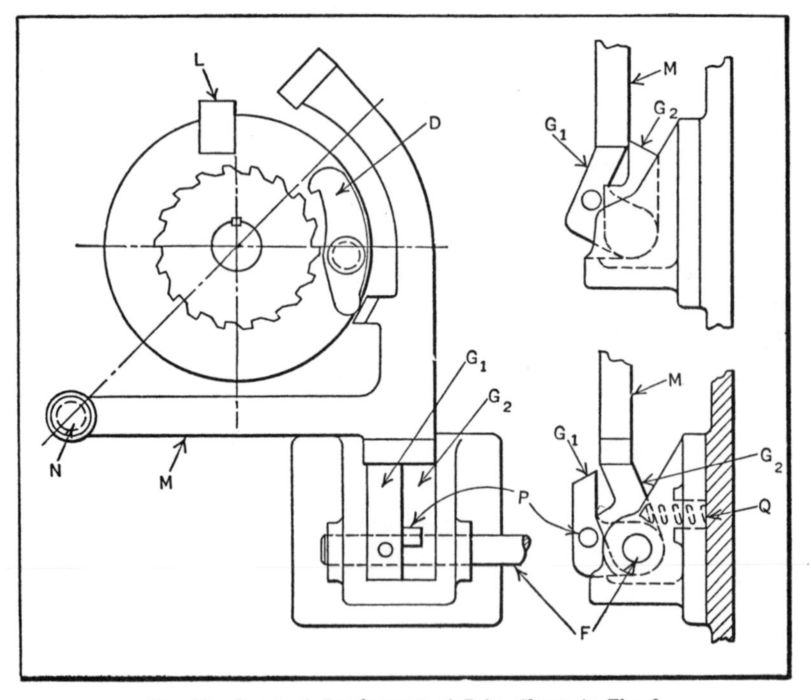

Fig. 12. Improved Development of Drive Shown in Fig. 9

high, and with a very slow feed, the trip motion shaft *F* did not operate quickly enough. Thus, latch *G* would not be ready to retain bar *H* when the latter member was lifted by the lug *L*. This meant that the disk would be given a second revolution before pawl *D* was disengaged. To avoid this difficulty, latch *G* was replaced by two latches, as shown at G_1 and G_2, Fig. 12. Latch G_1 is keyed to shaft *F*, but latch G_2 is free to revolve.

When shaft *F* is rocked clockwise, latch G_1 pushes latch G_2 from under lever *M* by means of the pin *P*. Lever *M* then occupies the position shown by the view in the upper right-hand corner. Thus lever *M* is raised slightly, but when it is released by latch G_2, latch G_1 occupies a position directly under lever *M*, thus preventing it from dropping the full distance. When shaft *F* is released and latch G_1 returned to the original position, the pressure of spring *Q* tends to return latch G_2, but cannot do so because the latch strikes the side of lever *M*. Lever *M* is then free to drop between the two latches and start the one-revolution drive. This arrangement prevents the one-revolution drive from functioning until shaft *F* has been tripped and returned to its original position.

Dwell of Sprocket-Chain Conveyor at Regular Intervals for Loading.—Endless sprocket-chain conveyors are used extensively for carrying lacquered or paint-sprayed parts through drying ovens. The work-holders, as a rule, are mounted at each link joint. In one application of this type, the chain travels a distance equal to the length of seven links and then dwells long enough to allow these links to be loaded by means of an automatically operated feeding device. This alternate movement and dwell of the chain is repeated continuously, so that the chain is always fully loaded as it passes through the oven.

The arrangement for obtaining this intermittent movement is shown in Fig. 13. It consists of driving sprocket *A*, ratchet wheel *B*, pawl *C*, and driving sleeve *D*. In each link joint of the chain is mounted a slender shaft, threaded at its upper end for a work-holder. The lower end of these shafts is provided with a roll, which, at a certain point in the chain travel, comes in contact with a rapidly moving endless leather belt (not shown). This causes the shaft, work-

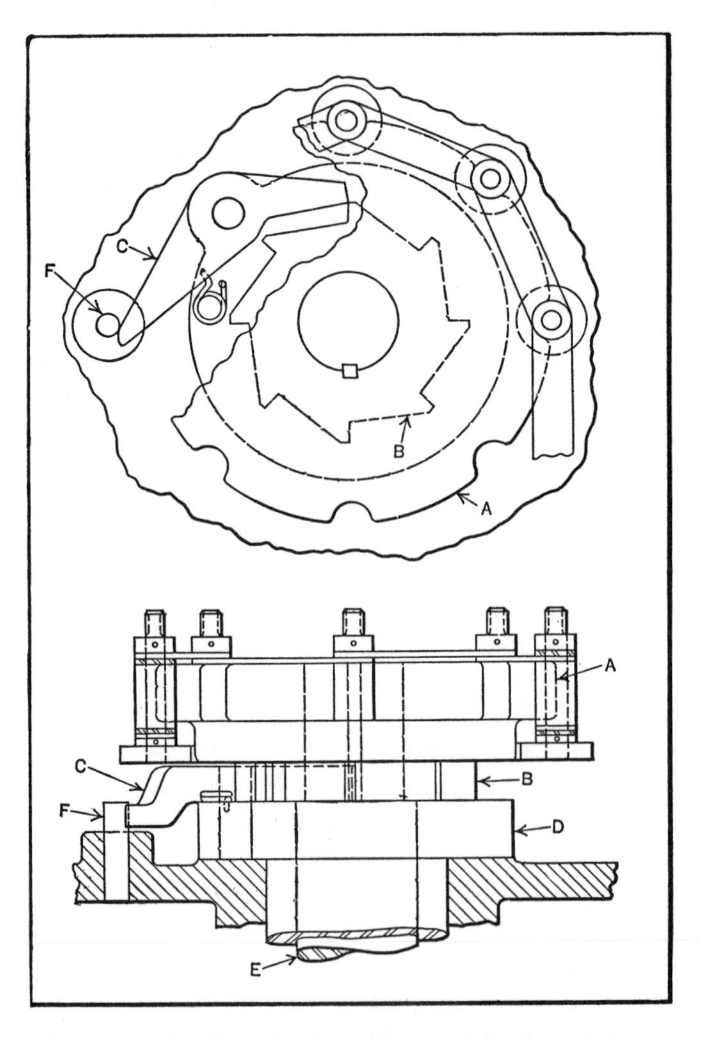

Fig. 13. Ratchet Mechanism that Disengages a Sprocket Wheel from its Driving Member at Regular Intervals to Obtain an Intermittent Movement of the Chain

holder, and work to rotate, so that the paint or lacquer being applied will be distributed evenly.

Both the sprocket and ratchet wheel are keyed to shaft *E*, which turns freely in the sleeve *D*. The sleeve is rotated continuously in a clockwise direction by another member of the machine, and at its upper end is pivoted the ratchet pawl. As indicated, the pawl has been forced out of engagement with the ratchet wheel by pin *F*. This causes the sprocket and chain to dwell long enough for seven work-holders to be loaded. As sleeve *D* continues to rotate, the outer end of the pawl passes by pin *F*, permitting the other end of the pawl to engage the next tooth and carry the sprocket wheel around to the position shown. Here, the pawl is again forced out of engagement with the ratchet, causing the sprocket wheel to dwell a sufficient time for seven more work-holders to be loaded. The movements described are repeated for each succeeding revolution of sleeve *D*.

Duplex Type of Intermittent Drive.— The mechanism shown in Fig. 14 has the component parts so arranged that it can be used for two distinct purposes. In the application to be described first, the continuously rotating shaft *A* transmits an intermittent rotating movement to the shaft *B*. In disk *C*, fastened to shaft *A*, is a pin *D* which contacts with the surface of the internal cam on the combination cam and lever *E*.

Lever *E* is pivoted on shaft *B* and is provided with a lug to which the spring pawl *F* is fastened. Two spring pawls are shown, but one or more may be provided, depending upon the movements imparted to lever *E*, as will be explained later.

The enlarged view, Fig. 15, illustrates the method of producing the oscillations in lever *E*. The movement of pin *D* on the internal cam surface of lever *E* forces the lever to one side until a point is reached, as shown by the dotted lines, where pin *D* passes over the ridge or shoulder in the

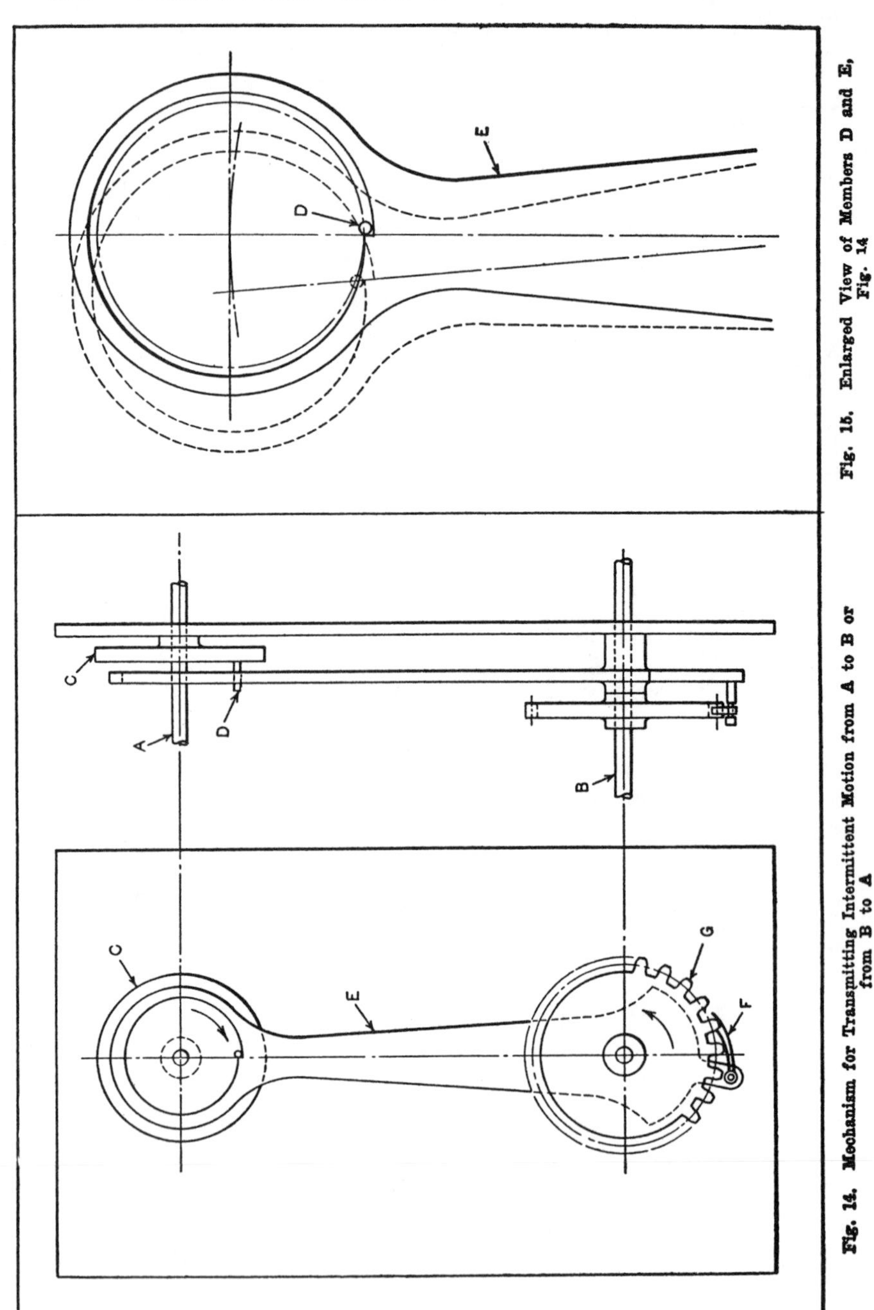

Fig. 14. Mechanism for Transmitting Intermittent Motion from A to B or from B to A

Fig. 15. Enlarged View of Members D and E, Fig. 14

cam surface. The location of pin *D* inside the cam produces a positive motion in the lever and eliminates the use of any form of spring.

If the stroke of the lever movement is sufficient to produce a movement in the pawl equivalent to one-half the pitch of the teeth in the gear or ratchet wheel *G*, then two spring pawls are required, as illustrated, the inner pawl being shorter by one-half the pitch of the teeth. If the lever stroke is equivalent to one-third of the tooth pitch, three spring pawls are required, each being shorter than the next outer one by one-third the tooth pitch. In this manner, the number of intermittent movements of gear *G* can be varied to conform with requirements.

In the second application, shaft *B* is driven continuously in a counter-clockwise direction. Through a friction clutch arrangement (not shown) shaft *B* drives shaft *A* in a clockwise direction. Lever *E* stops the rotation of shaft *A* at predetermined intervals, so that shaft *A* rotates intermittently.

Referring to Fig. 15, assume that pin *D* has reached the position shown by the dotted lines and that shaft *A* is being driven by shaft *B* through the friction clutch drive. As shaft *A* with its disk *C* revolves, pin *D* acts on the cam surface of lever *E*, causing it to pivot about shaft *B* in a direction opposite to that of the rotation of shaft *B* until pin *D* reaches its lowest position, as shown by the full lines in both Figs. 14 and 15. At this point the pin comes in contact with the shoulder on lever *E* and rotation of shaft *A* is stopped, due to the engagement of pawl *F* with the teeth on gear *G*.

It should be mentioned here that the drive between shafts *B* and *A* is designed to rotate shaft *A* at a faster speed than the driving shaft *B*. Slippage of the friction clutch in the drive from *B* to *A* occurs, of course, at this point and continues until shaft *B*, rotating in a counter-clockwise direction and carrying the spring pawls *F* around with it, causes

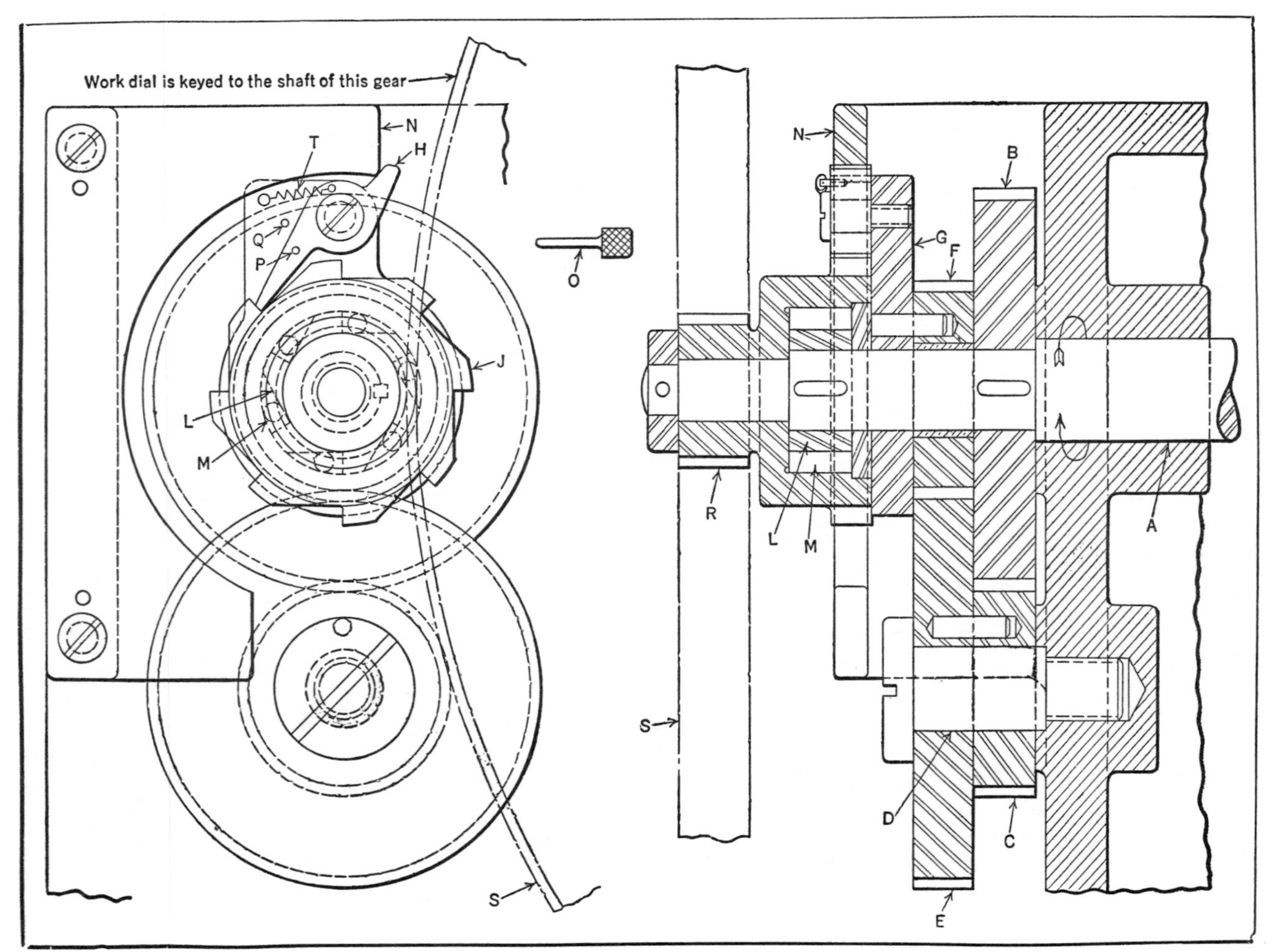

Fig. 16. Variable-velocity Rotary-motion Mechanism for Work-carrying Dial of Polishing Machine

lever *E* to pivot or swing around to the position shown by the dotted lines in Fig. 15. Pin *D* is released when contact with the shoulder on lever *E* is made at this point, and the cycle is repeated as described.

Combination Roller Clutch and Ratchet for Imparting Variable Rotary Movement.—The over-running or "free wheeling" feature of roller friction clutches is used to advantage for imparting a variable rotary motion to the work-holding dial of a polishing machine. In connection with this movement, a ratchet mechanism is employed. Similar pieces of two different diameters are polished on this machine. The work stations are so spaced that the large pieces will be close together on the dial, in order to reduce to a minimum the non-productive time of the wheel in passing from one piece of work to the other. However, the same dial and the same station spacing are employed for polishing the small-diameter work; hence, without the special variable-motion mechanism illustrated (see Fig. 16), an appreciable loss in production time would result, owing to the gaps between the work over which the wheel must pass.

With the mechanism shown, each station is advanced rapidly toward the polishing wheel until the polishing action commences. At this time, the rotary movement of the dial is immediately decreased to the desired speed. This speed of the dial continues until the part passes out of contact with the wheel, and the speed is then immediately increased, so that the succeeding dial station is moved rapidly to the wheel. Provision is made for imparting a constant rotary movement to the dial when polishing the larger work.

The mechanism for imparting the variable dial movement is relatively simple. The dial is driven by the shaft *A*, which rotates at a constant velocity. On this shaft is keyed the gear *B*, meshing with gear *C* on stationary stud *D*. Gears *C* and *E* are pinned together, the latter meshing with gear *F*, which is free to turn on shaft *A*. Pinned to gear *F* and also free to turn on shaft *A* is the plate *G*, to which is pivoted the pawl *H*. This pawl intermittently engages the ratchet wheel *J*, its engagement being controlled by the stationary cam *N*. Wheel *J* turns freely on shaft *A*, and its left-hand end forms the pinion *R* for rotating the dial, the pinion meshing with the bull gear *S* secured to the back of the dial. The inside of the ratchet wheel is bored out to receive the roller clutch arrangement, which consists of the core *L* and rolls *M*. The core is keyed to shaft *A*.

The ratio of the gears *B*, *C*, *E*, and *F* is such that plate *G* rotates four times as fast as shaft *A*. Hence, owing to the over-running feature of the roller clutch, if pawl *H* is in engagement with the ratchet wheel, the dial will be rotated at high speed through gear *R*. However, provision is made for automatically disengaging the pawl just as the polishing wheel comes in contact with the work. This is accomplished by means of the cam *N*.

For example, the pawl is shown engaged, the dial having been rotated at high speed to advance the work to the wheel. With the ratchet wheel in the position indicated, the wheel is just about to come in contact with the work. As the plate *G* continues its rotary movement, the cam *N* swings the pawl out of engagement with the ratchet wheel, allowing the roller clutch to pick up the motion and continue the rotation of the dial at one-fourth the preceding angular velocity.

This new angular velocity is constant and continues until the polishing wheel has passed over and commences to leave the work. At this point, the pawl leaves the lower part of cam *N* and coil spring *T* forces it into engagement with the ratchet wheel, so that the high speed of the plate is once more transmitted to the ratchet, thus rotating the dial at a corresponding velocity to advance the next station toward the polishing wheel. In this way, the two driving members *G* and *L* alternately transmit the required speeds to the pinion gear *R*. Cam *N* is designed to hold the pawl out of engagement with the ratchet wheel for one-half revolution of gear *R*. Hence, since plate *G* rotates four times as fast as core *L*, the angular movement of gear *R* for one-half turn will be four times that for the remaining half turn of the cycle.

When the larger work is being polished, a constant angular velocity of the dial is required. To obtain this condition, the holes *P* and *Q* in the pawl and plate, respectively, are aligned, and the pin *O* is inserted through them. Thus, the pawl is held out of engagement with the ratchet wheel to allow the roller to transmit the constant rotary movement of shaft *A* directly to the gear. There is sufficient friction in the various moving parts of the mechanism to prevent over-run of the roller clutch each time the pawl is disengaged. In applications where the inertia of the rotating part is likely to cause over-run, a simple brake can be mounted on one of the driving members.

This mechanism is designed to operate with the dial in a vertical plane, as the engagement of the rollers in the clutch is dependent upon gravity. However, the mechanism can be readily adapted to any other position of the clutch by inserting coil springs behind the rolls. Creeping movement of the clutch rolls in this unit does not accumulate and change the dial station location, since the creep for each cycle of the gear *R* is compensated for by the positive action of the ratchet.

Differential Ratchet for Imparting Slight Axial Movement to Feed-Screw.—In designing machines, it is sometimes necessary to make provision for transmitting an intermittent movement to a feed-screw from a reciprocating slide. For a comparatively large movement of the screw, an ordinary ratchet arrangement is suitable, but for an extremely small movement, such as that required for the feed-screw illustrated (see Fig. 17), special means must be provided.

The screw *G* provides the transverse feed for the wheel of a surface grinding machine used in grinding wood planer knives. It has the very short axial movement of 0.00045 inch for each cycle of the reciprocating slide *A*. To obtain the desired feed, two ratchet wheels *C* and *D*, operating on the differential principle are used. One wheel has 23 teeth and the other 24. Axial movement of both wheels is prevented by the bracket *B* fastened to the machine. This

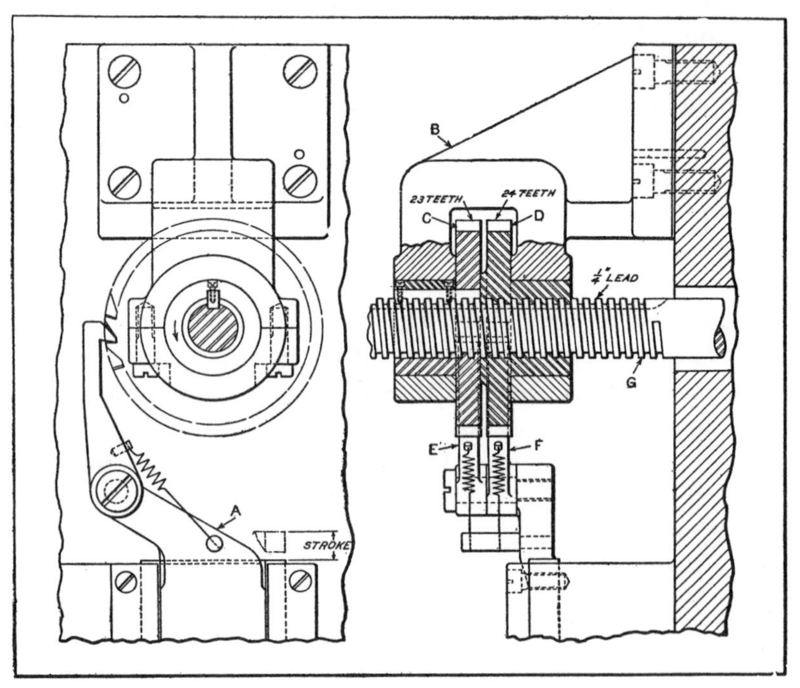

Fig. 17. Ratchet Mechanism for Imparting Slow Movement to Feed-screw

bracket also forms the bearings for the wheels. It will be noted that wheel *C* is provided with a feather key which engages the spline in the screw and that the screw is free to slide axially in this wheel. The bore in wheel *D*, however, is threaded to engage the screw. The angular movement is imparted to the ratchet wheels by their respective pawls *E* and *F*, pivoted to the slide *A*.

In explaining the action of this mechanism, let us assume, for simplicity, that both ratchet wheels have the same number of teeth. Thus, for every cycle of slide *A*, both the wheels, as well as the screw, would rotate together; consequently, there would be no axial movement of the screw. Now returning to the actual case, wheel *C* has one less tooth than wheel *D;* therefore, during one cycle of the slide, pawl *F* will rotate wheel *D* 1/24 revolution. Owing to the difference in the number of teeth, however, wheel *C* will be rotated by pawl *E* 1/23 revolution; and the axial movement of the screw will be equivalent to the difference between these two movements multiplied by the lead of the screw, or:

$$\left(\frac{1}{23}-\frac{1}{24}\right)\times\frac{1}{4}=0.00045 \text{ inch}$$

Perhaps it should be mentioned here that, in so far as the preceding description is concerned, both pawls could have been incorporated into one wide pawl encompassing both wheels. However, the requirements of the machine were such that a faster feeding movement of the screw was required for certain jobs. To accomplish this, the pawl *E* is swung to the left to clear wheel *C* and held in this position by a latch. A plunger is then released which locks wheel *C* to prevent its rotation. All movements for disengaging and locking the wheel are obtained by shifting one lever. However, the latch, plunger and operating lever are not shown. With the pawl *E* disengaged, one cycle of the slide will cause only ratchet wheel *D* to rotate, its angular movement being 1/24 revolution. The corresponding movement imparted to the screw is approximately 0.010 inch.

Adjustable Pawl Shield to Vary Ratchet Wheel Movement.— Alterations made in a certain product necessitated shortening a ratchet movement on the production machine. The new ratchet arrangement is shown in Fig. 18. Oscillating lever *A* is actuated by a cam (not shown) and carries the pawl *E*, which through the ratchet wheel *C*, transmits the required intermittent movement to shaft *F*. Adjustment of the angular movement of this shaft is obtained by means of the shield *D*. Bearing *B* was turned down on one end to serve as a support for the shield, which is held in a stationary position by a set-screw. Near the end of the back stroke of the lever, the shield lifts the pawl away from the ratchet wheel. Thus, part of the subsequent forward stroke of the lever is completed before the shield permits the pawl to engage the ratchet wheel, so that the angular

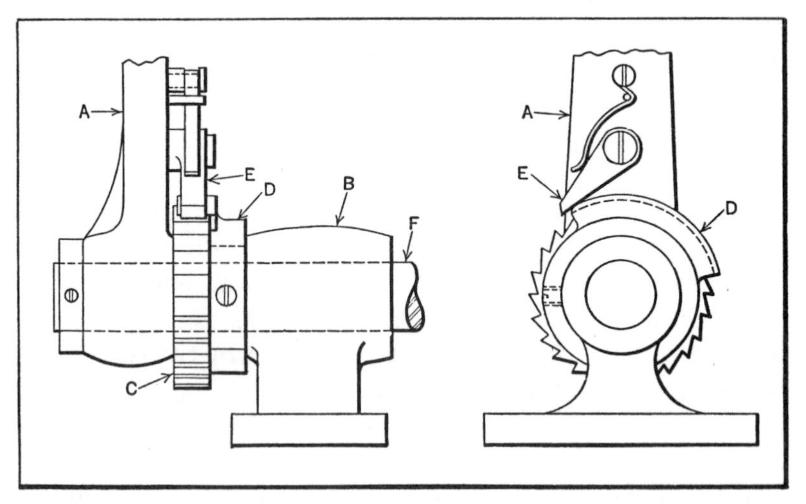

Fig. 18. Application of a Shield to a Ratchet for Reducing the Angular Movement Transmitted

movement of the shaft is shortened. By varying the position of the shield, the angular movement of the shaft will also be varied.

Combined Eccentric and Friction Ratchet for Automatically Varying a Feeding Motion.—On a special polishing machine, the work is passed under a set of oscillating brushes charged with abrasive. In the original design, the work-table was fed at a uniform rate by means of a toothed ratchet, which transmitted its movement through a shaft to the work-table. However, under certain light conditions, the surface of the polished work showed a series of marks corresponding to the movement of the work-table. Though it was considered impossible to eliminate the marks entirely, it was thought advisable to break up their symmetry, so as to render them less noticeable. This was accomplished by the use of a variable ratchet movement, the design of which is shown in Fig. 19.

The shaft *A*, which operates the work-table, carries the eccentric *B*, which is keyed to it. The eccentric *B* is en-

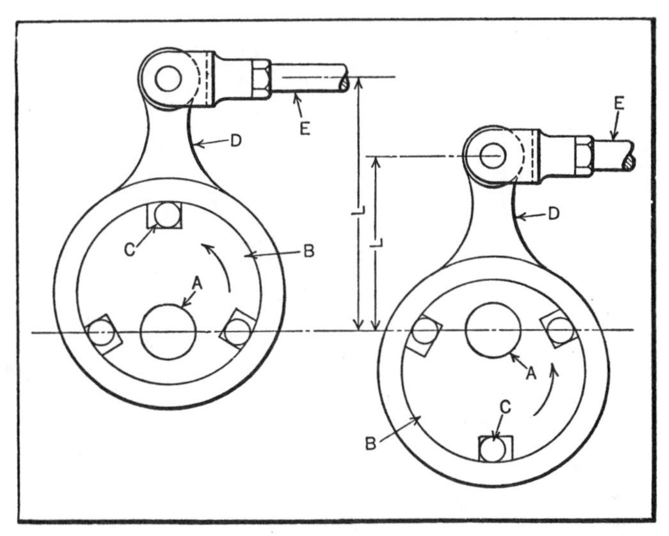

Fig. 19. Mechanism in which Eccentric Mounting of Ratchet Member B Varies Effective Length L of Ratchet Arm from Maximum to Minimum Once in Each Revolution of the Driven Shaft A

circled by the strap *D*, which is given an oscillating motion by the rod *E*. Eccentric *B* is grooved to receive the rollers *C*, forming a conventional type of roller clutch or ratchet which operates through the wedging action of the rollers *C* between the eccentric *B* and the strap *D*.

As the eccentric *B* revolves with shaft *A*, the effective length *L* of the lever arm changes constantly, the range of variation being controlled by the throw of the eccentric *B*. The view to the left shows the mechanism with *L* at its

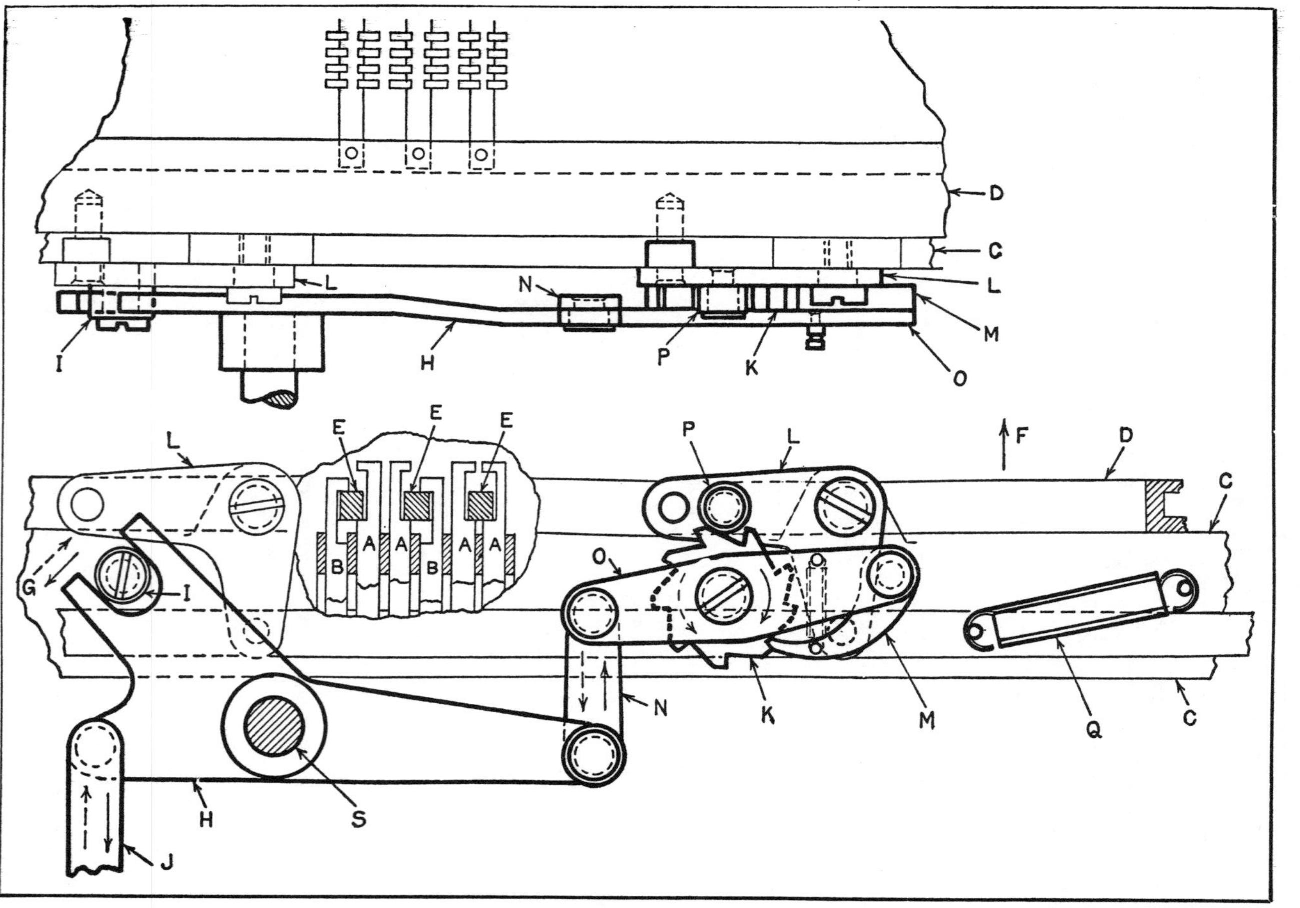

Fig. 20. The Parts Shown by Heavy Lines are Applied to a Type-casting Machine to Produce an Oscillating Movement of Frame D while the Frame C Moves through the Downward Cycle of an Oscillating Movement

maximum, while the view to the right shows *L* at its minimum. As the reciprocating movement of rod *E* is constant, the degree of movement imparted to shaft *A* is controlled by the length *L* of the lever arm. One cycle of variations is produced at each rotation of shaft *A*.

Mechanism for Oscillating a Part Mounted on a Moving Member.—Sometimes a designer finds it necessary to provide means for imparting a short, quick oscillating movement to a machine part on a member that is also in motion. The mechanism shown in Fig. 20 was designed to meet such a requirement. In this mechanism, the part or unit *D* is required to have a quick movement upward from the frame or part *C* in the direction indicated by arrow *F*, and then return to the starting position. This movement takes place while frame *C* is moving through the downward cycle of its oscillating movement, the direction of which is indicated by the full arrow at *G*.

The mechanism shown is a stop-pin bed used in a type-casting machine for the purpose of selecting groups of type. The bed consists of a multitude of flat stop-pins, slidable in a honeycomb holder. Some of these pins *A* are shown in the restored or cleared position, while others, such as those marked *B*, are shown in their depressed positions. These stops are arranged in rows, and in plan view present a two-dimensional field in which code patterns can be depressed.

Frame *D* is built above bed framework *C* with a number of bars *E* attached to it, and forms a sort of rigid grate which, when raised at right angles to the bed in the direction indicated by arrow *F*, restores all the depressed stops *B* to the clear positions *A*. When lowered into contact with the frame *C* as shown, the restoring unit, consisting of the members *D* and *E*, permits a new pattern to be depressed in the stop-pin bed to suit the new cycle.

Now, it is necessary to perform this clearing operation by raising and immediately dropping the unit consisting of members *D* and *E,* while the bed section is making a short, quick stroke in the direction *G*. This is effected by means of a single outside connection—a lever *H* acting with the bed frame through a roller and slot at *I*. The lever is propelled by a cam (not shown) through link *J*.

Under one of the several interconnected bellcranks *L* provided to secure true parallelism in the motion of members *D* and *E* is mounted a ratchet *K*. This ratchet is

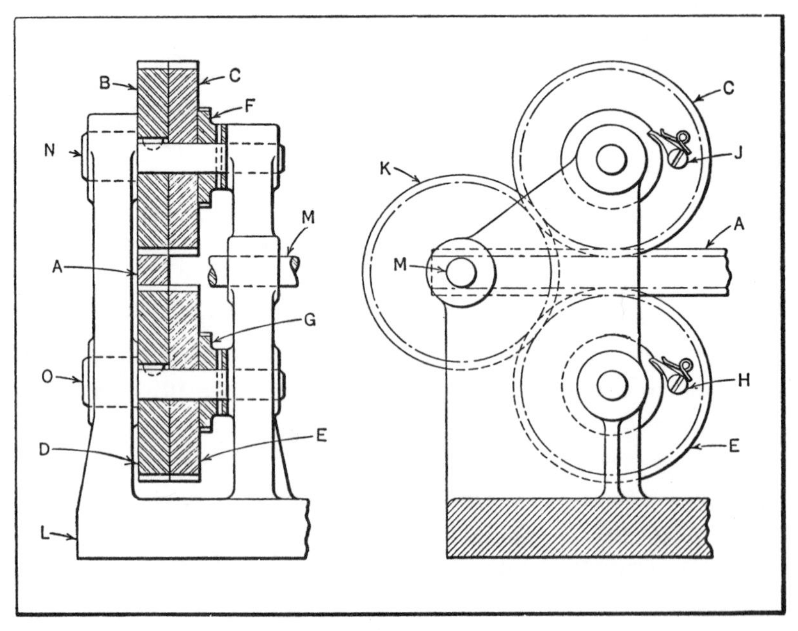

Fig. 21. Train of Gears Operated by a Rack for Imparting an Intermittent Motion in One Direction to Shaft M

mounted idly on a stud in the bed frame *C* and is made to act in combination with the roller *P* on the bellcrank *L*. A pawl *M* is arranged to receive its motion from the lever *H* through the connecting link *N* and rocking plate *O,* mounted idly on the same stud as the ratchet *K*.

It can be readily seen that during the stroke indicated by the dotted arrows there is skipping of the pawl *M* over the ratchet teeth, while during the stroke in the direction indicated by the solid arrows, the ratchet wheel is given positive indexing which forces the bellcrank *L* and the restoring unit upward. As soon as the crest of a ratchet tooth passes the roller, the frame unit is restored to its lower position on the bed frame by virtue of its weight, assisted by the tension in the spring *Q*.

The lay-out of the ratchet gear is such that, when the roller is in any of the ratchet tooth gashes, the roller clears the adjacent teeth by a slight amount. The stroke of the pawl is sufficiently in excess of the tooth spacing to cover this clearance and insure a full tooth spacing. The number of teeth in the ratchet is, of course, of no consequence, and such requirements as desirable size of the roller, easy camming of the roller by the back slope of teeth, available space for the whole device, etc., are factors governing the design of the ratchet.

Rectilinear Movement Converted to Intermittent Rotary Movement.—By means of the mechanism shown in Fig. 21, the reciprocating rack *A* imparts an intermittent rotary movement in one direction to the shaft *M* through the gears *B, D, C, E,* and *K,* and the ratchet wheels *F* and *G.* During each stroke of the rack, shaft *M* rotates one-half of a revolution and then dwells. The length of this dwell, as well as the velocity of shaft *M,* is controlled by automatic valves on an air cylinder (not shown) which actuate the rack.

Teeth cut in opposite sides of the rack engage gears *B* and *D,* keyed to shafts *N* and *O,* respectively. Gears *C* and *E* are free to turn on their shafts and mesh with gear *K* keyed to shaft *M.* Pawls, pivoted to gears *C* and *E,* engage ratchet wheels *F* and *G,* fixed to their shafts.

When rack *A* moves toward the left, gears *B* and *D* rotate in opposite directions, and pawl *J* simply rides over the ratchet teeth without imparting motion to gear *C.* Pawl *H,* however, engages ratchet wheel *G* and causes gear *E* and ratchet wheel *G* to rotate together. Now, as gear *E* is in mesh with gear *K*, gear *K* will rotate in a clockwise direction. When the rack reaches the end of its stroke, the automatic valves close for a predetermined time, thus holding the rack stationary and causing shaft *M* to dwell.

When the valves open, air is admitted to the opposite side of the piston and the rack moves toward the right. In doing so, pawl *H* rides over the teeth of ratchet *G*, and pawl *J* engages the teeth in ratchet *F*, causing gear *C* and ratchet wheel *F* to rotate together; and as gear *C* is in mesh with gear *K*, gear *K* will rotate in a clockwise direction as before.

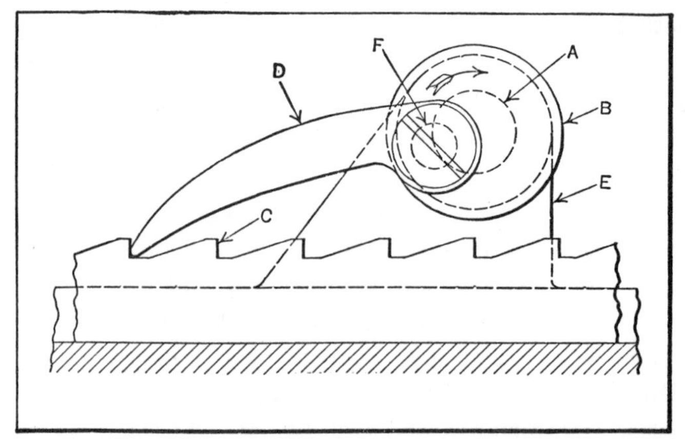

Fig. 22. Crank-driven Pawl that Imparts a Slow Movement at Both Ends of its Stroke to Prevent Jerking and Over-run

The movement of shaft *M* continues until the rack has reached the end of its stroke, at which time the automatic valves close once more to obtain the required dwell.

The angular movement of shaft *M* after each dwell depends upon the stroke of the rack and the ratio of the gears. In this case, all the gears have the same number of teeth; consequently, during one stroke, the travel of the rack must equal one-half the pitch circumference of gear *K*, or slightly more, to allow the pawls to engage properly.

Ratchet Pawl Having a Slow Movement at Both Ends of Its Stroke.—A feed-slide in a special tube cutting-off machine is given an intermittent movement by means of a pawl engaging evenly spaced teeth on the slide. To eliminate jerking at the beginning and over-run at the end of each slide movement, the pawl was actuated by means of a crank, as shown in Fig. 22. Here the teeth on the slide are indicated at *C*, one of which is engaged with the pawl *D*. The pawl turns freely on pin *F* in the crank disk *B*. The disk is integral with the end of the continuously rotating shaft *A*, which turns in the stationary bearing *E*.

As the shaft *A* rotates in the direction of the arrow, the pawl is carried toward the right until it engages the next tooth. At this time the pin *F* is diametrically opposite the position in which it is now shown. Continued rotation of shaft *A* causes the slide to start very gently toward the left, owing to the curvature of the path through which the pin travels. The velocity of the slide, however, increases as the pin approaches the bottom of disk *B*, and decreases as it approaches the position shown, the movement at the end of the stroke being barely perceptible. Consequently, the momentum of the slide is greater at the middle of the stroke and decreases at the end of the stroke.

Automatic Indexing Head with Self-Locking Mechanism.—Automatic indexing heads are used on many special machines having work-tables of the reciprocating type. One design of head particularly adapted for these machines, especially where an unusually large number of divisions is to be indexed, is shown in Fig. 23. The work is secured by some suitable means to the left-hand end (not shown) of shaft *A*. At the end of each indexing movement, the shaft is locked to prevent rotary movement of the work during the machining operation.

The indexing head housing *B* is fastened to the reciprocating machine table. Extending from one side of this head is the shaft *C*, to which are keyed the bevel gear *D* and the forked lever *E*. The forked end of this lever engages a stationary pin *F* secured to the machine, while gear *D*

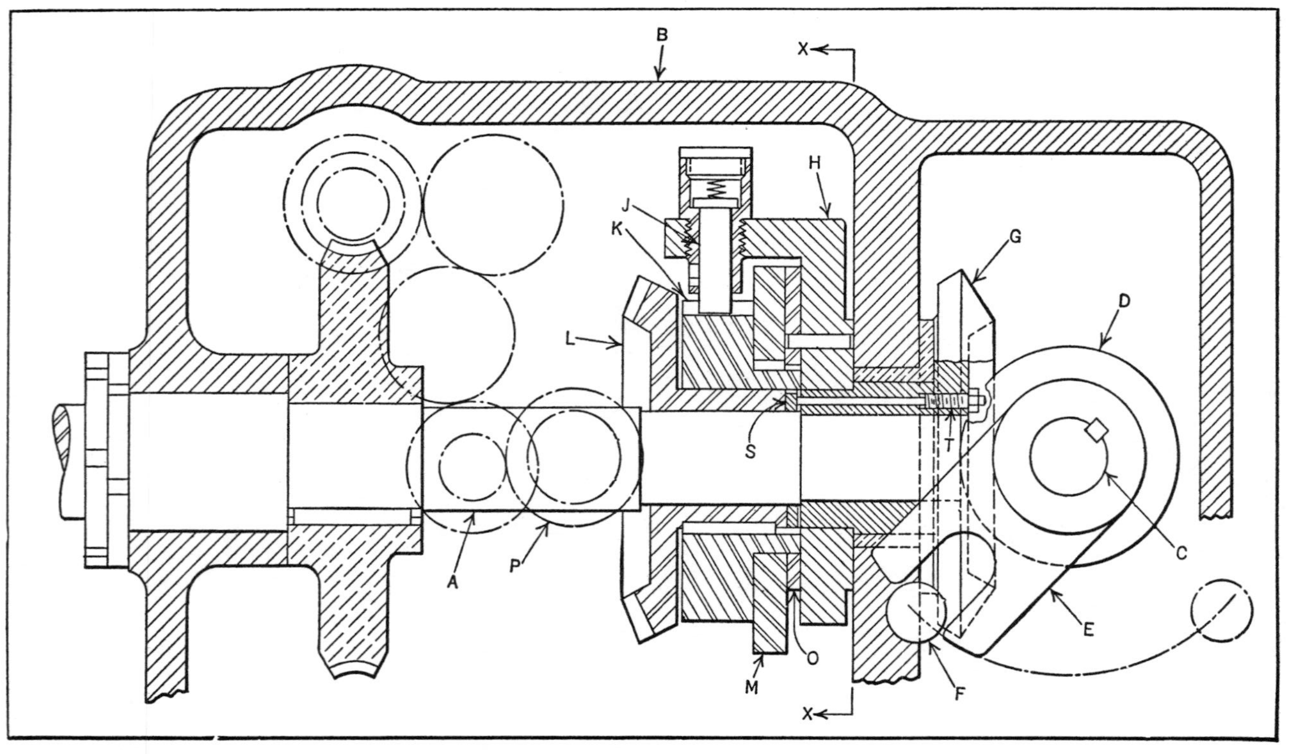

Fig. 23. Self-locking Indexing Head for Obtaining a Large Number of Divisions

meshes with gear *G*, keyed to the arm *H*. Arm *H* carries a spring-actuated pawl *J* that engages a ratchet wheel *K*, keyed to the bevel gear *L*.

A locking ring *M* is keyed to the ratchet wheel and, as indicated in Fig. 24, engages the plunger *N* in the housing *B*. Cam *O*, which is a thin plate pinned to the arm *H*, dis-

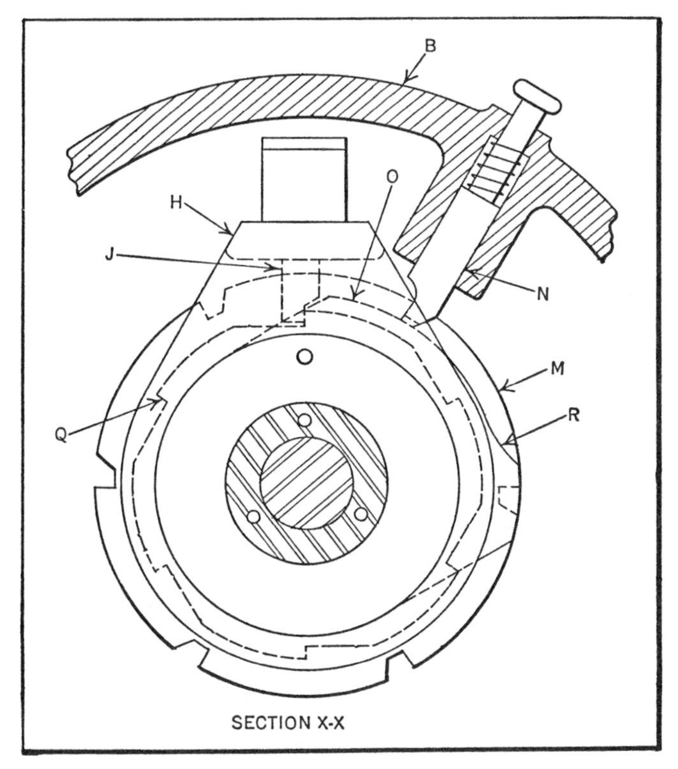

Fig. 24. Cross-section of Indexing Mechanism, Fig. 23, Showing Action of Locking Cam

engages the plunger *N* from the locking ring just before each indexing movement. Plunger *N* is wide enough to engage both the locking ring and the cam. To reduce the indexing movement so that a large number of divisions could be indexed, a combination spur and worm gear train was introduced, as indicated by the dot-and-dash lines in

Fig. 23. This gear train is operated by the bevel gears *L* and *P*.

Shaft *A* is indexed as the head moves toward the right just before the cutter engages the work, and is idle during the return stroke, while the machining is being done. Incidentally, if the cutter thrust is against the head, these movements should be reversed by mounting lever *E* and gear *D* on the other side of gear *G*. In the position indicated, the head has just completed the end of its indexing stroke toward the right. Now, as the head returns toward the left, the lever *E*, meshing with pin *F*, will be swung in a counter-clockwise direction, causing gear *G* and arm *H* to rotate about 63 degrees.

By referring to Fig. 24, it will be seen that during this movement of arm *H*, pawl *J* will be carried to the left and will engage ratchet tooth *Q*. However, just before pawl *J* engages this tooth, plunger *N* is disengaged from ring *M* by the lobe *R* on cam *O*.

On the return or indexing stroke of the table, lever *E* is swung in the opposite direction (clockwise), causing pawl *J* to rotate ratchet wheel *K*, with gear *L*, one-sixth revolution. This movement of gear *L* is transmitted through the spur and worm gear train, causing the shaft *A* to turn one division.

A washer *S*, Fig. 23, is provided to eliminate any backlash in the bevel gears and preserve the accuracy of the head. The backlash is taken up by tightening the screw *T*, thus bringing the gears into closer mesh. The number of divisions obtained with this type of head can be varied by changing the number of teeth in the ratchet wheel, by changing the gears or by varying the throw of lever *E*. This lever, with pin *F*, may be replaced by a rack and pinion.

Intermittent Motion for Feeding Wire to a Cutting-Off Machine.—Short pieces of twisted wire, approximately one inch long, are used in a certain product. Measuring and cutting off these short lengths by hand was found to be a slow and unsatisfactory process; hence the machine shown in Figs. 25 and 26 was designed to do this work automatically. It consists essentially of a mechanism for feeding the wire intermittently to two rotating shear blades.

All the working parts of the machine are mounted on a steel baseplate. Bracket *A* provides a support for the fixed shear blade *B* and also contains a double bearing for the shaft *C* (Fig. 25) on which a rotary shear blade head is mounted. It will be noted that bracket *A* is threaded to receive the two bronze bearing bushings *D* for shaft *C*. These bushings provide the necessary adjustment for setting the blades of the rotating shear head close to the stationary blade *B*. After this adjustment is made, the bearing bushings are locked in position by tightening the screws *E*.

Shaft *C* is driven by a 1/4-horsepower motor through reduction gearing, and drives the shaft *F* by means of helical gears. Shaft *F* runs free in the grooved roll *G* and in the ratchet wheel *H*. Keyed to shaft *F* is the feed-crank *K*, and connected to this crank is the link *L* (Fig. 26). This connection is made by the screw *M*, which also serves as a pivot on which link *L* oscillates. Secured to the head of screw *M* and to the link *L* is a spring, the purpose of which will be explained later. This spring, which has been omitted to avoid confusion, is of the "mouse-trap" type and is wound around the head of the screw *M*.

At the lower end of link *L* is secured the roller *N* and the feed pawl *O*. The pawl engages the ratchet wheel *H*, while roller *N* rides upon the periphery of cam *P*. Cam *P* is a running fit on the hub of feed-crank *K* (Fig. 25), and is prevented from turning by screw *Q* (Fig. 26). Roll *G* is keyed to the ratchet wheel *H*.

Operation of Wire-Feeding Mechanism.—The manner in which the feeding movement is imparted to the ratchet wheel and roll *G* is as follows: As shaft *F* revolves, the feed-crank *K* and its connecting members *M*, *L*, *N*, and *O*

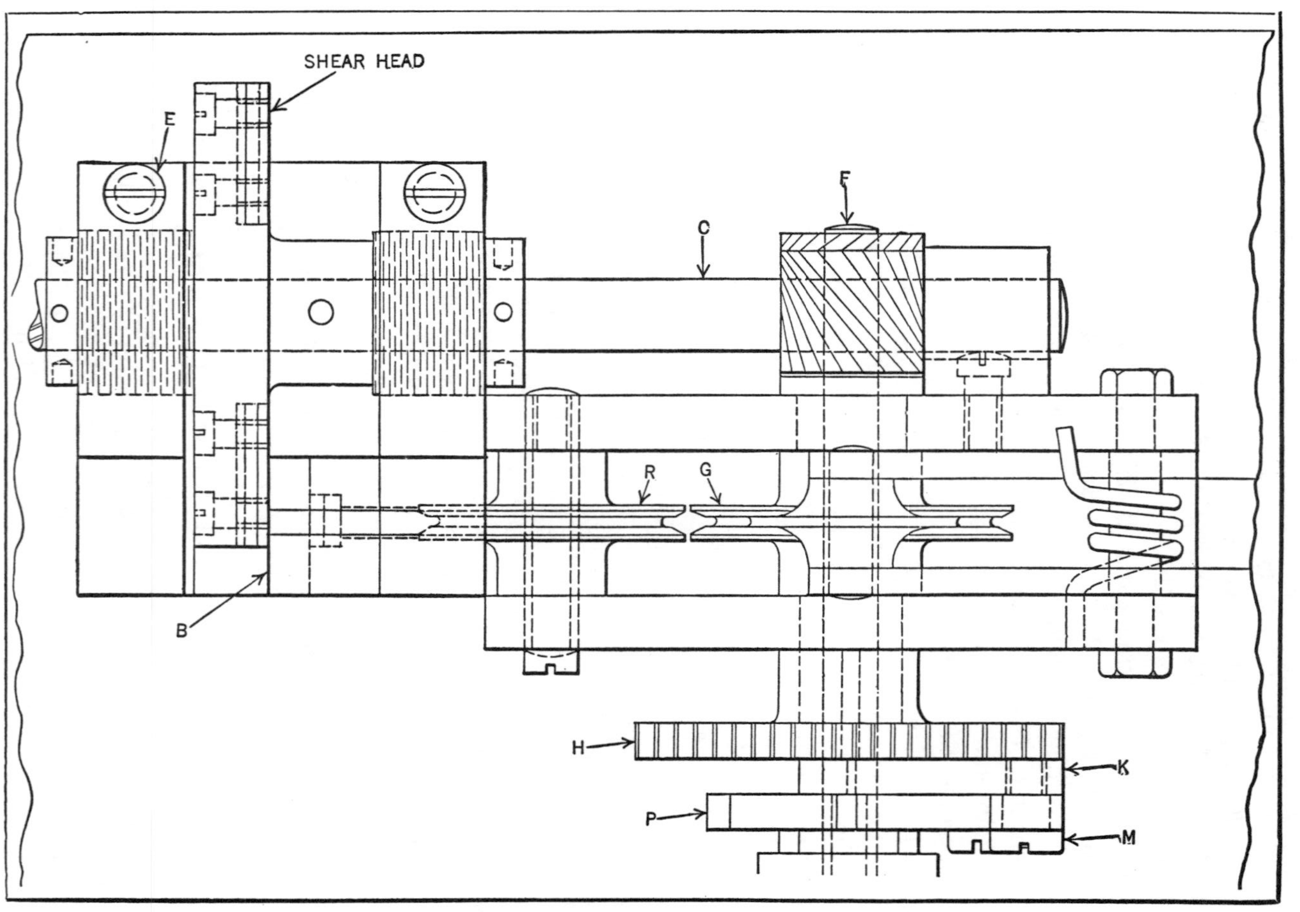

Fig. 25. Plan View of Intermittent Wire-feeding Mechanism Shown in Fig. 26

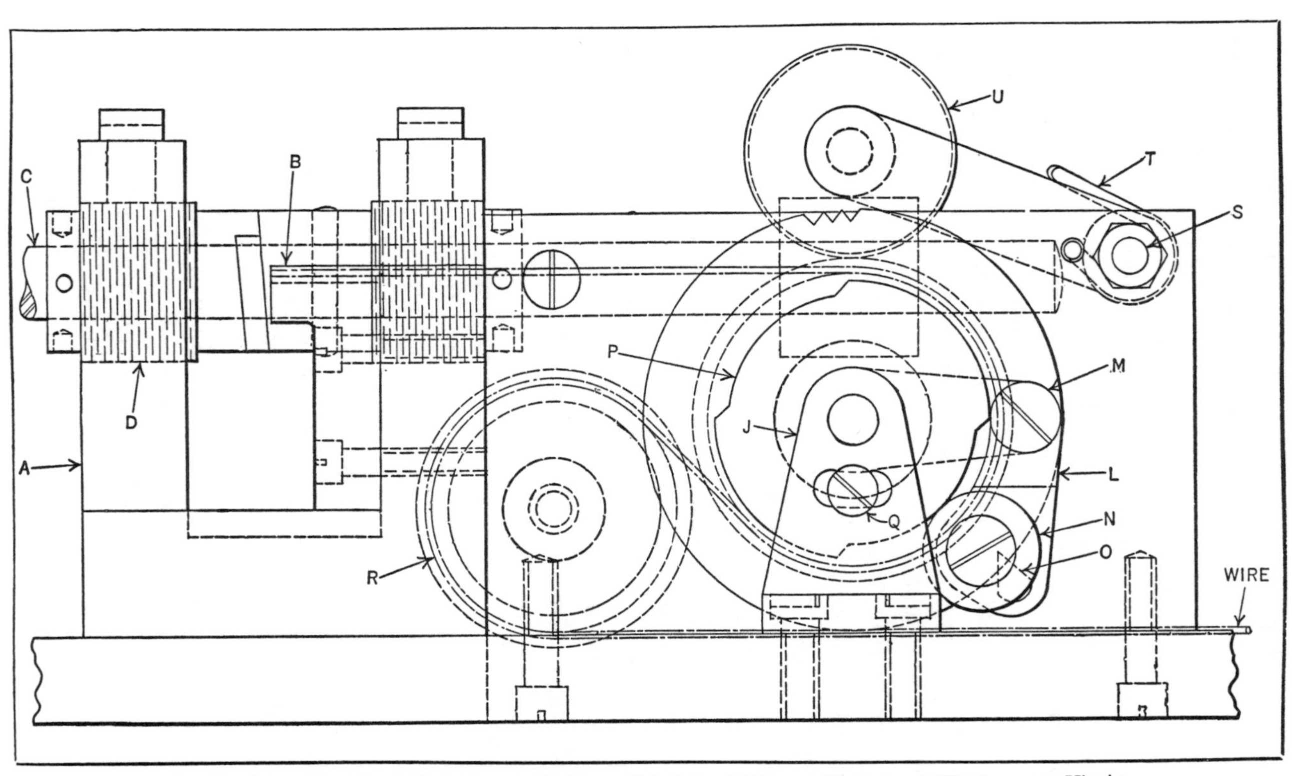

Fig. 26. Intermittent Feed Mechanism on Machine which Cuts off Wire at the Rate of 120 Pieces per Minute

are carried around cam *P*. Roller *N* is forced to maintain contact with the cam by means of the "mouse-trap" spring previously mentioned. The cam has two low places corresponding to the feeding intervals.

As roller *N* drops into these low places, the link *L* is pulled toward the center of the cam, carrying pawl *O* into engagement with the ratchet wheel *H* and thus rotating the ratchet wheel and roll *G*. This movement continues until roller *N* engages the high part on the cam and is forced outward, carrying link *L* outward also, and disengaging pawl *O* from the ratchet wheel.

In operation, the end of the wire is carried by hand under and over idler roll *R*, under and over feed-roll *G*, through a short piece of tubing (not shown) to keep it from buckling, and then over the edge of fixed shear blade *B*, where it is cut to length by the blades in the rotating head after the machine is started. Idler roll *U*, mounted on two arms pivoted on the stud *S*, serves to exert a pressure on the wire against roll *G* through the medium of spring *T*. This provides the necessary traction to pull the wire from the reel, which, although not shown, is located on the steel baseplate at the right.

Ninety-nine teeth were cut on the ratchet wheel, so that all the teeth would come into action. In this way, the wear is distributed over all the teeth. These teeth have a face angle of 15 degrees to permit the pawl to disengage readily under load. Any feeding movement from 1/99 of the circumference of the roll *G* to approximately one-half this amount can be obtained by substituting suitable plate cams.

As cam *P* has two low places on its periphery, it is obvious that two feeding movements take place for every revolution of shaft *F*. Through reduction gearing, shaft *C* operates at 60 revolutions per minute, cutting two wires at each revolution, or 120 wires per minute. Thus for a period of eight hours, the production is approximately 57,000 wires.

Indexing Mechanism with Interchangeable Turrets for Either Three or Four Stations.—A reciprocating slide on an automatic machine designed for drilling and counter-boring small fiber parts is provided with two turrets which are interchangeable. One turret has three equally spaced tool stations, while the other has four stations. The three-station dial is replaced by the four-station dial when the work requires the use of four different tools. By the elimination of the extra indexing movement through the use of the three-station turret whenever possible, an appreciable saving is realized.

Referring to Fig. 27, the tool-slide is shown at *A*. Upon the slide is mounted the permanent indexing dial *B* which is free to turn on the stud *C* fixed in the slide. The four-station turret *D* is shown mounted on the dial. The pin *E*, which is a drive fit in the dial and a slip fit in the turret, prevents the turret from rotating on the dial. The turret is indexed by means of the blade *F*, which is integral with the spring-actuated plunger *G*, the blade engaging the pins *H*, *J*, *K*, and *L* in the dial.

It will be noted that the four pins *H*, *J*, *K*, and *L* are in equally spaced holes in the dial, which correspond with the four turret stations. When this turret is replaced by the three-station turret (not shown), only three pins are used in the dial. In that case, two pins are located at *M* and *N*, pin *H* remaining in the position shown. The indexing of the turret is accomplished during the idle part of the stroke indicated.

In the position shown, the slide, moving toward the right, is approaching the working part of its stroke. On continuing this movement, the tool opposite the work performs its operation, after which the slide starts on its return stroke. As pin *H* leaves the heel of blade *F* on its movement to the left, the dial is prevented from reversing its movement by the pawl *P* pivoted to the slide. This pawl engages teeth cut in the periphery of the dial. Incidentally,

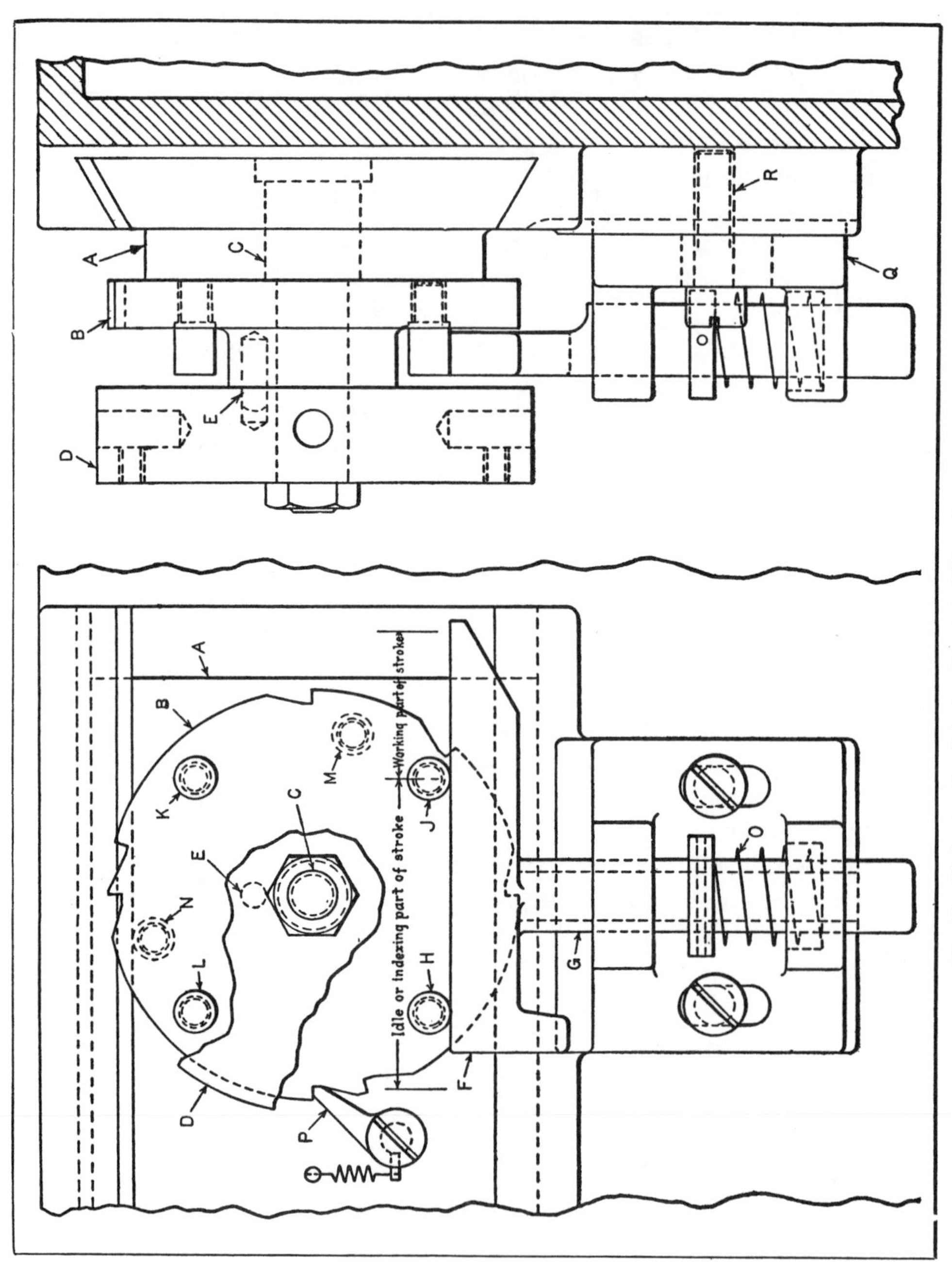

Fig. 27. Mechanism for Indexing Turret to Either Three or Four Stations

without this ratchet arrangement, the reversal of the dial would prevent the engagement of the blade with the succeeding pin, thus preventing the indexing of the dial. After pin *J* has left blade *F*, the latter is forced forward by spring *O*, a distance equal to about three-quarters the diameter of the pin.

Now, when the slide reverses its movement, pin *J* comes in contact with the left-hand end of blade *F*, so that further movement of the slide will cause the blade to rotate the turret. Continued rotation of the turret results in pin *K* coming in contact with the blade; and as the slide continues, the blade is pushed outward, so that pins *J* and *K* become located in contact with the long edge of blade *F*. The tool in the second turret station is now in position for performing its operation.

No means other than blade *F* are provided for locking the turret. The tool pins, pressing against the blade *F* as shown, serve to prevent the turret from rotating. This arrangement has been found entirely satisfactory for the class of work handled on the machine on which it is used. When the three-station turret is employed, it is necessary to move the bracket *Q* inward, so that the heel of blade *F* on plunger *G* will pass the center line of the right-hand pin when the slide is moved to its extreme left-hand position. To make this adjustment, it is only necessary to loosen the screws *R* which lock the bracket in place.

CHAPTER IV

INTERMITTENT MOTIONS OF THE GENEVA TYPE

The Geneva type of intermittent motion is based upon the principle of the Geneva stop which has been applied to watches, etc., to prevent winding the main spring too tightly. This stop mechanism, as the name implies, is intended to prevent rotation after a certain number of revolutions. This is not the case, however, when the principle of the Geneva stop is applied to intermittent gearing. Geneva wheels or mechanisms are used to transmit an intermittent motion to some driven member at regular intervals which may be repeated indefinitely, as, for example, when some part of a machine tool requires indexing or rotating through some fractional part of a revolution at certain intervals while the machine is in operation.

Geneva Motion Designed to Reduce Rate of Acceleration and Deceleration of Driven Member.—The Geneva stop mechanism is used frequently because of its simple design and serviceability. In the form generally used, the driving roll follows a circular path. With this arrangement, the disk begins its movement from a stationary position and comes to a stop without shock, but the acceleration and deceleration in the velocity of the disk occur at a rapid rate, producing a relatively high angular velocity in the rotating disk. In order to eliminate these disadvantages, a German inventor developed a modified form of Geneva stop mechanisms in which the driving roll that transmits intermittent motion to the cross or slotted disk is operated by a mechanism consisting of four articulated members, as shown at *A*, Fig. 1.

In this mechanism, the driving member *D* rotates on

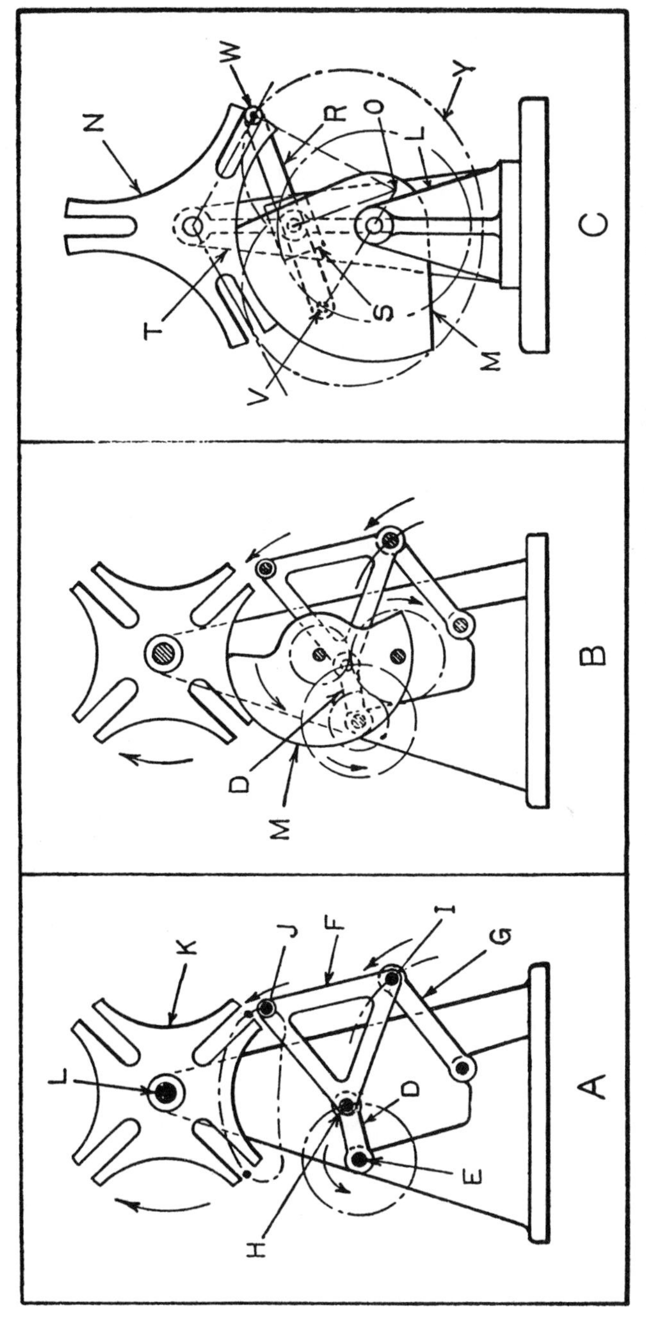

Fig. 1. Geneva Mechanisms Designed to Reduce Rate of Acceleration and Deceleration of Driven Member

axis *E*, and by means of rod *F*, gives member *G* a swinging motion. As the center of the connecting stud *H* describes a circle and stud *I* moves through only a part of a circle, all other points on the rod or member *F* describe curves of a distinct form. A stud at *J* supports the driving roll for the Geneva stop mechanism. When crank *D* completes a full rotation, roll *J* enters a slot in disk *K* and drives the disk to the next stopping position, after which it leaves the slot.

The difference between this mechanism and the older well-known arrangement is that the height of the curve followed by the roll on stud *J* is not so great; thus the angular velocity of the disk *K*, which depends on the distance of point *L* from the top of the curve, is considerably reduced. To prevent any unintentional movement of disk *K*, a blocking disk is necessary. For this purpose, a disk *M*, as shown in view *B*, is supplied. This disk is driven by intermediate gears from crank *D*. The addition of this blocking system, however, considerably complicates the mechanism. Another disadvantage of this drive is the bulky unsymmetrical design.

Another similar drive which functions through a turning block linkage is shown at *C*. The small fixed bracket *L* forms the bearing for the shaft of the driving crank *M*. This crank-arm also serves as the blocking disk for holding the driven disk *N* stationary during the dwelling periods. The crank-arm *M* is connected at *V* to the rod *R*, which slides in a block *S*, pivoted on the stand *T*. On the opposite end of rod *R* is mounted a roller *W*. As shown in the illustration, roller *W* describes a heart-shaped curve *Y*. The upper or spear-shaped portion of the curve *Y* is used for imparting the driving movement to the three-armed cross or driven disk *N*.

As roller *W* enters the slots in disk *N*, tangents to the path followed by the roller at this point must pass through the center of the disk and the center of the slots, which are radially located. The normal of curve Y is found by extending a line from V through the center of the shaft on which M is mounted, so that it intersects a line perpendicular to rod R, drawn from the center of the fixed stud on which block S is pivoted. A line from the point of intersection O to the center of roller W forms the desired normal to curve Y.

Combined Geneva and Intermittent Gear Movement.— In order to modify the operating characteristics of the

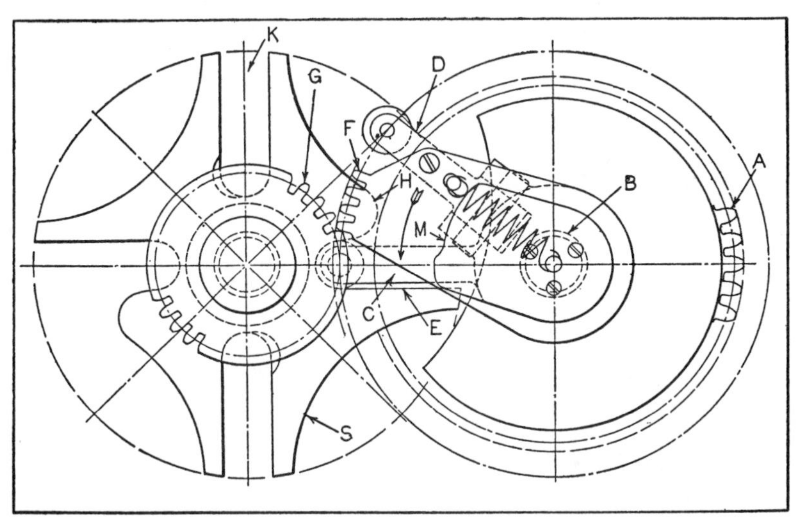

Fig. 2. Intermittent Drive Mechanism Designed to Accelerate and Decelerate Motion at Start and Finish of Driving Movements

well-known Geneva gear movement to adapt it for a particular purpose, intermittent gearing was incorporated in the design, as illustrated in Figs. 2 and 3. After laying out the design on the drafting board, a model was made which operated satisfactorily. The mechanism consists of a modified double driving arm Geneva wheel with intermittent gear segments. The gear segments are so placed that they transmit a practically uniform speed movement to the driven member from the instant the driving arm ends its accelerating movement until the second driving arm begins its decelerating movement in stopping the driven member. One advantage of the mechanism, in its application to an automatic machine, is that the driver requires a movement of only about 130 degrees to rotate the driven member 180 degrees. This leaves 230 degrees of the driver cycle

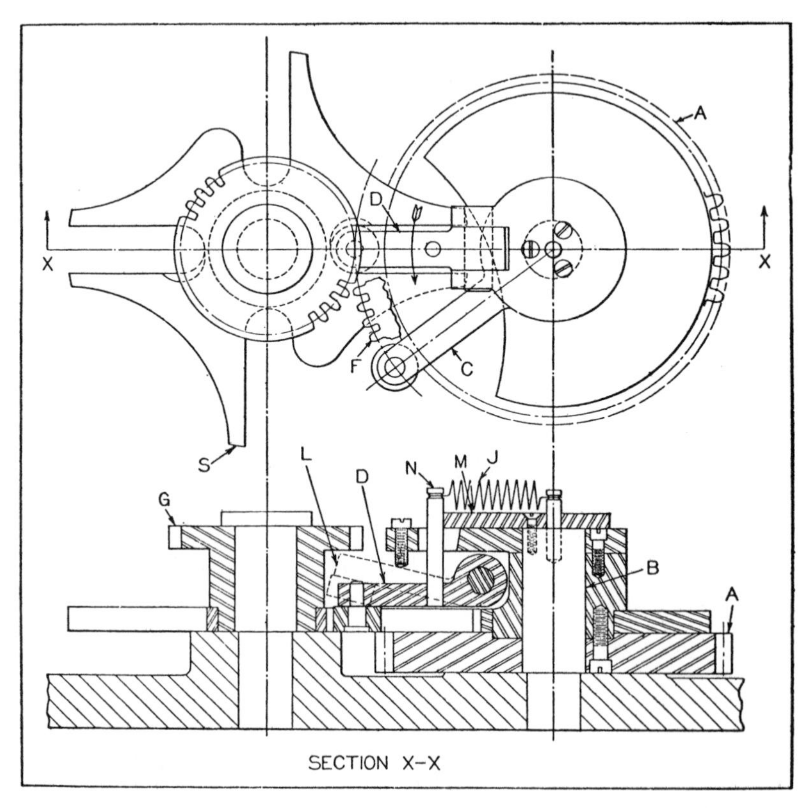

Fig. 3. Mechanism Shown in Fig. 2 with Various Members in Different Operating Positions

free to perform other useful work or operations while the driven member dwells.

Referring to the illustrations, *A* is the driving gear, which operates at a uniform speed. The whole driving unit is mounted on the stationary stud *B* and rotates in the direction shown by the arrow. The first driving arm *C* is integral with the driving member, while the second arm *D* is pivoted to it and suitably spaced from the first arm.

When arm *C* engages slot *E* in the driven spider *S*, it will start rotation of the latter member and accelerate its speed until arm *C* reaches the center line between the two members. At this point the intermittent gear segment *F* meshes with its mating segment *G* on the driven member. As the pitch line of the intermittent gears corresponds with the center line of the path in which the arm rollers rotate, the gears continue the motion of the driven member at approximately the same speed as was attained by the roller arm *C* at the instant it passed the line between the centers of the driving and driven members. The slot or arm on the far side of the spider *S* is shortened and so shaped at *H* that the roller cannot interfere with the uniform motion imparted by the gears as the roller recedes from the slot.

The ratio of the intermittent gears is such that the driven gear *G* will rotate 90 degrees while the gears are engaged, the remainder of the 180-degree movement being derived from the two driving movements of 45 degrees each, imparted by the accelerating arm *C* and the decelerating arm *D*. The latter arm, because of its pivoting feature (see Fig. 3) and the tension of the spring *J* is held out of engagement with its slot *K*, as indicated by the dotted lines *L*, until just before it reaches the center line, when the action of the lobe of the stationary cam *M* on the pin *N* forces the arm down into engagement with its slot. This engagement occurs at the instant when the intermittent gears pass out of engagement. The Geneva gear action of the arm *D*, in its further rotation, decelerates the driven member to a stop 180 degrees from the point where the accelerating arm *C* started its rotation.

Inverse Geneva Wheel Motion.—The term "inverse" is applied to an unusual form of the well-known Geneva mechanism for producing intermittent circular motion, because the driving and driven members rotate in the same direction, whereas with the usual form of Geneva motion, the rotations are reversed. The arrangement is such that the driving crank axis and the crank circle are entirely within the radius of the plate or driven member, and this produces a vastly different effect in the timing, acceleration, and the velocity of the plate. These effects are things to be considered in applying the mechanism to a machine design. In some designs, the effects produced may not be altogether desirable, while in others they may have distinct advantages and introduce an improvement.

The inverse Geneva stop or wheel motion was developed to fill the requirements of a particular type of drive for feeding strip stock into power press dies. Since its inception a variety of successful applications in automatic machinery have been made.

Typical forms of the inverse Geneva wheel are shown in Figs. 4 and 5, the former showing a three-station and the latter an eight-station plate. The essential parts are few and simple, consisting of a constant-velocity driving crank *C* and a variable-velocity driven member *D*, called the plate. The plate rotates in equal intermittent movements from station to station, stopping for a short interval of time at each station. As the rotation of the plate is caused by the motion of the crank-pin roller *E* in passing through radial grooves in the plate surface, the number of stations is dependent upon the number of grooves.

The smallest number of radial grooves with which a Geneva mechanism will function is three. The greatest number is infinite, being limited only by the diameter of the plate and the width of the grooves, both of which may theoretically be made to any proportions. In actual practice, however, the number of grooves required is not very great.

Working and Idling Angles of Driver Rotation.—By comparing Figs. 4 and 5 it will be seen that as the number of grooves increases, the working angle a of the driving

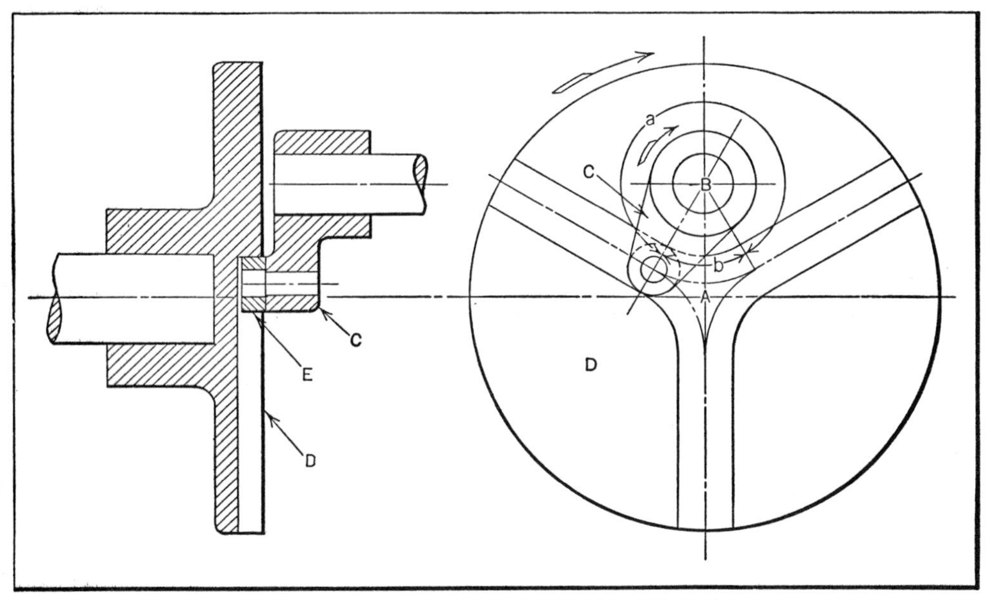

Fig. 4. Three-station Inverse Geneva Wheel Mechanism

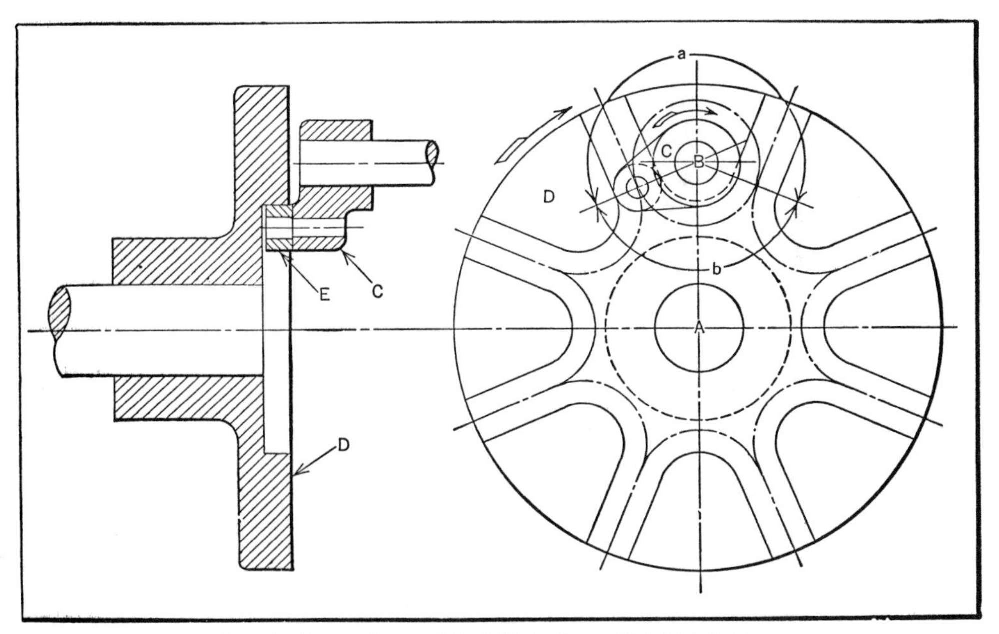

Fig. 5. Inverse Geneva Wheel Mechanism with Eight Stations

crank decreases and the idling angle b increases. These angles are determined as follows: Referring to Fig. 6, the angle s between two adjacent radiants on the Geneva plate is equal to 360 degrees divided by the number of radiants N. As the roller enters and leaves the grooves when the crank center line is at right angles to the radiants, two equal triangles are formed by the lines AEB and $AE'B$, from which it is seen that angle b equals $180 - s$. Then angle a equals $360 - (180 - s) = 180 + s$.

Fig. 6 also shows some of the practical points to be considered in the design of an inverse Geneva wheel or stop. In both Figs. 4 and 5 the inner ends of adjacent grooves are joined by a circular arc which is concentric with the crank circle. This arc is of little or no use, and to facilitate machining, it is preferable to connect the grooves with straight lines, as at h, Fig. 6. The corners should be broken by a small radius to permit the roller to enter the grooves more easily.

Locking the Driven Member.—In any sort of intermittent motion device it is desirable, and usually necessary, that some means be provided for locking the driven member in position while it is at rest. The locking feature employed in this mechanism is shown in Fig. 6. In this illustration, a circular arc lobe, machined concentric with the driving crank axis, is shown as an integral part of the crank at d.

During the idling period of the driving crank, this lobe is in contact with one of the locking segments e, which are made to project from the face of the plate and are machined to the radius of the lobe just described. This feature prevents any accidental rotation of the plate while it is in one of the rest positions. The angles f and g subtended by the arcs on e and d, respectively, are equal, and their magnitude is a matter that should be given careful consideration.

The angles should be made as large as possible in order to keep the plate locked during the entire time that it is

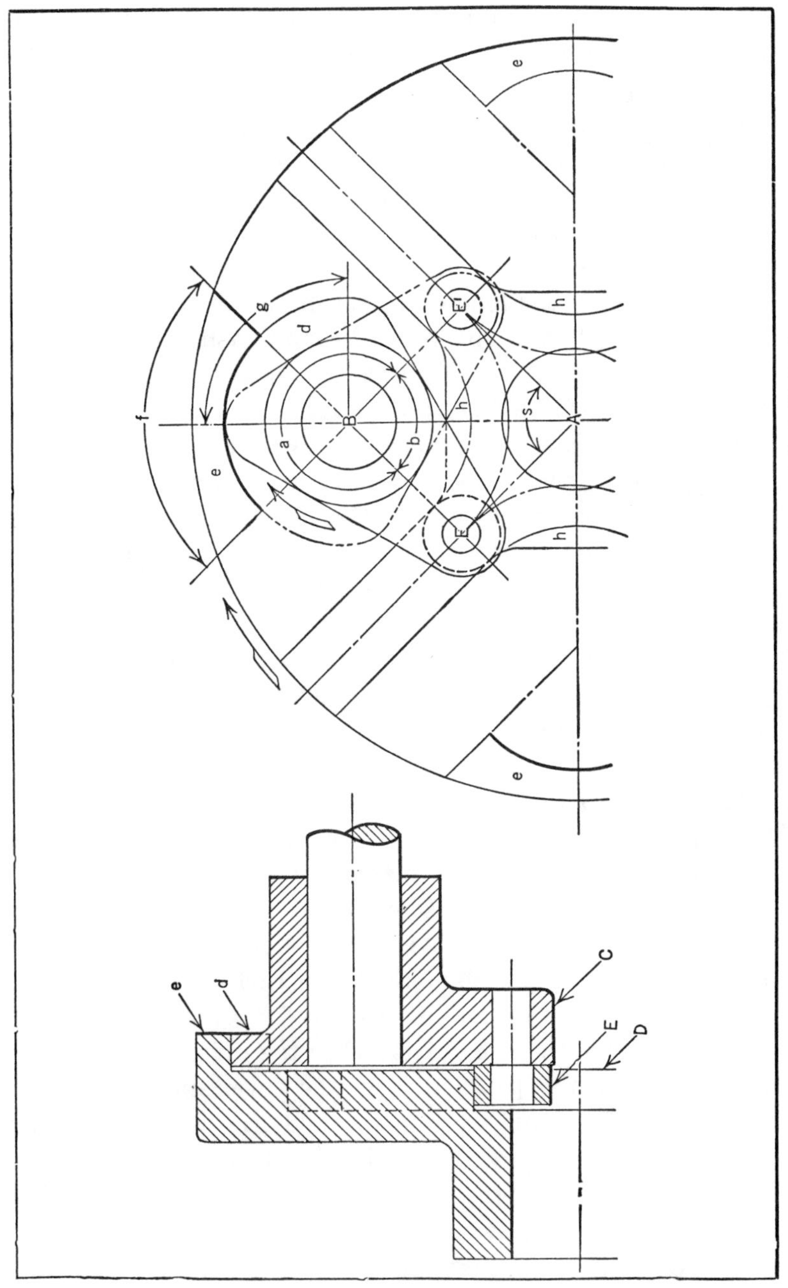

Fig. 6. Locking Arrangement Employed on Inverse Geneva Wheel Mechanism

not in motion, but making them too large will result in interference of the parts. The best results are obtained by making angles *f* and *g* each equal to angle *b*. This permits the locking action to begin the instant the crank roller leaves one groove and to end the instant it enters the next groove. This, it will be seen, is the case in Fig. 6. The heavy full outlines show the crank in the act of entering a groove, while the left half of the locking segment is just being cleared by the lobe on the crank so that rotation may begin in the plate. The light dot-and-dash outlines show the crank in the act of leaving the groove, with the lobe engaging the left half of the locking segment.

A locking segment *e* is placed midway between each two grooves, as shown. Their centers represent the relative positions of the crank during the idling interval between working periods. The radius of the locking segment is more or less arbitrary, but it will be limited by the radius of the plate. The locking segments do not add greatly to the cost of manufacturing the plate, because their shape is quite simple. In fact, the structural lines of the entire plate are made up of simple geometrical figures and are easily machined without the use of templets or masters.

Geneva Type of Work-Reversing and Transfer Mechanism.—Many types of automatic machines must be provided with means for turning over or reversing the position of the work at some point during its progress through the machine. This is accomplished very effectively in one case by the mechanism shown in Fig. 7. The work at *A* is turned over and transferred to position *B*. To accomplish this, the work *A* is fed in the direction indicated by arrow *C* to the position *D* in the reversing mechanism *E*. A plate *F* in this mechanism holds the work by means of pressure applied by two springs, only one of which can be seen in the illustration. Attached to the reversing mechanism is a gear *H* which is mounted on a shaft *J*.

The Geneva mechanism shown below gear *H* turns the

entire unit through an angle of 180 degrees in the direction indicated by arrow *K* each time it functions. This movement transfers the work from position *D* to *L*. When the next piece is pushed into position *D*, it comes in contact with pad *N* on plunger *M*, pushing it to the right. Thus the pad *P* on the other end of the plunger pushes the work into the position indicated at *B*. This action is repeated

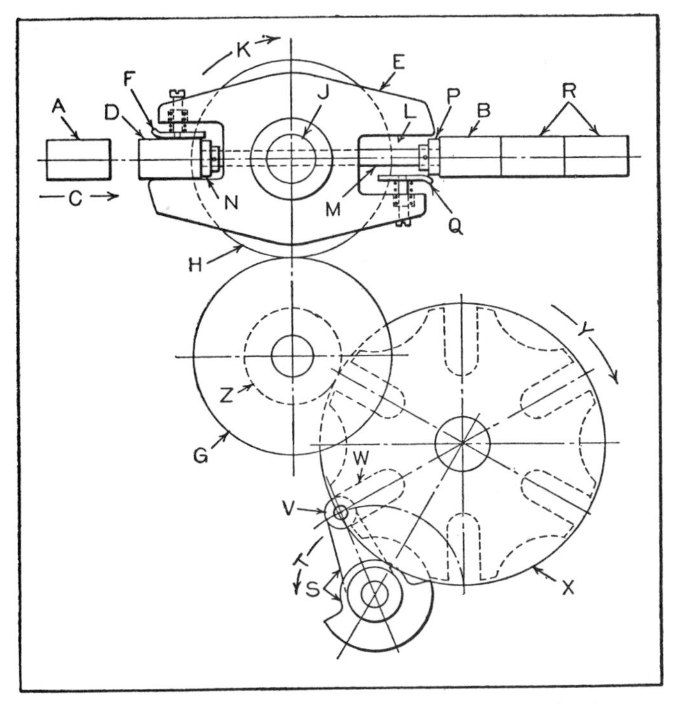

Fig. 7. Mechanism for Reversing Work D and Transferring it to Position B

at each cycle of the machine causing the work, which has been reversed, to be pushed along, as shown at *B* and *R*. The indexing is accomplished by means of a Geneva movement, in which the combination lever and locking segment *S* revolves in the direction indicated by arrow *T* through one complete revolution for each 180-degree indexing movement of member *E*.

Fig. 8. Mechanism for Producing Intermittent Rotary Motion for a Heavy Table

The roll *V*, at each revolution, engages one of the slots *W* in the plate, causing the large spur gear *X* to revolve in the direction indicated by arrow *Y* through one-sixth of a revolution. The speed ratio between gear *Z* and gear *X* is three to one. Thus gear *G* is revolved anti-clockwise one-half revolution, causing gear *H* to revolve one-half revolution in the direction indicated by arrow *K*. This completes one cycle in the operation of the automatic machine.

Segment Gear and Geneva Wheel for Intermittent Rotary Motion.— The mechanism shown in Fig. 8 was designed to give a large heavy table or turret an intermittent rotary motion. The drive shaft *A* carries a gear segment *B* which contains just enough teeth to cause gear *C* to make one revolution. On the same shaft with gear *C* is a crank carrying a roller *D* which engages slots in plate *E*. Plate *E* is so mounted on the table or turret that it is free to revolve.

Plate *E* carries pins *F* which are engaged by the rim or periphery of the circular segment *G* which is a part of the crank. This action locks the plate *E* in position while the roller is out of engagement with the slots. By varying the number of teeth in the pinion, the size of the segment, and the number of slots, the length of the dwell period can be increased or decreased. During the dwell period, while segment *B* is out of contact with the gear *C*, the roller *D* remains in the position shown. Very little shock occurs when the segment comes into contact with the gear. This mechanism is adapted for moving heavy loads, as the power is applied near the periphery of the table.

Locking Driven Wheel of Geneva Movement.—The modified Geneva movement shown at the left in Fig. 9, illustrated and described in Volume I, "Ingenious Mechanisms for Designers and Inventors" (page 74), provides positive locking of the driven member between the indexing movements. The locking is accomplished by having one or two of the rollers *R* engage the annular groove *G*. The roller *P*, carried by the driving disk *B*, is shown about to leave its slot, having completed the indexing of shaft *S*. The roller *R*, shown entering the groove *G*, serves to lock the disk and shaft *S* until the next indexing movement. This mechanism has the disadvantage of being rather large. Also, roller *R* is so located that the driver *B* must be of such a large diameter *D* that it would interfere with the shaft *S* if it were extended through the driven wheel. This construction necessitates placing the driven wheel at the end of the shaft, thus preventing the use of an outboard bearing.

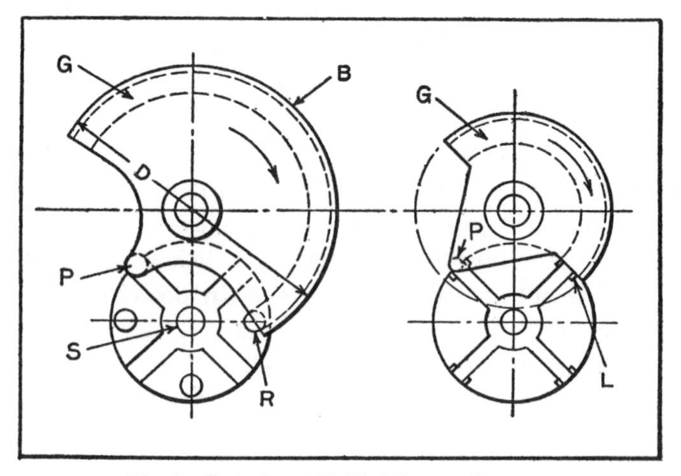

Fig. 9. Examples of Modified Geneva Movements

These objections have been overcome by the arrangement shown by the diagram at the right in Fig. 9, in which the locking of the driven disk is accomplished by lugs *L* which extend beyond the radial grooves of the driven member, so that they are engaged by the groove *G* in the driving member. Rollers can be substituted for the lugs *L*, but they are more expensive. It will be noted that the same diameter of driven wheel requires a driver of much smaller diameter than the mechanism shown at the left. It will be noted also that the lugs *L* are located in the most effective positions for locking the driven member, whereas the rollers *R* are so positioned that they lose about 30 per cent of their effectiveness. In other words, a given clearance between the locking members will permit more play or looseness in the case of the mechanism shown at the left. The loads on the locking pin required to resist a given torque will be about 50 per cent greater than on the lugs *L*.

Another advantage of the improved mechanism is that it is easier to make, especially in shops not accustomed to handling precision work, because the groove *G* extends for exactly half the driver area, or through an angle of 180 degrees, while the groove *G* of the other mechanism is somewhat over 180 degrees and must be carefully calculated and laid out. The rollers *R* must not only be accurately spaced between the grooves of the star wheel of the driven member, but they must also be accurately located at the correct distance from shaft *S*. On the other hand, the lugs *L* can easily be centered on the slots in the star wheel, and their location from the center is also easily accomplished. Still another advantage is that by cutting away a little material on the star wheel and modifying the arm of the driver that carries the roller *P*, either the driver or the driven disk can be assembled or dismantled without disturbing its mating part by sliding one part past the other.

Application of Geneva Wheel to Turret Indexing.— A well-known method of indexing the turrets of automatic machines is by the use of the principle of the "Geneva" motion. This has the advantage of giving a slow starting movement which gradually accelerates and then slows down before reaching the stopping point, thus securing rapid indexing and at the same time avoiding shock. An example illustrating the application of a Geneva wheel to turret indexing is shown in Fig. 10. In this case, the pin *A* engages the slots in the disk *B* to index the turret. The cylindrical portion of the pin carrier *C* engages concave portions of the disk *B* to locate the turret approximately; the automatic spring-operated latch D accurately locates the turret by engagement with notches in the large dividing wheel E. The turret is afterwards locked by a sliding steadyrest. This indexing mechanism gradually accelerates the heavy turret at the time of indexing and

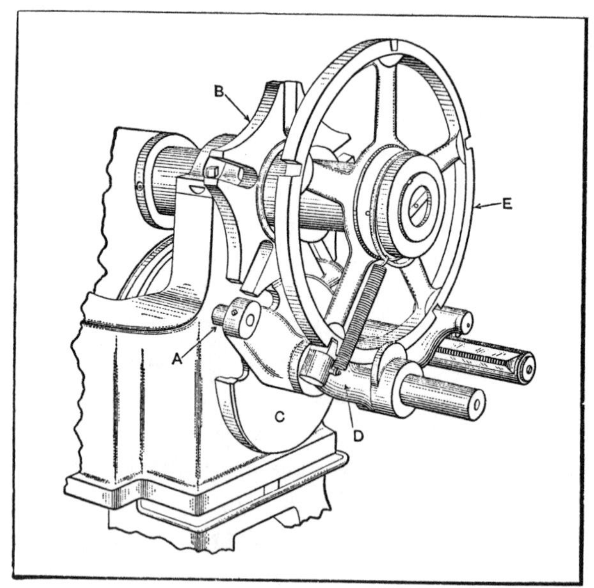

Fig. 10. Turret Indexing Mechanism of the Geneva Type

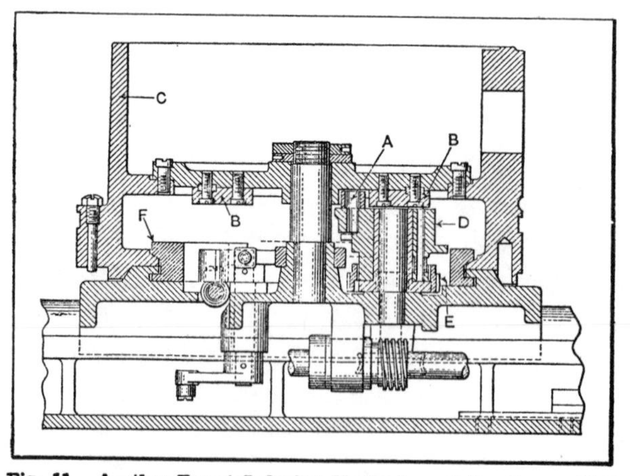

Fig. 11. Another Turret Indexing Mechanism of the Geneva Type

then gradually checks its momentum. Fig. 11 shows another application of the Geneva drive, in which a roller *A* engages slots formed between blocks *B* for indexing the turret *C*. The roller is carried on a sleeve *D* which is intermittently turned by gear *E*.

Graphical Analysis of the Geneva Mechanism.—In designing a Geneva mechanism for intermittently indexing a shaft or some other machine member through part of a revolution, it is frequently difficult for the designer unfamiliar with the mechanism to study its action. The following analysis is presented with a view to making the study of this mechanism easier. In the analysis it will be shown that a pair of imaginary arms connected by an imaginary link can be substituted for the Geneva mechanism and be kinematically identical with it. This is true for every point of the working range of the motion. The imaginary arms and connecting link will be of varying lengths at the different points of action, but at every point will be subject to the very simple laws covering the action of link work.

Fig. 12 shows in outline a typical form of the Geneva transmission at the beginning of the cycle. For simplicity, four slots are shown in the driven wheel *N*, although this analysis is equally applicable to any number of slots. The driving arm is shown at *M*, the center of the driving arm at *A*, the center of the driven wheel at *B*, and the center of the roller at *E*. The same mechanism is illustrated in Fig. 13 at an intermediate point in the cycle; here the letter *O* represents the imaginary driving arm, *P* the imaginary driven arm, and *Q* the imaginary connecting link. The imaginary arms and link are laid out as follows: Draw a line connecting *E*, the center of the roller, and *B*, the center of the driven wheel. Passing through *E*, draw the normal *ED*, and through *A* draw a normal to *ED*, intersecting at *D*. The imaginary driving arm is now length *AD*, the imaginary driven arm. *EB*, and the imaginary con-

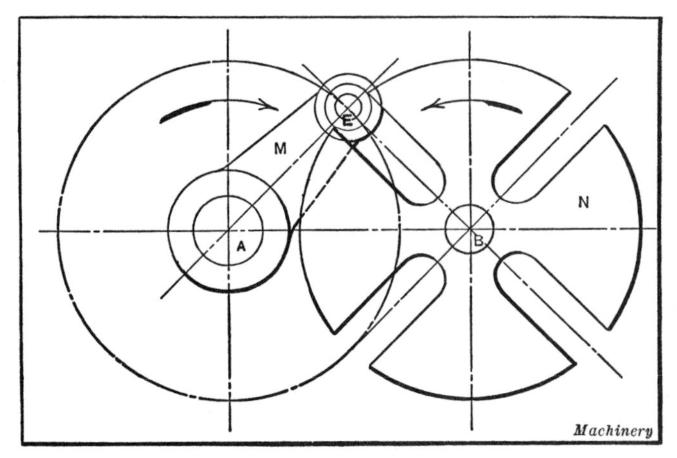

Fig. 12. Outline of a Typical Form of the Geneva Mechanism at the Beginning of a Cycle

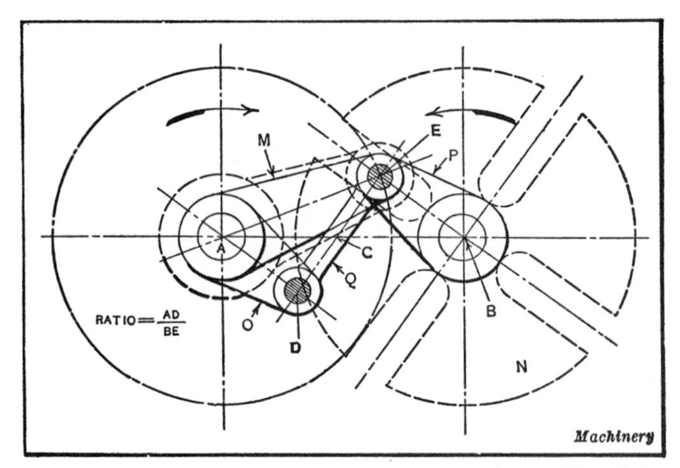

Fig. 13. Geneva Mechanism, with Imaginary Arms and Link that are Kinematically Identical with the Geneva Motion

necting link, *ED*. This imaginary linkage system kinematically replaces the Geneva mechanism for the point of the cycle at which *E* is in this illustration.

Fig. 14 shows the mechanism laid out with center *E* at a point still further advanced in the cycle. It will be noted from this illustration that the imaginary driving arm *O* has lengthened and that the arm *P* and link *Q* have short-

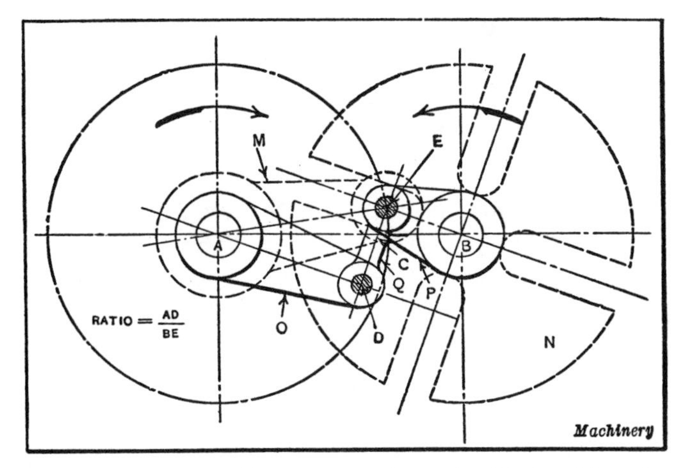

Fig. 14. Illustration Showing Geneva Mechanism Further Advanced along its Cycle

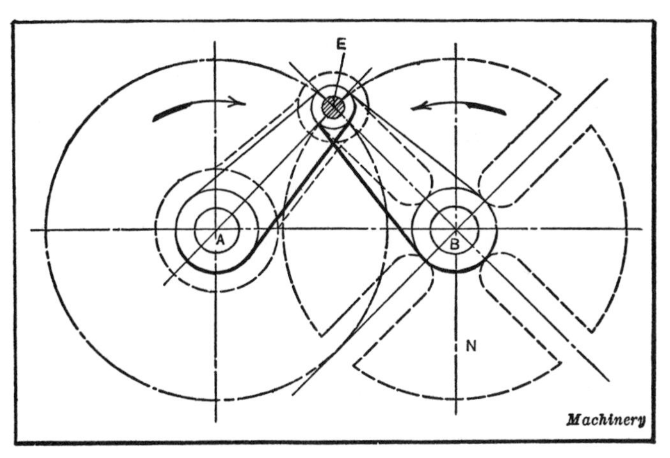

Fig. 15. Initial Position of Geneva Transmission, at which Point the Velocity Ratio is Zero

ened. In Fig. 16, the mechanism is shown at the middle of the cycle, where the imaginary driving arm attains its maximum length and coincides in length and position with the actual driving arm *M*. The imaginary driven link has shortened to the minimum, coinciding in position and length with line *BE* of the driven wheel, and the length of the connecting link is now zero. At this point the velocity ratio is at the maximum.

At the initial position of the Geneva transmission, which is illustrated in Fig. 15, the length of the imaginary driving arm is zero and the length of the driven link equal to *BE*, the maximum acting radius of the driven wheel. The length of the imaginary connecting link is also zero. In this case, the velocity ratio is evidently zero.

Determining the Velocity Ratios at Intermediate Points.— It is now in order to show the method of finding

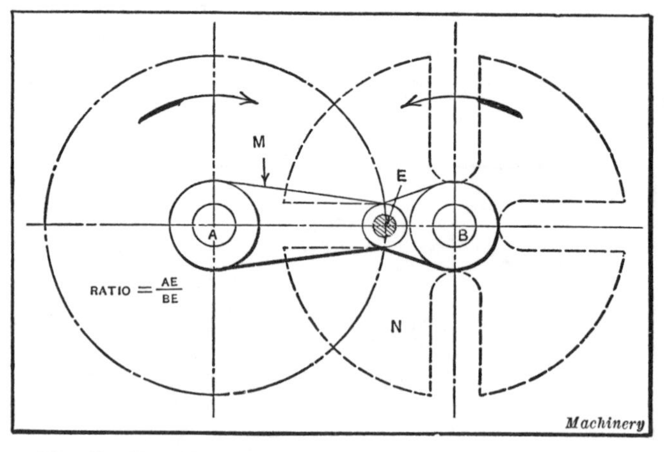

Fig. 16. Mechanism at Middle of Cycle, where Velocity Ratio is at the Maximum

the velocity ratio at intermediate points. In Fig. 13, the ratio is, by the law of leverages, $\frac{AD}{BE}$, as the connecting link is normal to both levers, and in Fig. 14 the velocity ratio is also $\frac{AD}{BE}$. (Compare with a pair of pulleys of radius *AD* and *BE*, connected by a belt *DE*). By laying out a number of lever systems, as in Figs. 13 and 14, the velocity curve of a Geneva mechanism can be determined for as many points as desired.

There is, however, a more direct method of determining the velocity ratio. In Figs. 13 and 14, the triangles ADC and BEC are similar by construction; therefore $AD:BE::AC:BC$. Hence $\frac{AD}{BE}$ equals $\frac{AC}{BC}$, and the velocity ratio is equal to $\frac{AC}{BC}$. Lines AC and BC are the segments into which the imaginary connecting link Q divides the line of centers AB.

Laying out the Velocity Curve.—To lay out the velocity curve, first divide the circumference EE_1, Fig. 17, into any number of parts, preferably equal. Then from B draw lines BX, BY, etc., through these points spaced out on the circumference. From these lines BX, BY, etc., draw normals 40-40, 35-35, etc., intersecting the points on the circumference and the line of centers AB. Now, assuming that A-15 along the line of centers measures 2.4 inches, and B-15, 6 inches, the velocity ratio at point 15 on the circumference equals $2.4 \div 6$ or 0.4. Compare this result with the velocity curve in Fig. 18.

Now prepare for laying out the velocity curve by erecting ordinates on the base line 0-90, Fig. 18, at equal distances apart. Lay off on each ordinate a distance corresponding to the quotient obtained by dividing $\frac{A\text{-}5}{B\text{-}5}$, $\frac{A\text{-}10}{B\text{-}10}$, etc., and connect these points. The resulting curve will be tangent to the base line at 0 and 90 and tangent to a line parallel to the base at the vertex, as shown. The velocity begins at zero, gradually increases to a maximum at the vertex of the curve, and then gradually diminishes till at the end of the cycle it again becomes zero.

Using the Velocity Curve.—If it is desired to find the point at which a velocity of 100 per cent occurs, bisect the line of centers AB, as in Fig. 19. On these segments draw the semicircles ADC and CEB. Through point E where one

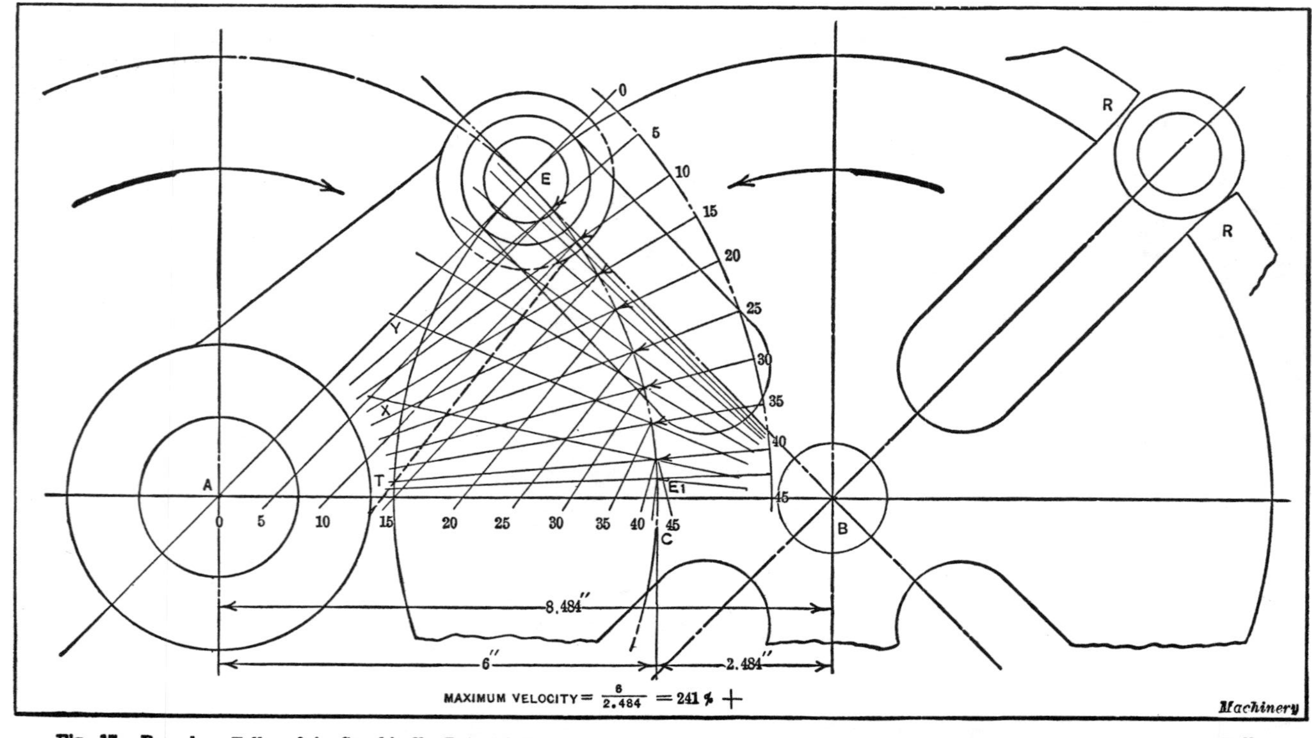

Fig. 17. Procedure Followed in Graphically Determining the Velocity of the Driven Member at Various Points along the Path of the Roller Attached to the Driver

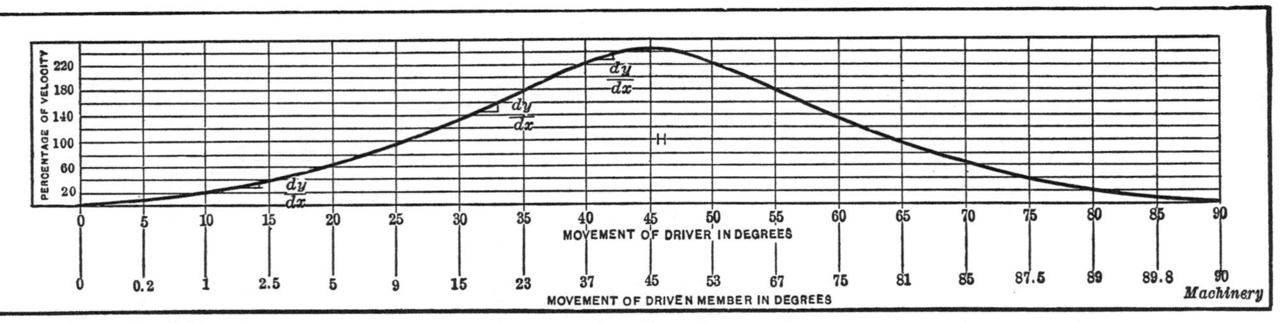

Fig. 18. Curve which Shows the Velocity Ratio Between the Driving and Driven Members of a Geneva Mechanism at Different Points along the Path of the Roller Attached to the Driver

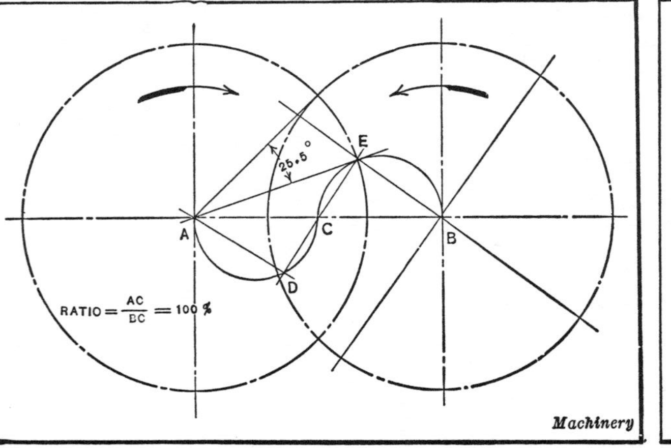

Fig. 19. Method of Determining at which Point a Velocity Ratio of 100 Per Cent is Obtained in the Movement of the Driving and Driven Members

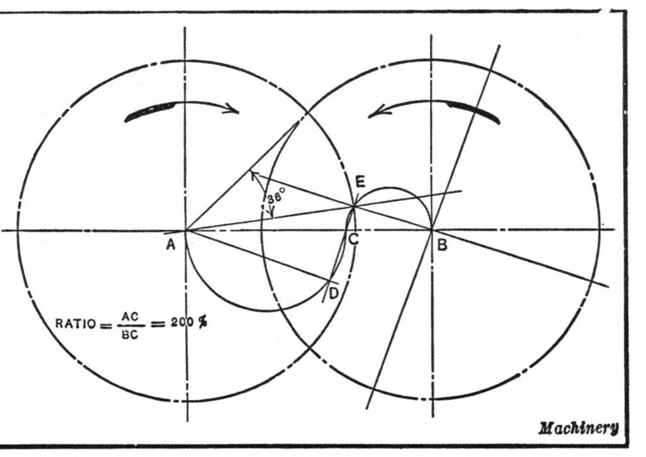

Fig. 20. Construction Used to Determine at which Point a Velocity Ratio of 200 Per Cent is Obtained Between the Driving and Driven Members

semicircle intersects the path of the center of the roller, draw the line *ED*, intersecting the line of centers *AB* through *C*. The angle *CEB*, being drawn on a diameter, will be a right angle. From *A* drop a normal *AD*, intersecting *DE*. The velocity ratio is $AC \div BC$, which equals **1**.

The construction for the 200 per cent ratio is shown in Fig. 20; in this case, the line of centers is trisected. Horizontal lines may be drawn intersecting the velocity curve in such a manner as to afford instant means for determining the velocity at any desired point. A ready method is to determine by the means suggested in the preceding paragraph, the height of ordinate *H*-45, Fig. 18, for a 100 per cent increase. Then divide *H*-45 into five equal parts, each of which will be equal to 20 per cent. Draw horizontal lines through these points of division and where these lines intersect the velocity curve, the velocity percentage will be known. The whole range of the velocity curve can be treated in this manner. It will be obvious that the velocity, after the driving member has moved 10 degrees, is found at the intersection of the ordinate 10 and the horizontal line marked 20 per cent.

To find the angular position of the driven wheel, prolong the lines *BX*, *BY*, etc., in Fig. 17, until they intersect arc *ET*. Measure angles *ABX*, *ABY*, etc., and lay off below the velocity curve, as shown in Fig. 18. The angular position of the driven wheel corresponding to any position of the driver can then be read off directly.

The only practical point in the design of the Geneva transmission that will be referred to here is to call attention to the desirability of enlarging the diameter of the driven wheel, as shown at points *R*, Fig. 17. This permits of operating a locking mechanism while the driven wheel is constrained by the driver, and makes the Geneva stop of very general application in automatic and semi-automatic machinery.

CHAPTER V

TRIPPING OR STOP MECHANISMS

Mechanisms of this general class may be used to stop a machine automatically either at the conclusion of a series of operations, possibly for stock renewal, or after a predetermined number of revolutions. Another function of a stop mechanism is to prevent the transmission of power to the machine whenever an abnormal operating condition would result in damage to the machine. These and other applications will be described.

Mechanism for Stopping Machine Automatically when Reel is Filled with Wire.—The mechanism shown in Fig. 1 is part of a machine for insulating electric wire. The purpose of this mechanism is to automatically disengage the machine clutch and thus stop the machine when the reel upon which the finished wire is being wound has been filled. This leaves the operator free to attend to other duties while the wire is being wound on the reel.

As the wire reel *A* gradually becomes filled, the roll *B*, resting on the layers of wire, is forced outward, causing the arm *C*, through a sliding clutch mechanism, to disengage the power actuating the reel. The roll *B* is held snugly against the wire by means of a weight (not shown) connected to the arm *C* by the cable *Y*.

The driving shaft *D* for the reel is supported in bearing *E* bolted to the machine base *G*. On this shaft is shrunk the clutch member *H* which engages the clutch teeth on sleeve *J*, to which gear *K* is keyed. Gear *K* meshes with gear *L* keyed to the shaft *M* on which the reel is secured. Sleeve *N* is a free fit in bearing *F*, and at its right-hand end has a turned collar. One end of coil spring *O* is placed

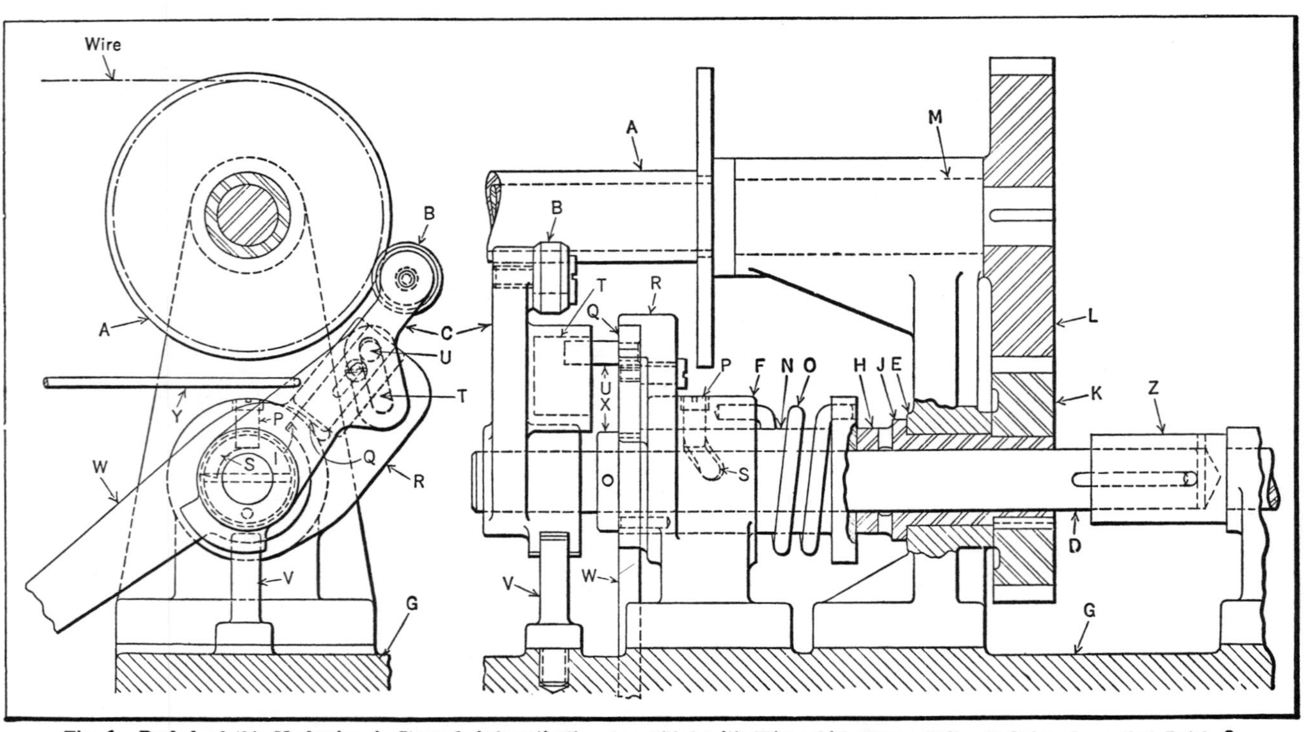

Fig. 1. Reel A of this Mechanism is Stopped Automatically when Filled with Wire which Forces Roller B Outward, so that Latch Q Releases Sleeve N, Permitting Torsion Spring O to Revolve Sleeve N. This Causes the Pin P in the Helical Slot S to Slide the Clutch Member H out of Mesh with the Clutch Sleeve J, thus Disconnecting the Drive from Shaft D to Reel A through the Gears K and L

in a hole drilled in the end of bearing *F*. This spring, when released as explained later, serves to rotate sleeve *N* in bearing *F*, so that the screw pin *P*, engaging a cam slot *S* in the sleeve, causes the clutch to move axially, disengaging the clutch members *H* and *J*.

The latch for releasing the spring is shown at *Q*. This latch slides radially in the guide *R* cast integral with bearing *F*. The movement of this latch is controlled by the pin *U* which is riveted to the latch and engages the deep cam slot *T* in a projection on arm *C*. Arm *C* turns freely on drive shaft *D* and is prevented from moving axially by the stationary pin *V* which engages a segmental groove in its hub.

When in its lowest position, the latch engages the projection *I* on the hand-lever *W*, which is a free fit on shaft *D* but is kept from moving axially relative to sleeve *N* and shaft *D* by collar *X* pinned to this shaft. Incidentally, lever *W* is pinned to the end of sleeve *N* and must therefore rotate with it. Sleeve *Z* provides for an axial movement of shaft *D*; the shaft slides in the sleeve, but is prevented from rotating by a key engaging a spline in the shaft. This sleeve is part of a shaft which is connected directly to the driving motor shaft.

The wire passing on to a nearly full reel is indicated in dot-and-dash lines in the end view. At this time, the last layer of wire on the reel has forced the roll *B* and arm *C* very nearly to their farthest right-hand position. During this movement of the arm, the cam pin *U* has been forced outward radially until latch *Q* is just about to leave projection *I* on the hand-lever *W*. As soon as another layer of wire is wound on the reel, the lever *C* will swing to the right a corresponding amount and cam slot *T* will raise pin *U*, so that latch *Q* will be entirely disengaged from projection *I*.

It should be mentioned that a torque has been developed by the coil spring *O* which, up to this point, has held the projection *I* tightly against the latch. As soon as the latch releases the lever, the energy stored in the spring causes the lever to swing in a clockwise direction, rotating sleeve *N* until pin *P* comes into contact with the opposite end of cam slot *S*. The action of the pin in this slot will cause the sleeve, together with lever *W*, shaft *D*, and clutch member *H*, to move axially toward the left and thus disengage clutch member *H* from the clutch teeth in sleeve *J*. By disengaging these clutch members, the power is thereby disconnected from the reel, causing the latter to stop, so that it can be replaced by an empty one.

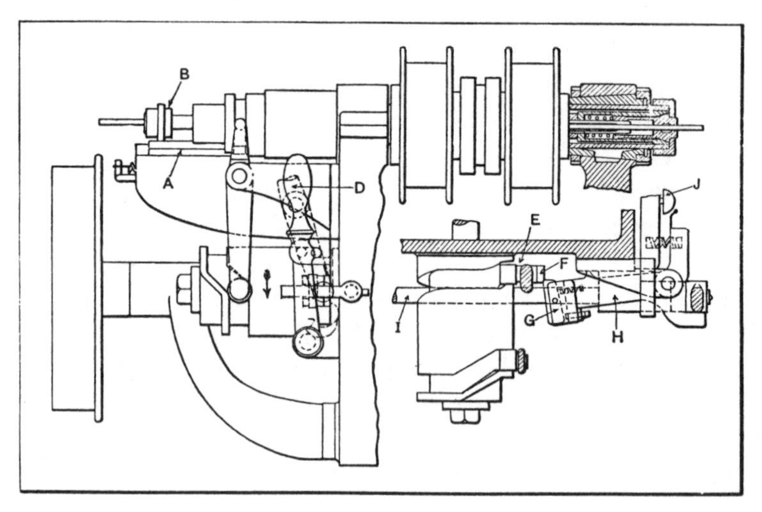

Fig. 2. Mechanism for Automatically Stopping the Machine when New Bar of Stock is Required

When the empty reel is in place, the hand-lever *W* is merely swung downward to start the reel rotating. This causes roll *B* to rest on the core of the reel and the cam slot *T* to allow pin *U* to move downward, so that latch *Q* once more engages the projection *I* and locks the lever in position. This downward movement of the hand-lever, of course, rotates the sleeve *N* so that the reverse action of the pin in the cam slot *S* occurs, moving the sleeve *N* toward the right and engaging the clutch members *H* and *J*, thus rotating the reel.

One outstanding feature of this arrangement is that

there is no excessive end thrust on the bearing, a condition which is typical of mechanisms of this type.

Stopping Machine for Stock Renewal.—A device for stopping the machine when the bar of stock has all been used is shown in Fig. 2; this result is accomplished by a mechanism controlled by the disengagement of the feeding device with the stock. The mechanism is so designed as to stop the machine with the jaws of the chuck open, so that a new rod of stock may be inserted; it is also devised so that the machine is not stopped nor the chuck opened until the length of stock projected by the forward movement of the feeding mechanism is acted on and severed from the remaining stock. This is accomplished by so constructing the stop mechanism that it is thrown into operative position when the feeding devices are disengaged from the stock, but does not operate to stop the machine until the feeding devices are again advanced.

In this construction, the slide *A*, connecting with the feeding tube by the grooved collar *B*, is drawn back by a spring when the stock passes beyond the feeding fingers, there being no friction to hold it; this operates lever *D* (shown dotted), the movement being made possible by the widened space in the cam groove at *E*. This movement allows the projection *F* to pass the latch *G* so that, on the next revolution of the cam, the lever *H*, carrying the latch, is rocked together with shaft *I*, which throws the driving mechanism out of operation and also sounds the gong *J* to notify the operator that a new piece of stock is needed.

Cork Cap-Disk Feeding Mechanism which Operates Only When Caps are in the Receiving Position.— The device shown in Fig. 3 is used in conjunction with a cap-feeding mechanism for inserting cork disks in the caps. As the caps are fed down the line the device places a cork disk in each cap, after which the caps continue on their way to other stations. The outstanding feature of the device is that it will not feed a cork disk *R* from the mag-

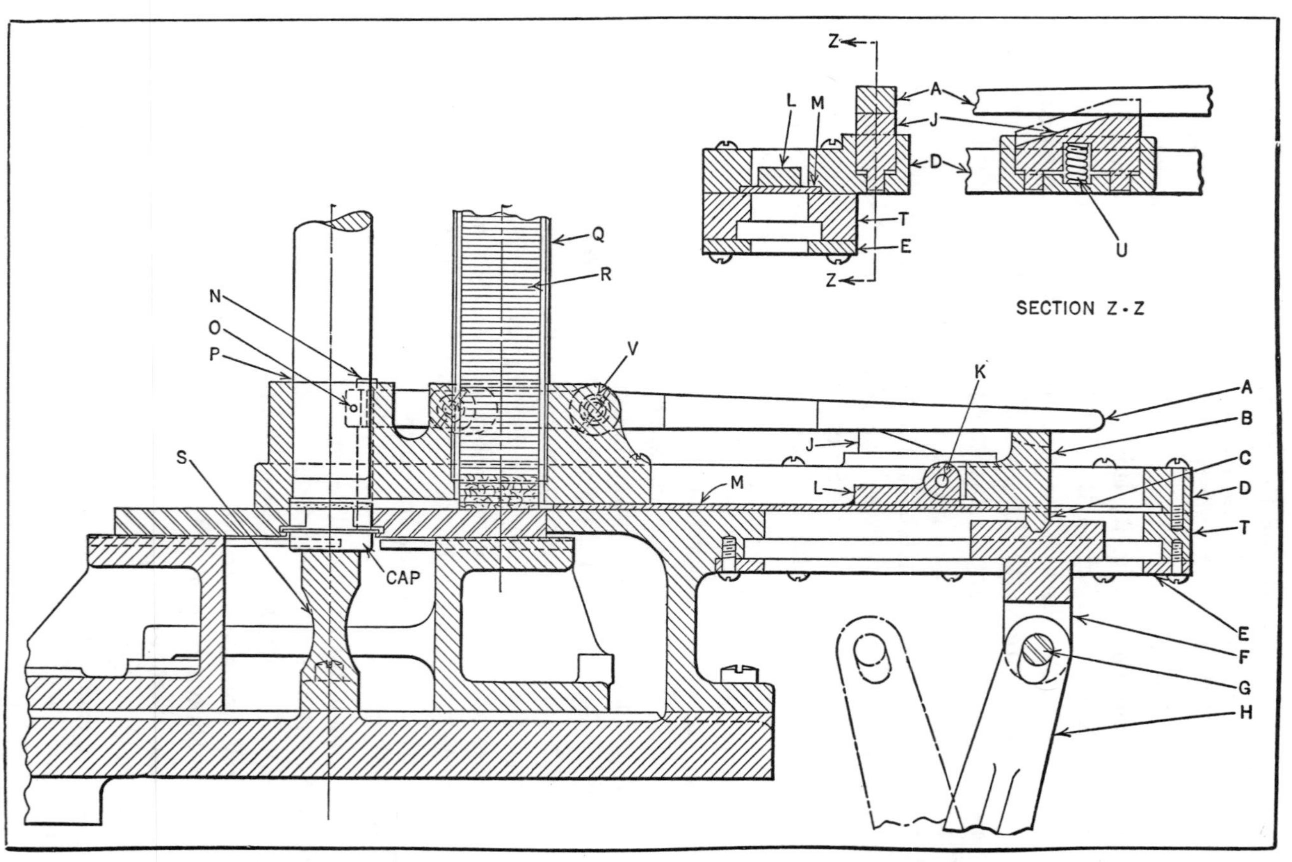

Fig. 3. Mechanism for Inserting Cork Disks in Metal Caps; an Automatic Stop Operates if a Cap is not in the Receiving Position

azine Q unless there is a cap ready to receive it. This is quite important, because it frequently happens that the flow of caps is interrupted. If the device continued to feed the cork disks, they would be wasted and in all probability, the mechanism would jam.

This cork disk feeding mechanism is synchronized with the cap-feeding mechanism, so that there is no chance for a misstep in production. The cork disks R are stacked in the vertical magazine or tube Q, which is kept in continuous agitation so that the disks assume a horizontal position, the upper ones falling down when those at the bottom are removed. A feeding finger M, slightly larger in width than the diameter of the cork disk, passes back and forth under the stack of disks, pushing them, one by one, under the plunger P. The plunger then forces the cork disk into the cavity provided for it in the cap. An anvil S is provided under the caps to take the pressure of the plunger.

The finger M slides in a groove provided for it between D and T. A hinged latch, consisting of parts K and L and the latch part B, is fastened to the feeding finger permanently. Below the feeding finger is the driving slide F, which functions in a groove between T and E. The driving slide reciprocates continuously under the action of the rocker arm H through the pin connection G. A groove cut in the top of the driving slide corresponds in shape to the projection C on latch B. When projection C rests in the groove on the top of the driving slide, the feeding finger M is reciprocated under the cork disk stack.

The mechanism is synchronized, so that a cork disk is in place ready to be forced down by the plunger just at the time when the cap begins to move from its place directly ahead of the cork disk feeding mechanism. A feeler bar N is placed at this point ahead of the cork disk feeding mechanism and, by its vertical motion, controls the feeding of the cork disks. The feeler bar is guided in a slot cut in the cap guide bar, and the opposite end rests in a slot in the

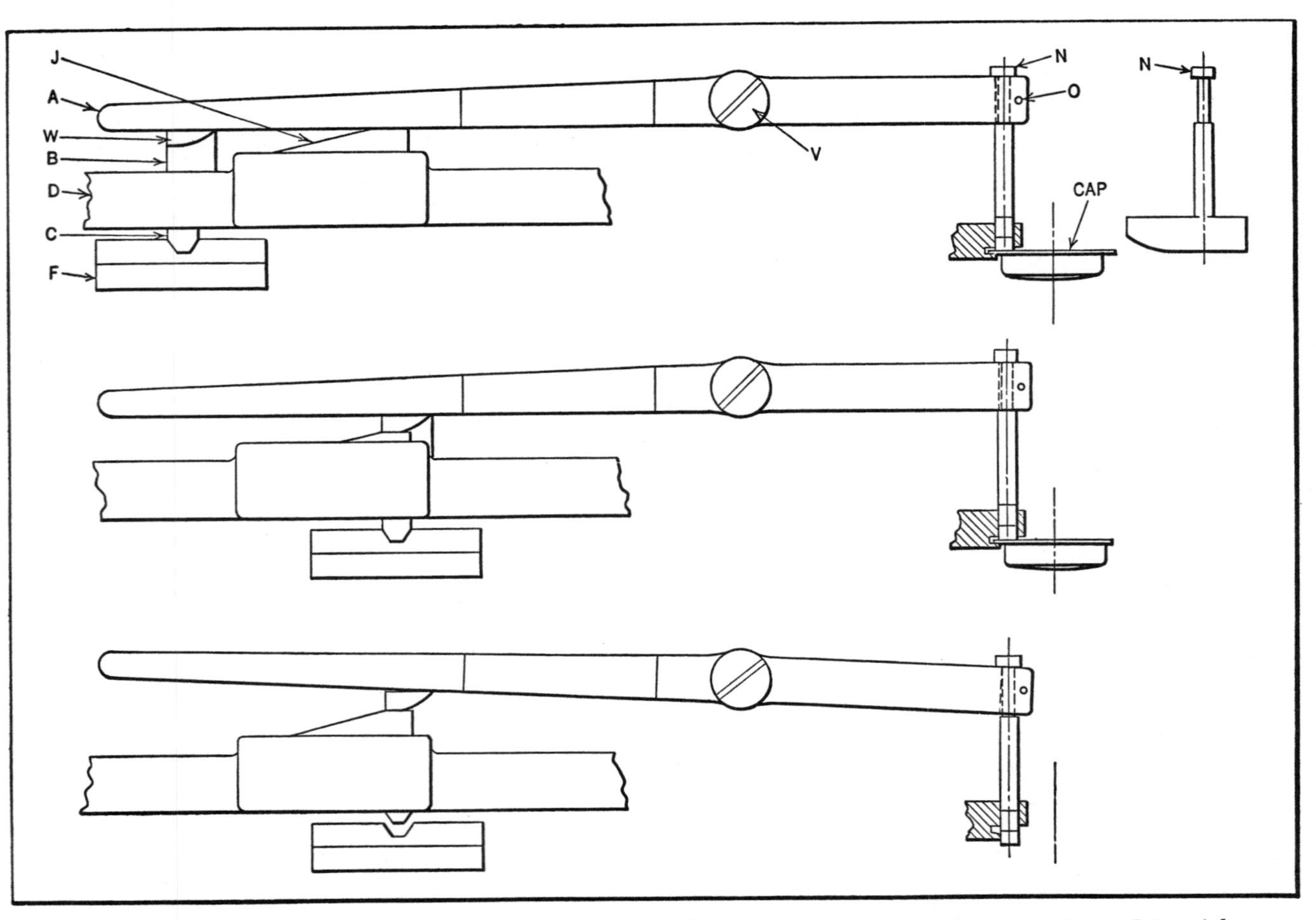

Fig. 4. Diagrams Illustrating Operation of Device that Stops Feed of Disks when Feed of Caps to Assembling Position is Interrupted

end of the lever *A*, where it is retained in place by a pin *O*. Lever *A*, in turn, is pivoted on a screw *V*, and is balanced about its pivot point so that the weight will be slightly greater at the end where the feeler bar is located. This insures the proper contact between the caps and the feeler bar.

The opposite end of lever *A* rests on a horizontal projection on latch *B* and on the tapered button *J*. The tapered button has a smooth vertical motion under the action of the light spring *U* in the well provided for it in *D*.

The operation of this device will be clearer by referring to the three views in Fig. 4, which show lever *A* in three different positions. The upper view shows the lever in the position assumed when a cap is under the feeler bar. The projection *C* on the latch is located in the slot on the driving slide *F*. At the same time, the tapered button *J* is in contact with the lever. The central view shows the position of the parts when the feeding finger has pushed a cork disk into place under plunger *P*. Lever *A* is in the same position as in the upper view, which indicates that another cap is under the feeler bar. The projection *W* on latch *B* has depressed the tapered button, so that the projection *C* engages the driving slide *F*. When there is no cap under the feeler bar, the parts assume the positions shown in the lower view. In this case, lever *A* has been raised from the tapered button so that as the projection *W* rides up the tapered surface it lifts projection *C* out of the slot in the driving slide and thereby stops the movement of the feeding finger *M*, Fig. 3.

Device that Prevents Engagement of Clutch Until Slide is in Operating Position.—The rotating spinning tool of a machine for spinning an inaccessible joint in kitchenware had to be of the expanding and contracting type to allow access to the work. The machine clutch was required to be disengaged and the rotary movement of the tool positively stopped while the tool entered the work, as

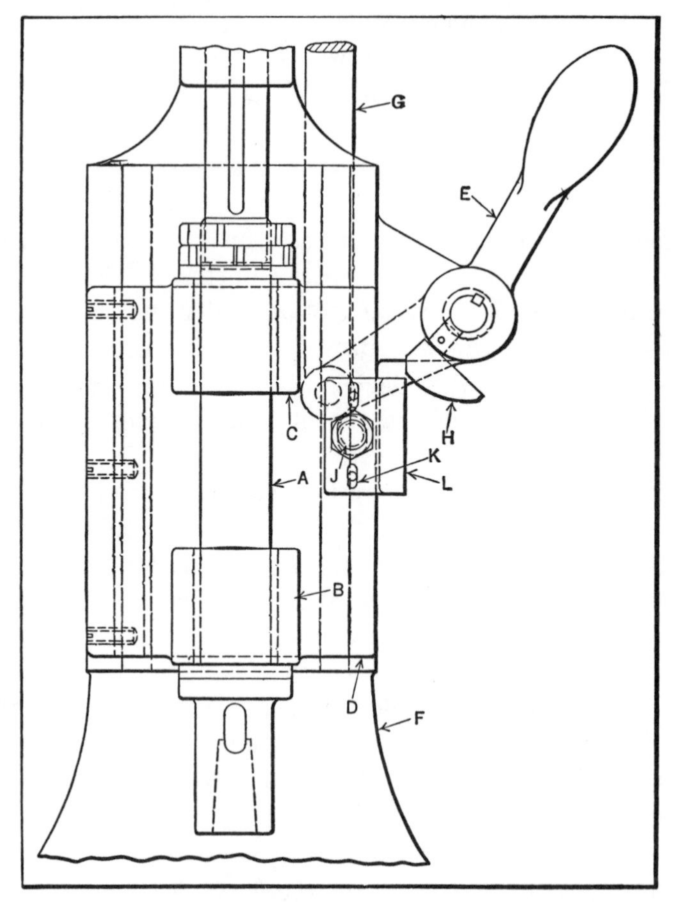

Fig. 5. Safety Device that Prevents Expanding Spinning Tool from Being Operated while it is Entering or Leaving Work

otherwise, the centrifugal force would cause the tool to expand and damage the work. To prevent the operator from accidentally leaving the clutch engaged at this time, the simple locking arrangement shown in Fig. 5 was devised.

The tool-spindle *A* rotates in bearings *B* and *C*, which are cast integral with a vertical tool-slide *D*. This slide is fed downward by a foot-pedal, which actuates a rack and pinion. The foot-pedal and rack and pinion are not shown in the illustration. The machine clutch is operated by lever *E*, pivoted to the bracket cast on the machine frame *F*. This lever is connected to the clutch mechanism by a link *G*. The interlocking arrangement consists merely of the dog *H* on the hub of lever *E* and the stop *L*, which is secured to the vertical slide by a nut on the stud *J*. This stop is of angular shape and has a slight vertical adjustment to accommodate similar work of different sizes. The adjustment is provided by the elongated holes for the aligning pins *K* and stud *J*.

The slide is shown in its working or lowest position, and the stop is down far enough to allow dog *H* to pass when the lever is swung in a clockwise direction to engage the machine clutch. In swinging the lever for this purpose, however, the dog is moved toward the left, thus blocking the return of the stop, with the slide, while the clutch is engaged. After the work has been spun, the operator must shift lever *E* back again to disengage the clutch before returning the slide to its upper position. Thus, when the slide is at any other point than the lowest point indicated, the dog will come in contact with the stop and prevent the lever from being swung clockwise to engage the clutch.

Roller Clutch with Tripping Device.—In designing special machinery, it is often necessary to provide a tripping clutch similar to that used on a power press. Such clutches may be required to have the added safety feature of not repeating should the operator fail to take his foot off the starting treadle. Special wire or pipe bending machines, special cutting-off machines, and single-action machines for such operations as gluing, notching, stamping, and scoring, are typical machines on which clutches of this kind are used. To meet the requirements of such machines, the clutch shown in Fig. 6 was designed. Although not new in principle, the design has been developed to a point where the device is light in weight, compact, and effective in action. Different applications may, of course, necessitate

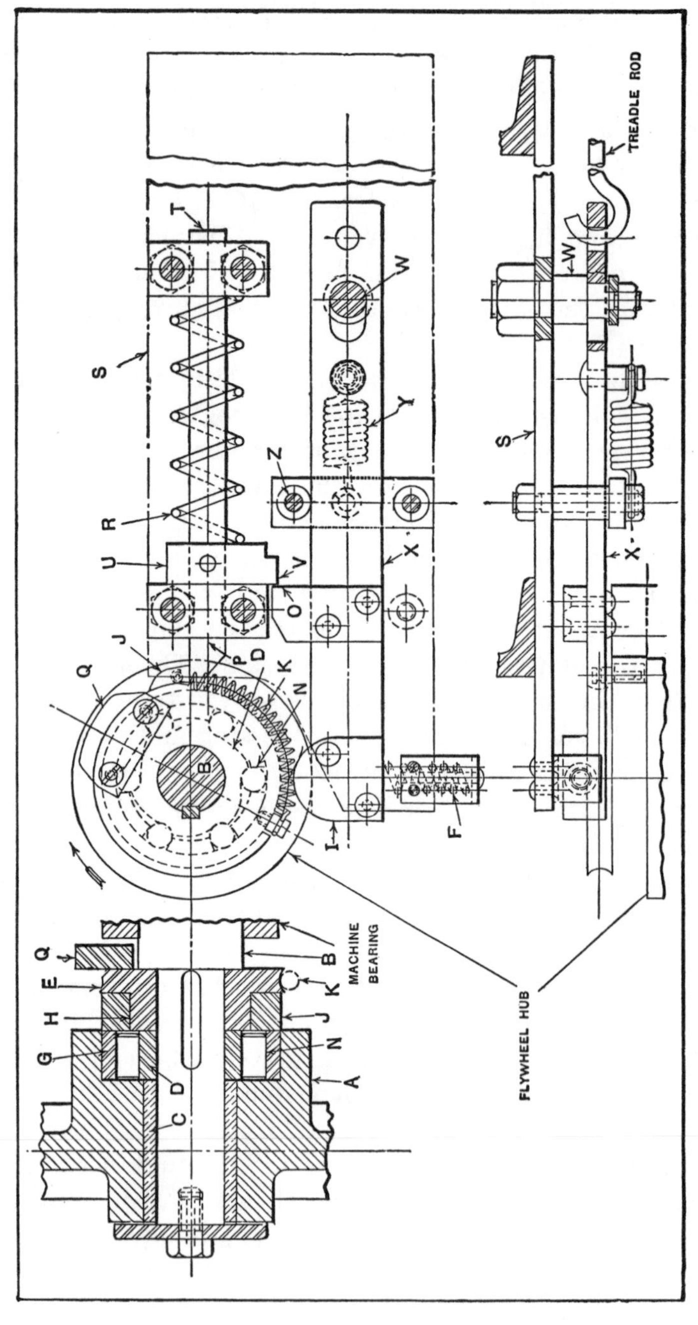

Fig. 6. Roller Clutch with Tripping Device Designed to Prevent Repeating

changes in the mounting, treadle action, and driving means.

When the machine is in operation, the flywheel *A* revolves continuously. Shaft *B* remains stationary until the operator depresses the foot-treadle. When the treadle is pressed down, shaft *B* makes one complete revolution and stops, regardless of whether the operator removes his foot from the treadle or keeps the treadle depressed. In order to cause the shaft *B* to make another complete revolution, the treadle must be allowed to return to its normal position and be depressed again.

Briefly, the action of the tripping device is as follows: When the foot-treadle is depressed, the treadle-rod pulls plate *X* to the right against the tension of spring *Y*. This movement brings latch *O* into contact with collar *U* at *V*, causing rod *T* to move to the right against the tension of spring *R*. The downward movement of the foot-treadle is continued until the latch *P* at the end of rod *T* is disengaged from the nose *J* of the friction roller cage, also shown in Fig. 7. The friction roller cage, being thus released, is revolved clockwise on the hub of collar *E* through the action of spring *K* fastened to pins *L* and *M*, Fig. 7. Referring to Fig. 6, it will be noted that collar *E* is keyed to shaft *B*.

Now as the roller cage revolves, it carries the rollers *N* with it, forcing the rolls to climb up the cam surfaces of the cam member *D*, which is keyed to shaft *B*. The rollers finally reach a point where they act as wedges between cam *D* and the hardened steel ring *G*, which is pressed into the hub of the flywheel *A*. Flywheel *A* then drives shaft *B* forward in the direction indicated by the arrow, the roller cage and collar *E* revolving with the shaft.

The shaft *B* is revolved but a fractional part of a revolution before the cam *Q*, secured to collar *E*, comes in contact with the cam *I* at the end of plate *X*, causing plate *X* to pivot about pin *W* against the tension of spring *F*. This action serves to disengage collar *U* from contact with latch *O* at point *V*, allowing spring *R* to force rod *T* back into the position shown, with the latch *P* ready to engage nose *J* of the roller cage when it has made a complete revolution.

As the rolls are prevented from moving forward by the roller cage when nose *J* is stopped by latch *P*, their wedging action between the ring *G* and the cam *D* is released and

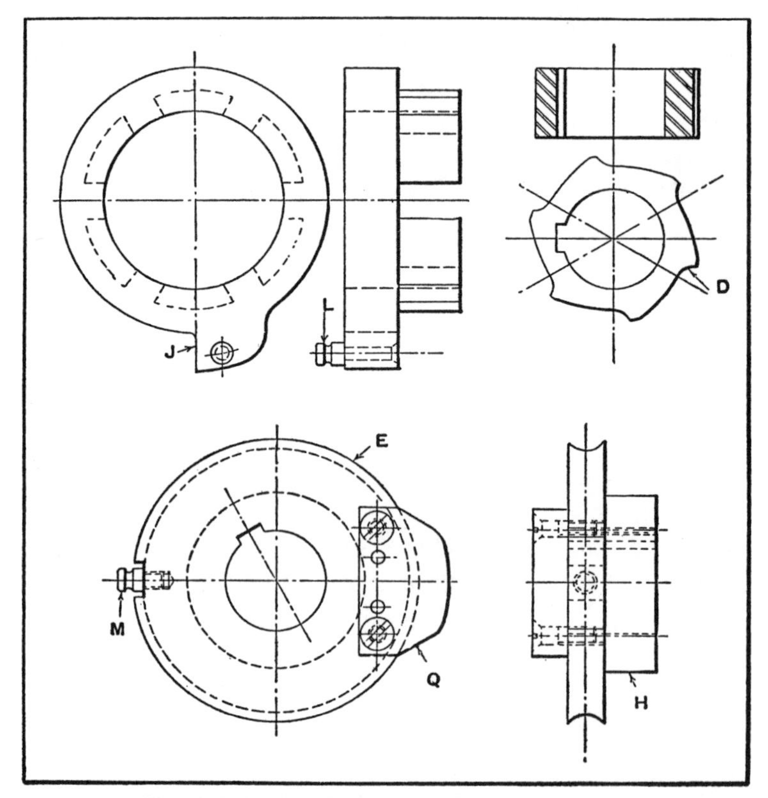

Fig. 7. Roller Cage J, Collar E, and Cam D of Clutch Shown in Fig. 6

shaft *B* stops, while the flywheel continues to revolve. All parts of the device are now in the positions shown in Fig. 6, ready for the tripping operation to be repeated. While the treadle-rod and the latches are shown in a horizontal position in the illustration, they are usually located in a vertical position, plate *S* being mounted on the machine frame, portions of which are shown in cross-section in the lower view. However, the construction is such that the latch can be located at any desired angle relative to the roller cage. To insure efficient operation of the safety device, care should be taken to see that the rod T slides freely in its bearings, and that spring R has a very snappy action. All wearing parts should, of course, be hardened.

Tripping Device for Bead-Chain Cutting-Off Machine.— Bead chain made of brass is used in large quantities for electric-light pull-sockets. This chain is wound on spools

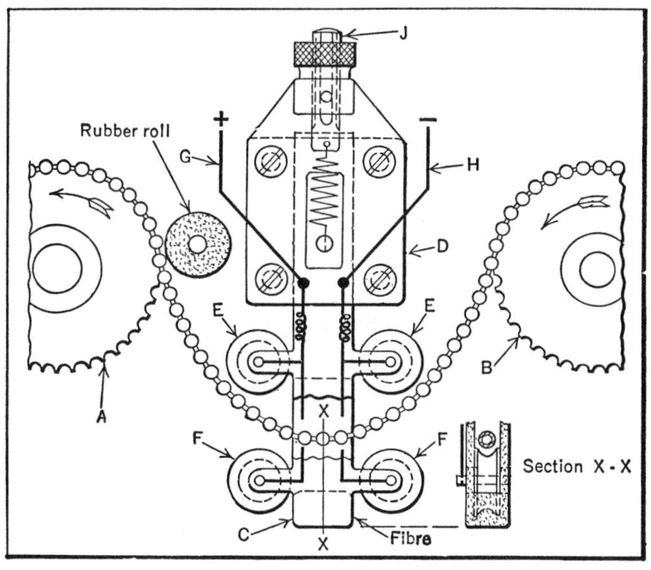

Fig. 8. Electrical Tripping Device that Stops Machine when Chain Breaks or when the Slack Varies

in bead-chain forming machines and then delivered to other machines where it is cut off to the required length for assembly in the sockets. In the cutting-off machines, the chain is passed over two sprockets A and B, Fig. 8. Creeping of the chain on the sprockets is one of the major troubles experienced with these machines; and if the machine continues running after creeping occurs, mutilation of the chain in another part of the machine results.

It was found that this difficulty could be overcome by

maintaining a certain amount of slack between the sprockets; and to obtain this condition, the tripping arrangement shown was incorporated in the machine. The tripping mechanism is so arranged that if the slack becomes appreciably greater or less than that indicated, the chain closes the electric circuit of a solenoid. This causes the core of the solenoid to release a clutch which stops the machine. The operator then gives the chain the required amount of slack. One of the advantages of this type of tripping device is that the chain is not required to lift or support any weighted latch member in order to close the circuit; the chain itself closes the circuit. In addition to this, if the chain breaks, the circuit is also closed, causing the solenoid to stop the machine.

The tripping arrangement consists chiefly of the fiber slide *C*, which is guided in the stationary block *D*. On the slide are mounted two sets of rolls *E* and *F*. At the end of the small pins on which the rolls turn, copper wires are soldered. These wires are connected to the two main wires *G* and *H* leading to the solenoid (not shown). It will be noted that slide *C* has a floating action, its weight being supported by the spring attached to screw *J*. This prevents excessive pressure of the upper rolls on the chain. Screw *J* can be adjusted so that the chain is normally half way between the upper and lower sets of rolls. The slide is made from fiber in order to insulate it from the machine, and the machine is separated from its foundation by layers of insulation to prevent grounding of the current.

The action of the device is as follows: If the chain creeps forward on sprocket *B*, the slack will increase until the chain rests on rolls *F*. This closes the circuit formed by wires *G* and *H* and operates the solenoid, which, in turn, releases the clutch and stops the machine. If the creeping of the chain is such that the slack is reduced, a similar action of the solenoid occurs, the chain being drawn against rolls *E*, in this case, and thus closing the circuit.

Quick-Tripping Mechanism for Clamping Device.—In designing a special clamping device, it was necessary to incorporate a quick-tripping mechanism which would provide for a rather slow releasing and an almost instantaneous clamping action. The mechanism designed for this purpose consists of a common plate cam *A* (Fig. 9), and an

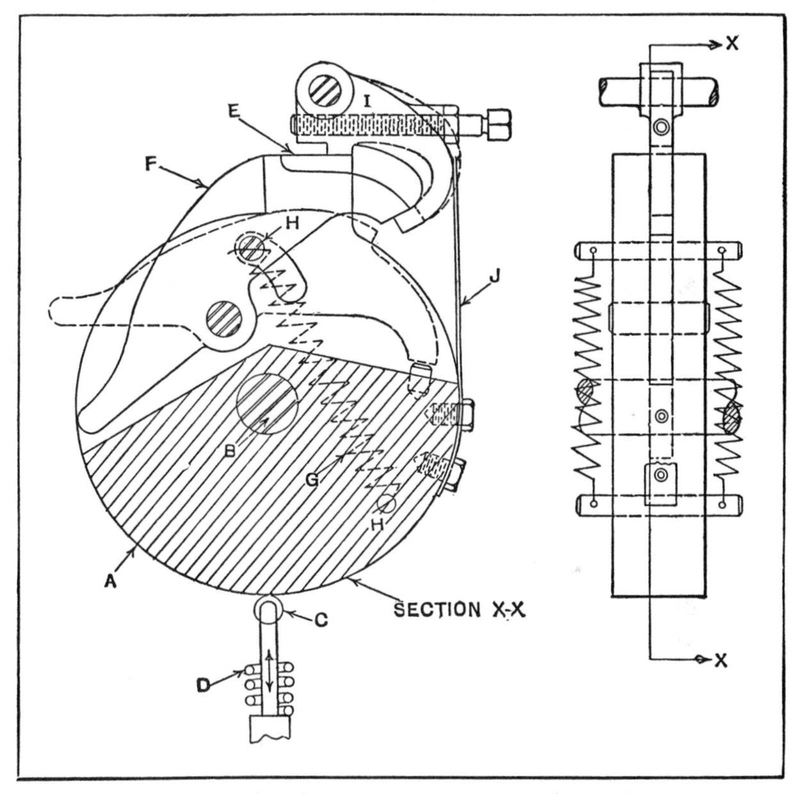

Fig. 9. Quick-tripping Mechanism for Clamping Device

involute cam *F* pivoted in a slot cut through the middle of the plate cam. These two cams actuate the cam-roll C which performs the required tripping and releasing operations.

The shaft *B*, to which the plate cam *A* is keyed, has a reversing motion. It revolves approximately 180 degrees

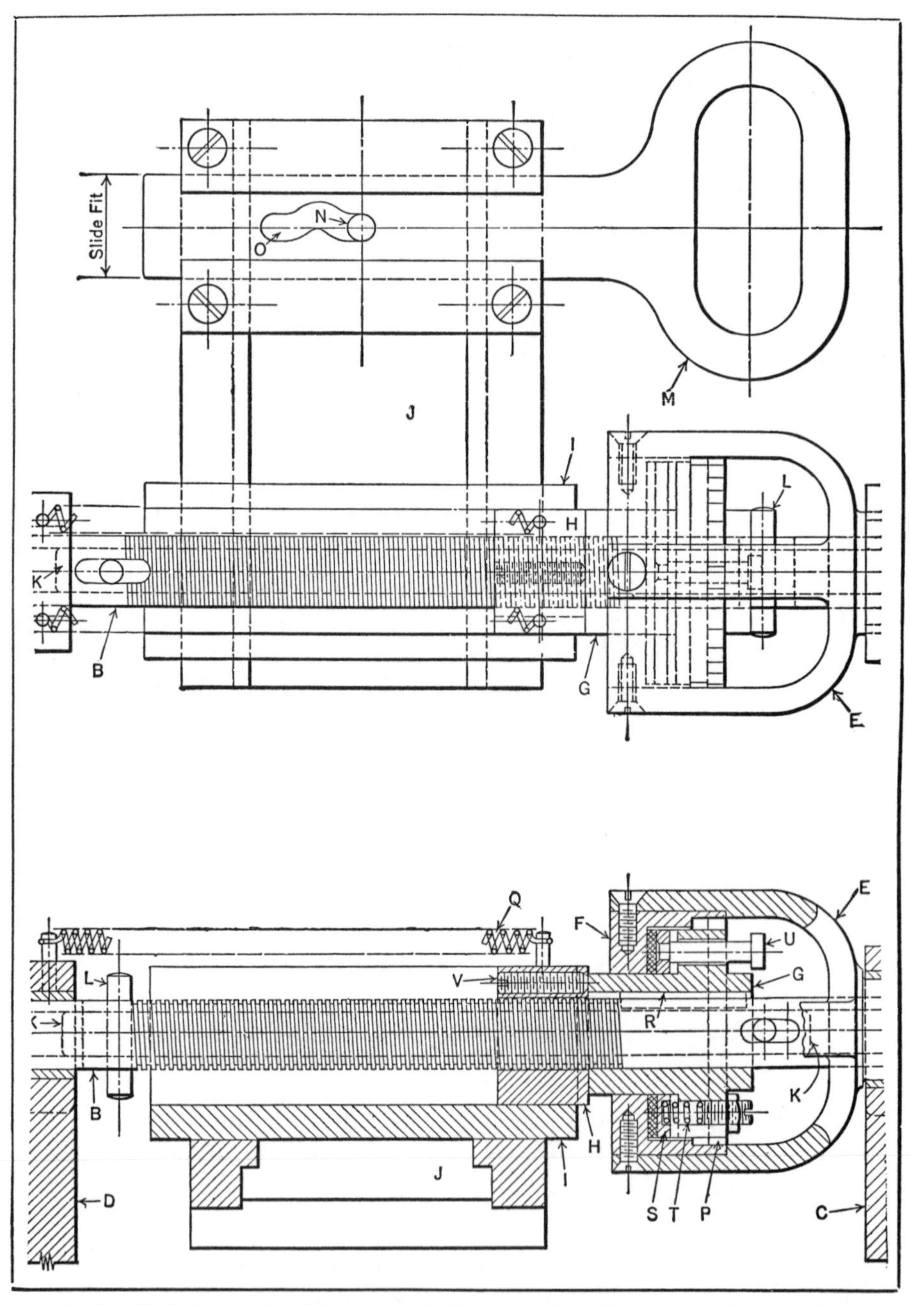

Fig. 10. Mechanism with a Lead-screw and Nut Arranged to Disconnect the Driving Clutch after a Given Number of Revolutions

in a counter-clockwise direction, reverses, and then returns to the starting position, this oscillating movement being continuous. As the plate cam *A* revolves in a counter-clockwise direction, the roll *C* is gradually depressed by the cam *F*, causing the clamping device to be released slowly. When the cam *F* reaches the point where the roll *C* is in contact with the lobe *E* of the cam *A*, the clamp is fully released, and continued rotation of the cam brings the end of the adjusting screw of latch *I* into contact with roll *C*, causing the latch to release cam *F* and allowing the springs *G* to draw the cam into the position shown by the dotted lines.

At this point the direction of rotation of shaft *B* is reversed by a mechanism on the machine, which is not shown in the illustration. As the cam assembly turns in a clockwise direction, the roll *C* reaches the end of the dwell surface of lobe *E* and is returned into contact with the concentric portion of cam *A* with a very sudden action through the tension exerted by spring *D*. This sudden movement of roll *C* engages the clamping device almost instantly. As the cam continues to rotate, the roll *C* comes into contact with the protruding end of cam *F*. As the tension of spring *D* is sufficient to overcome the tension of springs *G*, the cam *F* is returned to the position shown by the full lines, where it is held by latch *I*, which is under the tension of the flat spring *J*. The mechanism is thus set in position for a repetition of the cycle just described, upon the reversal of the shaft *B*.

Mechanism for Stopping a Machine After a Given Number of Revolutions.—The mechanism shown in Fig. 10 is designed for use on either a hand- or a power-driven machine. The object of the device is to control the number of pieces fed into an assembly from a magazine, by automatically stopping the machine at the end of the count. The device is applicable to any kind of machine in which a shaft or the complete machine is required to be stopped after a given number of revolutions. The machine or shaft remains idle until the work is removed and the handle *M* is moved to the starting position.

A pulley (not shown) drives the main tubular shaft *B* through clutch members *E, F,* and *G.* Shaft *B,* in turn, drives the machine. The pulley is keyed to the hub of clutch spider *E.* The circular barrel *F,* having thirty internal teeth, is positively secured within the four arms of spider *E* and revolves with the pulley. The connecting and disconnecting circular clutch member *G* between *B* and *F* has thirty external teeth designed to mesh with the teeth in *F* at *P.* Member *G* is a sliding fit on shaft *B,* but is prevented from rotating on it by key *R.*

The concentric spring-pad ring *S,* mounted on three sliding pins *U,* is backed up by three compression springs *T.* On the face of this pad is riveted a piece of brake lining or fabric which provides a frictional engagement between *F* and *G.* This friction clutch is adjusted to allow two revolutions before the speeds of the two revolving members become synchronized and the teeth become positively engaged at *P.*

Within the shaft *B* is a sliding rod *K* having crosswise holes near each end, through which two pins *L* are driven. These pins project through slots in the sides of the shaft. Shaft *B* has either a buttress or a square thread with a lead and a horizontal length that is sufficient to provide for the largest number of revolutions required. The half-nut *H* is a sliding fit in the channel *I,* mounted on the slide *J.* The threaded hole in nut *H* is elongated on one side an amount equal to a little more than double the depth of the thread. The thread is also cut away until only about one-half of the threaded circumference is left for engagement.

The pin *N* in the cross-slide actuates the transfer channel *I,* either engaging the half-nut *H* with the lead-screw or disengaging it, by its action on the cam slot *O* in the sliding portion of handle *M.* Two coil tension springs *Q* are attached to the half-nut, their opposite ends being positively fixed to the left-hand bearing on the machine. The half-nut is shown just making contact with the front of clutch member *G*. As shaft *B* continues to revolve, the nut advances until the clutch teeth at *P* are disengaged. The member *G* then comes in contact with pin *L* as both *G* and *K* are moved forward, disconnecting the clutch pad *S* from frictional contact with *F*. The machine is thus stopped, allowing the clutch barrel to run idle.

When it is desired to start the machine, the handle *M* is pulled forward, causing the nut to become momentarily disengaged from the lead-screw, so that it returns instantly to its left-hand position, and is in mesh again with the lead-screw thread. On its return movement, nut *H* strikes the pin *L* at the left and causes the friction pad *S* to engage *F*. This, in turn, causes *G* to rotate in synchronism with *E*, so that the teeth at *P* are engaged by means of rod *K* and pins *L*. This starts the machine, and the nut begins to travel on the lead-screw toward the clutch, where it repeats the stopping operation. Cam slot *O* is designed to lock the nut channel in its proper position while the nut is traveling on the lead-screw.

When the nut is disengaged from the lead-screw by handle *M*, it is returned by springs *Q* and caused to strike pin *L* at the left, thus being held momentarily in contact with the pin. The nut cannot start back until it has caused the clutch to engage and thus commence the count. The springs *T* under the pad *S* act as buffers against the sudden return impact of the nut on pin *L*, and allow the clutch to engage more smoothly. Screws at the back of the compression springs *T* provide means for adjusting the spring pad to give the proper synchronizing friction and buffer action. Screw *V* in the half-nut is adjusted to give the exact number of revolutions required. Any desired number of revolutions within the range of the lead-screw can be obtained by using a split washer of the right thickness on the contacting face of nut H or by screwing pins of the required length into the face of the nut.

For a large number of revolutions which would require a lead-screw of excessive length, if arranged as illustrated, the nut can be operated on an independent screw in a channel at one side of the clutch, using speed reducing gears between it and shaft B. In this case, the nut is made wide enough to surround shaft B and long enough to lead properly in its guiding channel. However, with a lead-screw 12 inches long, having 18 threads per inch cut on shaft B, over 200 shaft revolutions can be obtained before the clutch is disengaged. This is sufficient for most counting and machine stopping operations.

Stopping Spring Fatigue Testing Machine at Time of Breakage.— An old punch press is used as a fatigue testing machine for shock absorber springs. The equipment operates twenty-four hours a day and a small counter indicates the number of times the spring is compressed. In the past, when a spring broke during the night, the machine continued to operate and a wrong number of compressions was recorded.

The problem was solved by installing a photo-electric relay in such a position that the light beam passes beneath the bottom of the plunger when the plunger is in its lowest position. With this arrangement, the spring intercepts the light beam and prevents it from falling upon the photo-tube under ordinary circumstances. When the spring breaks, however, it collapses and the beam passes over it to the photo-tube, which actuates a relay and stops the operation of the machine.

Power Press Stop Mechanism which Disengages Clutch when Magazine Feed Jams.—When parts become jammed as they are fed from a magazine to the dies of a press, it is likely to prove disastrous, not only to the die members, but to the press members as well. The most practical method of preventing damage in such a case is to stop the press instantaneously at the top of its stroke by some automatic means. This is done by the mechanism shown in Figs. 11 and 12, which makes it possible for one operator to tend three presses running at about 75 revolutions per minute.

This arrangement, as applied to a battery of power presses equipped with automatic feed mechanisms, has proved highly satisfactory. In addition to eliminating damage to press and die members, this mechanism also provides a valuable safety feature in that the movement of two levers is required to start the press; hence, both of the operator's hands must be on these levers at this time and out of the danger zone.

Only the magazine feed-slide and tripping mechanism are shown. They are mounted on the baseplate *A*, which is secured to the press at a point adjacent to the magazine. The feed-slide, which is indicated at *B*, reciprocates in guides on the block *C*, this block being fastened to the baseplate. The reciprocating movement of the feed-slide is obtained through link *D*, rocker arm *E*, lever *F*, and cross-head *G*. Rocker arm *E* pivots about the stationary sleeve *J*, and lever *F* is pivoted to arm *E* at *K*. The lower end of lever *F* enters a slot in the side of sleeve *J* and is held against the side of plunger *L* by the spring *M* (Fig. 12). With this arrangement, arm *E* and lever *F* oscillate normally as an integral unit. Thus, link *D*, which is attached at its upper end to the press ram, oscillates arm *E* and lever *F*, imparting an oscillating movement to the cross-head *G*, which operates in the vertical guides *H* on the feed-slide.

The oscillating movement of the cross-head causes the feed-slide to reciprocate and push the work from the magazine into the die. However, should jamming of the feed-slide occur, the lower end of lever *F* would immediately swing away from the side of plunger *L* and cause the tripping levers *N*, *Z*, and *P* to operate, as will be explained

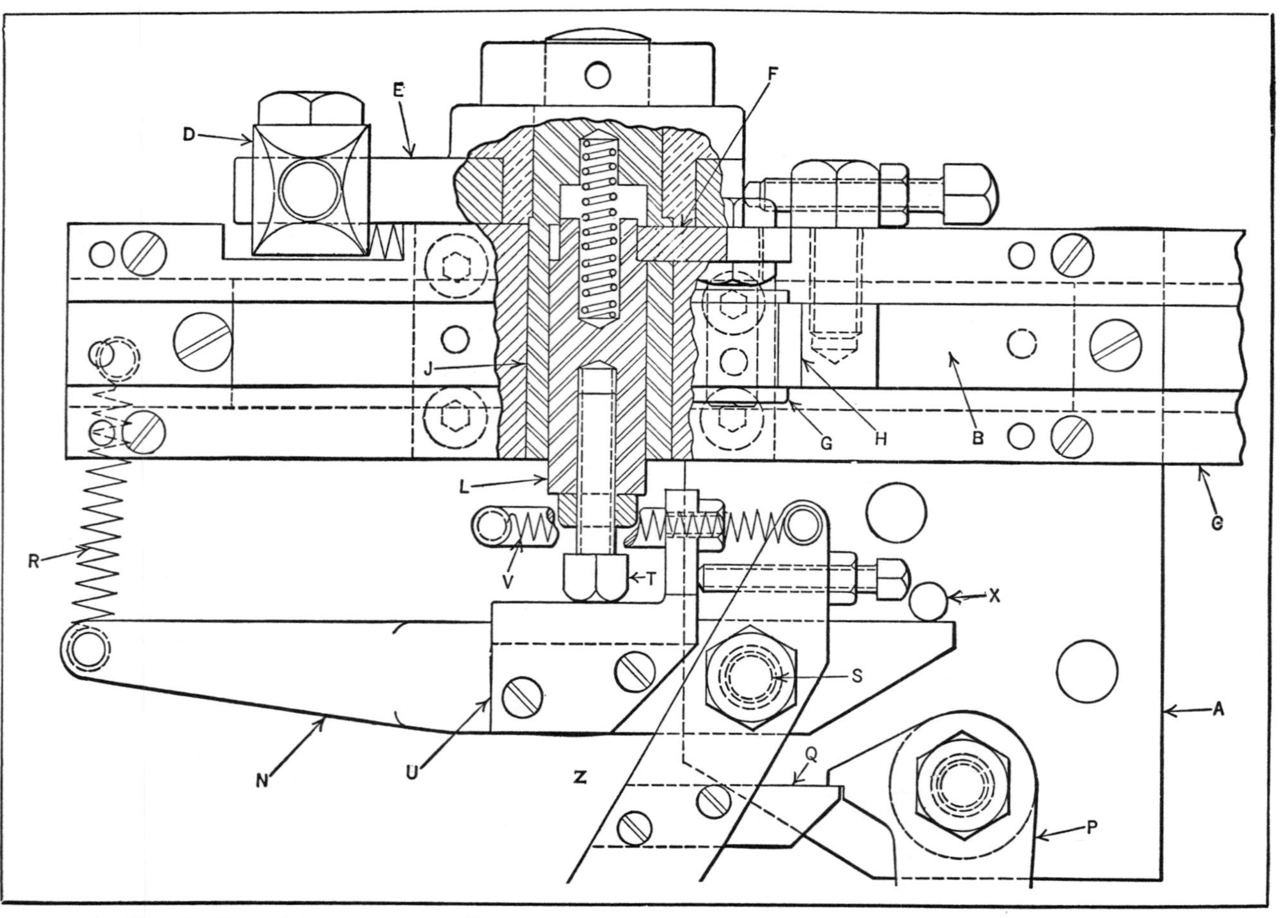

Fig. 11. Plan View of Power Press Stop Mechanism (shown in Fig. 12) which Disengages Clutch when Magazine Feed Jams

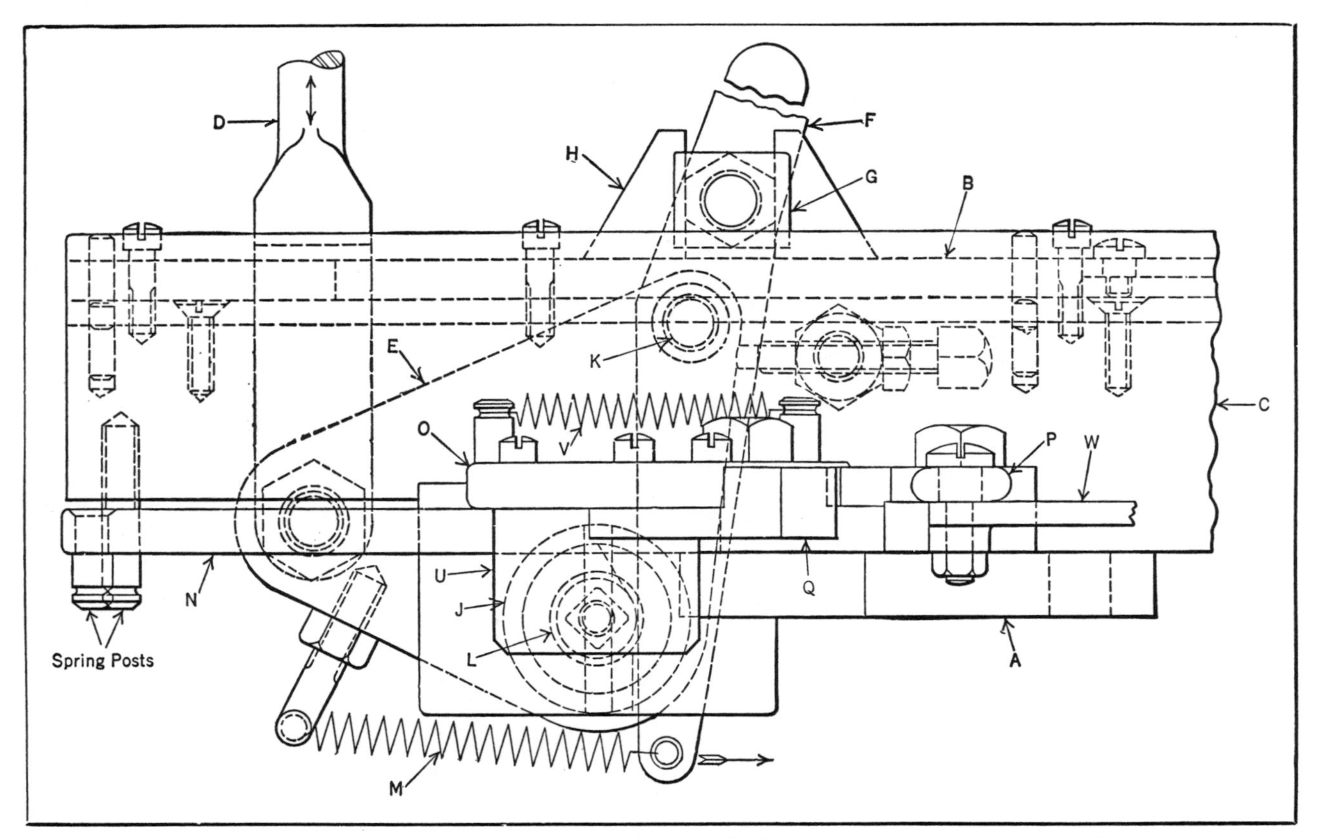

Fig. 12. Tripping Mechanism which Automatically Disengages the Press Clutch when the Magazine Feed Jams

later. Levers *N* and *Z* are pivoted at *S*. Bracket *U*, secured to lever *N*, is held against the adjusting screw *T* in plunger *L* (Fig. 11) by means of the coil spring *R*, thus holding plunger *L* in engagement with lever *F*. Latch *Q*, on lever *Z*, is normally held in engagement with lever *P* by the spring *V*. Link *W* (Fig. 12) is connected to the clutch mechanism, and when moved toward the right disengages the clutch and stops the press ram at the top of its stroke.

The Tripping Action when a Jam Occurs.— Assuming that the feed-slide *B* has jammed, the movement of the slide will be shortened or discontinued altogether, so that the lower end of lever *F* will leave plunger *L*, allowing the plunger to be forced further into the sleeve *J* by the lever *N*. As levers *N* and *Z* swing together (by resetting and when jamming occurs only), the latch *Q* leaves the notched end of lever *P*, allowing this lever and link *W* to move toward the right under the action of a spring (not shown) and stop the press at the top of the stroke. The cause of the jamming can then be removed.

To restart the press, the operator grasps levers *Z* and *P*. Lever *Z* is swung toward the right, carrying lever *N* back into contact with the stop-pin *X*, and thus allowing the shoulder at the end of plunger *L* to once more engage the lower end of lever *F*. After this, lever *Z* is swung slightly to the left to clear the notched end of lever *P*. Lever *P* is then swung to the left until its notched end engages the latch *Q*, thus throwing in the press clutch.

As the work is fed to the dies on the upward stroke of the ram, jamming of the feed-slide usually occurs at this time; hence, the press is stopped at the completion of this stroke, thus preventing damage to the press tools. This arrangement also insures the safety of the operator, as both hands are occupied with levers *Z* and *P* when starting the press.

Mechanical Device Stops Press if Punch Breaks.— Electromagnetic devices for stopping a punch press by releasing the clutch when the machine or dies fail to function properly are sometimes used (see Ingenious Mechanisms, Vol. I, page 148). Serious damage to the dies or press is often prevented by such devices. Electromagnetically controlled devices have numerous advantages over the all-mechanical type, but there are times when such equipment is not desirable, especially when it is possible to install a non-magnetic device that is cheaper and practically as simple as the circuit-closing mechanism required for operating the electromagnet. Such a device is shown in Fig. 13. It is designed to stop the press and prevent the work from being

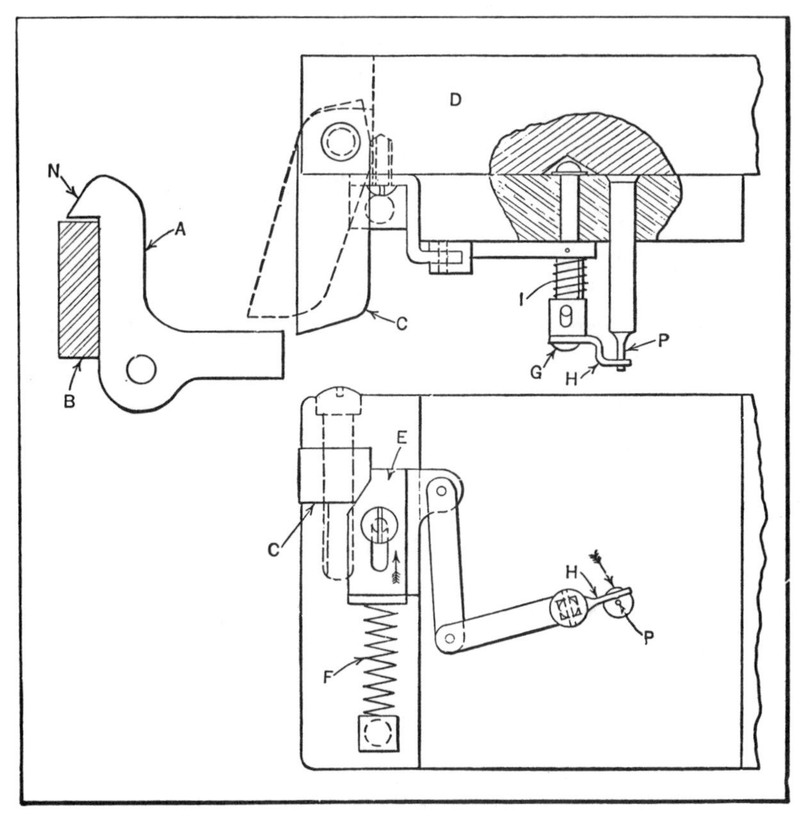

Fig. 13. Mechanical Device for Stopping Press if Piercing Punch is Broken

spoiled in the event that the small piercing punch *P* becomes broken.

Referring to the illustration, latch *A* is attached to the frame of the press in such a position that the hook *N* will snap over the hand trip-lever *B* or an auxiliary member that operates parallel with the trip-lever when the latter is depressed sufficiently to engage the press clutch. If punch *P* is broken, the dog *C*, which swings freely in an extension of the punch-holder *D*, is pushed outward into the position shown by the dotted lines. On the next down movement, dog *C* trips latch *A* and releases the clutch.

The mechanism by which dog *C* is pushed outward is clearly shown in the illustration. The pressure exerted by the spring *F* is just sufficient to overcome the friction of the moving parts and the weight of dog *C*. Thus the pressure of the finger *H* on the small punch is not great enough to deflect the punch. The direction in which the spring pressure acts is indicated by the arrows.

The cam *E* is used to operate dog *C* in preference to a direct connection with the levers, because any jamming effect on the dog will not be transmitted through the levers to the small punch. Finger *H* is backed up by a spring *I*, and moves up and down the punch, the knob *G* striking a depression in the stripper plate. The movements are so timed that finger *H* does not touch the strip stock. It is evident that the basic principles here described can be employed in a great variety of lever arrangements that may be designed to suit different conditions.

Automatic Brake Mechanisms.—To safely hoist, hold, and lower a load, hoisting machinery is usually equipped with so-called safety, automatic, or retaining brakes. These brakes permit a load to be lifted freely by the motor, and lock the brake by the gravity action of the load as soon as the lifting torque of the motor ceases to act in the hoisting direction. The load is retained by the brake in any position, and only when the motor runs in the lowering direction is the acting power of the brake diminished, allowing the load to descend. The speed at which the load drops is regulated and determined by the lowering speed of the motor, while the brake, in the meantime, absorbs by friction the greater part of the potential energy of the dropping load, and generates heat in the brake.

Fig. 14 represents what is known as the *Weston* brake, which is the typical form of a very large class of automatic brakes used on hand and electric cranes to control the load. A pinion *A* mounted loosely on the shaft has formed on one hub a spiral surface normal to the shaft, and on the opposite end a faced surface to present to the friction disks *e*. A collar *D*, fast on the shaft, has a spiral surface which engages that of the pinion hub, and is backed up by a split washer or other device to resist end-pressure along the shaft. A flange *B*, loose on the shaft, has a faced surface similar to that on pinion *A*, and carries a ratchet to engage with a pawl *C*. A series of friction disks *e* is placed between the faced surfaces on *A* and *B* in such a manner that the disks in contact with *A* and *B* are keyed by sliding feathers to *B* and *A*, respectively, as shown at *X* and *Y*.

This gives each disk a motion opposite to that of its neighboring surfaces, and each two surfaces in contact having opposite directions of rotation form one friction surface of the brake. Thus the brake shown has five friction surfaces and four washers or disks. These disks are made of various materials; alternate disks of steel and brass, or steel and fiber are frequently used, and also polished saw steel for all the disks. The shaft revolves in the direction of the arrow on the right to hoist, and with the arrow on the left to lower; ratchet teeth are formed to permit the rotation of the flange *B* when hoisting, and prevent it when lowering; pawl *C* is counterweighted to throw it into engagement with the ratchet; flange *B* is backed against a shoulder on the shaft, so that all the end-thrust is taken by the shaft between this shoulder and the split collar *E*, and the brake is self-contained.

Action of the Weston Type of Automatic Brake.—The action of this brake is as follows: Suppose a load acts on the pinion *A* (Fig. 14) tending to revolve it in the direction of the left-hand arrow, and the shaft begins to turn in the direction of the right-hand arrow. *D* being fast on the shaft will revolve opposite to *A*, which will cause the spirals to slip or bind slightly and thrust *A* toward *B*, thus clamping the disks *e* between *A* and *B*, the end-thrust of *D* and *B* being taken by the shoulders on the shaft. In this manner

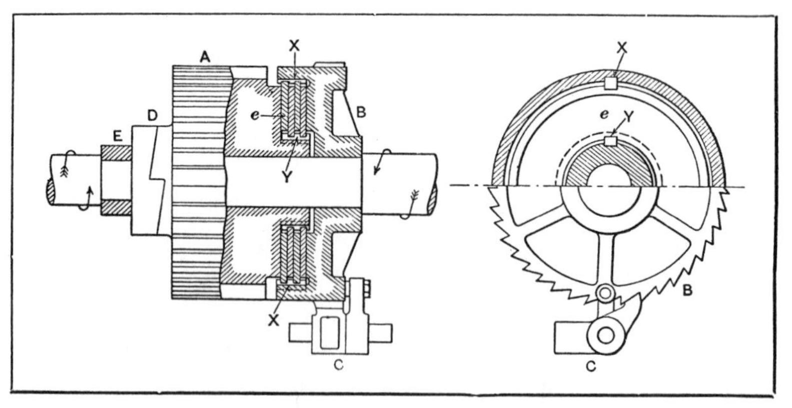

Fig. 14. Weston Type of Automatic Brake for Use with Hoisting Machinery

the whole mechanism consisting of *D*, *A*, *e*, and *B* is locked solidly together, and is made fast upon the shaft; thus the pinion *A* is driven and the load raised.

To lower, the shaft is turned in the direction of the left-hand arrow, carrying *D* with it, and since *A* (at the beginning) is clamped tightly to *B* through the disks *e*, and *B* is prevented from rotating by the pawl *C*, *D* is given motion relative to *A* in the direction of releasing the spirals, and hence the thrust upon *A*. As soon as this thrust is relieved, *A* turns freely in the direction of the left-hand arrow under the influence of the load, and, overhauling the shaft with its collar *D*, brings the spirals again into contact, reestablishing the locked condition and holding the load suspended. A further motion of the shaft results in a repetition of this cycle, and the act of lowering the load consists of an in-

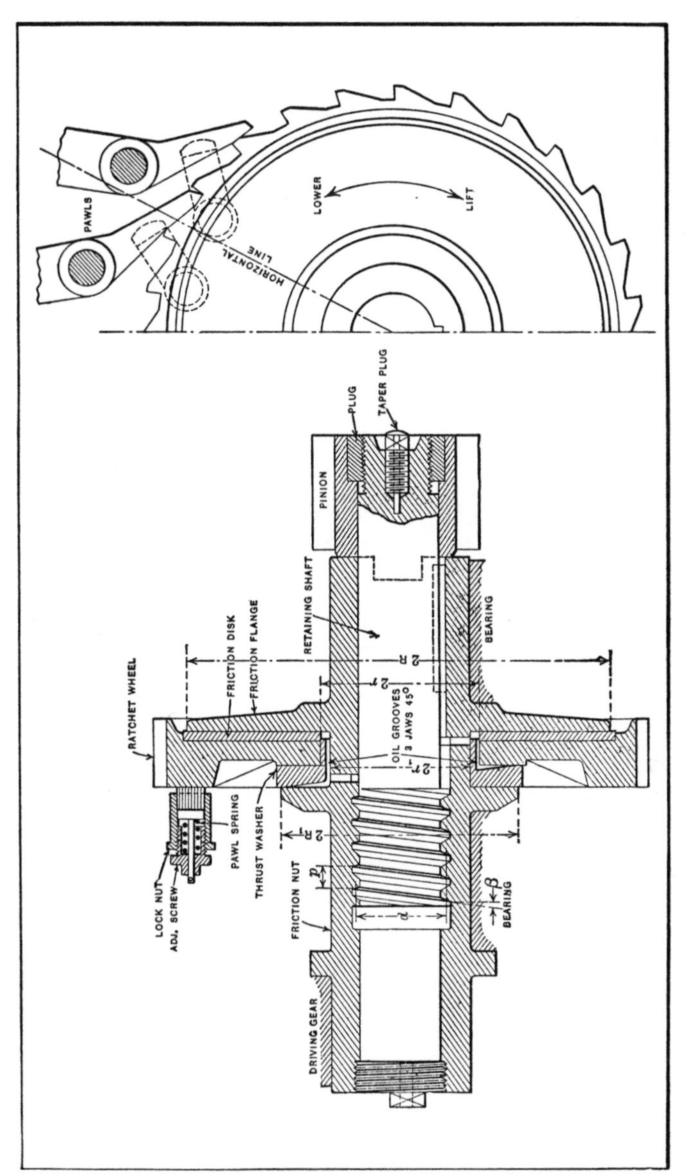

Fig. 15. Another Design of Weston Load Brake for Cranes and Other Hoisting Machinery

finite number of such repetitions in a unit of time, the motion of the load resulting from each cycle being infinitesimal, thus making the motion of the descending load uniform.

Special Type of Weston Brake.— Another type of Weston brake, embodying exactly the same principle as that shown in Fig. 14, is shown in Fig. 15. The ratchet is free to revolve when hoisting, but is held by two silent pawls from turning in the lowering direction. The friction nut is geared to the motor and the retaining shaft with gear pinion leads to the hoisting drum. The retaining shaft and friction nut are threaded either right- or left-hand, according to the hoisting direction. The friction flange is keyed to the retaining shaft and mates with the friction nut by means of three jaws which have about 15 degrees angular play. The friction flange drives the pinion direct through tongued and grooved projections between the pinion and flange. Any tendency of the load to revolve the retaining shaft when the motor is at rest causes the friction flange with friction disk to be pressed against the ratchet wheel and the thrust washer of the nut, due to the action of the threads. The friction of this washer against the ratchet wheel, which, as already explained, does not turn in the lowering direction, is sufficient to hold the load. Upon starting the motor to lower, it turns the friction nut and relieves a certain amount of pressure on the washers, until the pressure is overcome so far as to permit the load to revolve the friction flange in unison with the speed of friction nut, or motor. In hoisting, the jaws of the friction nut and flange engage, thus relieving the brake of all friction.

CHAPTER VI

OVERLOAD RELIEF MECHANISMS AND AUTOMATIC SAFEGUARDS

Certain types of machines or other forms of mechanical apparatus are likely to be subjected to excessive overloads resulting possibly in breakage of one or more parts unless provision is made to prevent, automatically, any dangerous overloading. These overloads are due to some abnormal operating condition and the function of the relief or release mechanism is to automatically disconnect the machine or driven member from the source of power, thus safeguarding it against excessive strains and serious damage. These overload relief mechanisms may be classed as a form of tripping or stop mechanism designed especially to safeguard a machine or its parts against excessive strains and breakage.

Automatic Overload Release for Worm-Gear Drive.—A machine for cutting coal in mining is subjected to such strains, jerks, and shocks that some overload protection is essential. A cast-iron safety washer which has been applied on mining machinery for many years has certain disadvantages which have been overcome by the improved overload release to be described. The safety washer is used in conjunction with a worm-gear drive, as shown by the left-hand sketch, Fig. 1. This washer is placed over the worm and is held by a nut and a short section of pipe. The idea is to make this washer strong enough to hold the worm in place under normal loads. If the load is excessive, however, the thrust of the worm will break the washer, thus releasing the worm from its driving key.

An overload release designed to eliminate certain disadvantages of the safety washer is shown by the sketch at the right of the illustration. The worm is held in the running position by the coil spring *A*, the tension of which may be adjusted by nut *B* on the worm-shaft. A jaw

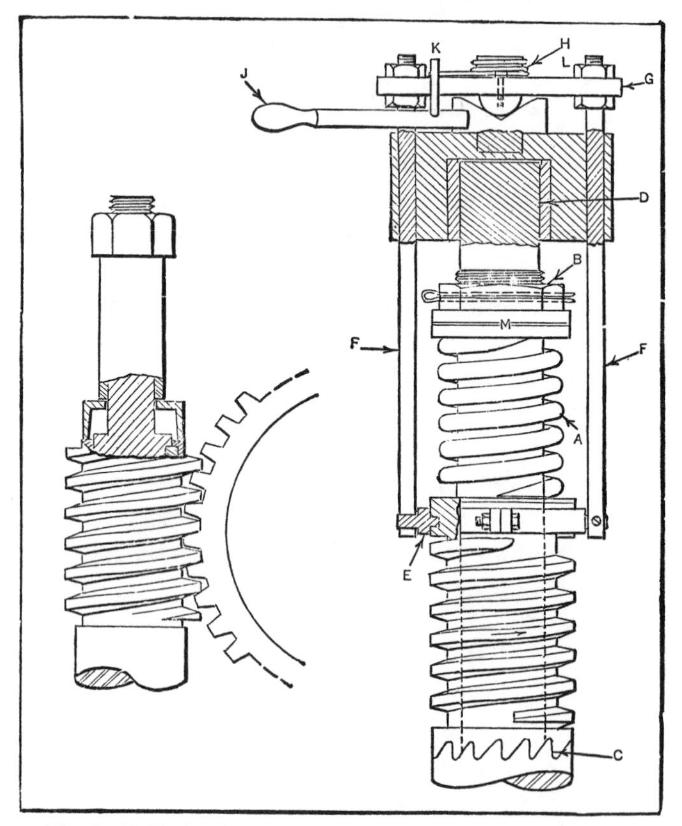

Fig. 1. (Left) Cast-iron Safety Washer Release; (Right) Improved Type of Overload Release

clutch *C* is used instead of a driving key, and the extended worm-shaft has an outer bearing at *D*.

The action of the mechanism is as follows: During normal load, the parts are in the relative positions shown. If there is an overload, the worm thrust compresses spring A so that clutch *C* is disengaged. This axial movement of the worm is transmitted through ring *E* and rods *F* to plate G, causing the small coil spring at *H* to swing clutch handle *J* from position *K* to *L*, which locks the worm in the out or disengaged position. To reset, lever *J* is simply returned to its former place, which permits the driving and driven parts of clutch *C* to come into engagement.

It is important to have the ball thrust bearing *M* between spring *A* and nut *B*, because when the clutch is in the disengaged position, the worm-shaft and nut must necessarily continue to turn, while the worm is idle. The teeth of clutch *C* should be rounded at the edges to prevent damage at the moment of disengagement under load.

This mechanism is quick and positive, and can be applied to various other drives, especially when a machine is likely to encounter some obstruction due to careless adjustment or operation. It can be utilized to provide overload protection when a machine is running in one direction but not in the other. For example, many machines are geared for a higher speed during the return stroke, and overload protection is desirable for this reversal or backward movement. Various other applications will be apparent to designers. It is obvious that every machine subject to overload should have its safety device, for the same reason that motors need fuses and circuit-breakers.

Worm-Gear Equipped with Friction Drive that Prevents Overload.—A friction release is incorporated in the worm-gear shown in Fig. 2. This gear was designed for use in a wrapping machine in which failure of any part to function would merely result in slippage of the drive gear. The same principle, however, has many applications in special and automatic machinery.

Instead of making the gear from one piece, it is constructed from three pieces, namely, a hub, a ring on which the teeth are cut, and a friction disk. These are assembled, as shown at the left, by six bolts. Originally helical springs were placed between the disk and the bolt heads, as shown at *A*, but in this particular application, it was found that

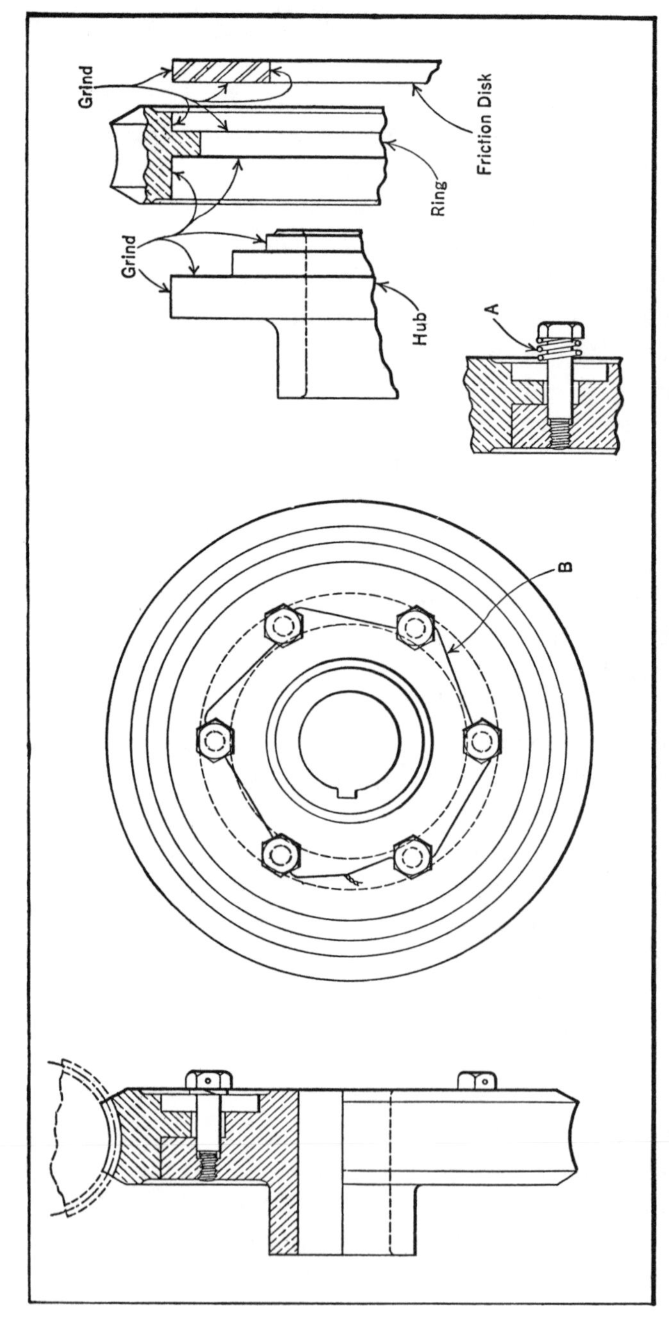

Fig. 2. Worm-gear with Friction Clutch that Prevents Overloading the Driving or Driven Units

spring lock-washers were satisfactory if the studs were not screwed up too tight. The use of helical springs, however, is recommended when slippage must occur at any accurately specified stress. After the proper adjustment has been made, the bolts are restrained from turning by the wire *B* which passes through holes in the heads of the bolts. To insure concentricity, it is best to grind the surfaces as indicated at the right, allowing just enough clearance to offer a free-running fit. These units are then assembled, after which the teeth are cut just the same as in any regular gear.

Before adopting this design, the gear was tested by means of a prony-brake mechanism, comprising a pinion drive, a brake-shoe, and an arm that worked in conjunction with an ordinary weight scale. The precision with which the drive could be made to release was quite surprising. A prony-brake mechanism is recommended for adjusting units for a given load that must be maintained closely. The hub and gear ring of the worm-gear are made of bronze, and the friction disk is made of steel.

Another Overload Release of Friction Type for Gears or for Other Rotating Members.– The release clutch shown in Fig. 3 is of the friction type and has proved very satisfactory in protecting parts of machine drives against overloading. The device can be built directly into a spur or worm gear and requires no additional space; hence it can be easily incorporated in a drive where no provision was originally made for such a device. The clutch is of simple design and very economical to build, since standard gears requiring only a little extra machining can be used. The friction disks used are standard Ford parts, costing less than five cents each.

Gear *A* is provided with six equally spaced holes *B* containing the pins *C*. These pins engage notches in the friction disks *E* and act as drivers. Other disks *F* are provided with lugs *H* which engage corresponding notches drilled in hub *J*. Disks *E* and *F* are free to slide on each other. They are held tightly against the web of gear *A* by hub *J* and collar *L*, the required pressure being transmitted to the hub and collar by the springs *M* on the shoulder-screws *N*. With this arrangement, gear *A*, pins *C*, and disks *E* comprise the driving member of the clutch,

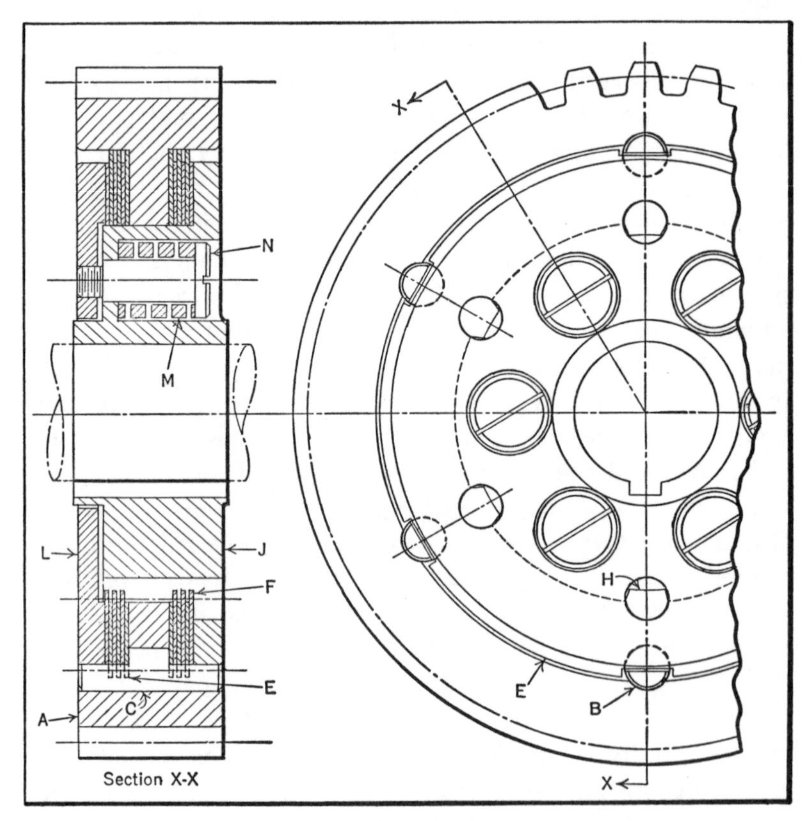

Fig. 3. Gear Equipped with Friction Drive for Stopping the Driven shaft when Excessive Loads are Applied

and disks *F*, hub *J*, springs *M*, screws *N*, and collar *L*, the driven member.

Any excessive torque applied to the driven shaft will cause the friction disks to slide on each other, thus stopping the rotation of the driven shaft until the excessive torque is removed. The magnitude of the torque transmitted is directly proportional to the total axial spring pressure applied on the clutch disks and their coefficient of sliding friction, and can be controlled by proper spring adjustment.

Overload Friction Release for a Large Gear Drive.—Each machine operating the large valves for filling and emptying the locks of a certain ship canal has embodied within the main spur gear an overload friction release.

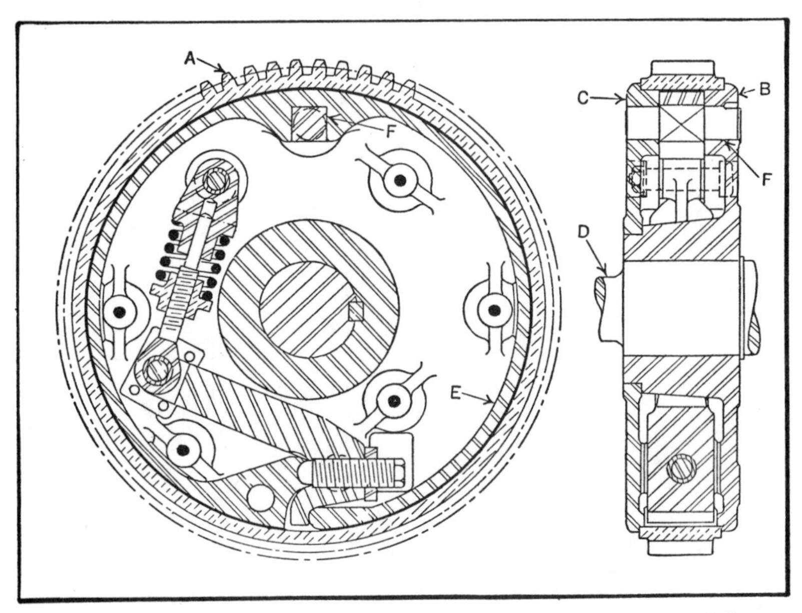

Fig. 4. Overload Friction Release with Adjustment for Controlling Point of Release

The object of this device is to prevent damage to the valve-lifting mechanism in case the valve gate should be suddenly stopped by some obstruction.

The main spur gear unit consists of a manganese bronze rim *A* (Fig. 4), about 37 inches in diameter, with teeth cut on its outer face. The gear is free to rotate in a groove formed by the cast-steel casing *B* and the cover *C*, which are bolted rigidly together and keyed to the gear-shaft *D*. In the casing is an internal brake-band *E* of cast steel,

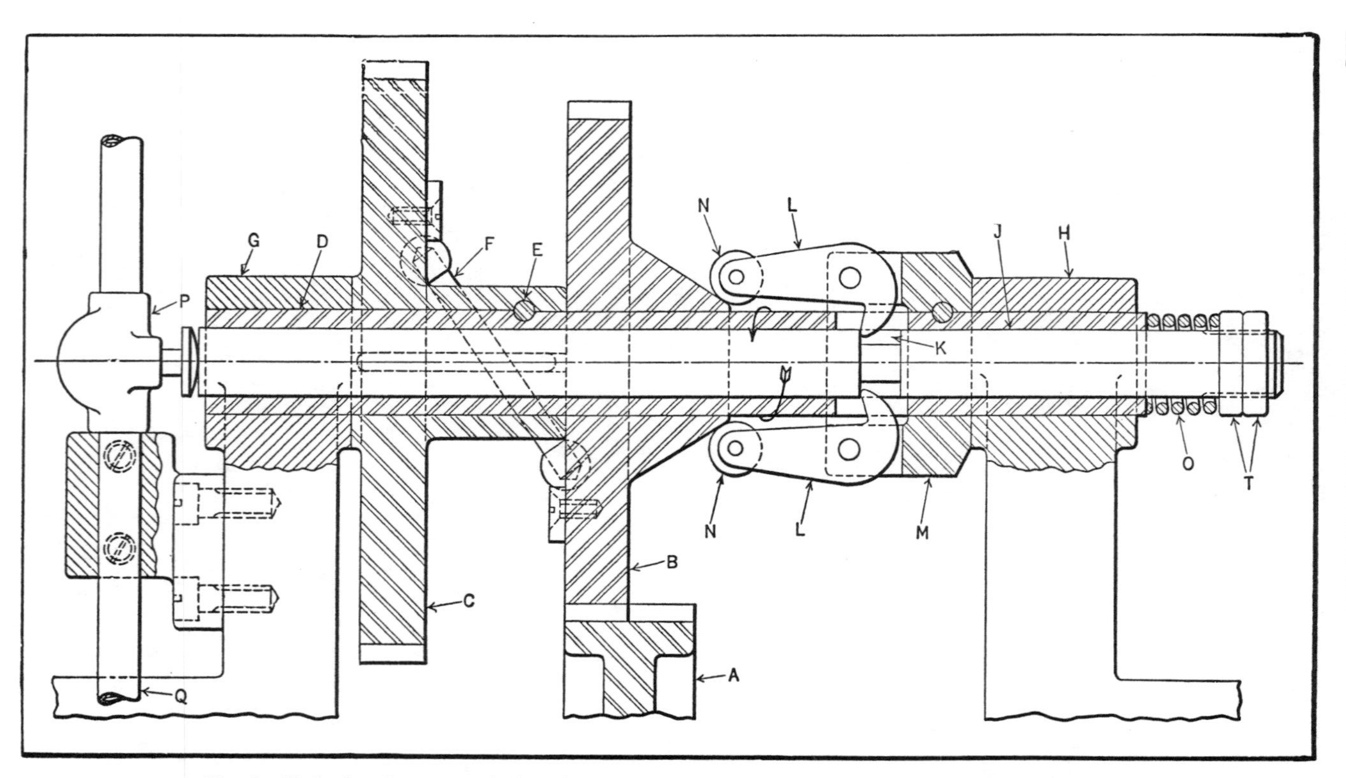

Fig. 5. Mechanism for Pneumatically Disengaging Clutch at Remote Point when Machine is Overloaded

which has an asbestos lining secured to it with copper rivets. The band is pivoted to the casing by the pin *F*.

A spring-actuated lever is provided to expand the brake-band and press it against the rim. Thus the torque is transmitted from the gear teeth through the brake-band to the casing and then to the shaft *D* to which the valve-operating drum is keyed. The spring mounting is adjustable, so that the proper load can be applied to the lever and the load regulated to compensate for wear on the band and rim. With this arrangement, the gear unit acts as a whole. However, should a log or other foreign material obstruct the valve, slippage would occur between the bronze rim and the brake-band and thus prevent damage to the machine or valve.

Pneumatic Overload Relief Mechanism for Automatically Disengaging Clutch at Remote Point.—A conveyor system is employed in a certain plant for delivering gravel over a relatively long distance to a washing and screening machine. Too large a quantity of material fed into the machine is likely to cause damage; to prevent this, an overload relief mechanism is provided on the machine for stopping the conveyor, the power for which is applied at some distance from the point of delivery of the gravel. With this arrangement, the relief mechanism opens a valve in a compressed air line when the machine is overloaded and delivers air to a cylinder, the piston of which disengages the conveyor clutch.

This mechanism is shown in Fig. 5; the air cylinder and conveyor clutch are omitted, as their design is generally known. The driving gear *A*, which rotates at a constant speed, transmits the required rotary movement to the machine through gears *B* and *C* and another gear (not shown). Gear *C* is secured by pin *E* to the hollow shaft *D*, supported in the stationary bearings *G* and *H*. Gear *B* is a running and sliding fit on this shaft, but when the machine is not overloaded, is caused to rotate with gear *C* by the bar *F*.

Balls secured to the ends of this bar engage corresponding ball sockets in gears *B* and *C*.

Plunger *J* which operates the valve *P* in the compressed air line *Q* when the machine is overloaded is a sliding fit inside of the hollow shaft. Plunger *J* contains a groove *K* which is engaged by the two fingers *L*. These fingers are pivoted in the collar *M*, which is pinned to the hollow shaft. At the left-hand ends of fingers *L* are rollers *N*, which rest on the tapered hub of gear *B*. The fingers *L*, gear *B*, and plunger *J* are held normally in the position shown by the coil spring *O*. The collar-nuts *T* provide the necessary adjustment for setting the tension of spring *O* to hold gear *B* in the position indicated when the machine is not overloaded.

In operation, gear *A* rotates shaft *D* in the direction indicated by the arrow. When the machine is running under a normal load, gear *B* maintains the axial position shown. However, if the load becomes excessive, the pressure against the ball ends of bar *F* will increase so that the bar will push gear *B* toward the right. As this movement occurs, the tapered hub of gear *B* opens fingers *L*, which causes plunger *J* to be forced toward the left and the button in valve *P* to be depressed.

In this way, air is admitted to the line *Q* leading to the clutch-operating cylinder, which causes the piston to disengage the clutch and stop the conveyor. When the excessive load on the machine is relieved, gear *B* once more returns to the normal position shown, causing plunger *J* to move toward the right and close the air valve. Springs provided on the air cylinder then return the piston and thus re-engage the conveyor clutch.

It should be mentioned that a small hole is drilled in the air cylinder head at the pressure end to permit the air to escape when the piston is actuated by the springs. This allows the air to leak out of the cylinder fast enough to permit the springs to return the piston when the air valve *P* is closed, but not so fast that full line-pressure will not operate the piston. This leak hole has another important advantage in that it prevents the operation of the clutch cylinder through leakage which might occur in the valve *P*.

Overload Relief for Oscillating Lever.—A release mechanism that was designed for a feed slide subject to jam-

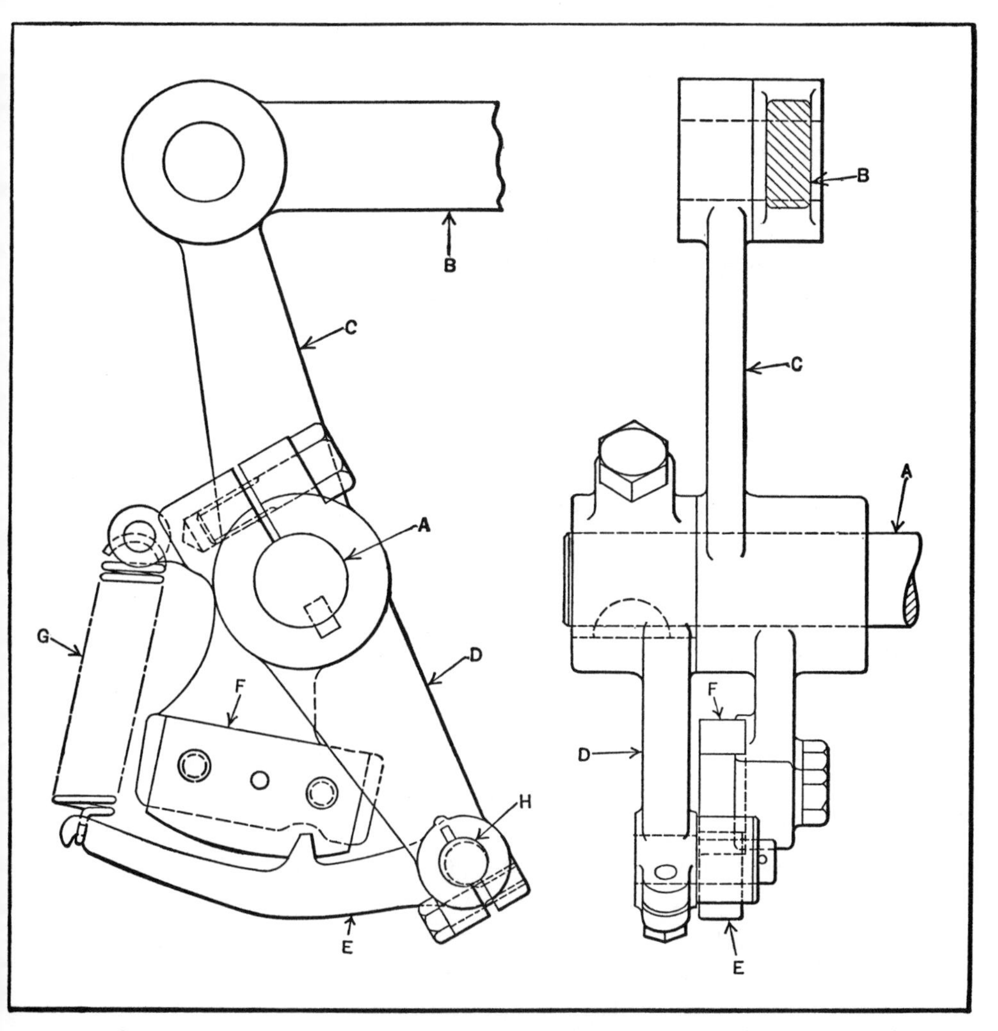

Fig. 6. Arrangement for Automatically Disengaging a Driving Lever from its Shaft when the Load Becomes Excessive

ming but that can also be applied to various types of movements is shown in Fig. 6. Oscillating shaft *A* transmits a reciprocating movement to link *B* connected to the feed slide (not shown) through the lever *C*. Lever *C* is a slip fit on the shaft, but is prevented from turning by a locking arrangement consisting of lever *D*, locking bar *E*, locking plate *F* secured to a projection on lever *C*, and spring *G*. At the outer end of lever *D*, which is keyed to the shaft, is pivoted the bar *E*. A tooth in this bar engages a notch in plate *F* and is held in this position by the spring *G*.

Normally, the entire mechanism is locked together and rocks back and forth with the shaft. However, if link *B* becomes overloaded, lever *C* will stop oscillating and shaft *A* will merely turn in the hub bore of this lever. Lever *D*, being keyed to the shaft, will continue to oscillate and cause the tooth on bar *E* to ride out of the notch and slide along the now stationary plate *F*. The tooth will continue to slide back and forth along this plate and in and out of the notch until the overload on link *B* is removed. When this is done, the tooth will engage the notch and the entire mechanism will once more function as a unit. An eccentric stud *H* is provided so that the angular position of lever *C* can be adjusted to vary the position of link *B* at the beginning and end of its stroke.

Overload Slip Arrangement for Feed-Screw.—The overload slip mechanism, Fig. 7, is so designed as to allow for the application of varying loads. The slide *A* of this mechanism is operated by means of a threaded sleeve or feed-screw *B* in the threaded hole *C*. The rod *D* passes through sleeve *B* and has a handwheel pinned to one end. The hub *E* of the handwheel has a cam-shaped end which is in contact with a similar cam face on the end of sleeve *B*. The opposite end of rod *D* is threaded and fitted with lock-nuts *F*. When the handwheel is turned until the screw *N* comes in contact with button *O*, any additional movement of the handwheel will cause the cam face on hub *E* to ride up on the cam surface on sleeve *B*, compressing the spring *P*. When the cam load reaches the high point, spring *P* causes rod *D* to return to its original position. Varying pressures from zero to maximum can be obtained either by adjusting nuts *F* to vary the loading of spring *P* or by increasing or decreasing the angle on the cam faces of the handwheel hub and sleeve *B*. Both the spring pressure and the angle of the cam faces can, of course, be adjusted when this seems desirable.

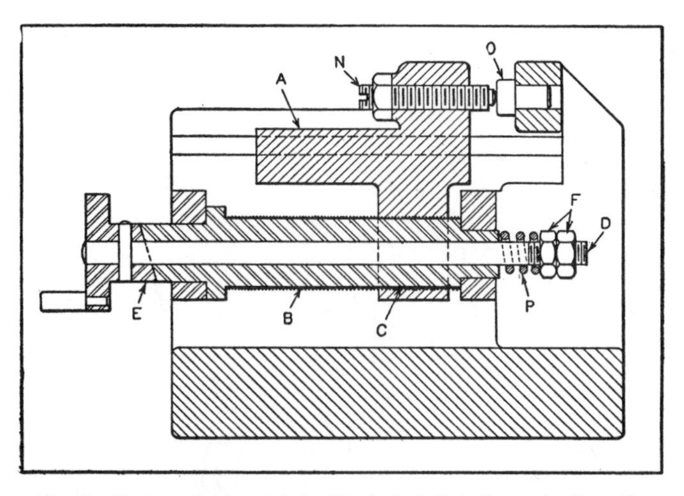

Fig. 7. Feed-screw Operated by Handwheel that Ceases to Turn the Feed-screw when the Slide Meets Obstruction or is Overloaded

Ratchet Feed with Automatic Overload Safety Stop.— A ratchet feed mechanism provided with a safety attachment that protects the mechanism from breakage in case the feed becomes jammed, and that also serves to stop the machine when this occurs is shown in Fig. 8. The attachment is so designed that the feed can be reengaged as soon as the obstruction has been removed. Previous to the installation of this attachment, a shear pin was used to protect the feeding mechanism from breakage. The shear pin arrangement merely protected the feed mechanism and did not prevent the loss in production that resulted from operating the machine while the feed was jammed; in addition, it required the suspension of production while the shear pin was being replaced.

Referring to the illustration, the feed-shaft *A* is given an intermittent rotary movement by means of the ratchet wheel *B* and the pawl *C* which is carried on lever *D*. Lever

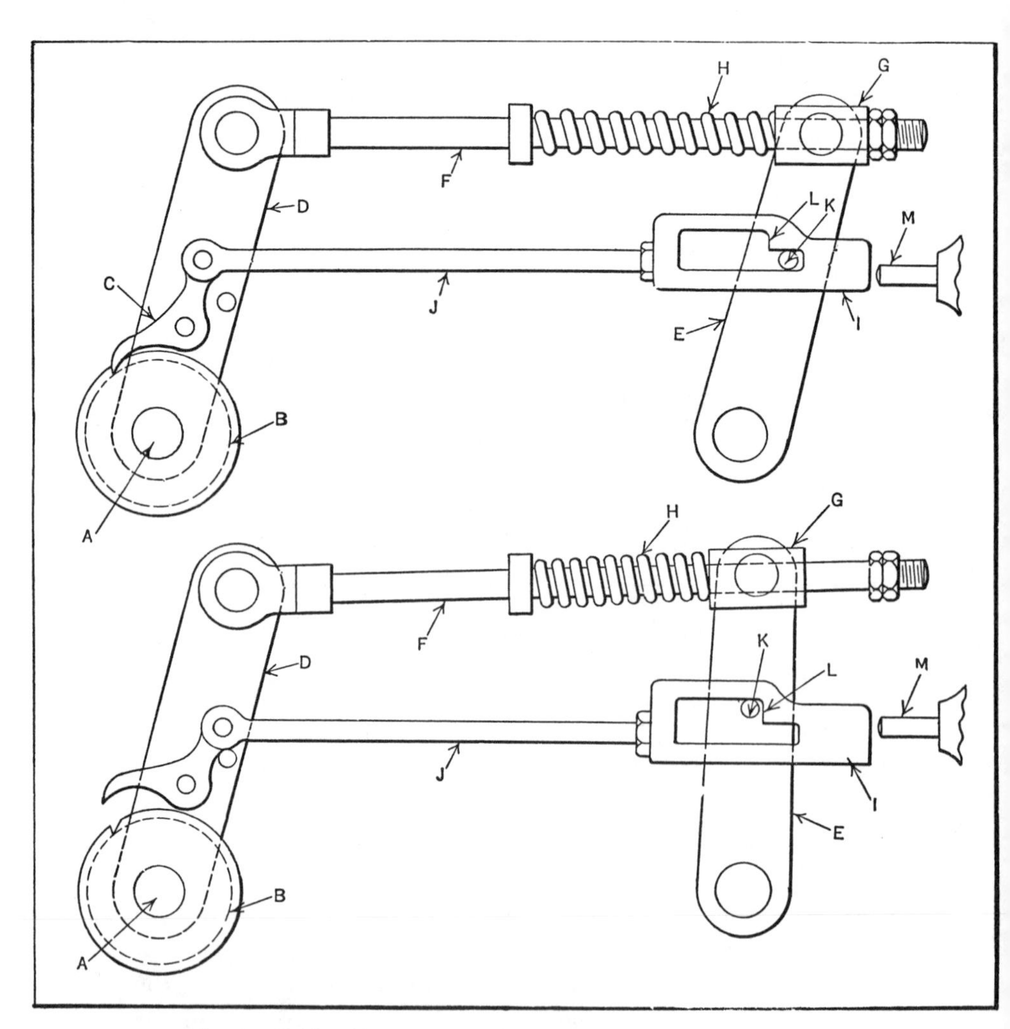

Fig. 8. Ratchet Feed with Attachment for Protecting Mechanism and Stopping Machine if Feed Jams

D receives its movement from lever *E* through the connecting-rod *F*. Lever *E* carries the swinging yoke *G* through which rod *F* passes. Rod *F* carries the spring *H*, which is compressed when the load on shaft *A* exceeds a predetermined limit. Pawl *C* is connected to the plate *I* by the connecting-rod *J*. Plate *I*, which has an irregular-shaped hole, rests on pin *K* carried on lever *E*. Under normal conditions, connecting-rod *J* carries no load, merely riding between levers *D* and *E*.

The upper view of the illustration shows the mechanism in its normal operating position. The oscillating movement of lever *E* is transmitted to shaft *A* by pawl *C*, which is held in engagement with the ratchet wheel *B* by a spring (not shown). Should the movement of shaft *A* be prevented, the continued movement of lever *E* would simply result in compressing spring *H*, thus preventing the breaking of parts. The forward movement of lever *E* carries pin *K* into the larger portion of the irregular hole, as indicated in the lower view. On the return stroke of lever *E*, the pin *K* engages the shoulder at *L* on plate *I*, thus causing pawl *C* to remain out of engagement with ratchet wheel *B*, and preventing further movement of shaft *A*, although the machine may not be stopped immediately. In this position, plate *I* extends beyond lever *E* sufficiently to push over the rod *M* far enough to open the electric switch that controls the driving motor, and thus cause the machine to come to a stop.

When plate *I* is lifted and disengaged from pin *K*, the machine is again ready to start, but if the resistance of shaft *A* is greater than the tension of spring *H*, the machine will again be stopped on the first stroke of lever *E*. In the actual construction, two pairs of levers *D* and *E* are used, the mechanism being located between them. In order to show the mechanism more clearly, however, the outer levers have been omitted in the illustration.

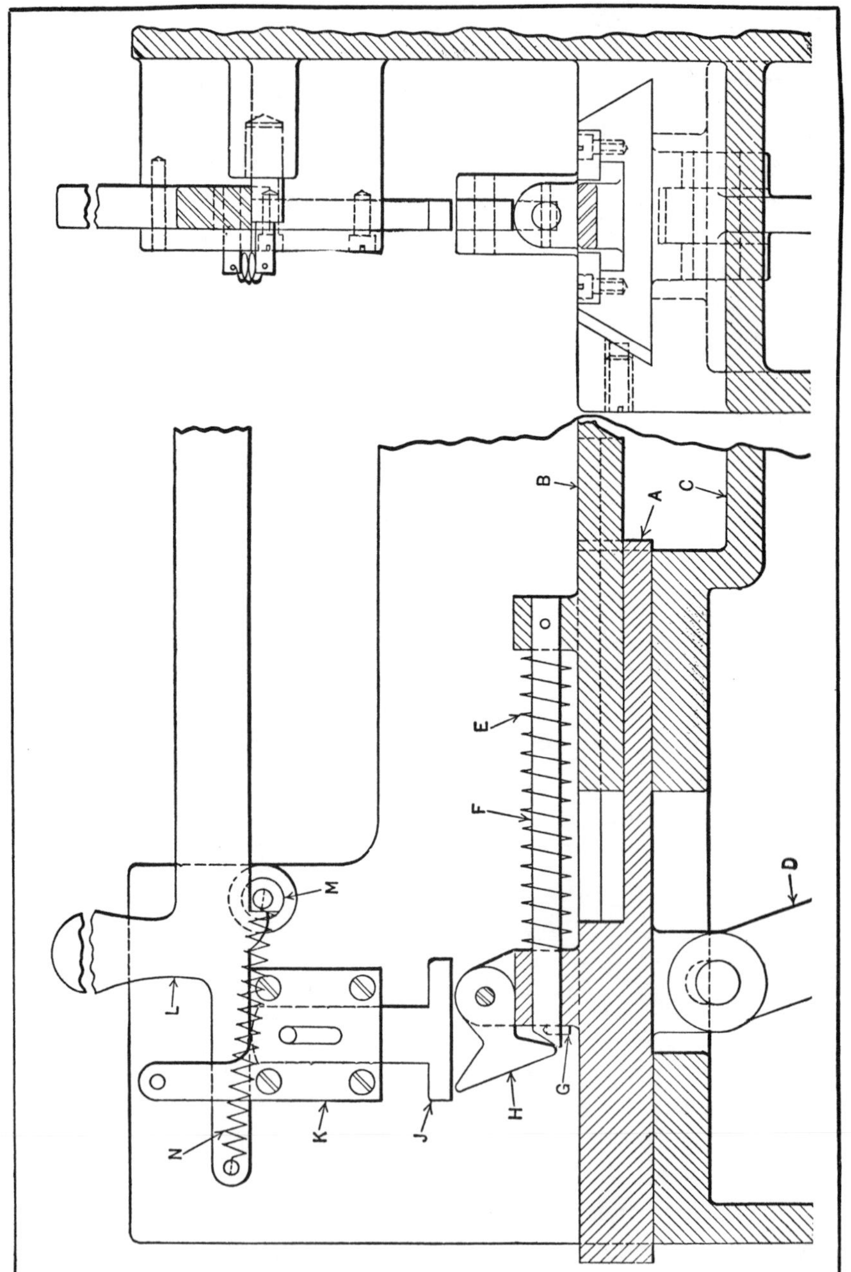

Fig. 9. Mechanism for Stopping Slide Automatically when it Becomes Overloaded

Mechanism for Instantly Disengaging Clutch at Point of Overload.— Many mechanisms designed to disconnect the power drive to the machine when it becomes overloaded function only at one point in the operating cycle. With the mechanism shown in Fig. 9, however, the machine clutch through which power is transmitted to the slide is disengaged instantly, at the exact point in the slide movement at which the overload occurs.

This arrangement is incorporated in a machine for assembling metal caps on electric fuse plugs, a number of the plugs being capped simultaneously. The capping tool slide actually consists of two slides *A* and *B*, slide *B* being superimposed upon slide *A*. Slide *B* carries the capping tools and is normally held in one position relative to the main slide *A* by means of a stiff coil spring *E*. If the tool-carrying slide *B* meets with an obstruction, it telescopes into the main slide, actuating a latch *H*, through rod *F*, which causes a spring-operated hand-lever *L* to shift and disengage the machine clutch. Although not shown here, a band brake operated by the same hand-lever prevents over-run of the machine members after the clutch is disengaged. Slide *A* is reciprocated in a dovetail guide in the machine frame *C* by the oscillating lever *D*. This lever is actuated by another member of the machine (not shown).

Rod *F*, together with stop-pin *G*, limits the telescoping movement of the slides, in addition to tripping the pawl *H* when the slide meets an obstruction. Pawl *H* is pivoted at the top of the main slide lug, and when swung upward engages latch *J* sliding in the guide *K* on the machine frame. The upper end of this latch, when the latter is raised by pawl *H*, serves to disengage the clutch lever *L* from the stationary pin *M* in the machine frame.

The coil spring *N*, secured to pin *M*, then forces the lever *L* toward the right, disengaging the machine clutch and applying the band brake. All these movements take place at practically the same instant that the overload occurs, so

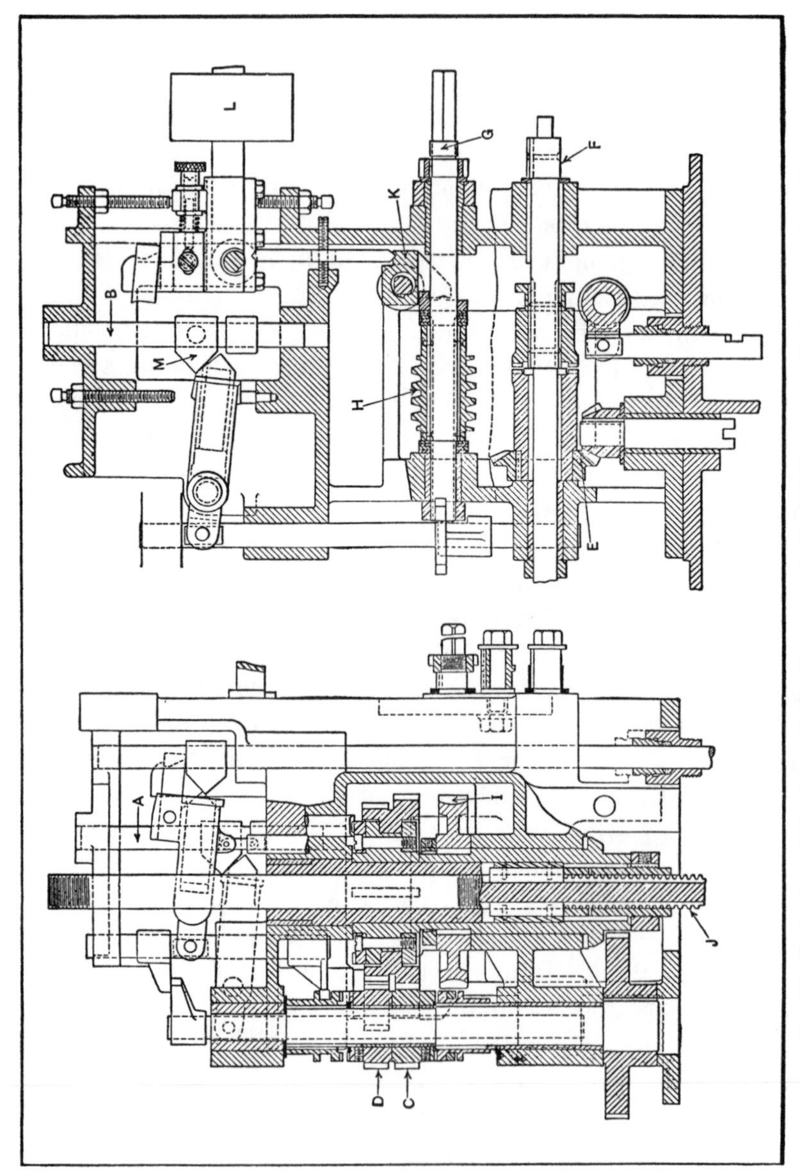

Fig. 10. Mechanism which Automatically Disengages the Feed if Pressure on Tool Becomes Excessive

that no further movement of slide *B* results after meeting the obstruction. Slide *A* is then moved to the left by hand and the obstruction removed, after which the machine is started by shifting lever *L* back to its original position.

Disengagement of Feeding Mechanism when Pressure on Tool is Excessive.—The mechanism Fig. 10 is part of an "automatic" of the vertical type. The rods *A* and *B* carry dogs which engage stops on the frame of the machine and, under normal conditions, trip the advance and return feed movements respectively. The rapid-traverse motion, which retracts the tools quickly and can also be utilized to bring them forward rapidly to the point of cutting, is operated through gears *C* and *D* controlled by clutches. The advance feed for cutting is through bevel gears *E* and change-gears connecting shafts *F* and *G*, the worm *H* on the shaft *G* driving worm-wheel *I*. These two trains of mechanism give the desired advancing and retracting movements through connection with screw *J*. A feature of this feed is in providing means for automatically tripping it whenever the pressure on the cutting tool becomes excessive. This is accomplished by providing a thrust bearing for the worm *H* which, through bellcrank *K*, is held in place by an adjustable weight *L*. When the pressure is sufficient to raise the weight, the mechanism operates to trip the latch *M* and engage the return motion the same as if the regular tripping point had been reached.

Spring-Plunger Release Mechanism for Preventing Damage to Reciprocating Parts.—Mechanisms for automatically preventing damage to reciprocating parts are sometimes required when there is the possibility of such an occurrence resulting from the jamming or overloading of the machine. A mechanism of this kind is shown in Fig. 11. The reciprocating movements of the parts to which this mechanism is applied are obtained by means of the lever *A*, which swings back and forth as indicated by arrows *E*.

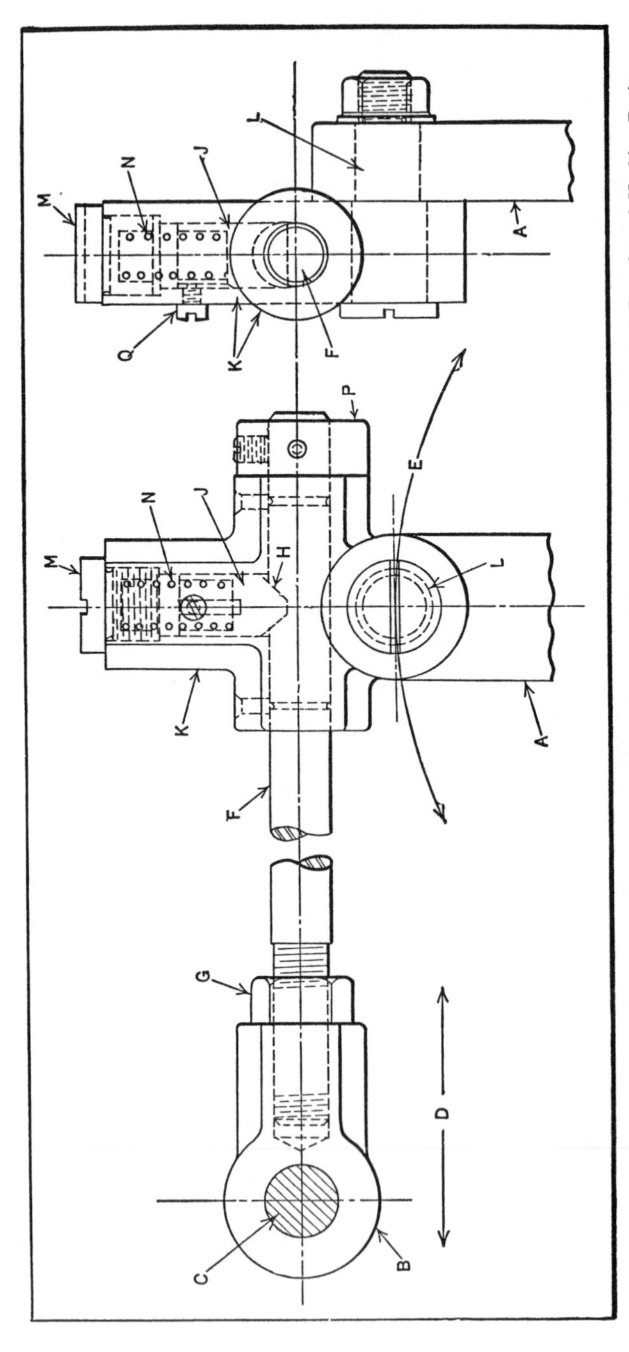

Fig. 11. Release Mechanism for Reciprocating Lever Designed to Prevent Breakage Due to Overloading or Jamming of Machine Parts

Lever *A* is mounted on a shaft (not shown) which is operated by a cam-and-roll mechanism. At *B* is shown a connecting link, through the center of which passes a short shaft *C* by means of which the rocking action of the lever *A* is transmitted to a vertical reciprocating slide. This slide travels back and forth in the directions indicated by arrows *D*, being propelled through the medium of the connecting-rod *F*, which is adjustable in the link *B*. Rod *F* operates in a vertical position instead of in the horizontal position shown in the illustration. This rod is securely locked in place by nut *G*, the adjustment being obtained by making a complete revolution of rod *F*, so that the crosswise notch at *H* will be in the proper position to receive the locking plunger *J*.

Ordinarily, the movement of lever *A* is imparted directly to shaft *C*. The safety unit is provided to prevent breakage in case the machine becomes jammed or overloaded. This safety device consists of the cast-iron housing *K*, which is free to pivot on stud *L* in lever *A*. Screw *M* holds spring *N* in place, so that the plunger *J* is forced into the notch *H* in rod *F*. Under normal operating conditions, the unit acts as a non-yielding driving block. In the case of an overload on shaft *C*, spring *N* yields sufficiently to permit plunger *J* to snap out of notch *H*, with the result that lever *A* and block *K* will reciprocate without imparting any motion to rod *F*, thereby preventing the driven parts from being broken or damaged.

When the obstruction is removed, plunger *J* automatically springs back into notch *H* and the normal operation of the machine is resumed. The collar at *P* is provided to insure a positive driving movement for rod *F* on the down stroke. Thus, the driving movement imparted to rod *F* is interrupted only on the upward stroke. There is a pin-shaped end on screw *Q* that enters a keyway in plunger *J* and thereby prevents the latter member from turning in block *K*.

CHAPTER VII

REVERSING MECHANISMS OF SPECIAL DESIGN

Reversing mechanisms may be designed to act at a fixed point in the cycle of movements or to vary the point or time of reversal. A reversal of motion in some cases may also be accompanied by a change of velocity. This chapter deals with reversing mechanisms of the different types mentioned and includes only special designs not found in Chapter VI of Volume I (pages 161 to 197).

Compact Reverse Mechanism of Rapid-Acting Parallel Worm Type.—In a certain type of can-seaming machine, the work is controlled by a mechanism having a continuous reciprocating movement. This mechanism provides a traverse movement of constant velocity. The reversals are positive and practically instantaneous. A compact design was essential in this instance, because the mechanism was used in making an alteration to a machine where the small space available made it impossible to use a long-throw cam. A nut and feed-screw provided with the usual dog-operated reversing mechanism was considered, but was rejected owing to the lost motion attending each reversal.

The mechanism is mounted on the machine frame *A*, Fig. 1, and consists essentially of the two worms *B* and *C* and the follower-roll *D*. Both the worms have right-hand threads and are rotated at a constant velocity in opposite directions by means of gears *E* and *F*, mounted on their respective worm-shafts. These gears, in turn, are rotated by gear *G* on the shaft *H*, which is driven by another member of the machine. Follower-roll *D* is free to turn on its bearings in cross-slide *K*, which moves laterally in the

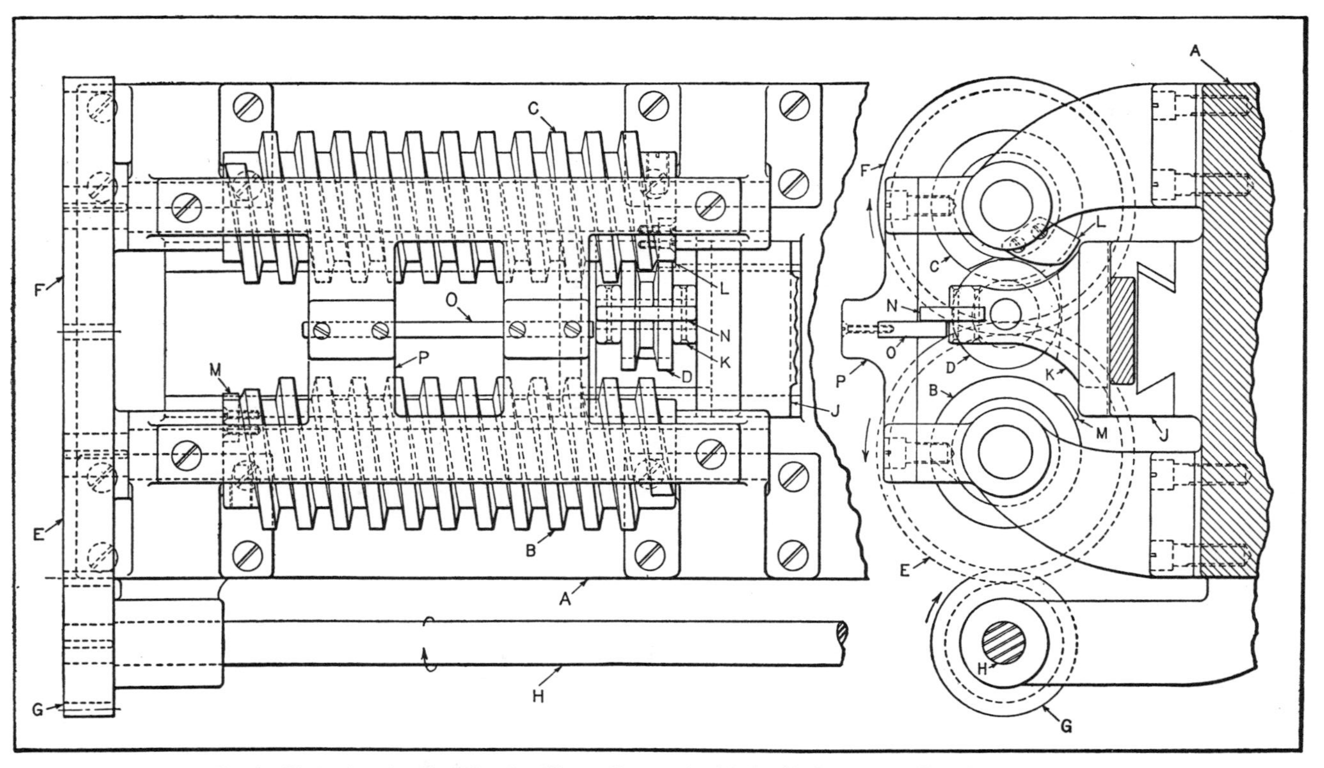

Fig. 1. Mechanism for Obtaining Rectilinear Movement with Rapid Reversal at Each End of Stroke

slide *J*. Slide *J* is mounted on a dovetail guide on the machine and transmits the required movement to the work. This slide is given a reversal of its movement at each end of its stroke through the action of cam lugs *L* and *M*, secured by screws to opposite ends of worms *C* and *B*, respectively. The roll is held in engagement with each worm by the insert *N* which rides along one side of stationary bar *O*, depending upon which worm is engaged with the roll. Bar *O* is held in the stationary position by the top plate *P*, secured to the bearings of the worm-shafts.

When the follower-roll has reached the position indicated, slide *J* is at the end of its right-hand stroke. It will be noted that insert *N* on the roll cross-slide has just passed the end of bar *O*. Now as the worm continues to rotate, the cam lug *L* comes in contact with one flange of the roll and forces it over into engagement with the beginning of the thread on worm *B*, holding or locking it in this angular position until the worm thread has carried the left-hand end of the insert *N* past the right-hand end of bar *O*. Worm *B* then reverses the movement of the follower with slide *J*, carrying it toward the left until, at the end of the stroke, cam *M* comes into contact with the other follower-roll flange, which forces the latter over into engagement with worm *C*.

The roll is held in this position until its movement toward the right carries insert *N* past the end of bar *O*, the bar preventing disengagement of the roll and worm during the remainder of the stroke. Thus worm *C* returns the roll and slide *J* to the position shown, where the reversal of the slide *J* is repeated. The reversals of slide *J* are effected rapidly and with absolutely no shock. The pressure between insert *N* and bar *O* is insignificant, owing to the relatively small angle of the worm thread. However, in order to insure a long life, as well as to increase the efficiency of the unit, both of these members are hardened and ground on their wearing surfaces.

Mangle Gear Mechanism for Changing Direction of Rotation.—The mangle gearing mechanism shown in Fig. 2 is designed to drive, from a continuously rotating shaft *P*, a shaft *S* a portion of a turn backward and forward. The pinion shaft *P* is driven through universal joints which permit it to move back and forth in the slot in guide *B*. To

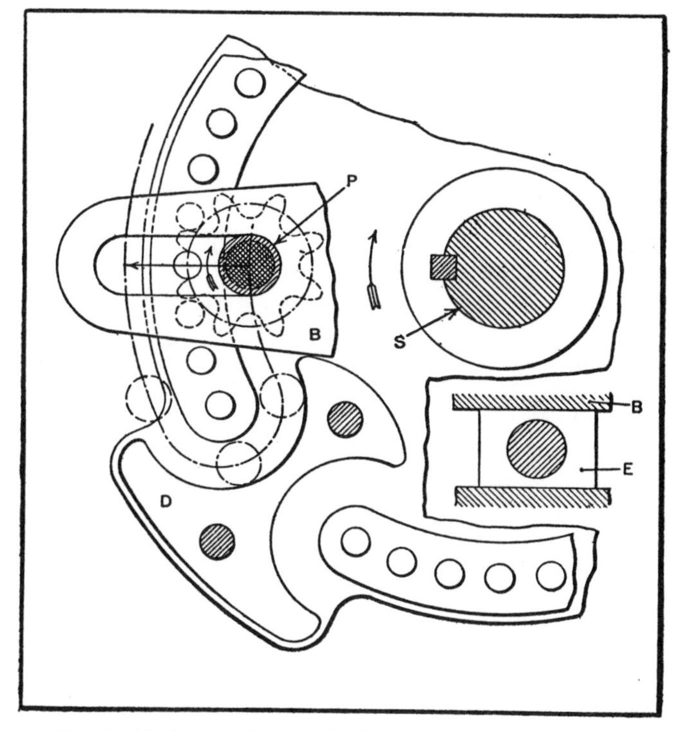

Fig. 2. Pin Type of Mangle Gearing for Reversing Rotation of Driven Shaft S

the shaft *S* is keyed a center plate to which is attached concentrically the mangle gear proper, consisting of a ring of cast iron or steel fitted with a number of pins which act as gear teeth and which mesh with the teeth of a gear of the sprocket type.

To the center plate is attached a reversing dog or guide *D* into which the end of the sprocket shaft passes, restricting the movement of the sprocket, and causing it to move from one side of the pin ring to the other as the sprocket and ring rotate together. After the passage of the sprocket from one side of the ring to the other, the ring turns in the opposite direction; that is, its motion is reversed, but its velocity remains unchanged, except, of course, during the passage of the sprocket to the other side of the ring while it is in contact with guide *D*. In passing through the guide, the end of the sprocket shaft travels from one end of the

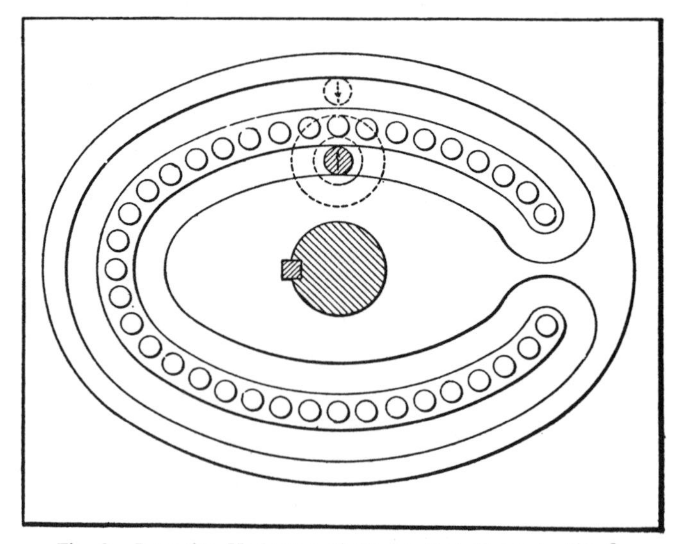

Fig. 3. Reversing Mechanism Similar to that Shown in Fig. 2, but Designed to Produce Variable Velocity

slot in the fixed bracket *B* to the other. The ends of the slot prevent the sprocket from being forced out of mesh with the pins. When the center plate has rotated through its full complement of a turn, the sprocket is again transferred to the side of the ring with which it was previously in mesh through the action of the opposite guide. Instead of rotating directly in the bracket *B* as a bearing, the shaft may rotate in a sliding block *E*, such as shown in the small cross-sectional view.

In Fig. 3 is shown a variation of the type of mechanism

just described. This design is arranged to vary the velocity of the driven shaft. The end of the pinion shaft may be fitted with a ball journal bearing.

Mangle Gearing for Reversing Rotation of Shaft After One Complete Turn.—The pin type of mangle gearing mechanism shown in Fig. 4 is designed to reverse the driven shaft after it has made a complete turn. With this

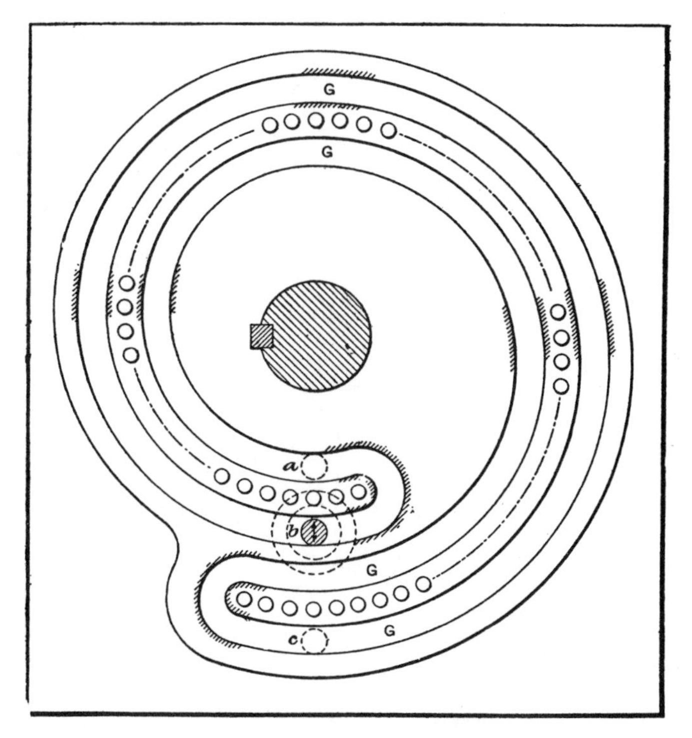

Fig. 4. Pin Type of Mangle Gearing Arranged to Reverse Driven Shaft after One Complete Turn

mechanism, the driven shaft has a somewhat variable motion. The smaller the lead of the spiral in relation to the distances of the pins from the shaft center, the less will be the velocity variation. Fig. 5 shows one arrangement for driving the pinion shaft *b*, Fig. 4, which provides for the required oscillation of the pinion or sprocket shaft.

It is obvious that an infinite number of velocity combinations are possible by varying the shape of the mangle gear shown in Fig. 4. The continuous groove G serves to keep the sprocket in mesh with the pins. The end of the pinion shaft b may either rotate directly in contact with the groove or it may bear against the groove side through a ball journal bearing attached to the shaft.

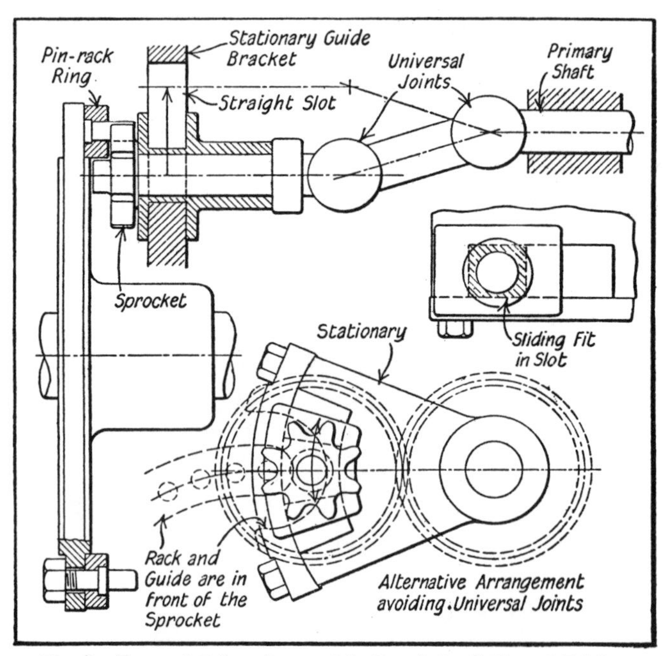

Fig. 5. Types of Drives Arranged to Permit Oscillation of Driving Shafts such as the One Shown at b, Fig. 4

Perhaps the most common method of driving the pinion of a mechanism of this kind is by bevel gears, the bevel gear on the pinion shaft serving as a universal joint. This method has the objection, however, that owing to the oscillation of the shaft, the pinion occupies different angular positions, not only during the oscillation but when it is driving. To allow for the change in the angle, the teeth of the pins must be barrel-shaped. Generally they are permitted to assume this shape through wear. The driving method shown in Fig. 5 is preferable to the bevel gear drive.

Shaft-Reversing Mechanism Giving Higher Velocity in One Direction.—When a shaft-reversing mechanism is required in which the velocity in one direction must or can be greater than in the other, the driver and driven elements may be ordinary gears or gear segments, such as shown in Fig. 6. With this type of mangle gearing, the velocity of

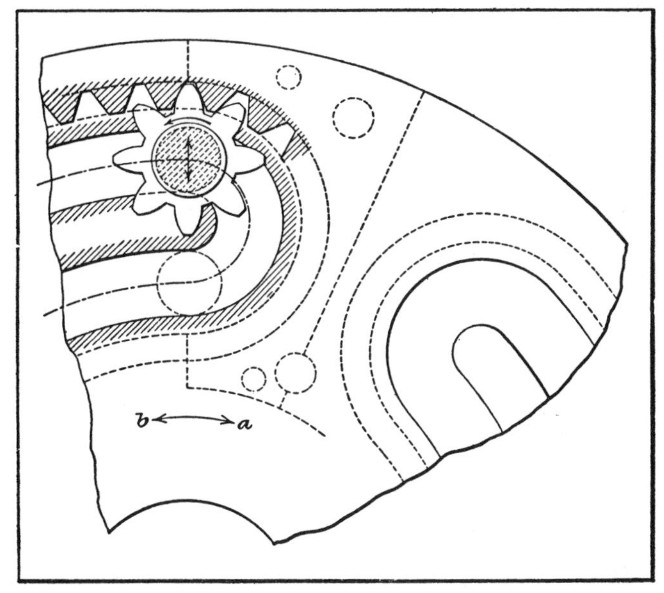

Fig. 6. Shaft-reversing Mechanism that Gives Higher Velocity in One Direction

the driven shaft is greater in the direction indicated by arrow *a* than in the direction *b*. The velocity is uniform, however, in each direction. The pinion shaft in this case is guided wholly by a groove in the center plate, into which the end of the pinion shaft projects.

In Fig. 7 is shown a double-edge rack segment-form mangle gear for obtaining reversal of the driven shaft. The round end *c* of the gear ring can be half of a pinion, having a boss or hub by which it is located in a drilled hole in the plate.

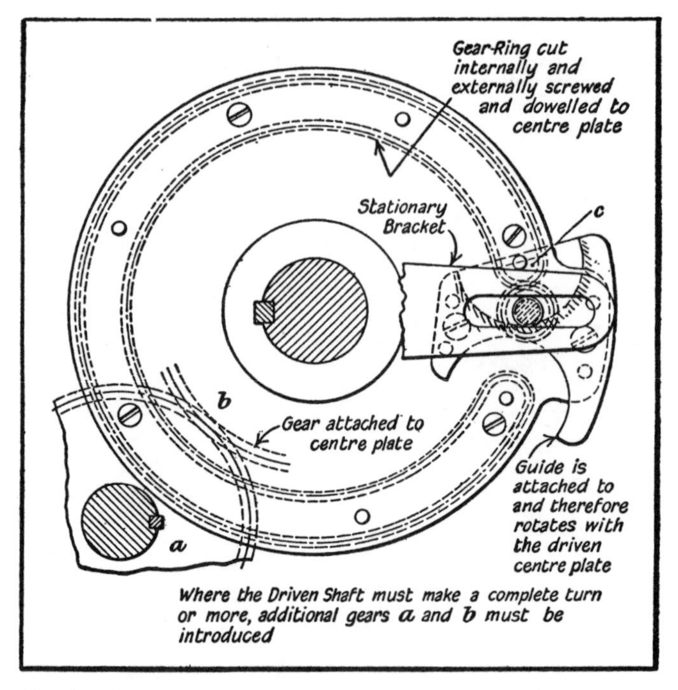

Fig. 7. Mangle Gear Shaft-reversing Mechanism with Additional Gears for Obtaining a Complete Turn or More than One Turn

Shockless Reversing Mechanism which Varies Point of Reversal.—Some mixing machines of the agitator type, employed for mixing liquid or plastic materials, require a reversing movement of the agitators; at the same time, however, the point at which reversal occurs must advance uniformly. These combined movements may be obtained by means of the mechanism shown in Fig. 8.

Here the drive shaft *D*, rotating at a uniform speed, imparts the required movement to the shaft *G* through the action of a combination planetary and elliptical gear train. All three shafts *D*, *L*, and *G* rotate in stationary bearings, and owing to the ever changing radii of the elliptical gears at the tooth contact, an alternating accelerated and retarded movement is imparted to shaft *L*. This movement, in turn, is transmitted by spur gears *J* and *K* to the ring

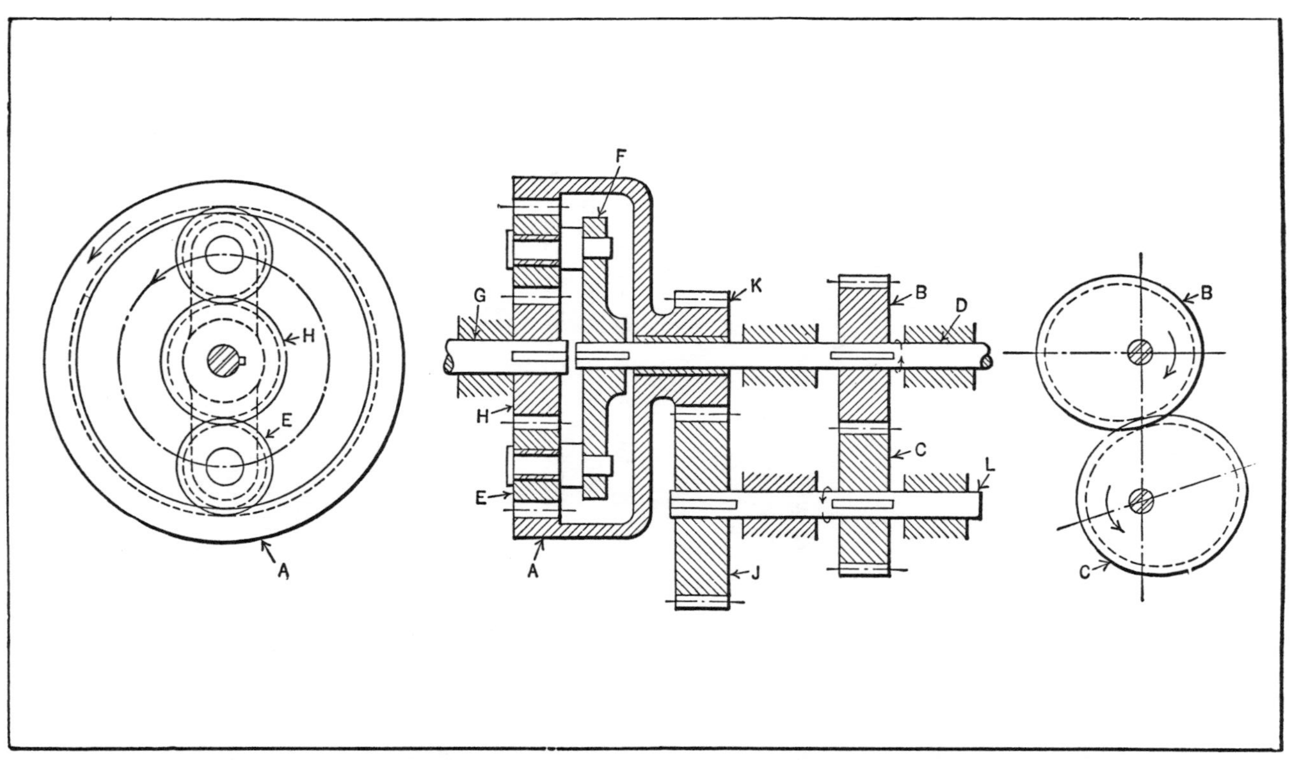

Fig. 8. Planetary Gear Train which is Given a Variable Movement by Elliptical Gears to Reverse the Movement of Shaft G

gear *A*. Now assume that the elliptical gears *B* and *C* have rotated into the positions indicated in the end view at the right, and that the ratio of spur gears *J* and *K* and also the ratio (momentarily) of the elliptical gears is such that the velocity of the centers of pinions *E* is one-half that of the pitch line velocity of the ring gear. Then, according to the principle of epicyclic gear trains, the pinions will simply roll about and not rotate the spur gear *H*.

Now if the shaft *D* rotates in the direction of the arrow, the ratio of the elliptical gears at the tooth contacts will gradually change so that the movement of shaft *L* and ring gear *A* will be retarded. Therefore, as the pitch line velocity of the ring gear decreases, the velocity of pinions *E* relative to the spider *F* will also decrease, and the lag of these pinions will cause the gear *H* and shaft *G* to rotate in the same direction as shaft *D*. This movement of shaft *G* will be accelerated until the elliptical gear *B*, whose engaging radius is gradually diminishing, has rotated through an angle of 90 degrees. At this point the ratio of the elliptical gears is at its minimum and as they continue to rotate, the ratio increases. This has the effect, through the movement transmitted to the pinions, of retarding the angular movement of gear *H* and shaft *G* until the elliptical gear *B* has passed through another 90-degree angle. At this time, the elliptical gears are once more (momentarily) in a position where the pinions roll about but do not rotate gear *H*; and, on further rotation of the elliptical gears, the velocity of the pinions will be gradually increased with respect to spider *F*, thus reversing the angular movement of gear *H* and shaft *G*. This movement of shaft *G* will be accelerated during a 90-degree movement of the elliptical gear *B* and then retarded through the next 90 degrees, at the end of which time the point of reversal has again been reached and the mechanism has passed through a complete cycle.

The movement transmitted to the pinions during the first half revolution of the elliptical gears is slower than the movement transmitted during the second half; and since the velocity of these pinions governs the amount of angular movement of shaft *G*, then the angular movement of this shaft, in a counter-clockwise direction, is less than that in a clockwise direction. Therefore, the point of reversal of the shaft *G* will vary or advance about the shaft center an amount equal to the difference in these two angular movements. By varying the ratio of the spur gears *J* and *K*, the advance of the reversal points may be increased or diminished to suit the requirements; or in case no variation of the reversal point is required, the same procedure may be followed. This type of mechanism, owing to its retarding and accelerating movements, is particularly desirable where reversal must take place without shock.

Oscillating Motion Converted to Variable Reversing Motion.— In a special electrical switch testing machine, an oscillating motion of one shaft is converted to a reversing motion in another shaft, the latter alternating at each reversal between the two speeds of 60 and 30 revolutions per minute.

The shaft *X* (Fig. 9), on which the segment gear *A* is keyed, is the oscillating member. The shaft *T*, to which the irregular motion is transferred, turns in the machine bearings (not shown) and serves as a pivot for the arm *B*. Gears *O* and *P*, located under this arm, are keyed on shafts *U* and *Y* and are connected by the three gears *S*, *V*, and *G*. The concentric grooves *E* and *D*, milled in the segment gear, are joined at both ends to form one continuous groove and serve as a guide for the cam-roll *C* in the end of arm *B*. Dogs *R* and *N*, which engage projection *Q* on the arm, are fastened securely to the segment gear. Latches *J* and *M* swing on shoulder-screws, and normally bear against pins *I* and *K*, due to the tension of the coil springs.

In the position shown in the illustration, the segment gear *A* is oscillating in the direction of the arrow, and the dog *R*, against lug *Q*, is about to swing the arm *B* around shaft *T*. A further upward movement of dog *R* will throw gear *O* out of engagement with the segment gear. However, just before the teeth of gear *O* have become disengaged, a partial engagement of the teeth in gear *P* and segment *A* takes place. While gears *O* and *P* are being shifted, roller *C* swings up to the beginning of the groove

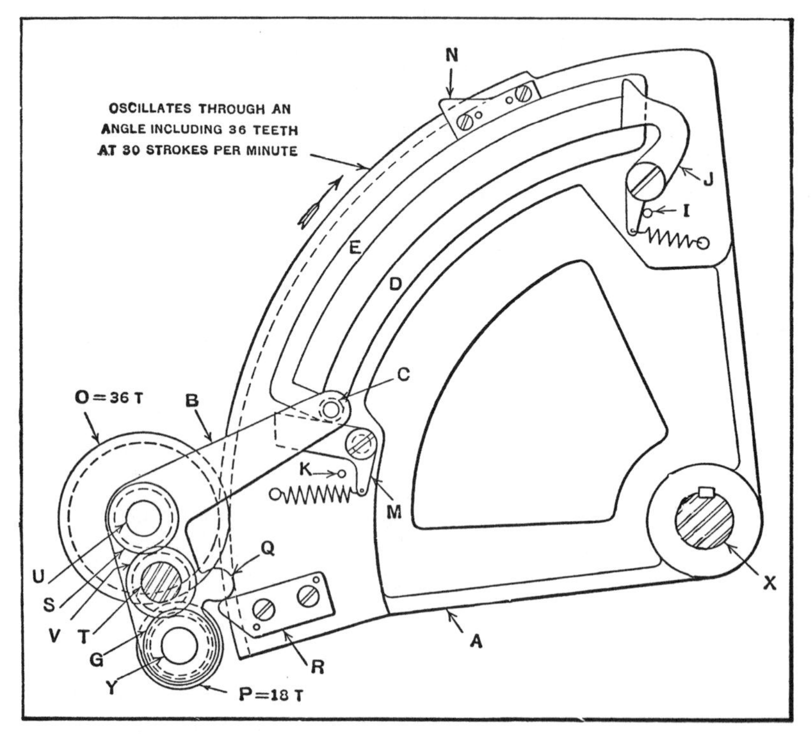

Fig. 9. Mechanism for Converting Oscillating Motion into Reversing Motion

E. When the roll reaches this position, the oscillating segment *A* has come to the end of its upward stroke and is about to return. The latch *M* closes the end grooves and prevents the roll from dropping back to groove *D* when the segment reverses.

The roll now follows groove *E* and serves to hold gear *P* in mesh with segment *A* until dog *N* comes in contact with the lug *Q*. This disengages gear *P*, after which gear

O is engaged with segment *A* again. In the meantime, roll *C* has forced latch *J* to one side and is swung down to the end of groove *D*, being prevented from coming out of this groove by the return of latch *J*. The roll, running in groove *D*, serves to hold gear *O* in mesh during the return stroke of the segment. This completes one cycle of the movements.

Because of the difference in the number of teeth between gears *O* and *P*, as noted in the illustration, and the arrangement of the gear train, the uniform oscillation of segment *A* will result in one clockwise revolution of shaft *T* for every up stroke of the segment, while the down stroke will result in two counter-clockwise revolutions of the shaft. With some slight modifications in the design, shaft *T* may be made to revolve at varying speeds other than described and in the same direction instead of reversing. This may be done by varying the number of teeth in the gears and adding an idler between any two of the gears *S*, *V*, or *G*.

Mechanism for Reversing Tap Spindles in Drill Head.—When more than one tap is used in a drill head, the problem of reversing the taps is often simplified by having one tapping spindle drive on the "in feed" and another spindle drive on the "return feed." The arrangement of the gearing for such a drive is shown diagrammatically in Fig. 10. In this case, the large drill head carries a number of drilling spindles (not shown in the illustration), in addition to the four tapping spindles, *A*, *B*, *C*, and *D*.

The drill head slides up and down on column *E*, being kept in alignment by an external projection that slides in a vertical track. The drive is obtained from a vertical shaft *F* at the end of which is keyed the pinion *G*. This pinion is in mesh with the gear *H* which drives the drilling and tapping spindles. The drive for the drill spindles is very simple and is not shown in the illustration. The drive for the tapping spindles begins with the clutch shaft *I*, which is driven from the gear *H* through the pinion *J*. The shaft *I* carries a sawtooth double clutch *K* which can be engaged with either the upper member L or the lower member M.

When the downward feed of the head begins, an arrangement of levers similar to the belt-operating mechanism on a planer causes the clutch K to engage the upper member L. The drive to the tapping spindles is then through the idler

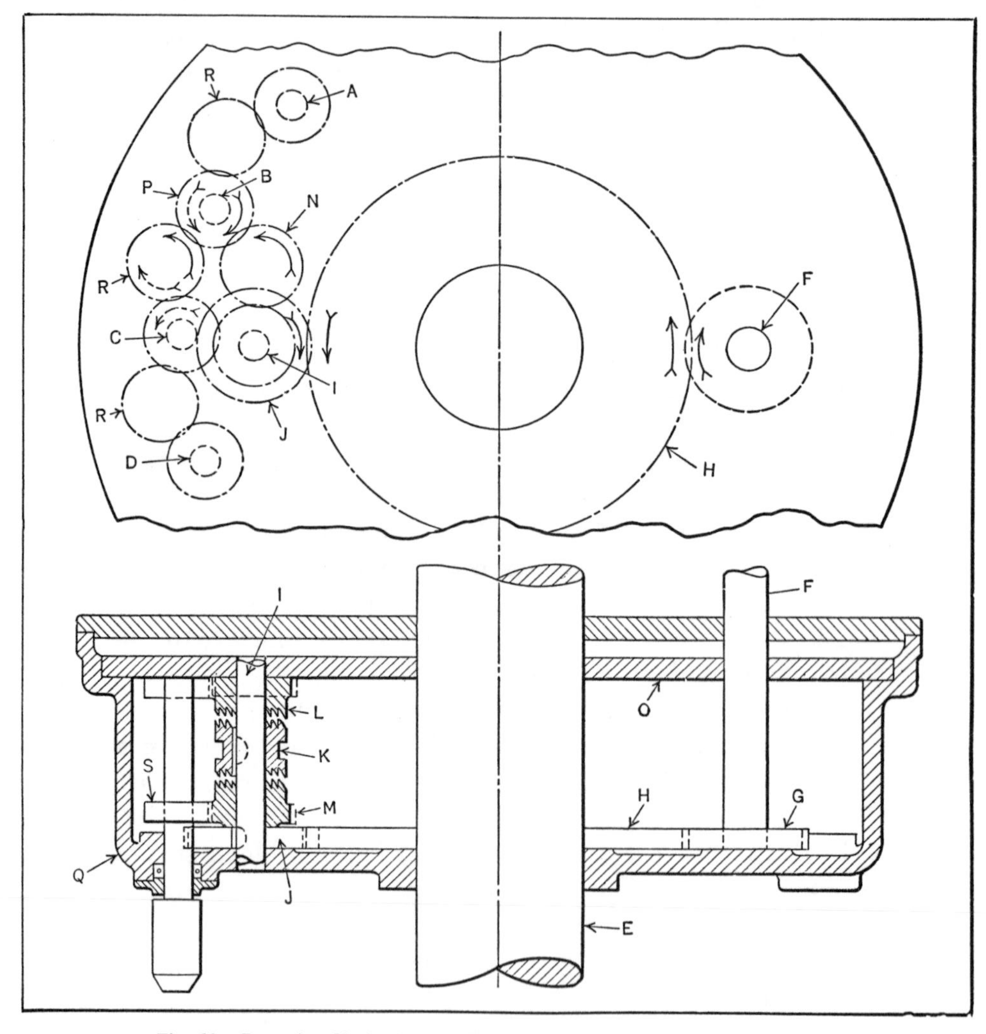

Fig. 10. Reversing Mechanism for Taps Used in Multiple-spindle Drill Head

gear *N*, which is mounted on the top plate *O*, and thence to the gear *P* on the top of the tapping spindle *B*. The tapping spindle is then revolved in the direction required for tapping. The other three spindles *A*, *C*, and *D* are driven in the same direction through the idler gears *R* which are mounted on the bottom of the gear-case *Q*.

As soon as the head begins to travel upward, the clutch *K* comes into engagement with the lower clutch member *M* and drives directly through gear *S*, which is fastened to the bottom of the tapping spindle *C*, revolving it, together with the other spindles, in the opposite direction. The full-line arrow-heads show the direction in which the meshing gears revolve when tapping, while the dotted arrows indicate the direction in which the gears revolve when the spindles are reversed on the "out feed."

Rotary Reversing Mechanism for Varying Angular Movement and Dwell of Driven Shaft.— Wire-forming machines of the four-slide type usually have various ingenious mechanical movements incorporated in their design that are applicable to machines used for other purposes. For example, in one four-slide, wire-forming machine, there is a reversing movement for a feed-slide shaft that has unusual features. This mechanism is designed to give the driven shaft a short dwell at each point of reversal. Besides, provision is made for varying the angular movement of the driven shaft without altering the dwell. The movements are transmitted from another shaft which oscillates continuously at a constant angular velocity.

On the oscillating driving shaft (not shown) is an arm to which is connected the link *A*, Fig. 11. This link, in turn, is pivoted to the sector or segment *B*. Sector *B* is free to oscillate on the stationary stud *C* and is provided with a sliding gear sector *F* which meshes with the driven gear *G*. Sector *B* is also provided with adjustable split stops *D* and *E*. These stops are used for regulating the angular movement and the dwell of the driven shaft *H*, to which gear *G* is keyed. The stops are clamped in place by bolts on the dovetailed periphery of the sector *B*.

An important part of the mechanism is a friction stop or brake on shaft *H*, which is necessary to prevent over-run of this shaft at the point of reversal. The brake arrangement, however, being of simple and well known design, is not illustrated.

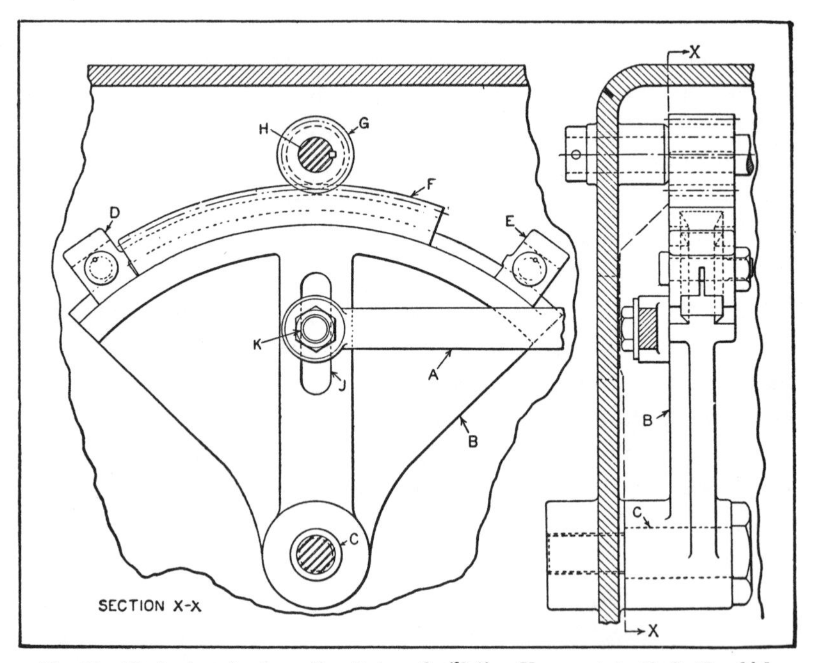

Fig. 11. Mechanism for Imparting Rotary Oscillating Movement to Shaft H, which Permits Varying Angular Movement and the Length of Dwell at Each Reversal

When the machine is in operation, the sector *B* is oscillated by link *A* at a constant angular velocity. In the position shown, the sector has moved toward the right to its central point, rotating gear *G* in a counter-clockwise direction. This motion continues until the sector has reached its farthest position at the right. The sector then reverses its movement and the rack *F* and gear *G* remain stationary until the end of the rack comes into contact with the stop *E*. At this time, continued movement of the sector toward the left will carry the rack segment toward the left, rotating gear G in the opposite direction. The rotary movement of this gear continues until the sector comes to the end of its movement toward the left. Now as the sector once more moves toward the right, gear G dwells until stop D comes in contact with the sliding gear segment. Thus, gear G and shaft H are given a rotary reciprocating motion with a short dwell at each point of reversal. Owing to the different kinds of jobs adapted to this machine, a variation in the angular movement of shaft H is frequently required, the dwelling period remaining constant. This is obtained by the combined adjustment of the link and the stops. The extent, however, to which the angular movement can be increased is limited by the length of the sliding gear segment.

Suppose, for example, a greater angular movement of the shaft were required. In this case, the stud K would be adjusted to a lower point and stops D and E would be moved farther apart to avoid reducing the time periods of the dwells. In making these adjustments, a few trials are usually necessary in order to obtain the proper positions of stud K and the stops. Obviously, the same arrangement can be used to reduce or increase the dwell within certain limits, the angular movement of the driven gear remaining the same.

CHAPTER VIII

DRIVES OF THE CRANK TYPE FOR RECIPROCATING DRIVEN MEMBERS

The special designs of crank mechanisms described in this chapter are for transmitting motion to slides or other parts having a reciprocating action. These drives may be arranged to produce some special movement, such, for example, as arresting the motion of the slide momentarily during some part of the stroke or providing a quick return movement to reduce the idle period; or the design may be special in that provision is made for adjusting either the length or position of the stroke while the machine is operating.

Crank Motion that Causes Slide to Dwell at Center of Stroke.— The crank mechanism Fig. 1 is incorporated in a certain carton wrapping machine for changing the position of the carton as it passes through the machine. This mechanism imparts a reciprocating movement to the work-slide *A*, with a dwell at the center of its stroke in each direction. The slide is reciprocated in the stationary guide *D* through link *B* by the crank *E*. This crank is a free fit on shaft *F*, but rotates with *F* whenever spring-actuated plunger *H* engages one of the notches cut in the flange on the crank. A pin *J* in the plunger projects through and below arm *G*. When this pin engages the cam-block *K*, which is secured to the machine frame, the pin and the plunger are moved radially outward. This causes the plunger to disengage the notch and allows the crank and slide to dwell while the shaft continues to rotate.

The shaft and arm *G* are rotated in the direction indicated by the arrow. In the position shown, pin *J* has engaged cam-block *K* and has withdrawn the plunger from the notch *L*. At this time, the slide is at the center of its stroke; and since the plunger is all that locks the crank *E* to the shaft, the shaft will turn in the bore of the crank hub and allow the crank and slide to dwell. A spring-actuated V-plunger is provided at *N* to hold the slide securely in the "dwell" position.

As the shaft and arm continue to rotate, pin *J* leaves cam-block *K* and allows the end of the plunger to ride on

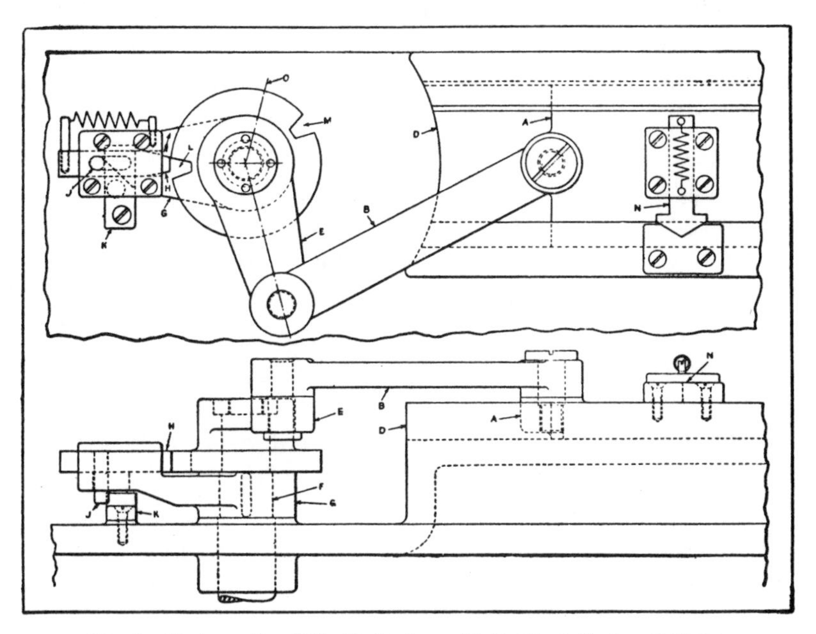

Fig. 1. Reciprocating Slide Mechanism with Dwell at Center of Stroke

the periphery of the crank flange until it drops into notch *M*. When this occurs, the crank is once more locked to the shaft so that continued rotation of the shaft will cause the slide to move toward the left and return to the position shown. At this time, the center line of the crank will coincide with center line *O* and cam-block *K* will have forced pin *J* outward, thus disengaging the plunger from its notch and allowing the crank and slide to dwell.

The withdrawn plunger then rides along the periphery

of the crank flange until it drops into notch L and locks the arm to the crank again. The rotating arm now rotates the crank, causing the slide to move this time toward the right and then back to the position shown. At this point, the cam-block once more disengages plunger H. This completes one cycle of movements, which is repeated for each revolution of shaft F.

It will be noted that the angular movement of the crank is different for each half of the slide cycle, owing to the angular position of the connecting-rod B. This results in a variation of the time interval for each succeeding dwell and stroke. Fortunately, however, this variation is permissible. In other applications, where the dwell and stroke must have the same time interval, the well-known Scotch yoke crank movement could be used instead of the crank and connecting-rod shown. In this case, the notches would be located in the flange diametrically opposite each other.

Planetary Type of Crank Motion for Obtaining Dwell.— In attempting to bend a stranded copper cable into a U-shape by means of a kind of wing die, it was found, while experimenting with a punch press, that a distinct stop or dwell was required at a certain point in the bending stroke to permit the copper to set. If the dwell was omitted, a springing back of the metal occurred, resulting in variations in the form of the bent section. This dwell had to take place before the end of the stroke, because the latter part of the stroke was utilized to eject the formed piece.

A special machine was designed to actuate the slide from which the bending die receives its motion. This operating slide also requires a dwell at the top of the stroke to allow time for inserting unbent parts into the die, so that the machine can be operated continuously instead of using a single-stroke clutch and tripping device. The planetary type of crank motion used causes a crankpin to follow, during the dwelling periods, an arc having a radius equal approximately to the length of the connecting-rod, so that the crank end swings without transmitting motion. Fig. 2 shows the general arrangement.

The housing *A* contains shaft *C*, which is driven through worm-gearing and carries a crank disk *D*. An internal gear *H* having 120 teeth is bolted to housing *A* and meshes with a 24-tooth planetary pinion *J* attached to the eccentric crankpin *E*; consequently, when crank disk *D* revolves,

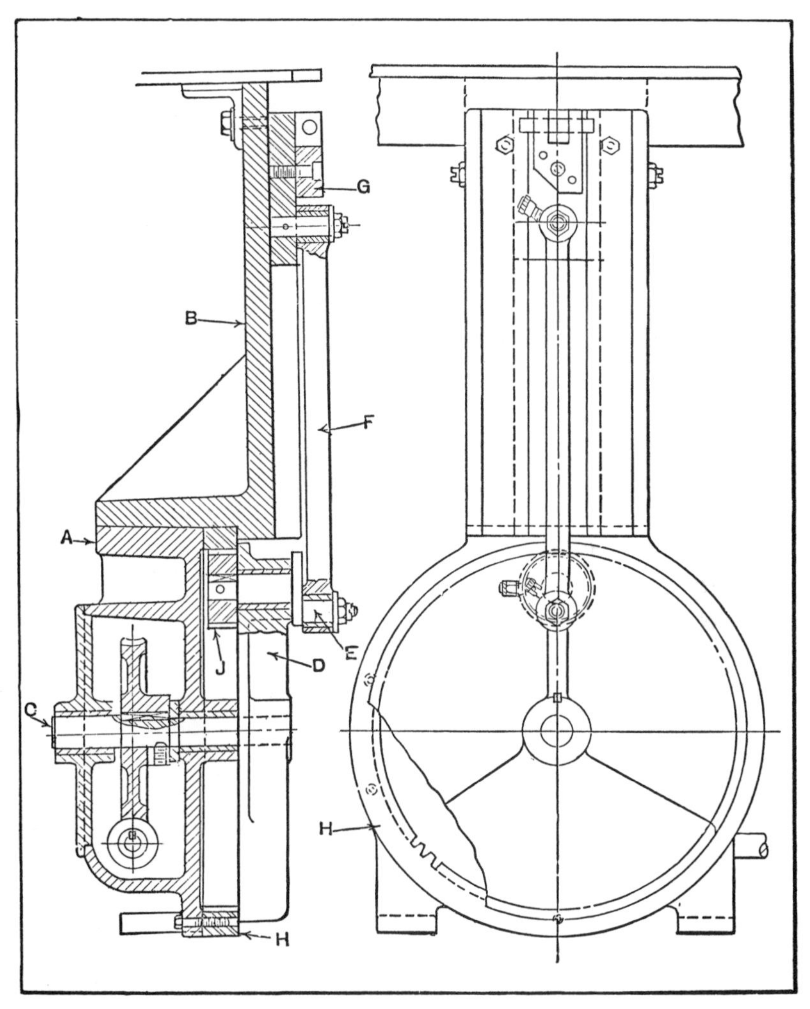

Fig. 2. Planetary Type of Crank Motion for Obtaining Dwell of Bending Die

pinion *J* and the crankpin revolve around their own axis and also around shaft *C*. These combined rotary movements modify the motion imparted to slide *G* and cause the axis of the pin *E*, to which connecting-rod *F* is attached,

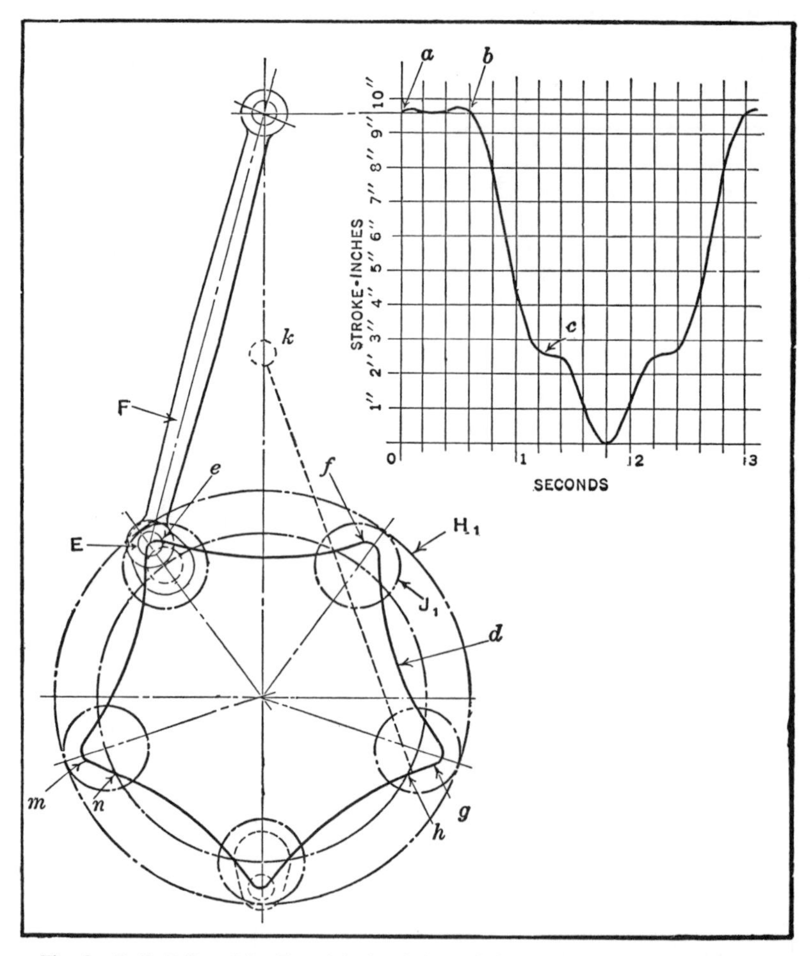

Fig. 3. Path Followed by Eccentric Crankpin, and Chart Showing Dwelling Periods

to follow the path indicated by the heavy line *d*, Fig. 3.

The curve in the upper right-hand corner of Fig. 3 illustrates how the action of the driven slide is changed during one complete cycle. The cycle begins at a point representing the top of the stroke of the slide. The vertical dimensions on the chart represent the stroke, in inches, and the horizontal dimensions, the time in seconds. One revolution is represented as 3 seconds, because the machine is designed to run about 20 revolutions per minute.

The relative positions of the internal gear *H* and the pinion *J* (Fig. 2) are indicated in Fig. 3 by dot-and-dash pitch circles H_1 and J_1. The radius of the eccentric crank *E* is considerably less than the pitch radius of the pinion, which causes the axis of the eccentric crankpin to describe a five-lobed curve *d*. The dwell of the driven slide at the top of the stroke occurs between points *a* and *b* on the chart and during about 6/10 of a second. This dwell is due to the fact that the length of the connecting-rod equals the radius of an arc which approximates that part of the crankpin path from *e* to *f*. As the lower end of the crankpin swings from *e* to *f* it transmits only a slight movement, and there would be none at all if this portion of curve *d* were a perfect arc with a radius equal to the connecting-rod center-to-center length. It is the dwell at this point that is utilized for removing the work and inserting unbent blanks.

The pause during the down stroke to allow the metal to set after bending occurs between points *g* and *h* where the curve *d* is practically tangent to the arc of the connecting-rod. This pause or dwell is represented on the chart at *c*, and at this time, the upper end of the connecting-rod is at *k*. An unnecessary dwell is made during the return stroke between points *m* and *n*, which correspond to *g* and *h*, but this slight delay in the upward movement does not affect the practical working of the mechanism. The entire device is located under a table about 2 feet square, which indicates that it is quite compact.

Oscillating Crank-and-Toggle Mechanism for Rapid Reciprocation of Slide.—In a metal ribbon crimping machine, four complete cycles of a slide are obtained from an oscillating arm as the latter passes through one cycle. This arrangement, which is shown in Fig. 4, has the advantage of simplicity of design and an unusually smooth action. The slide *A* that controls the crimping tools is mounted in guides *B*, cast integral with the machine frame *C*. Arm *D* is the driving member and is keyed to the shaft *E*, which oscillates at a constant angular velocity. This arm transmits

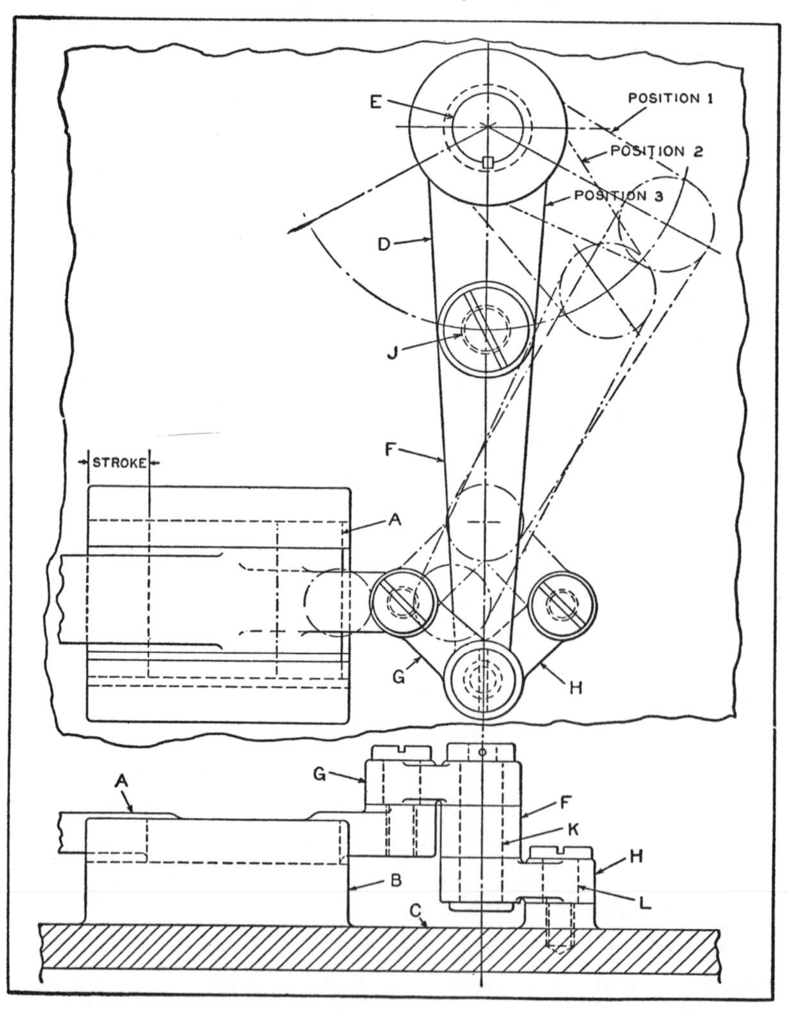

Fig. 4. Reciprocating Slide of a Metal Ribbon Crimping Machine, Operated by Oscillating Arm and Link Mechanism

the movement to the slide through a toggle arrangement consisting of links *F*, *G*, and *H*. Link *F* is pivoted at its upper end to arm *D* by pin *J*, and at its lower end to links *G* and *H* by pin *K*. The outer end of link *G* is pivoted to the slide, and the outer end of link *H* is pivoted to the shoulder screw *L* in the machine frame.

Three positions of the arm and links are shown. At Position 1, the toggle links *G*, *F*, and *H* are at their highest points; hence, slide *A* has been drawn to its farthest point at the right. As the arm swings downward to Position 2, these links assume a horizontal position, causing the slide to move to its farthest position at the left. The arm then continues its movement until it arrives at Position 3, where link *F* has forced the toggle links down to their lowest position, causing the slide to be carried back to the position indicated. Thus, during this one-quarter cycle of the arm, slide *A* has passed through a complete cycle. Consequently, as a repetition of these slide movements occurs during each quarter cycle of the arm, the slide will complete four cycles for each cycle of the arm *D*. An added advantage of this toggle arrangement is the unusually high working pressure that is delivered at the end of the stroke toward the left at the point where the pressure is needed most.

Auxiliary Crank that Assists Crankpin Past its Dead Center.— One method of overcoming the dead center condition in transmitting rotary movement to a shaft by means of a crank is shown in Fig. 5. Two of the outstanding advantages of this drive are its positive action and its low cost. The driven crank is actuated by a similar crank keyed to the driving shaft. By incorporating an auxiliary or "dummy" crank, the driven crankpin not only is helped past its dead center positions, but the angular velocity of the driving and driven shafts is held constant. In addition to this, the torque transmitted to the driven shaft is uniform at its various angular positions.

The shaft-to-pin center distance is the same for all three

cranks. The driving crank is indicated at *A*, the driven crank at *B*, and the "dummy" crank at *C*. It is important to note that the connecting-rod is of solid construction and connects all three crankpins. With this arrangement, the position of all three cranks is the same at any part of the machine cycle.

In the full outline, the cranks and connecting-rod are approaching the dead center position. When they reach

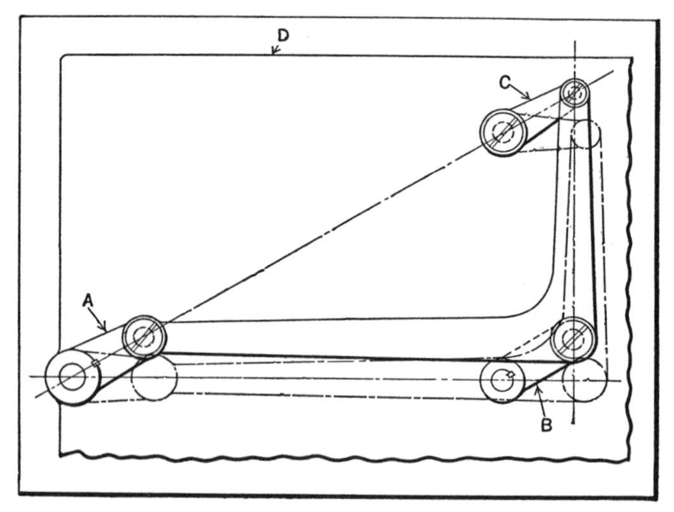

Fig. 5. Arrangement for Preventing Crankpin from being Stopped on Dead Center

this position, they will coincide with the dot-and-dash outline. Here it is obvious that the crankpin in the "dummy" crank has passed its dead center and can continue its movement unrestricted. Now, owing to the rotary action of crank *A*, crank *C* will swing downward, and as a result, crank *B* will be forced past its dead center. The same action occurs in reverse order when crank *C* is on its dead center relative to crank *A*. That is, crank *B*, having passed its dead center, will serve to force crank *C* past its dead center position. Incidentally, the location of crank *C* can be varied to suit existing conditions, although it should not be located too close to a straight line passing through the driving and driven shafts.

Auxiliary Crank for Quick Return Movement.— The cam-operated turret-feed mechanism of an automatic screw machine is shown in Fig. 6. The advance feed is obtained by the cam *A* operating through the segment lever *B* to feed the turret-slide *C*. The return motion is accelerated by the revolution of the crank *D* which brings the turret

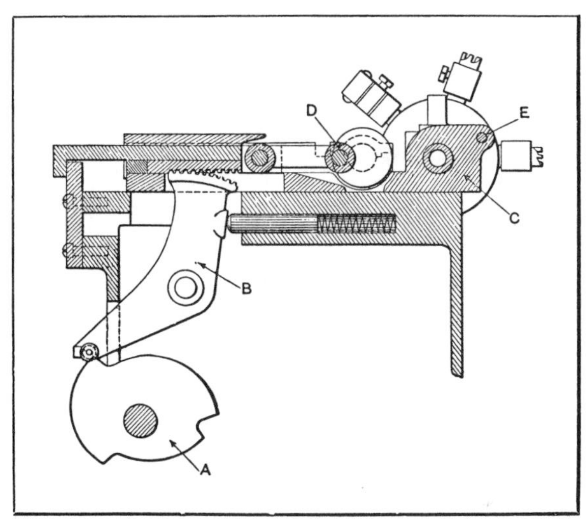

Fig. 6. Turret Slide is Withdrawn Quickly when Crank D Rotates

back quickly, a distance equal to the throw of the crank.

In the operation of a machine for high-speed work, it becomes important, both in securing the desired speed and in avoiding objectionable shocks, to move and reverse the lightest parts. For this reason, machines having turrets of the "revolver" or "barrel" type, in which each spindle can be fed independently, are especially adapted to high-speed work. In such machines, each tool carrier is connected successively with a reciprocating feed slide, and only the feed slide with one of the tool carriers connected with it requires to be reciprocated for the feed and return movements. In order to "speed up" this type of machine still further, the use of an auxiliary slide has been resorted to. This auxiliary slide alone is moved during that part of the quick-return movement required to retract each tool, and even this slide is disconnected for the remainder of the return movement, thus avoiding the shock which would result from rapid movement of the slide.

Fig. 7 shows an application of such an auxiliary slide with its disconnecting means. The turret *A* carries a series of tool spindles which are successively indexed to come into operative positions and be engaged by the block *B*. A main

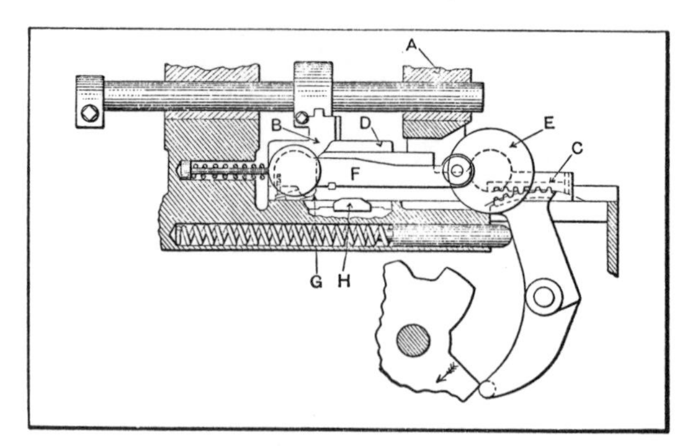

Fig. 7. Another Application of a Crank Motion for Obtaining a Quick Return

slide *C*, on which is an auxiliary slide *D*, is mounted on the bed of the machine. A crank *E* which is also on the main slide is connected to the auxiliary slide by the two-part connecting-rod *F*, one end of which is connected with the main slide and the other with the auxiliary slide. A latch at *G* holds these two parts together except during the quick-return motion which is obtained by revolving the crank disk *E*; then the latch is disengaged by passing over the cam *H*, which thus breaks the connection with the auxiliary or supplemental slide for the remainder of the crank throw and gives the quick-return movement and the quick-advance movement up to the point of cutting.

Rapid Return Movement Obtained by Roller Clutch and Crank Arrangement.—In designing machinery, it is frequently possible to make use of a roller friction clutch for reducing the time consumed during the idle part of the production cycle. This application is exemplified by the simple crank motion of the Scotch yoke type shown in Fig. 8. It is employed for actuating a slow-moving slide in a machine for forming plastic materials.

The slide indicated at *A* is reciprocated vertically. The crank is composed of the core *B*, integral with drive shaft *C*; the member *D*, which is bored to provide a running fit for the core; and the rollers *E*. The roller *F* on the stud that is secured in the projection on member *D* serves as the crankpin and engages a slot extending across the slide. As the crank rotates in the direction of the arrow, the slide is given its upward or working stroke, the movement being comparatively slow. During this stroke, the weight of the slide causes the rollers *E* to grip both the core and the member *D* tightly, so that both members rotate positively together. When the roll *F* has passed the center line *G*, the weight of the slide, which has caused the rolls to wedge tightly on the upward stroke, releases the gripping pressure of the rolls between the core and member *D* and allows the latter to rotate one-half revolution, returning the slide to its lowest position at a relatively higher velocity.

The downward stroke is the idle one, and its velocity in this particular case is unimportant in so far as the timing of the slide movements is concerned. This condition made it possible to use this crank. At the bottom of the stroke, the rolls *E* once more pick up the motion and move the slide upward at the slow speed required for the operation. It is estimated that with this design, an approximate gain of 30 per cent in production time is obtained over the time that would be required if a crank of the solid type were used.

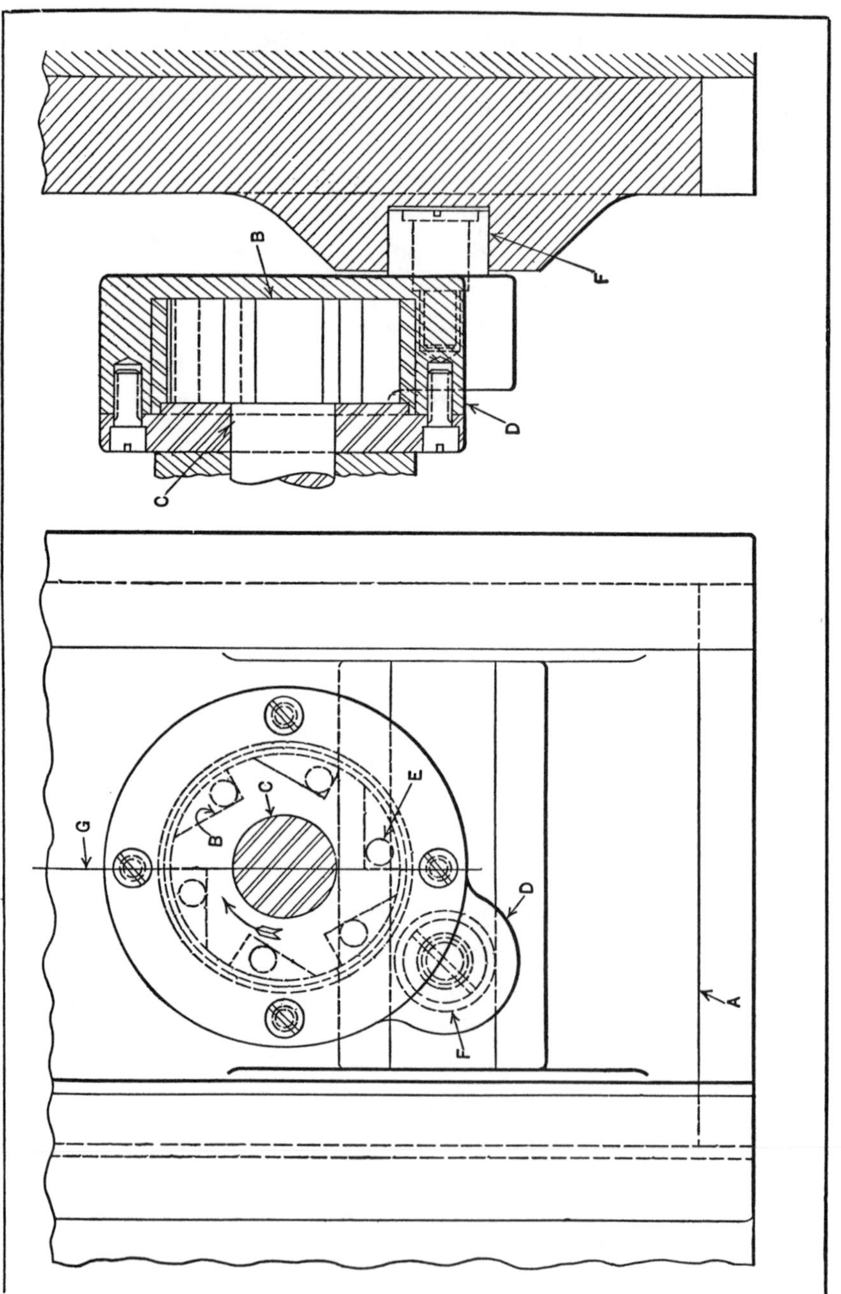

Fig. 8. Crank-operated Slide with Quick-return Movement

Quick-Return Crank Motion with Adjustment for Varying Velocity of Stroke.—Slotting machines, as a rule, are provided with some means of varying the cutting speed to suit the different materials to be machined. However, in reducing the cutting speed, the production is also reduced a corresponding amount, because the velocity of the entire cycle of the machine is slowed up. This objection was overcome in the case of one slotting machine by using a crank

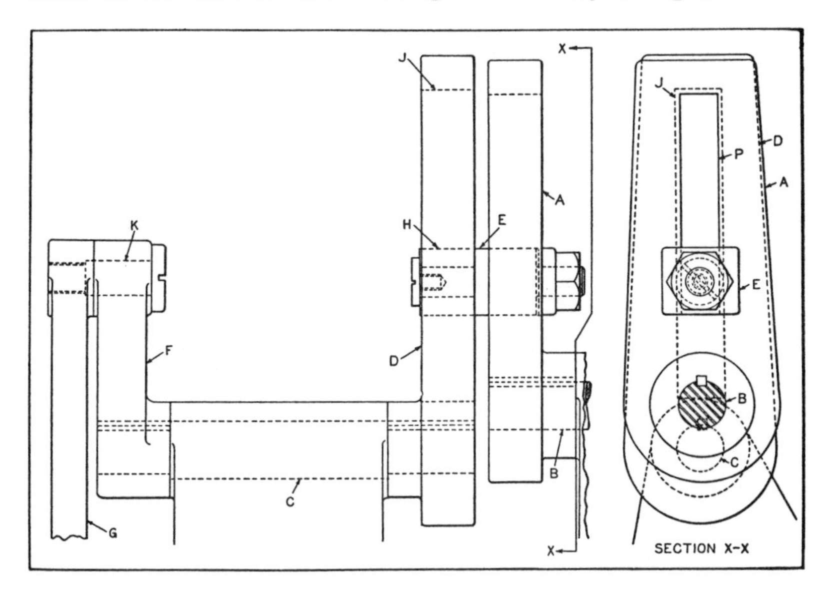

Fig. 9. Quick-return Crank Motion Mechanism with Means for Varying Velocity of Stroke

motion for actuating the slotting ram, the principle of this motion being shown in Figs. 9 and 10. With this arrangement, the velocity of the working stroke can be varied within certain limits without changing the time taken for the ram to pass through its cycle. Therefore, varying the cutting speed of the tool in this way does not change the rate of production, because, as explained later, the loss in velocity during the working stroke is compensated for by increasing the velocity of the return stroke.

The crank mechanism consists chiefly of the arm *A*,

Fig. 9, keyed to driving shaft *B;* the jack-shaft *C,* keyed to the arm *D* in which slides a cross-head *E* fastened to and adjustable along arm *A;* and crank *F,* also keyed to jack-shaft *C* and connected to the slotting ram by the connecting-rod *G.* It will be noted that jack-shaft *C* is offset from the driving shaft *B.* Consequently, as arm *A* rotates arm

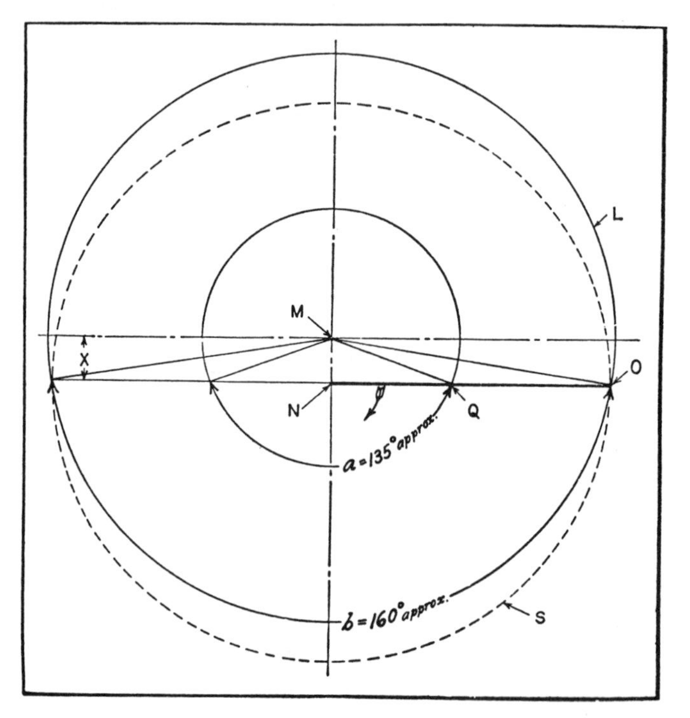

Fig. 10. Diagram Indicating Operating Principle of Mechanism Shown in Fig. 9

D and crank *F,* the pivot block *H* on cross-head *E* slides back and forth in the slot *J.* Arm *D* and crank *F,* therefore, will be given an irregular rotary movement; that is, the crankpin *K* will travel faster in its circular path *S,* Fig. 10, when below the horizontal center line of shaft *C* than when above this center line. This action will be more clearly understood by referring to Fig. 10. Here let circle *L* represent the path of cross-head *E,* point *M* indicating the center of driving shaft *B*. Let *N* indicate the center of shaft *C*, and let the heavy line represent the arm *D* with the cross-head at *O*.

Now, if the arm *D* is horizontal, as indicated by the heavy line, the cross-head, with arm *A* and shaft *B*, will rotate in the direction of the arrow only 160 degrees, in order to rotate arm *D* one-half revolution. Thus, during this movement, which corresponds with the return stroke of the ram, arm *D* and crank *F* rotate faster than arm *A*. In completing their revolution, however, the cross-head and arm *A* rotate 200 degrees to turn arm *D* and the crank the remaining half revolution. Hence, during the latter movement, which corresponds with the working stroke of the ram, crank *F* rotates more slowly than arm *A*. Thus, a slow working stroke and a rapid return stroke are obtained.

If it is required to reduce the velocity of the working stroke, the cross-head is adjusted inward in slot *P*, Fig. 9, to a new position, say, to *Q*, Fig. 10. In this case a 135-degree movement of arm *A* is required to rotate crank *F* through its return stroke, and a 225-degree movement to rotate the crank through its working stroke. Thus, the velocity of crank *F* is increased during the return stroke and reduced during its working stroke. Crank *F* and arm *A*, however, complete their cycle in the same time, so that the reduction in velocity of the working stroke does not affect the production rate of the machine. Incidentally, a greater range in the velocity variation of the crank can be obtained by increasing the offset *X* between shafts *B* and *C*. This change will, of course, affect the length of the slots in the arms.

Adjusting Operating Position of Reciprocating Slide without Stopping Machine.—Occasionally it is necessary to provide means for adjusting the operating position or point of reversal of a slide having a fixed length of stroke without stopping the motion of the slide. A parallel to this requirement is found in a vertical shaping machine, in

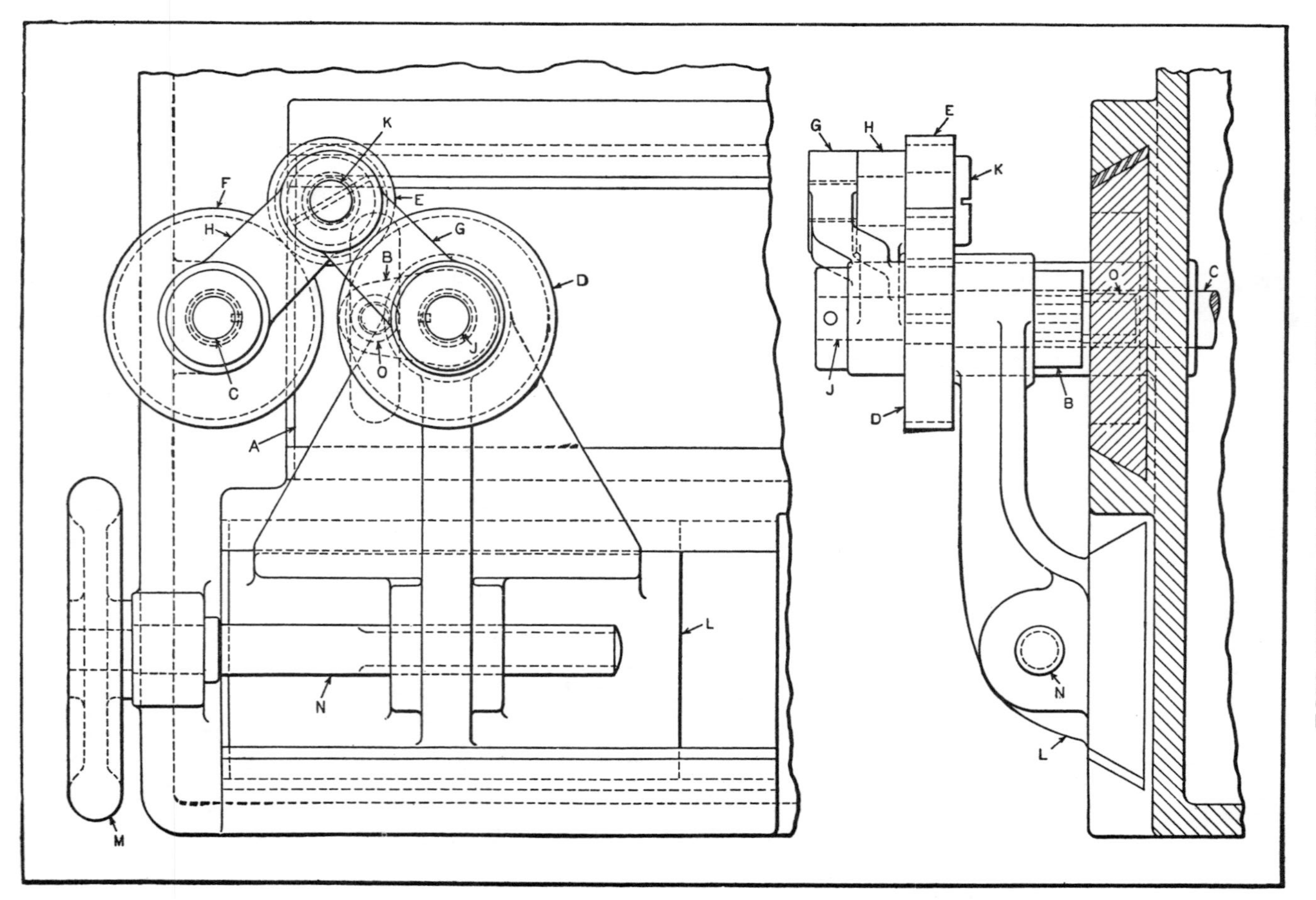

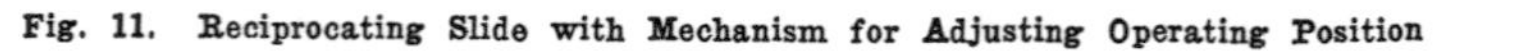
Fig. 11. Reciprocating Slide with Mechanism for Adjusting Operating Position

which the reversal point of the tool-slide is varied manually. A mechanism for obtaining this variation is shown in Fig. 11, the tool-slide being indicated at *A*. This slide is driven by the crank *B* keyed to shaft *J*. Shaft *J* is driven by shaft *C* through gears *D, E,* and *F*. Roll *O,* mounted on a stud in crank *B,* engages a groove in the tool-slide and operates on the principle of the Scotch yoke.

The center distances between gears *D* and *E* and between gears *E* and *F* are maintained by the links *G* and *H,* respectively. These links are a free fit on the gear-shafts *J* and *C*. Gear *E* and link *H* are also a free fit on screw *K*. Shaft *J* turns freely in a bracket cast integral with the adjusting slide *L,* and this slide is actuated by the handwheel *M* on the feed-screw *N*. Screw *N* engages a nut cast on slide *L*. Slide *A* is shown in its extreme left-hand position. Assume that both the left-hand point of reversal and the right-hand point of reversal are required to occur farther toward the right. To effect this change, the operator merely turns handwheel *M* the required amount or until slide *L* has carried shaft *J* a corresponding distance toward the right. In doing this, the links tend to straighten out, yet the gears remain in mesh and continue the rotation of the crank. The range of variation for changing the point of reversal is controlled by the diameters of the gears. If a larger idler gear *E* is used, the slide will have a greater range of adjustment.

Adjusting Crank Throw of Wire-Forming Machine while Machine is Running.—In the operation of a wire-forming machine, difficulty was experienced in holding the parts to a uniform shape, due to variations in the hardness of the low grade of wire used. These variations in hardness necessitated frequent adjustment of the forming dies to prevent excessive variations in the depths of the formed portions. As stopping of the machine for this purpose seriously affected production, it was decided to provide means for making the necessary adjustments while the ma-

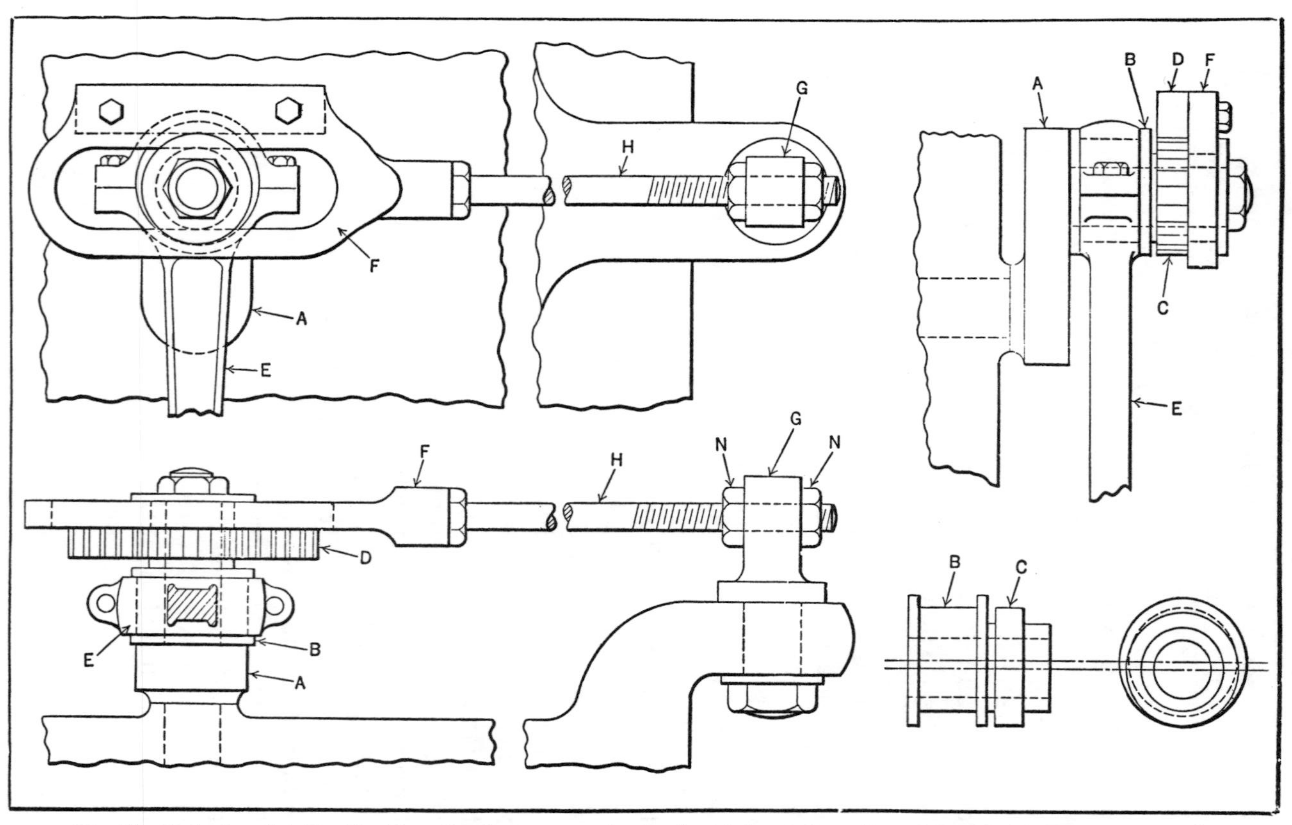

Fig. 12. Mechanism that Provides Means for Shortening or Lengthening Stroke of Connecting-rod E while Machine is in Operation

chine was in operation. This was satisfactorily accomplished by applying the stroke-changing mechanism shown in Fig. 12. With this mechanism, the length of stroke of connecting-rod *E*, which operates one of the dies, can be changed while the machine is in operation, simply by adjusting the nuts *N* on rod *H*.

The length of stroke of rod *E* is varied by means of the eccentric bushing on the pin of crank *A*. Crank *A* is of the conventional open-end type, except that the crankpin is longer than would ordinarily be required. The bushing *B* on the crankpin is turned eccentric with the bore to fit the bearing in the connecting-rod *E*. The hub of bushing *B*, which is machined concentric with the bore, carries the gear *C*, as shown in the view in the upper right-hand corner of the illustration. The yoke *F* is carried on the hub of bushing *B* outside of gear *C*. This yoke carries rack *D* which meshes with gear *C*. Rod *H* is fastened to yoke *F* and is threaded on its outer end where it passes through stud *G*. Stud *G* is located in a fixed position, but is free to turn or swing. All three assembly views show the crank *A* in its upper position. The bushing *B* is shown adjusted for the maximum length of stroke.

As the crank *A* rotates in either direction, the crankpin carrying bushing *B* moves in the slot in yoke *F*. This produces a rotating movement of bushing *B* on the crankpin as a result of the action of rack *D* and gear *C*. The number of teeth in gear *C* is such that a half turn of crank *A* produces a half turn of gear *C* and bushing *B*. Thus the throw of eccentric bushing *B* is reversed in relation to the crankshaft as the crank *A* reverses its position. This causes the throw of eccentric *B* to be added to the throw of crank *A*, thus increasing the stroke of connecting-rod *E*. This condition exists only in the opposite positions of the crank *A*, as the bushing *B* is constantly changing its position throughout the cycle. As the nuts *N* on rod *H* are changed, the rela-

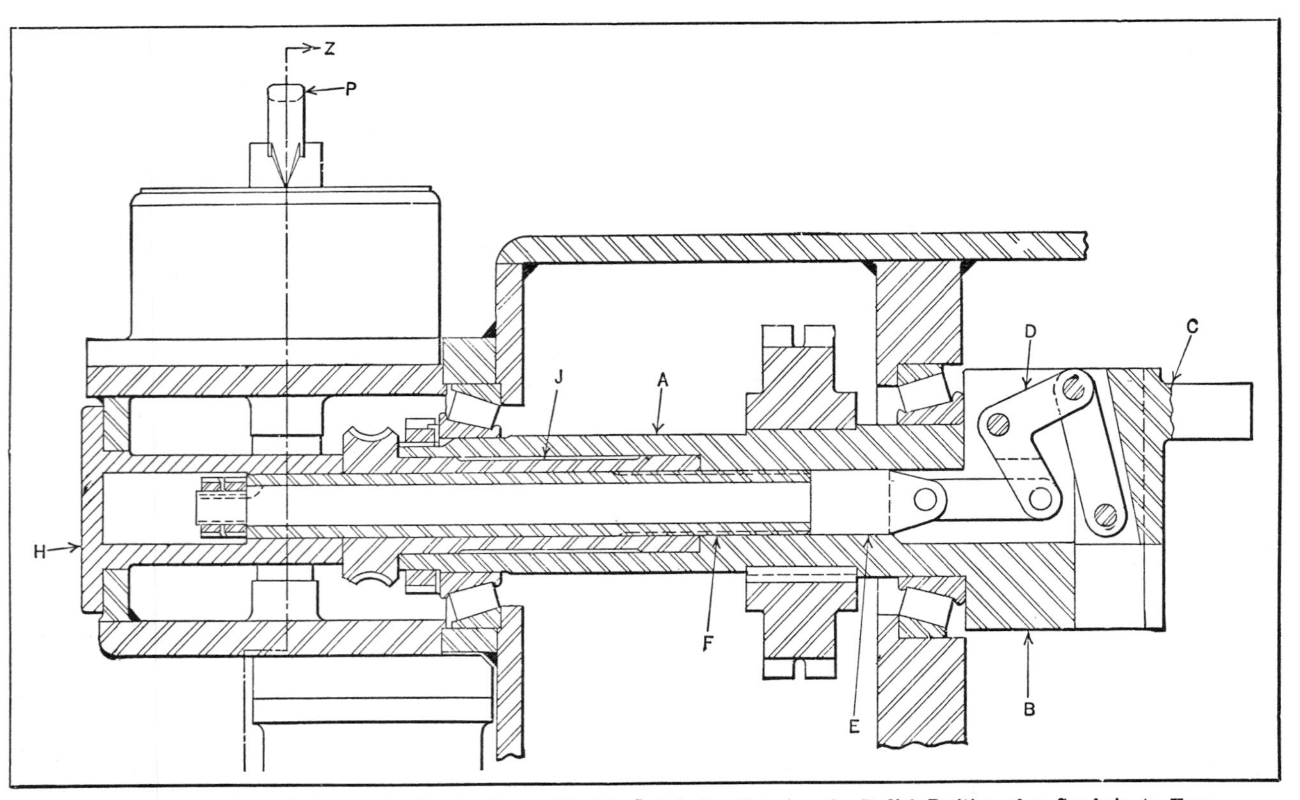

Fig. 13. Motor-driven Mechanism Combined with an Electric Switch for Changing the Radial Position of a Crankpin to Vary the Crank Throw

tive position of the bushing *B* is changed, causing a change in the throw of crank *A*.

Electric Control that Varies Throw of Crank While Machine is Running.—The mechanism illustrated in Figs. 13 and 14 provides a rapid adjustment of the throw of a crank while the machine is in operation. The crankshaft indicated at *A* serves to impart a reciprocating motion to another member of the machine, and any throw of the crank between zero and the maximum is instantly available.

The crankshaft is mounted on tapered roller bearings. Crankhead *B* is integral with the shaft and carries a sliding block of which the crankpin *C* is an integral part. This block is connected by means of link *D* with the draw-bar *E*, which is free to slide axially in shaft *A*. Sleeve *F* is threaded at its right-hand end and has rack teeth on it that mesh with gear *G* (Fig. 14).

Sleeve *F* is keyed to stationary cap *H* to prevent it from turning, but is free to slide axially in this cap. The worm-wheel nut *J* is a running fit in shaft *A*, and is threaded to fit sleeve *F*. A reversing motor rotates the worm-wheel nut through worm *K*, and thus moves the sleeve *F*, with draw-bar *E*, axially, so that, by means of link *D*, the radial position of the crankpin is changed.

Mechanical and Electrical Mechanism for Regulating the Crank Throw.—The apparatus for controlling the radial movement of the crankpin is shown in Fig. 14. It is contained in a separate housing, and consists of a special electric switch designed to control the reversing motor. This switch has a disk *L*, which is connected by a bushing to gear *G* and is provided with two semicircular contact segments *M*. The segments are insulated from disk *L*. Member *N* is connected to a handle and its pointer *P*.

The links *Q*, hinged to member *N*, are each equipped with a contact blade *R*. The links, with their blades, are held against the contact segments by a coil spring, as indicated. This spring also serves to hold the pointer against the

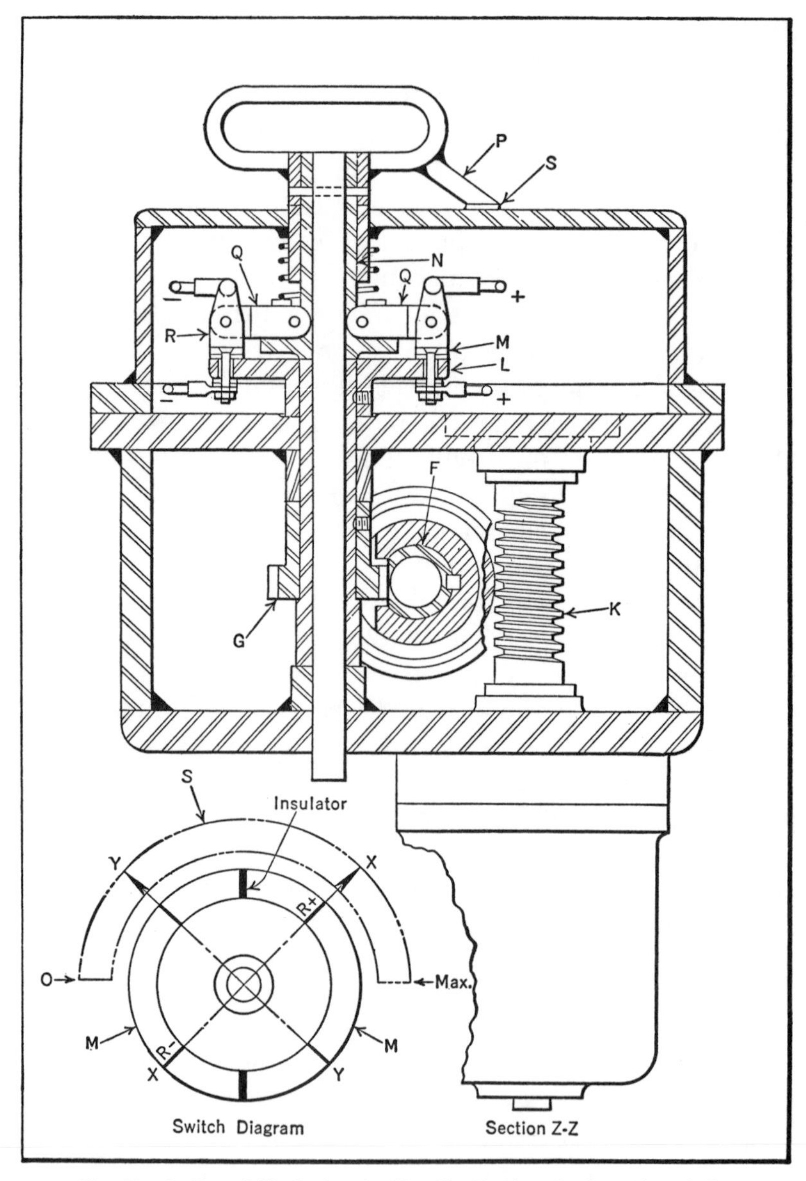

Fig. 14. Section of Mechanism in Fig. 13, Showing the Operation of the Electric Control Switch

graduated dial S and prevents its shifting after an adjustment of the crankpin has been made. To further prevent shifting, the finger is formed like a knife-edge and rests in radial grooves in the dial, which also serve as the graduations.

The operation of the switch will be understood from the switch diagram. Point $R+$ represents the positive contact blade and $R-$ the negative blade. The contacting segments M are connected with the motor and are separated from each other by insulators. When the $R+$ and $R-$ blades are on the insulators, the motor is idle, as the circuit is open, and the throw of the crank is indicated by the position of the finger on the dial S. If the finger is moved toward the right, so that the blades $R+$ and $R-$ coincide with line X-X, the blade $R+$ is in contact with the right-hand segment and $R-$ with the left-hand segment, and the electric circuit is closed. Consequently, the motor will start and shift the crankpin, as already explained. In the meantime, through the axial movement of sleeve F and the resulting rotation of gear G, the disk L turns clockwise until the contact blades engage the insulators. At this point the circuit is broken and the motor stops, leaving the crankpin in a radial position corresponding to the position of the pointer on the dial S.

If the pointer is moved toward the left, say on line Y-Y, the motor will run in the opposite direction and move the crankpin back toward its former position. In this way, to either shorten or lengthen the throw of the crank, the operator merely swings the pointer handle so that the pointer engages the graduation on the dial corresponding to the required throw. This adjustment, besides being rapid, is made with a minimum amount of effort, as the motor does the actual work of shifting the crankpin. In designing the switch, great care should be taken to thoroughly insulate the electrical contacts.

CHAPTER IX

RECIPROCATING MOTIONS DERIVED FROM CAMS, GEARS, LEVERS AND SPECIAL MECHANISMS

In designing the driving mechanisms for some parts having a reciprocating motion, cams, gears or levers are substituted for a transmission of the rotating crank type. Examples of these different forms of reciprocating drives will be described. As with the crank type of drive, the object of using cams or combinations of levers may be to vary the stroke in some way or the object may be to obtain a mechanical movement essential to meet a particular operating requirement.

Double Lever Mechanism to Provide Strokes of Unequal Length Synchronized During Part of Stroke.— The mechanism shown in Fig. 1 fulfills an unusual requirement in a simple manner. Two slides of a wire-forming machine were required to operate with different lengths of travel, the slide having the longer travel being arranged to operate in synchronism with the other during a portion of its stroke. Adjustability, both as to the length of travel and the period of synchronization, was also required on the slide with the longer travel.

Referring to the illustration, bearing *H* carries the shaft *A*, which is given an oscillating motion by a cam-operated lever (not shown). The motion of shaft *A* is transmitted to lever *B*, which is keyed to it. Rod *E* transmits the motion of lever *B* to one slide, and rod *D* transmits the motion of lever *C* to the other slide. Lever *C* oscillates on stud *K*, carried on lever *B*, and has gear teeth cut on the end.

The gear teeth on lever *C* mesh with teeth cut in disk *L*, which is carried free on the hub of lever *B*. Disk *L* carries

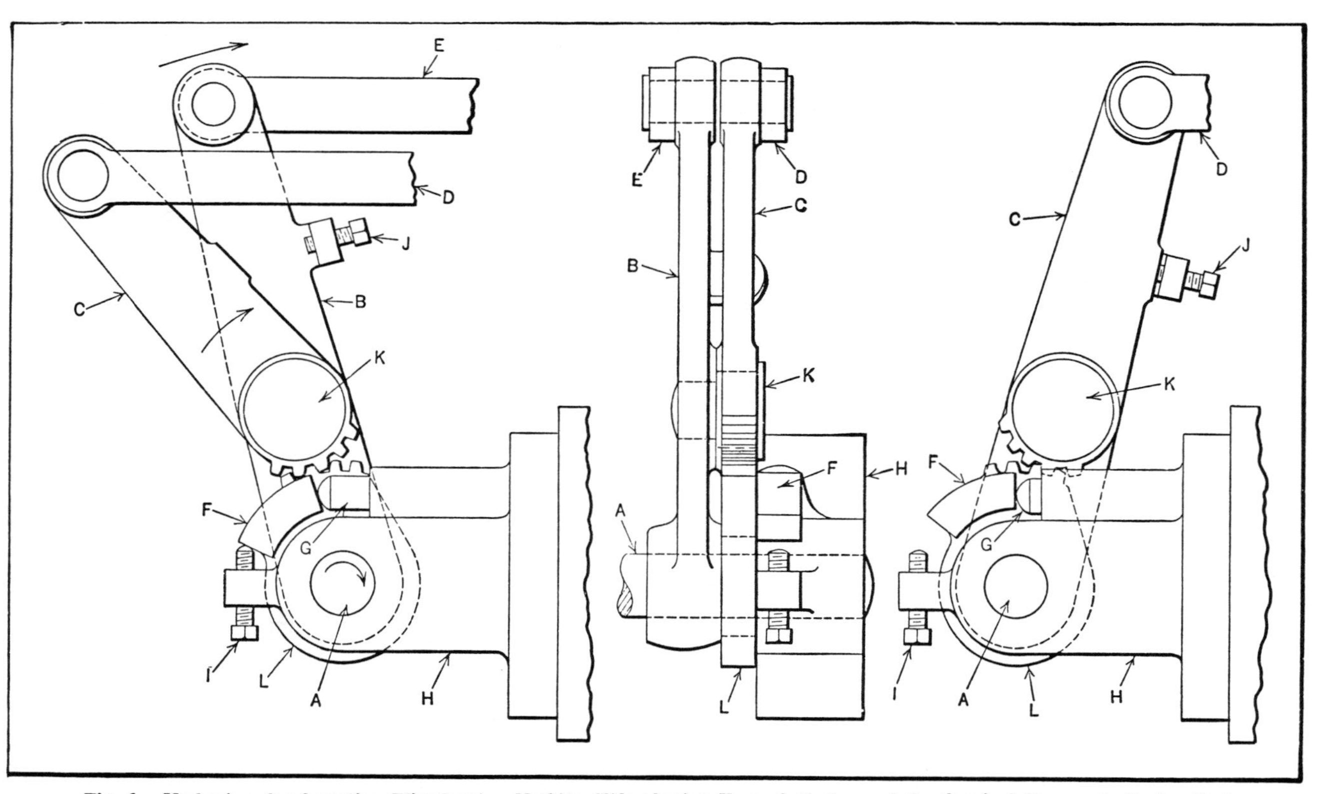

Fig. 1. Mechanism for Operating Wire-forming Machine Slides having Unequal Strokes and Synchronized Movements During Part of Stroke only

a pad *F*, which is in constant contact with plunger *G* in bearing *H*. Plunger *G* is backed up by a stiff spring. The action of plunger *G* against pad *F* tends to hold the pad down against the screw *I*.

As the shaft *A* rotates in the direction indicated by the arrow in the view to the left, lever *B* is carried in the same direction, but disk *L* is restrained from movement by the pressure of plunger *G* against pad *F*. This causes the gears to operate, so that the lever *C* is swung on stud *K* in the direction of the arrow. Rod *D* is thus given the combined movement of lever *B* and lever *C*, which continues until lever *C* makes contact with the stop-screw *J* on lever *B*. At this point, lever *C* is restrained from further rotation on stud *K*, and continued movement of lever *B* causes disk *L* to be carried around with it, due to the locking action that takes place between the gears and screw *J*.

As soon as disk *L* turns with lever *B*, levers *B* and *C* revolve around a common axis—the center of shaft *A*—and they move in synchronism from that point on. The view to the right shows the levers *B* and *C* in their extreme forward position, while an end view of the mechanism in the same position is shown by the central illustration. On the return stroke, synchronism is maintained until pad *F* again makes contact with screw *I*, at which time the movement of lever *C* is increased by the action of the gears. Screw *J* controls the period of synchronization, while screw *I* controls the travel of lever *C*.

Slide which Always Dwells During Initial Movement of Parallel Slide.—The mechanism shown in Fig. 2 provides a dwell or delay in the movement of slide *B* while slide *A* enters upon the first portion of its cycle. On the return stroke, slide *B* dwells in the same manner while slide *A* begins its movement back to its original position.

This requires a delay arrangement that will operate at each end of the cycle, so that the first slide will remain stationary in each position for a given length of time. The dwells could, of course, be obtained by means of cams. However, the mechanism shown is simple and more compact than a cam arrangement. In this mechanism, disks *N* and *D*, with their contacting pins, are arranged similarly to the tumblers employed on a combination lock. Referring to the illustration, the slide *A* moves a given distance at the start before slide *B* moves in the same direction. At

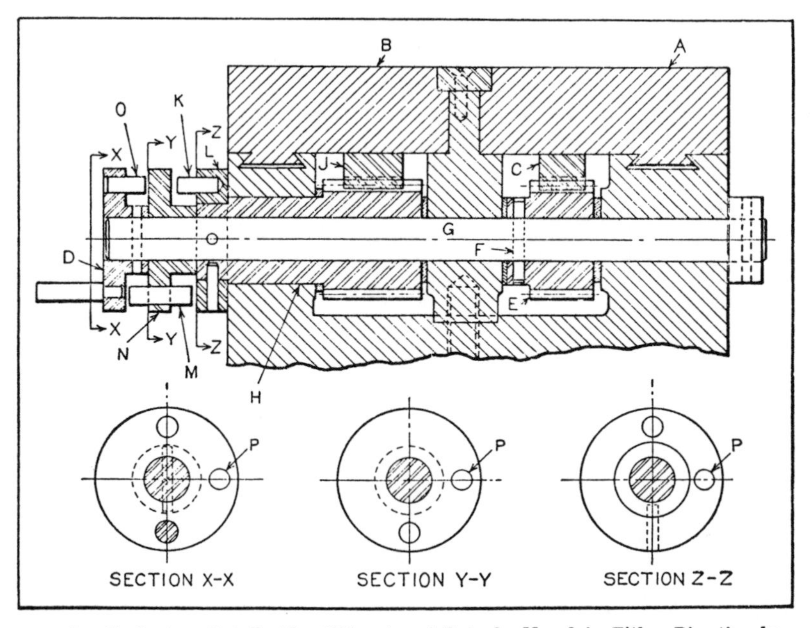

Fig. 2. Mechanism that Enables Slides A and B to be Moved in Either Direction by Turning Handwheel D, Slide B Always Dwelling for a Certain Period During the Initial Movement of Slide A

the end of the stroke, slide *A* moves in the reverse direction the same distance as at the start before slide *B* begins its return movement.

These motions are obtained in the following manner: Slide *A*, through rack *C*, is connected directly to the disk or handwheel *D* by gear *E*, pin *F*, and shaft *G*. Shaft *G*, however, is allowed to rotate freely in the combination gear and bushing *H*. Gear-bushing *H* is connected to handwheel *D* through pin *K* in collar *L* which comes into contact with pin *M* in the free-running collar *N*. The opposite end of pin *M* comes into contact with pin *O* in handwheel *D*. Thus pin *O* makes one revolution minus an amount equal to the thickness of the pin before it comes in contact with pin *M*. The opposite end of pin *M* can then make one revolution less the thickness of the pin before coming in contact with pin *K* which moves gear-bushing *H*, causing slide *B* to move. On the return stroke, the reverse action takes place.

This means that two revolutions of disk *D*, less the thickness of two pins, can be obtained before slide *B* follows the movement of slide *A*. Less than this amount of movement

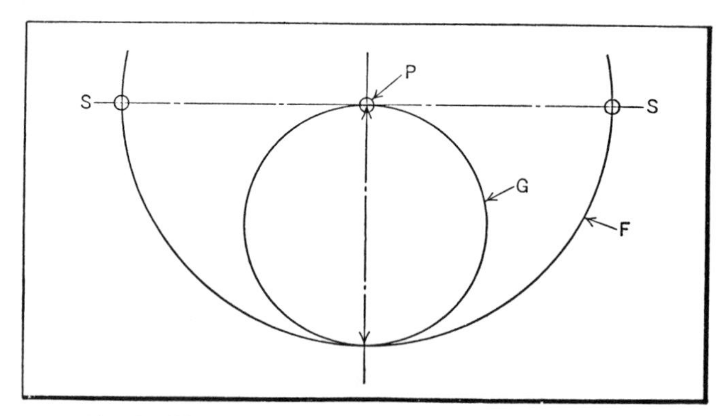

Fig. 3. Diagram Illustrating Application of Hypocycloid to Reciprocating Mechanism

can be obtained by placing two pins in each of the three disks at such angular positions as to give the required movements. Thus, in the case shown, the total movement of *A* in advance of *B* is 1 1/4 revolutions minus the thickness of three pins if pins *P* are inserted. This movement lends itself very readily to operations that require the withdrawal of a certain tool from the work before the entire carriage is withdrawn.

Mechanism for Converting Rotary into Reciprocating Motion by Application of Hypocycloid Principle.—The principle of the hypocycloid, as illustrated in Fig. 3, has been applied very effectively in the mechanism shown in Figs. 4 to 6. This mechanism is designed to convert rotary motion into reciprocating motion. The hypocycloid SS, Fig. 3, is generated by the point P in the generating circle G as it rolls on the inside of the circle F. The hypocycloid thus generated by point P is a straight line when the

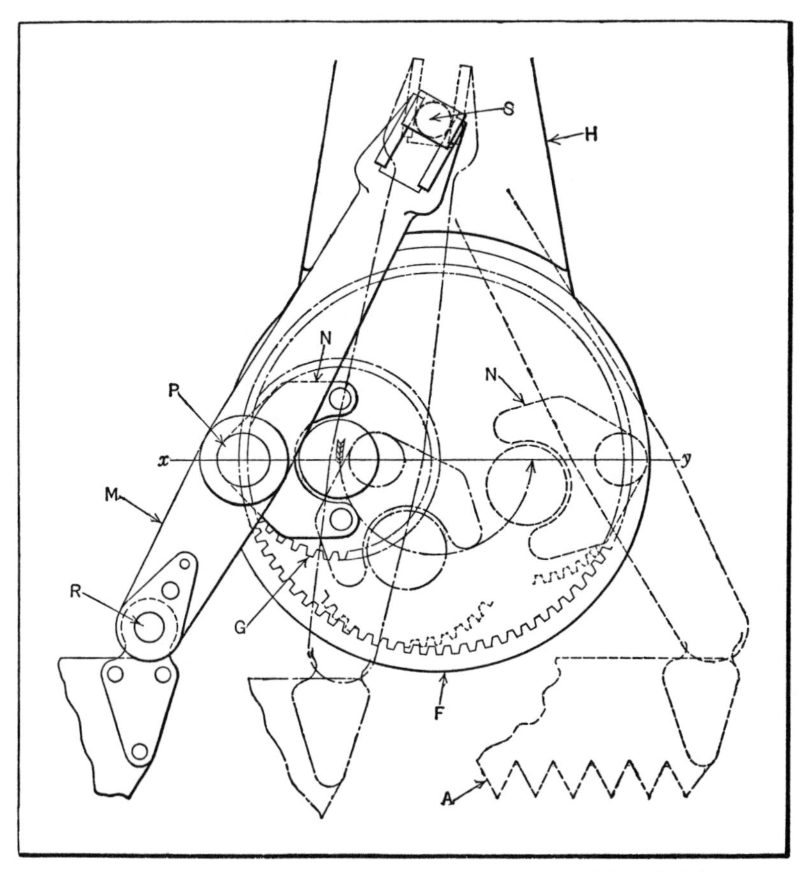

Fig. 4. Saw-reciprocating Mechanism Based on Hypocycloid Principle

diameter of the generating circle G equals the radius of the circle F. A mechanism designed on this principle will give a long stroke with a minimum number of small strong parts arranged in the most compact form.

In the principal application to be described, the circle F

becomes the pitch circle of a fixed internal gear, and the generating circle G becomes the pitch circle of a pinion that rolls around on the inside of the internal gear. The point P, which is at all times located at the exact intersection of

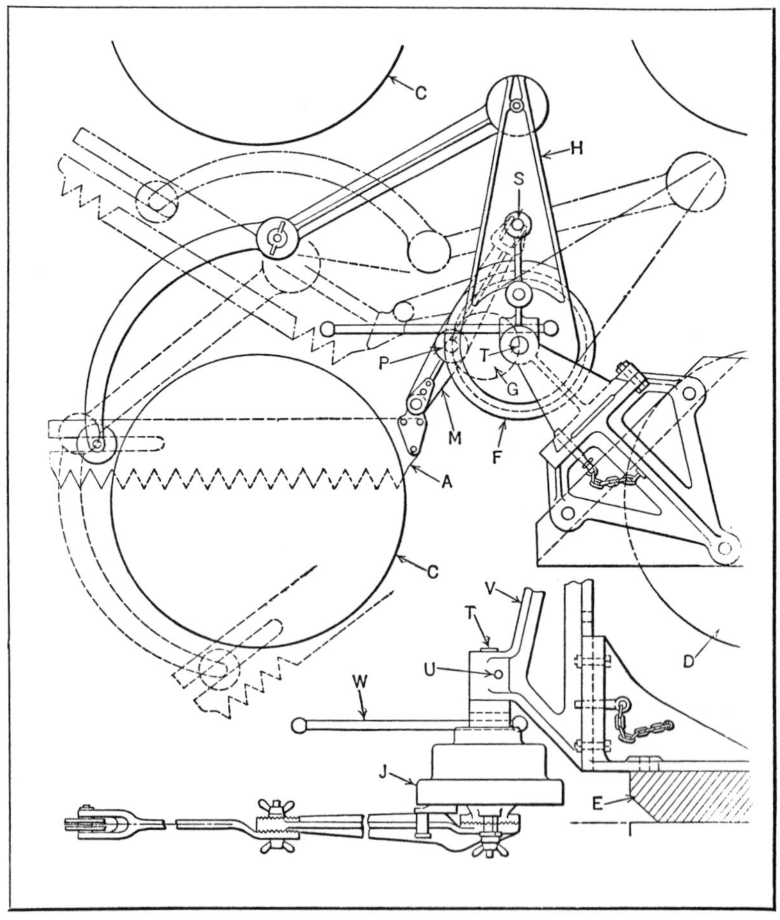

Fig. 5. Assembly of Saw Operated by Mechanism Shown in Fig. 4. Dotted Lines Show Starting and Finishing Positions of Saw in Cutting off Closely Spaced Piles

the pitch line G with the center line SS of the internal gear, becomes the stroke pin. As the pinion rolls, this pin moves back and forth on a straight line from S to S, which is the equivalent of the pitch diameter of the internal gear.

Hypocycloid Principle Applied to Saw-Reciprocating Mechanism.—The arrangement shown in Figs. 4 to 6 forms the operating mechanism of a saw for sawing off piling. The saw *A*, Fig. 4, has a stroke of 12 inches, and the internal gear *F* has a pitch diameter of 8 inches. The complete machine is made of Duralumin, and, without the motor, weighs 43 pounds.

The specific problem was to design and build a one-man portable machine for sawing off piles *C* (Fig. 5) at low tide. These piles were 18 inches in diameter and were spaced 30 inches apart, center to center. Plan and elevation views show the assembled machine attached to a 2- by 12- by 48-inch timber *E*, supported on pile *D*, which was hand-sawed. The saw and the guide arms are shown in three positions by dotted lines to indicate how the reciprocating members clear the adjacent piles. Fig. 4 shows the reciprocating mechanism to a somewhat larger scale. Sectional elevation and inverted plan views of the power-driven parts are shown in Fig. 6.

The direction of the stroke is determined and fixed when assembling stroke pin *P* and the pinion *G* in the internal fixed gear *F*. In Figs. 4 and 6, the line *x-y* is at right angles to the extension arm *H* of the frame *J*. The internal gear *F* is made as a separate piece only for convenience in cutting the teeth and to provide a bottom bearing for the driving pinion *K*, Fig. 6. The outside end *L* of pinion *K* was squared and connected to an air motor which runs at a speed of 800 revolutions per minute. This gives the saw about 72 strokes per minute.

The stroke pin *P* and its bracket *N*, riveted to the rolling pinion *G*, are shown in three positions in Fig. 4 to illustrate how the pin *P* follows line *x-y*, carrying with it the forked link *M* which has pin *R* pivoted on its short end. The long end of link *M* is forked around the squared fulcrum pin *S*, Fig. 6, which swivels in the hub *Q* of the extension arm *H*. The shoulder-stud *T*, Fig. 5, is supported in frame *V* and fixed in position by pin U. A vise handle W in frame J permits the operator to swing the whole assembly on stud T to adjust and feed the saw.

Referring to Fig. 6, the motor-driven gear X is a running

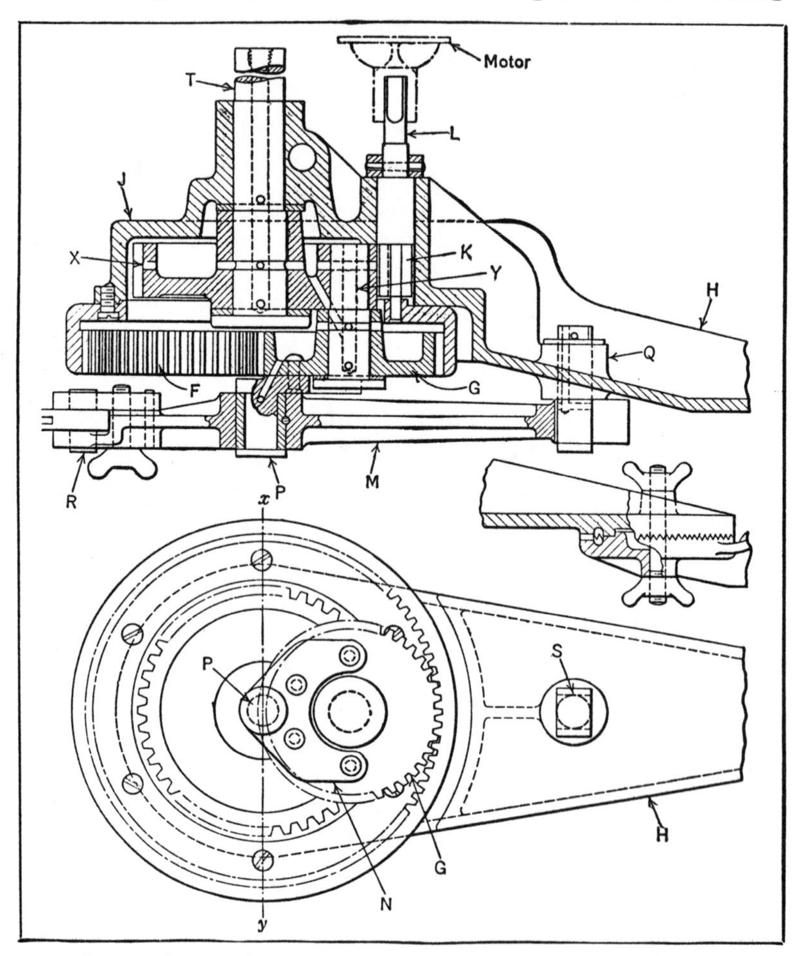

Fig. 6. Cross-section and Plan Views of Saw-reciprocating Mechanism

fit on stud T, the head of which serves as a support for the gear. The crankpin Y, fixed in gear X, has pinion G mounted on its head. Suitable thrust washers are provided for both gears X and G. The stroke-pin bracket N is riveted to pinion *G*, with the center line of the pin *P* located on the pinion pitch line. The pin *P* turns freely in a bushing made in halves to facilitate assembling. The halves of this bearing are pinned securely to the forked link *M*. Holes and grooves for providing ample lubrication from one grease cup screwed into the top of stud *T* are shown.

Uniform Reciprocating Motion.—A uniform reciprocating motion often is required in machine design, and the

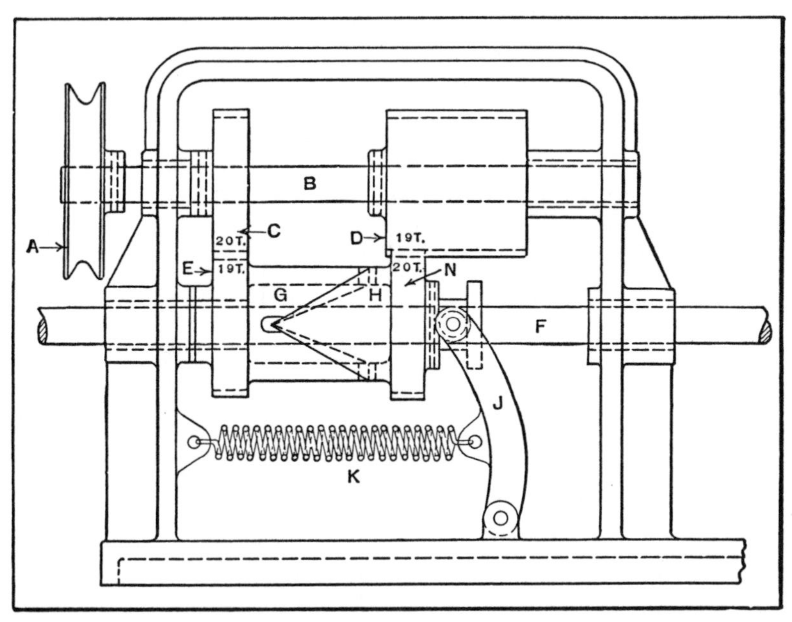

Fig. 7. Mechanism which Imparts an Even Reciprocating Motion to a Rotating Shaft

mechanism to be described produces such a movement. A belt drive to pulley *A*, Fig. 7, rotates shaft *B*, which drives gears *C* and *D*. Gear *C* meshes with and drives gear *E*. Cam *G* is integral with gear *E* and is opposed to the mating cam *H*, which is integral with gear *N*. Cam *H* and gear *N* are attached to shaft *F*, which rotates and also receives a reciprocating motion. Cam *G* and gear *E* are free to revolve around this shaft.

Gear *C* has twenty teeth, and gear *E* nineteen teeth,

whereas gear *D*, which drives gear *N*, has nineteen teeth, and gear *N* has twenty teeth; consequently, gear *E* and cam *G* are driven somewhat faster than the mating cam *H* and gear *N*, so that there is a differential motion between the two. The result is that cam *G* forces cam *H* and shaft *F* to the right at a constant speed until the point of the driven cam passes the point of the driving cam, when the return stroke begins. It will be noted that spring *K*, acting

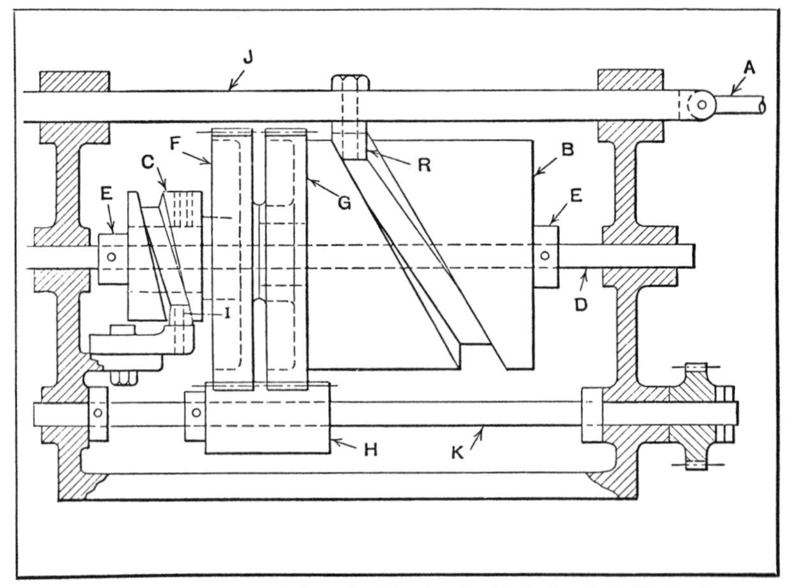

Fig. 8. Double Cam and Gear Combination for Producing Variable Stroke

through lever *J*, holds cam *H* in contact with cam *G* during the return movement. Gear *D* is made wide enough to permit gear *N* to continue in mesh during the entire stroke. This mechanism, with more or less modification to suit the purpose, could be applied to various classes of machinery.

Variable-Stroke Mechanism.—The purpose of the mechanism shown in Fig. 8 is to impart a variable-stroke motion to rod *A*. This is accomplished by two cams *B* and *C*, mounted on shaft *D*. These cams are a free running fit on shaft *D* and are held in place by two collars *E*. Cam *C* is keyed to the hub of gear *F*, which has 100 teeth. Gear *G*, which has 111 teeth of modified form and is of the same diameter as gear *F*, is keyed to the hub of cam *B*, which transmits motion to rod *A* through roll *R* and bar *J*.

The gears *F* and *G* are both driven by the wide-faced pinion *H*, secured to shaft *K*. The cam roll *I* is mounted on a bracket secured to the machine frame, and remains in a fixed position. As the gears *F* and *G* revolve, the former gains eleven teeth on the latter at each revolution, thus shortening and lengthening the throw of slide bar *J* and rod *A*, the stroke being lengthened when the relative angular positions of the cams are such that they impart motion in the same direction, and shortened when the motion imparted by the cams is opposed.

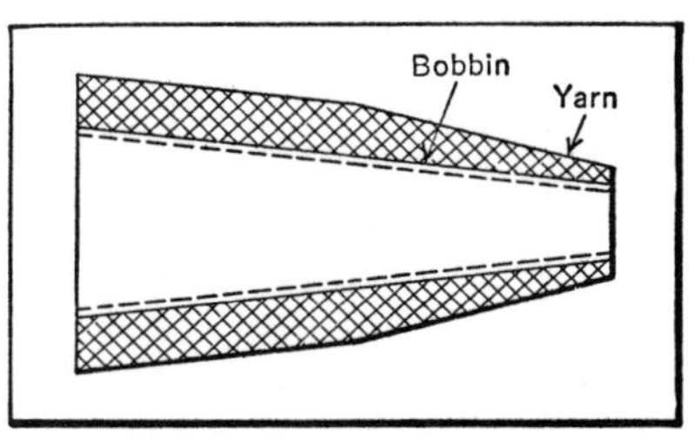

Fig. 9. In Winding this Bobbin, the Yarn is Guided by Means of the Mechanism Shown in Fig. 10

Varying a Reciprocating Movement at One Point of Reversal.—In a certain textile machine, the member that guides the yarn as it is wound on conical bobbins is given a reciprocating movement of uniform length until several layers of yarn have been wound. Then the length of this movement is gradually diminished so that when completely wound, the yarn on the small end of the bobbin forms a cone of greater taper than the bobbin itself, as shown in Fig. 9.

The mechanism for producing this movement is shown in Fig. 10, the member for guiding the yarn being indicated at *A*. This member slides on the stationary guide *C* and receives its motion from the reciprocating cross-head *G* through the bellcrank lever *M*, pivoted to the cross-head at *H*. The cross-head slides on stationary bars *E* and *F*, and is reciprocated by means of cam *K* on shaft *L*.

On the lower arm of lever *M* is a roll *m* which engages a

channel cut in the bar *O*, pivoted at *n*. Another roll *P* at the free end of bar *O* engages the groove in the cam *Q*. This cam controls the angular position of bar *O*, and is rotated at the required speed by the worm and worm-gear *S* and *r*. It will be noted that the path of cam *Q* is con-

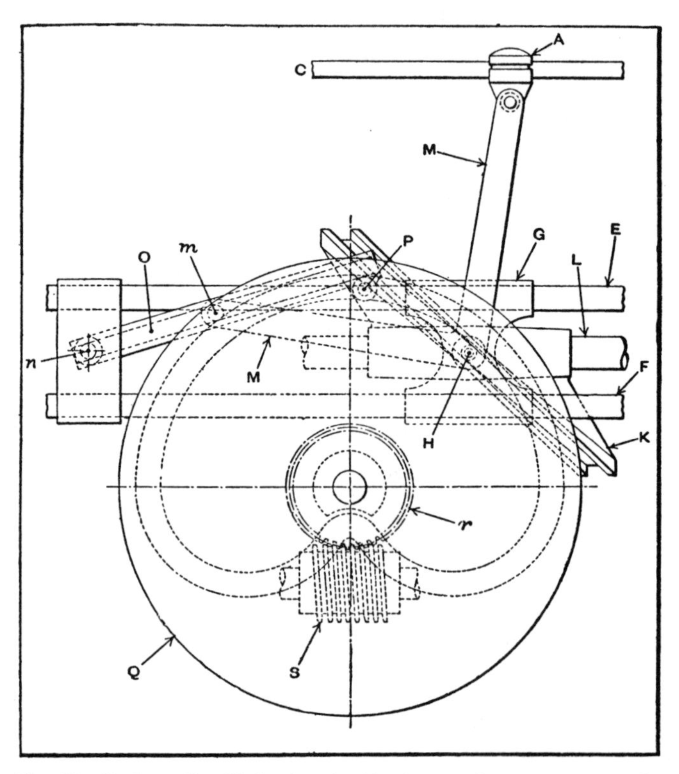

Fig. 10. Reciprocating Mechanism for Varying the Length of the Stroke of Member A which Guides the Yarn as it is Wound on the Bobbin Shown in Fig. 9

centric with its shaft for 180 degrees. Hence, while roll *P* is passing over this part of the cam, bar *O* will remain stationary and the length of the stroke of member *A* will remain constant. This is clearly shown in the diagram, Fig. 11, where the length of the stroke at this time is indicated at S_1. It will be seen that this stroke is equal to the movement of the cross-head *G*, Fig. 10, plus the movement of the upper end of lever *M* resulting from its engagement with bar *O*. It is during the cam dwell that the first layers of yarn are wound along the length of the bobbin and parallel to its conical surface.

At the end of the dwell, however, roll *P* moves toward the center of the cam, swinging the bar *O* downward and thus changing the angular position of the lever. As a result, the stroke of member *A* is gradually diminished until

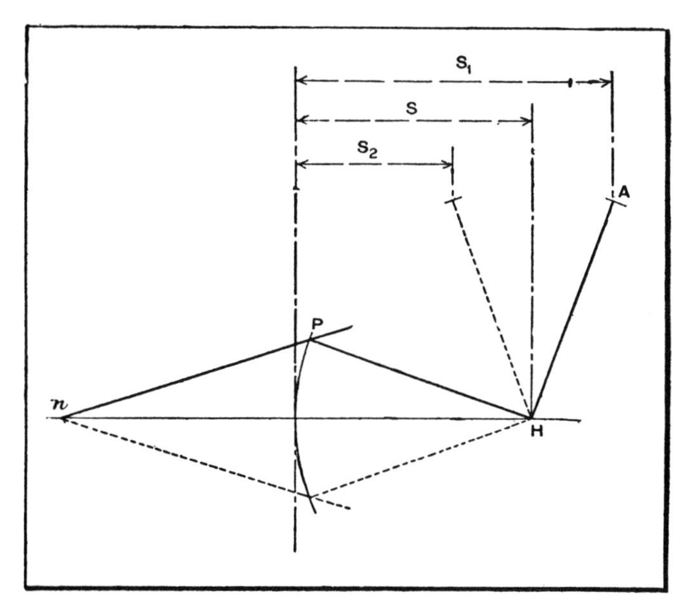

Fig. 11. Diagram Showing how the Oscillation of the Bellcrank Lever Shortens the Stroke in the Mechanism Illustrated in Fig. 10

the channel bar and the lower arm of the lever are in line. In this position (momentarily), the linear speed of member *A* and cross-head *G* are equal and their movement is indicated at *S* in the diagram, Fig. 11.

As roll *P*, Fig. 10, continues toward the center of the cam, the stroke of member *A* decreases still more until the channel bar and lever have assumed the position indicated by the dotted lines in Fig. 11. The stroke now is equal to S_2 and at this time, the bobbin is completely wound. Re-

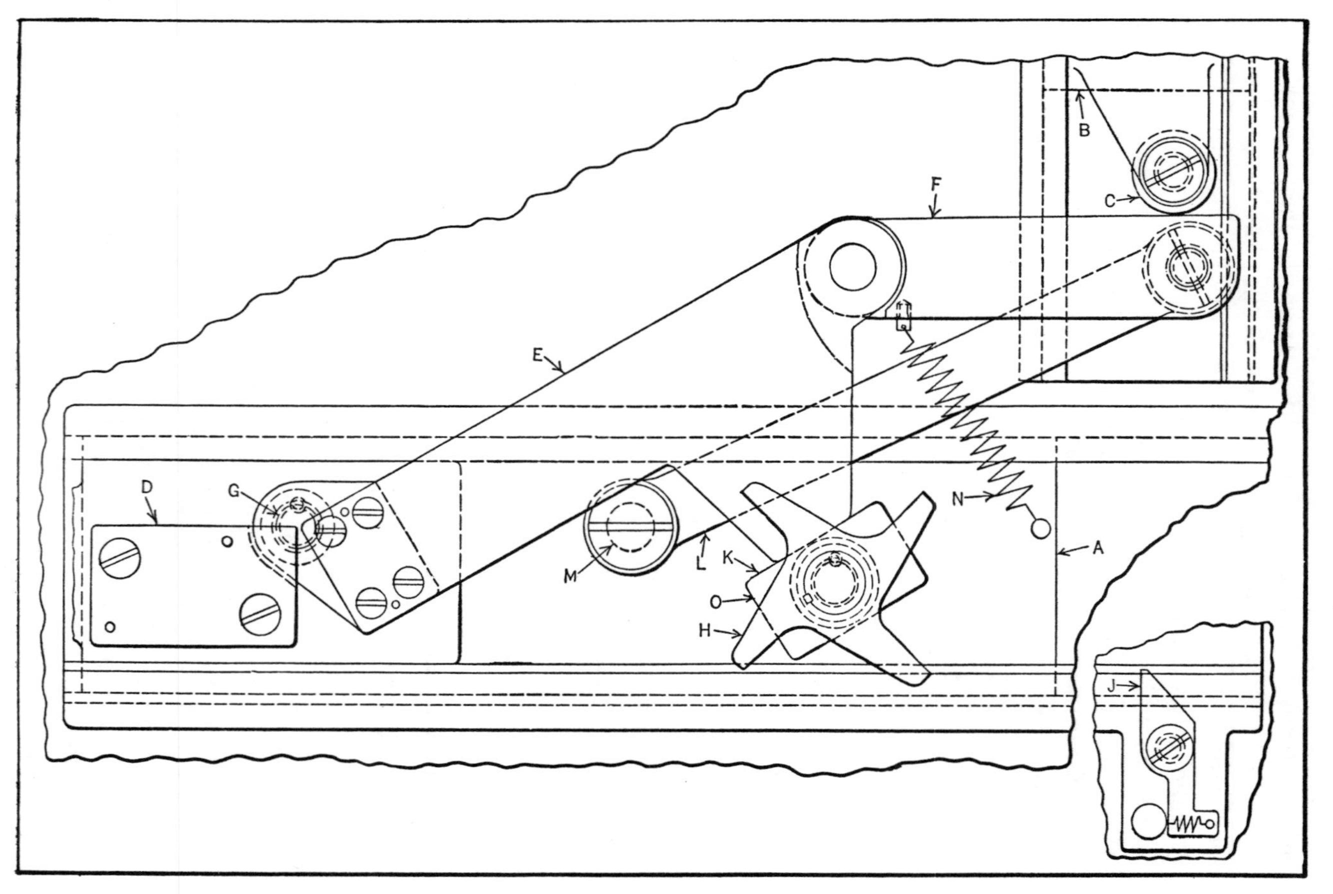

Fig. 12. Mechanism by Means of which the Horizontal Reciprocating Slide A Alternately Imparts a Long and a Short Stroke to Slide B

ferring to the diagram, it will be seen that the stroke is shortened only at one end. Consequently, as each successive stroke is shortened, the length of each successive layer of yarn is decreased a corresponding amount. Hence, the wound yarn at the small end of the bobbin forms a cone having a greater taper than the bobbin itself. This increased taper depends on the contour of cam *Q* and also on the rotary speed of the cam. This type of mechanism is used in many winding machines other than textile machines.

Alternately Imparting Long and Short Stroke to Slide.— The cam mechanism, Fig. 12, imparts a long and a short stroke alternately to a slide, a dwell occurring at both ends of each stroke. This slide serves to change the position of the carton of a carton-stapling machine, relative to the stapling tools.

Perhaps the most interesting feature of this arrangement is the fact that only one point of reversal is varied to obtain the two different strokes. The cam is made in three parts, consisting of block *D*, arm *E*, and bar *F*. It is mounted on the slide *A*, which is given a constant reciprocating movement by another member of the machine (not shown). The required movement is imparted to slide *B* by contact of the roll *C* with the cam.

Block *D* is secured to slide *A* and causes slide *B* to dwell at the left-hand end of the stroke. Arm *E* is pivoted at *G* to slide *A*, and its angular position is varied every other stroke by means of the star-wheel *H* and the pawl *J*. Star-wheel *H*, together with the block *K*, is keyed to a shaft that is free to turn in its bearing in slide *A*, while pawl *J* is pivoted to the machine frame. Contact between block *K* and arm *E* is maintained by spring *N*. Bar *F* maintains a horizontal position on both the long and short stroke of slide *B*, and it was to obtain this condition that the link *L* was incorporated. This link is pivoted to slide *A* at *M* and is connected to the bar *F*. With this arrangement, bar *F* remains in a horizontal position when arm *E* changes its angular position, thus maintaining the dwell at the right-hand end of the cam.

Reciprocating Slide with Cam Mechanism for Operating Tool or Drill Slide.— The reciprocating slide *A*, Fig. 13, is driven by an eccentric connected to rod *D*. This slide has a cam mechanism by means of which motion is applied to the slide *B*. The motion of slide *B* is employed for feeding metal-cutting tools to the work. It can be applied to

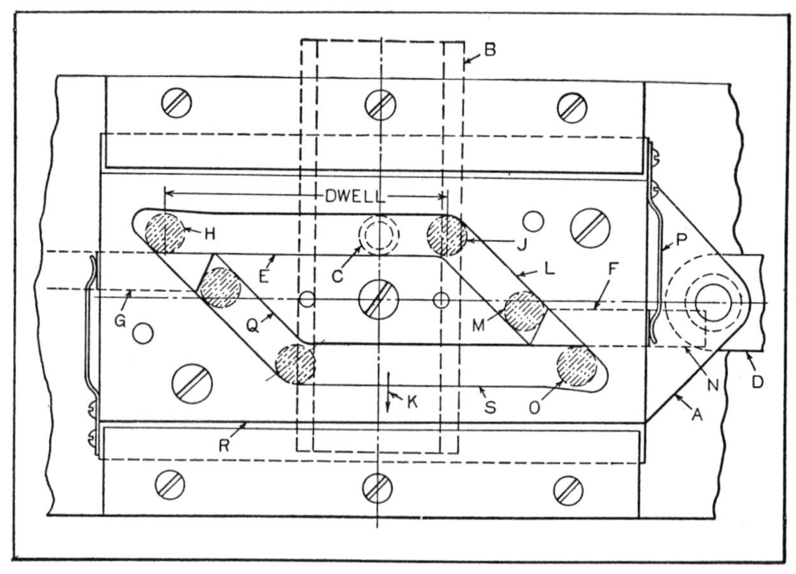

Fig. 13. Reciprocating Slide Mechanism for Operating Tool-slide

the spindle of a drill press, for example, to advance and withdraw the drill. By making suitable changes in the cam slot the drill can be given a rapid approach and reduced feeding movement, followed by a rapid return to the starting position and dwell. The movements required for this operation can be obtained by changing the cam-plate *Q* and guide plate *R*. These plates are secured in place by screws and dowels.

The cam-roll *C* is mounted on the driven slide *B*, which can move only in a direction at right angles to the movement of slide *A*. The cam slot *E* in slide *A* has latches or slides *F* and *G* which project into the slot and prevent the cam-roll from reversing its direction of travel in the cam slot. When slide *A* is at its extreme right-hand position, the cam-roll will be located in the cam slot as indicated at *H*. The roll remains stationary while the slide *A* moves to the left until it reaches the position indicated at *J*, thus allowing the slide *B* to dwell. Any further movement of slide *A* to the left will cause roll *C* to move down the inclined portion *L* of the cam slot, moving slide *B* in the direction indicated by the arrow *K*. When the cam-roll reaches the point *M*, it forces the latch *F* back to the position indicated by the dotted lines at *N*. As soon as the roll reaches position *O*, latch *F*, under pressure from spring *P*, snaps back to the closed position. At this point, the eccentric that reciprocates slide *A* has reached its highest point and the slide commences to travel in the reverse direction. On the return stroke, the same cycle is repeated, causing the slide *B* to be returned to its original position when the cam-roll reaches the position indicated at *H*. Within reasonable limits, the dwell positions of the cam slot, as at *S*, can be changed to produce any sequence of movements or dwells required.

Combined Reciprocating and Elevating Movement.—The device illustrated in Fig. 14 is used for skimming dirt and oxides from the surface of molten lead in a galvanizing vessel. Pieces to be galvanized are dipped in this vessel, and in order that they shall have a smooth and bright surface, all foreign matter must be removed from the lead before the pieces are withdrawn. The skimming is done by means of the reciprocating blade *G*. The blade is in contact with the lead on the stroke from right to left. The return stroke, however, is made with the blade in an elevated position, as shown in the end view. With the blade in the latter position, the pieces to be coated can be readily placed in or withdrawn from the vessel.

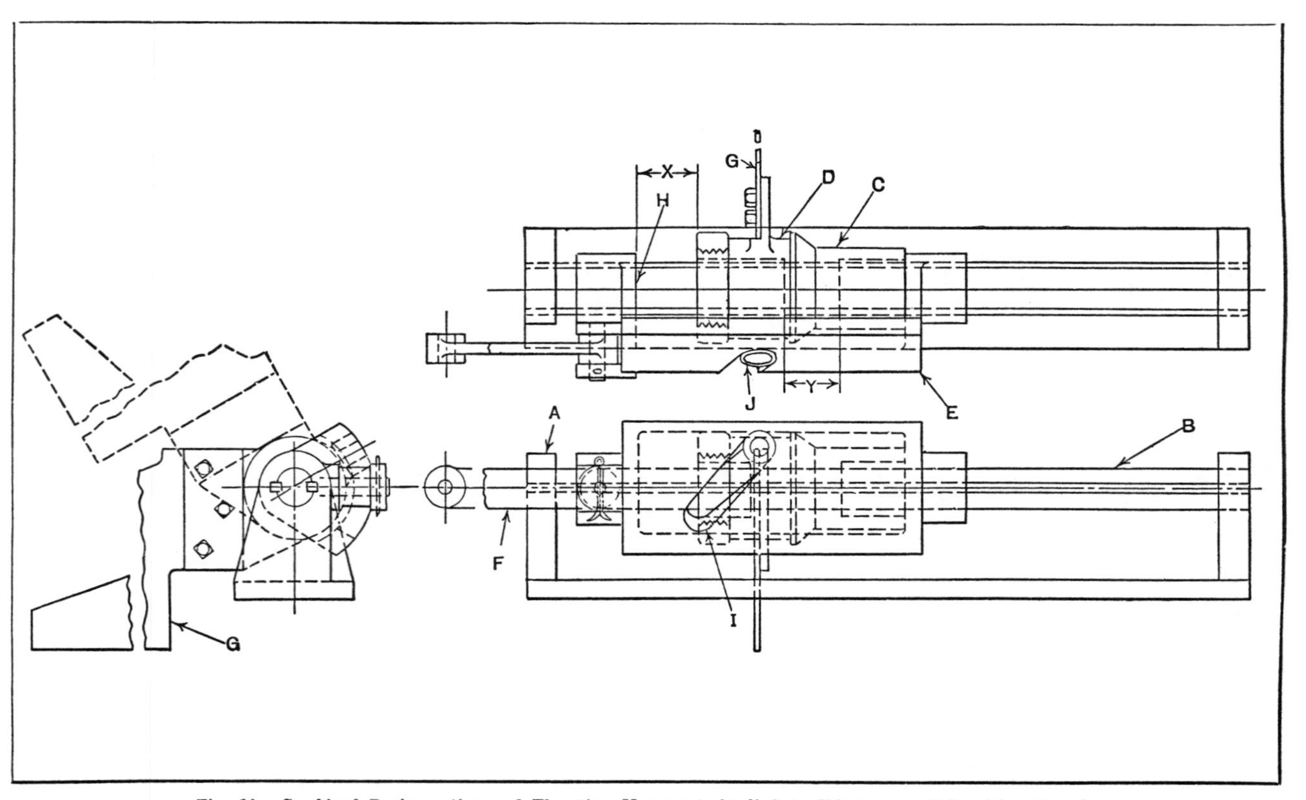

Fig. 14. Combined Reciprocating and Elevating Movement Applied to Skimmer on Galvanizing Vessel

On the bracket *A*, secured to the side of the vessel, is fastened a stationary shaft *B*. Sliding on this shaft is a sleeve *C* which, in turn, forms a bearing for the bushing *D*. On one side of this bushing is an extension to which the skimmer is fastened, while on the other side is mounted a cam roller *J* which engages an angular slot in the carriage *E*. The carriage slides on shaft *B*, and is given a reciprocating motion by a crank (not shown) through the connecting-rod *F*. Both members *C* and *E* are prevented from rotating by keys in shaft *B*.

Referring to the plan view of the illustration, it will be seen that the skimmer blade is at its farthest position to the left. The carriage *E* now moves toward the right, and after traveling a distance *X*, the surface *H* on the carriage boss comes in contact with the end of the sleeve *C*. While the distance *X* is traversed by the carriage, the sleeve *C* is stationary and the roll *J* is forced downward due to the angularity of the cam slot *I*. This movement of the roll will cause the blade *G* to rise above the molten lead.

The carriage *E* continues to move to the right with the blade in its elevated position until the end of the stroke is reached. On the return of the carriage, the gap shown at *X* will be on the other end of the sleeve *C*. As the width of this gap decreases, the cam roll will ride to the top of the slot *I*, causing blade *G* to enter slightly past the surface of the lead. The blade, held in this position, skims the surface of the lead as it continues its stroke to the position shown in the plan view.

There must be sufficient friction between the shaft and the sleeve so that the latter will remain stationary while the roll *J* raises the skimmer blade. This friction is obtained by counterboring both ends of the sleeve until the length of its bearing on the shaft is shortened to a distance *Y*. The location of this short bearing surface is such as to cramp the sleeve enough to obtain the desired friction. In

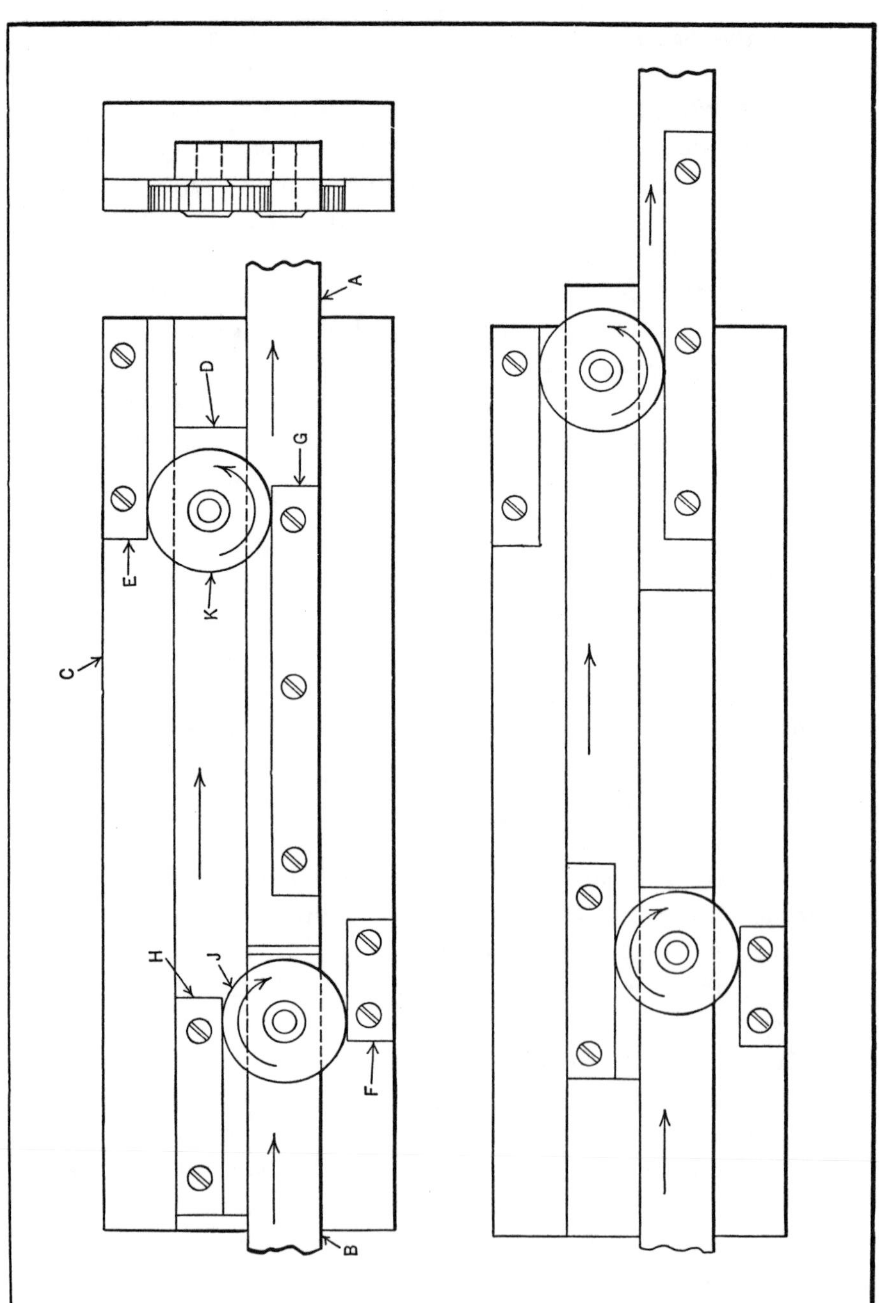

Fig. 15. Rack and Gear Mechanism for Increasing the Stroke Imparted by a Slide to Four Times its Original Travel

this simple way, a very effective and dependable frictional grip is obtained.

Quadrupling the Travel of a Slide.—In the mechanism illustrated in Fig. 15, the reciprocating slide *A* has a stroke four times as long as the slide *B* from which it receives its motion. This is effected through a series of racks and pinions, the pinions moving in a straight line and meshing with two opposite racks, one of which is fixed and the other free to slide. Obviously, the rack that is free to slide will move twice as far as the center of the pinion. This design is advantageous when a compact arrangement is required, and by using more than two gear combinations, the stroke imparted by slide *B* can be increased to any length.

The mechanism is mounted on the stationary block *C*, which is grooved to receive the three reciprocating slides *A*, *B*, and *D*. On each of the slides *A* and *D* is secured a rack, as indicated at *G* and *H*. Two stationary racks *E* and *F* are fastened to block *C*. On the ends of the slides *B* and *D* are the pinions *J* and *K*, each of which meshes with a fixed and a sliding rack. Full lines are used to represent the pitch lines of the gears and racks.

Now it will be seen that if slide *B* is advanced toward the right, say 1 inch, slide *D* will move 2 inches in the same direction through the action of pinion *J* meshing with the racks *F* and *H*. The same combination of gearing exists at the right-hand end of the block. Consequently, if slide *D* moves 2 inches, the stroke imparted to slide *A* will be 4 inches. At the end of this 4-inch stroke, slide *A* will be in the position shown in the lower view.

Intermittent Trigger Slide Having a Positive Working Stroke and a Swift Return.—A reciprocating slide having a trigger action is employed on an automatic nut-tapping machine for feeding the nut positively and at a relatively slow speed from the magazine to the tapping position. There, the slide dwells during the operation, after which the trigger action releases it, so that it returns swiftly to the magazine for another nut.

The mechanism is mounted on the machine frame *H*, Fig. 16. It consists of the dovetail slide *A*, on which is guided the auxiliary slide *B*, both slides being actuated by the oscillating segment gear *C*. Gear *C* oscillates at a constant velocity and receives its motion from another mem-

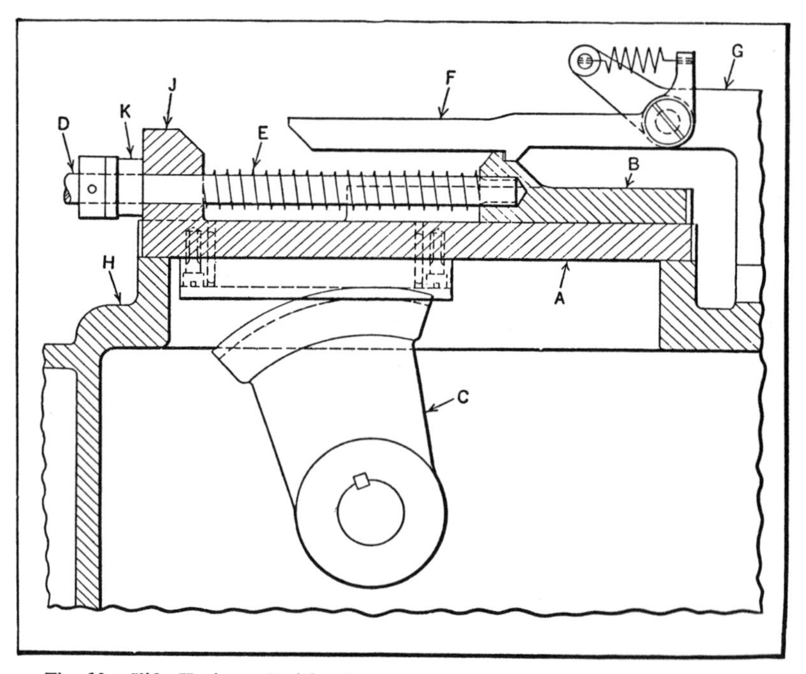

Fig. 16. Slide Having a Positive Working Stroke and a Rapid Return Movement

ber of the machine (not shown). Secured in slide *B* is the round rod *D* which is connected at its left end to the work-carrier (not shown). Spring *E*, which imparts the rapid return movement to the work-carrier, is mounted on rod *D*, and the spring tension is released for the return stroke by the trigger *F*, pivoted on the stationary bracket *G*.

Gear segment *C*, in the position shown in the illustration, has carried both slides to their extreme left-hand position, and in doing so, has caused rod *D* to transfer a nut from the magazine to the tapping position. At this point, trigger *F* has engaged the projection on the auxiliary slide *B*. The gear segment now reverses its movement and carries slide *A* toward the right. During this movement, slide *B* remains stationary, since it is held by the trigger. Near the end of the stroke of slide *A* toward the right, the projection *J* engages the end of the trigger and lifts the latter away from the projection on slide *B*; consequently, slide *B* is released and under the action of spring *E* is carried toward the right until the rubber bumper *K* comes in contact with projection *J*. Thus, with the return of slide *B*, the work-carrier and rod *D* are returned swiftly to the magazine for another nut. This completes the cycle of the mechanism.

Slide which Dwells at One End of Its Stroke.—The screw shells on the plugs attached to electric extension cords are spun in place on the plugs in an automatic machine. The assembled shell and plug is delivered to the machine from the magazine by means of a feed-slide having a dwell at one end of its stroke. The slide is designed to have a positive action. The dwell occurs when the slide has carried the plug to its spinning position and continues until the shell has been spun and the finished plug ejected from the machine.

Referring to Fig. 17, the feed-slide *A* is mounted in guides on the machine frame *B*. The driving lever *C* is equipped with a roll *D* which engages a cam slot *E* in a projection on the slide. The illustration shows the empty slide in the position it occupies after being carried back toward the left to the magazine (not shown). As the lever reverses its motion, the slide is returned with a plug to the position indicated by the dot-and-dash outline, the lever rotating through angle *b*. This is the position of the slide while the plug is being spun. The dwell of the slide that permits this operation is obtained as the lever continues its movement along the curved portion of the cam slot, the latter being concentric with the lever shaft at this time.

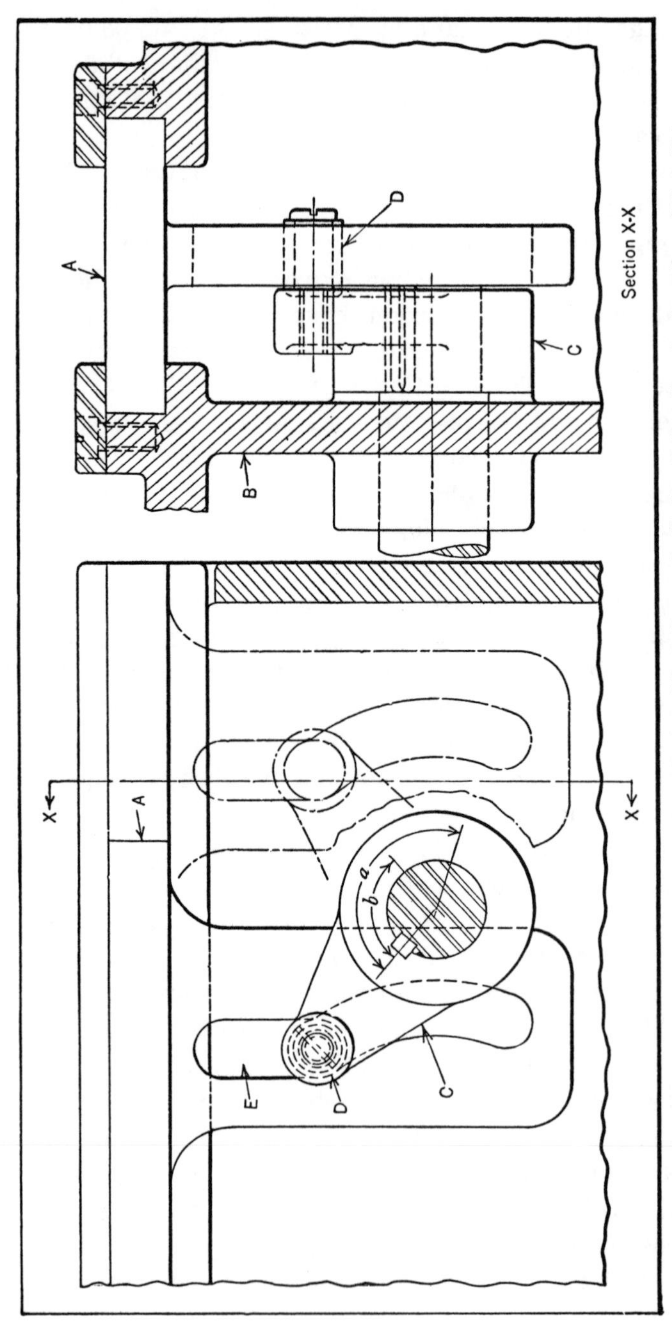

Fig. 17. Slide which Dwells at One End of its Stroke while Roller on Oscillating Driver Travels along Arc of Cam Slot

The dwell continues until the lever has moved through the angle *a* minus *b* and back to the position shown by the dot-and-dash outline. At this point the completed plug is ejected from the slide by a device not shown. The con-

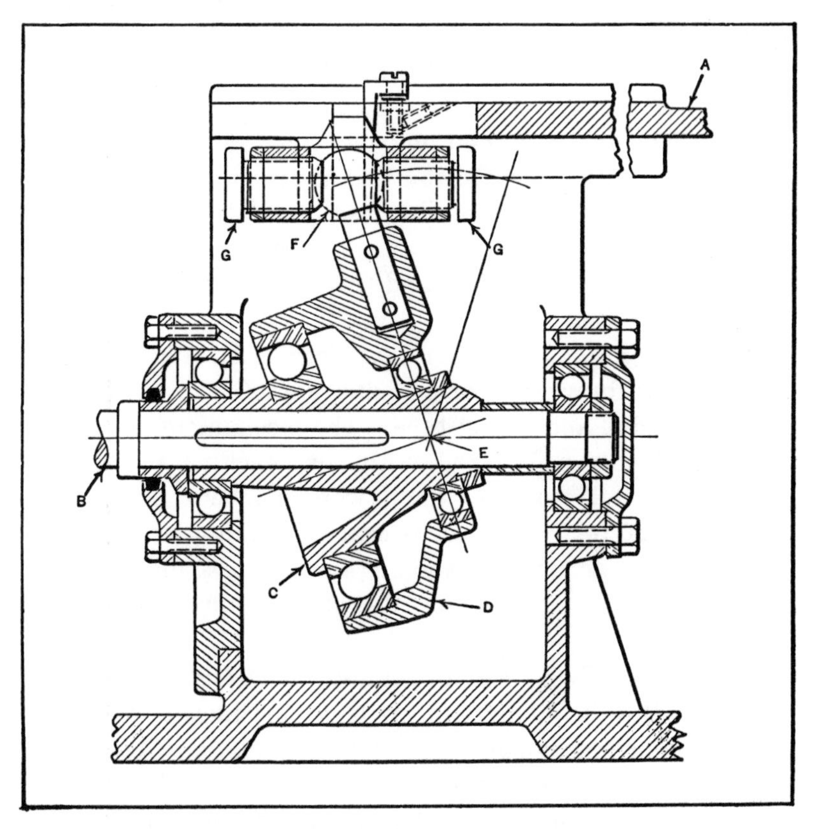

Fig. 18. Reciprocating Mechanism of a High-speed Machine for Operating Slide Parallel to Driving Shaft

tinued movement of the lever returns the slide to the magazine ready to pick up another plug.

Slide with Reciprocating Movement Parallel to Driving Shaft.—In a special high-speed machine used to shear and form fiber shields for electrical switches, it was required that slide *A* (see Fig. 18) have a reciprocating motion parallel to driving shaft *B*. It was also essential to operate slide *A* without lost motion due to wear, and the design here shown has proved satisfactory in this respect.

The reciprocating motion is obtained from an angular eccentric sleeve *C*, secured to driving shaft *B*. As this sleeve revolves, it imparts a swinging motion to part *D* about center *E*. This motion is transmitted to slide *A* through a rod which is fixed to *D* and has a ball-shaped end *F*. The spherical end engages concave seats in bronze screws *G*, held in a cross-head that is free to slide vertically far enough to provide for the rise and fall resulting from the circular movement of ball *F*. Screws *G* provide adjustment to eliminate play, and the vertical cross-head slide has adjustable gibs.

The two ball bearings that support the driving shaft and also the two between sleeve *C* and part *D* are of the combination radial and thrust type. The mechanism is enclosed in a bracket cast integral with the machine proper and forming a well so that the lower members are always in a bath of oil. This reciprocating mechanism operates smoothly and accurately, and requires little attention other than to add oil to the well at intervals of approximately two months.

Obtaining Two Reciprocating Motions from One Movement.— A change in a wire product necessitated changing the mechanism of a wire-forming machine so that the reciprocating motion originally used would be replaced by two similar movements of lesser magnitude in the same period of time. Fig. 19 shows how this was accomplished, using the same source of power. Originally the required reciprocating movement was furnished by rod *A*. In the new arrangement, this rod actuates rod *K*, causing it to move forward and back while rod *A* is moving in one direction.

Rod *A* is given a reciprocating motion from a distant source of power for transmitting the oscillating motion to the lever *B*, which is fastened to gear *C*. Gear *C* and lever *B* are free on stud *D*, and oscillate in unison. Gear *C* transmits motion to gear *E*, which carries the lever *F*, both of which are free on stud *G*. Lever *F* carries the pin *H*,

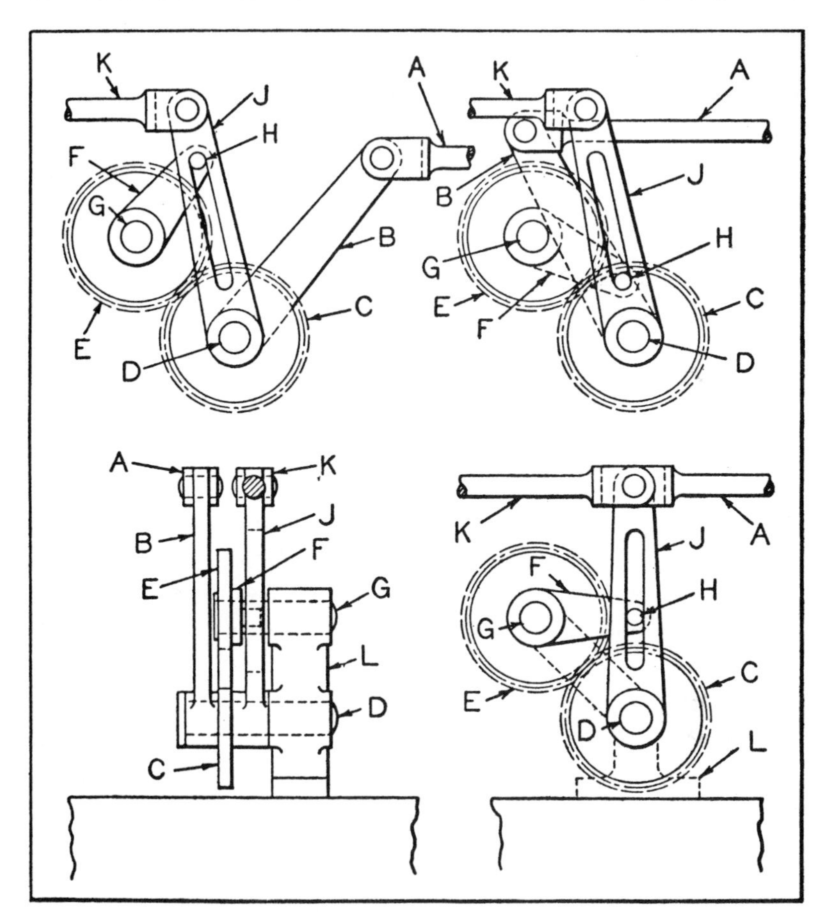

Fig. 19. Diagrams Illustrating Operation of Reciprocating Mechanism

which travels in a slot in lever *J*, transmitting motion to rod *K*. The assembly is supported by the bearing *L*.

In the two upper views, rod *A* is shown at its extreme right position and its extreme left position, representing half its cycle of operation. It will be noted that in both these views, lever *J* occupies the same position, having passed through one complete cycle and returned to its original position. Starting its movement from the position shown by the upper left-hand diagram, lever *B* is moved to the left by rod *A*, causing gear *C* to make a partial revolution. Gear *E*, meshing with gear *C*, is thus given a partial revolution in the opposite direction. As lever *F* is fastened to gear *E* and moves with it, pin *H* is moved downward in the slot in lever *J*, causing the latter to move to the right until pin *H* reaches the horizontal center line of stud *G*, at which time lever *J* is at its extreme right-hand position, as shown by the lower right-hand view.

Continued movement of rod *A* produces a further downward movement of pin *H*. As pin *H* passes the center line, it acts against lever *J* in the reverse direction, moving it to the left. As rod *A* reaches its extreme left position, lever *J* is also at its extreme left position, having completed its cycle, whereas rod *A* has completed but half its cycle. As rod *A* returns to its extreme right position, lever *J* again passes through its cycle. The magnitude of the movement of lever *B* may be determined by comparing its positions (see two upper views), while the movement of lever *J* will be understood by reference to the upper left-hand and the lower right-hand diagrams.

Slide with Dwell at Ends of Stroke and Quick Return.— A mechanism designed to give an intermittent movement to a reciprocating slide is shown in Fig. 20. For every revolution of the shaft *I*, the slide *J* rises at a comparatively slow speed until it reaches the position shown by the dotted outline; the slide then dwells at this point for a certain period of time, after which it returns to its original position. These movements are secured through the action of the lever arm *D* on the latch *C* and on the projecting lug *K* of the slide. The end of the arm, rotating at a uniform speed, engages the lug *K* and raises the slide until the latch *C* catches on the lower lug *G*.

The slide is held in this dwelling position until the arm trips the latch, when the slide drops down on stop-pin *B* to the position shown, thus completing the cycle. A further motion of the arm raises the slide. The member of the machine on which the end *H* of the slide acts (not shown) returns under the action of a coil spring, carrying the slide

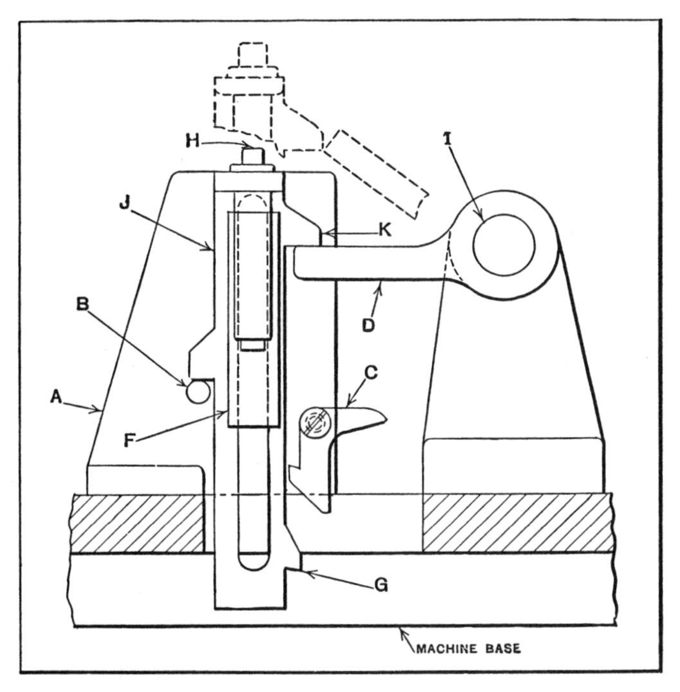

Fig. 20. Slide Mechanism Producing an Intermittent Motion and a Rapid Return

back also. It will be noted that the angle through which the arm must turn to raise the slide *J* the required height is governed by the over-all length of the arm *D* and the location of the shaft *I*. The slide is confined in its path by the T-shaped gib *F* on the bracket *A*.

Slide which Dwells During Every Other Cycle of Driving Slide.—On a certain bread-wrapping machine the mechanism, Fig. 21, controls the action of the bread-shifting slide.

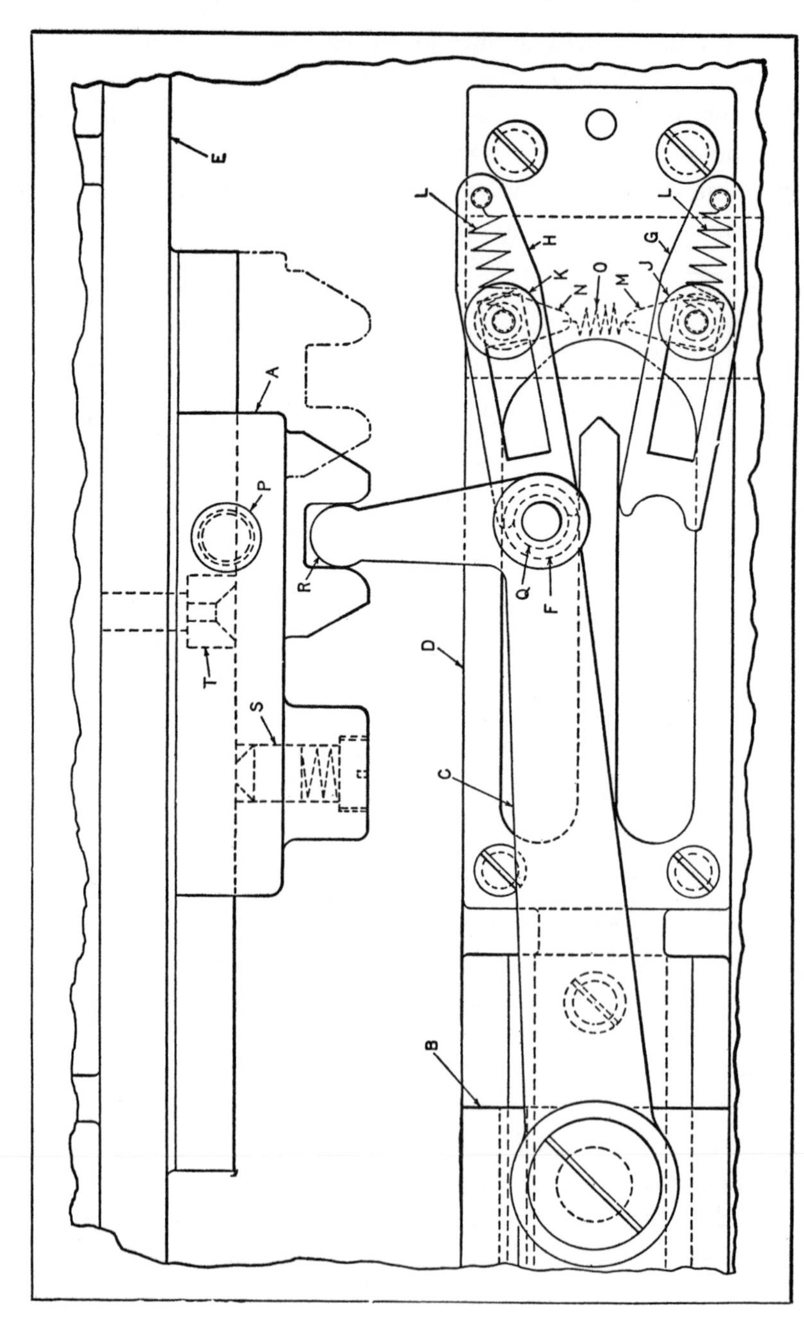

Fig. 21. Mechanism Driven by Constantly Reciprocating Slide B which Allows Slide A to Dwell During Every Other Cycle

This slide is operated by another sliding member and dwells during every other cycle of the driving member.

The bread-shifting slide is indicated at *A*. This slide transmits its controlling motion to the loaf-holding member (not shown) through stud *P*, and is actuated by the continuously reciprocating slide *B* through the connecting-rod *C*. Rod *C* is automatically disengaged from slide *A* after every other cycle of slide *B* by the switching arrangement mounted on the base *D*, which is secured to the machine frame *E*. Thus, base *D* serves also as a guide for the slide *B*. In the top of the base is machined a U-shaped groove with which the roll *F* on the connecting-rod engages.

Two spring-actuated switching arms *G* and *H* are pivoted to the base by the pins *J* and *K*. These pins are free to turn in the base and have a square shoulder near their upper end on which the arms slide. The arms are held normally in the position shown by the coil springs *L*. On the lower ends of the pins are secured fingers *M* and *N*, connected by the coil spring *O*. The tension of spring *O* serves to return the arms to their normal positions.

Slide *B*, together with connecting-rod *C*, moves slide *A* through part of its stroke toward the right. As these members continue their movement in this direction, the roll *F* enters the curved portion of the U-groove, withdrawing the projection *R* on the connecting-rod from its recess in slide *A*. This causes slide *A* to stop. In the meantime, however, the end of arm *H* engages the shoulder on the roll stud *Q* and is forced back toward the right. Thus, when the roll has reached its extreme right-hand position, the energy stored up in spring *L* forces the roll past the dead center. At this point, the slide *B* reverses its motion, and as it moves toward the left, the roll travels in the lower part of the groove. During this stroke of slide *B*, and also during its return stroke, the projection *R* on connecting-rod *C* remains disengaged from slide *A*. Hence, the latter dwells during this cycle. However, when slide *B* returns,

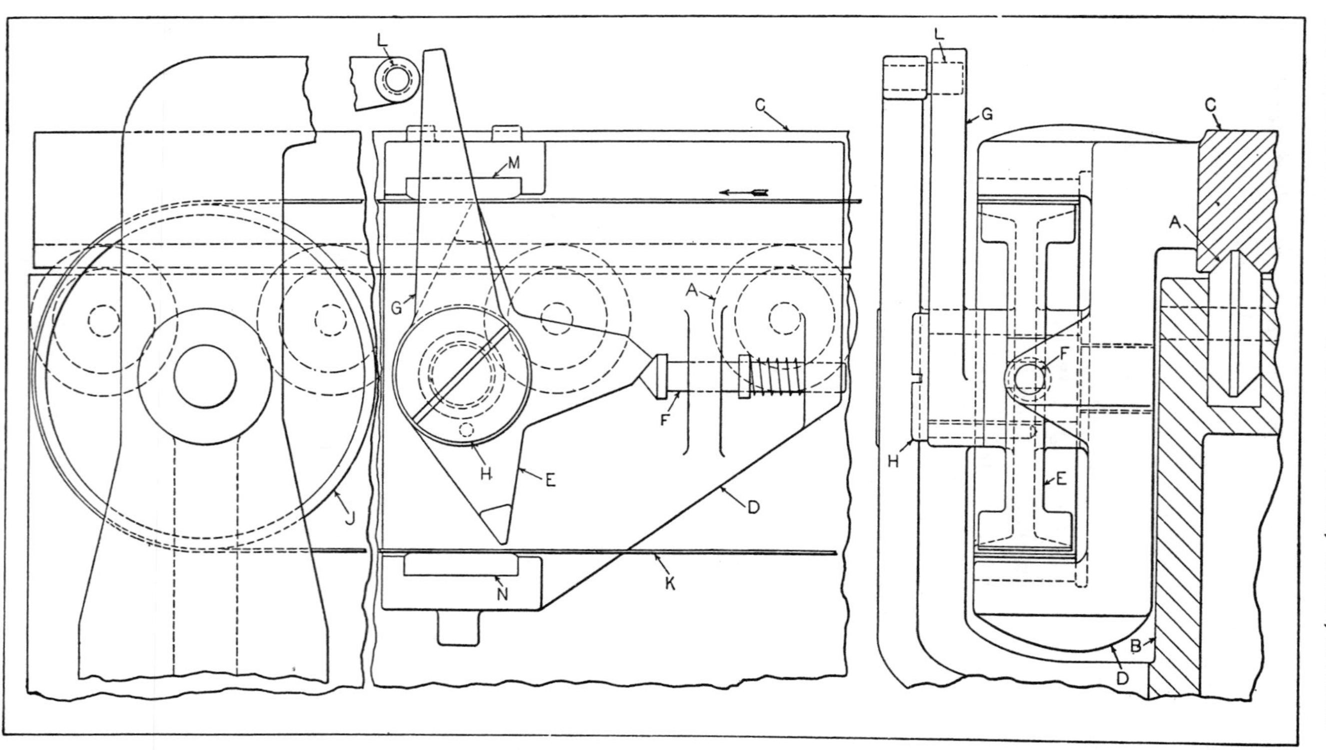

Fig. 22. Mechanism Used on Machine Table to Alternately Engage the Upper and Lower Sides of a Horizontal Belt, thus Obtaining a Reciprocating Motion

the roll stud *Q* engages arm *G*, so that when the roll reaches its extreme right-hand position in the groove, the arm forces the roll past the dead center and into the opposite section of the groove.

In entering this section of the groove, the projection on the connecting-rod again engages the recess in slide *A*, so that this slide is carried with slide *B* toward the left. It also returns with slide *B* to the position indicated by the dot-and-dash outline. At this point, the connecting-rod projection is again disengaged, as already described. Thus, slide *A* has a dwell equivalent to a complete cycle after every other cycle of the machine. In order to have slide *A* stop at exactly the same position every time it dwells, the spring-actuated plunger *S* was provided. The end of this plunger merely rides along the top of the guide for slide *A* until the slide reaches its dwelling position. When this occurs, the plunger drops into the depression in the bushing insert *T*, thus locking the slide securely during the idle stroke of slide *B*.

Reciprocating Motion Obtained by Alternately Engaging Upper and Lower Sides of a Steel Belt.—The turned shafts used in a certain type of machine tool are finished by polishing with abrasive cloth. This work is done on a machine in which the shaft is revolved between centers while the abrasive cloth, in a suitable holder, is held in contact with the shaft and moved back and forth by a reciprocating table. The long reciprocating movement of the table is obtained by alternately engaging and disengaging opposite sides of a horizontal steel belt. The mechanism secured to the table for automatically engaging and disengaging the belt is shown in Fig. 22.

The reciprocating table *C* is mounted on the beveled rollers *A*. An apron *D*, cast integral with the table, carries the reversing tumbler *E*, the spring-actuated plunger *F* and the dog lever *G*. Tumbler *E* and lever *G* are pinned together, but are free to turn on the stud *H*, secured in the

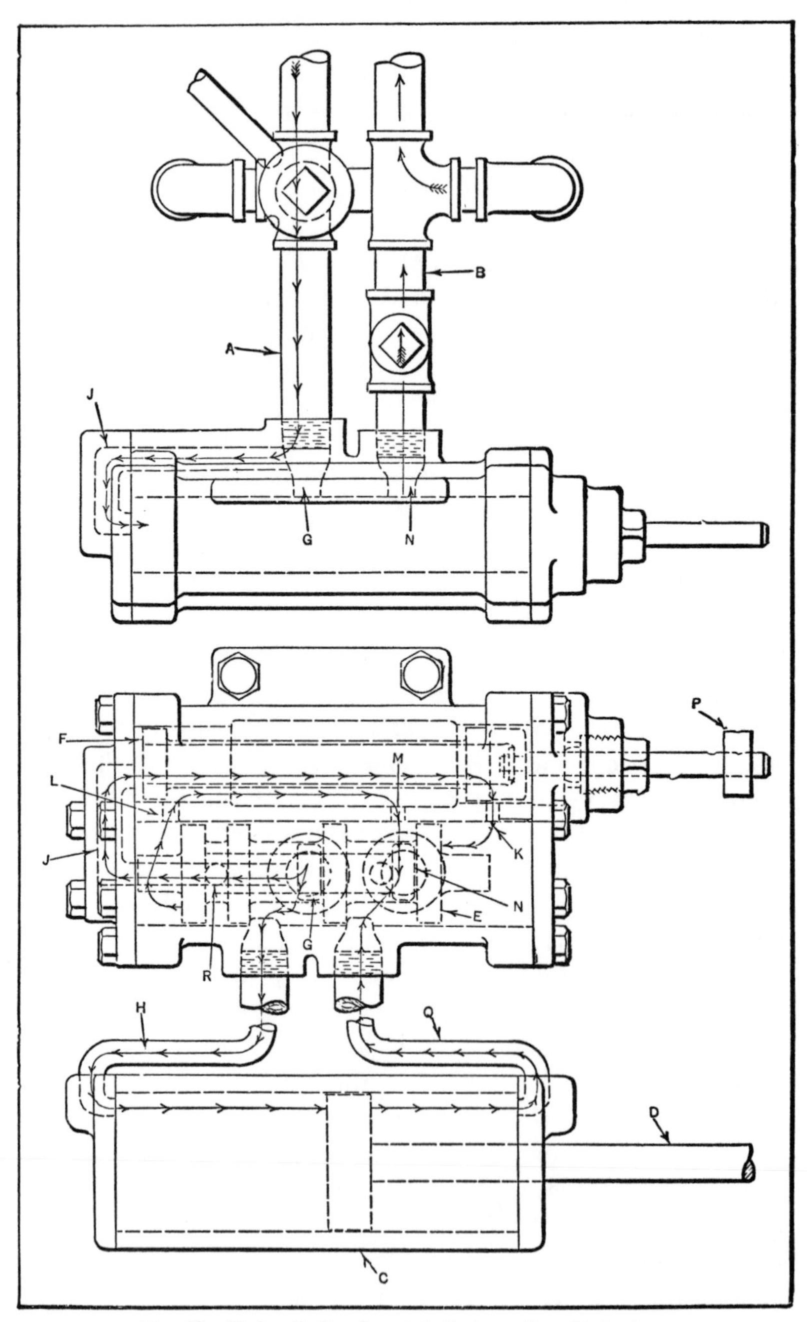

Fig. 23. Hydraulically Operated Reciprocating Mechanism

apron. The belt drums are mounted on shafts supported by brackets which are secured to the machine bed. Only the left-hand drum *J* is shown. This is the driving drum which rotates at a constant velocity. The other drum is the idler, and its bracket has a horizontal adjustment for taking up the slack in the steel belt *K*. On an extension on each of the brackets is a pin *L*, which engages the lever *G* at each end of the table stroke, causing the movement of the table to be reversed.

The spring on plunger *F* is made heavy enough to maintain engagement of tumbler *E* when the table is moving toward the right. As shown, the action of plunger *F* causes the top side of the belt to be gripped tightly between the top prong of the tumbler *E* and the block *M* secured to the apron. Consequently, the table must travel in the same direction as the top side of the belt, as indicated by the arrow. At the end of the movement in this direction, the lever *G* engages stationary pin *L* which swings the lever in a clockwise direction, thus disengaging the top prong of tumbler *E* from the belt.

Although the belt is then disengaged from the table, the resulting momentum causes the table to continue its motion in the same direction, so that the lower prong of the tumbler engages the lower side of the belt. This causes the belt to be gripped between the lower prong of tumbler *E* and the apron block *N*. As the lower side of the belt is moving toward the right, the movement of the table will be reversed. Wear resulting from use will eventually destroy the gripping action of the prongs and blocks, but this condition can be easily corrected by placing shims under the blocks *M* and *N*.

Hydraulic Reciprocating Mechanism for Machine Tools.— The hydraulic control valve (Fig. 23) is applied to hydraulically operated grinders. This valve controls the flow of oil to and from the hydraulic cylinder that contains the piston or plunger for operating the work-table.

A one-way pump supplies oil at a constant pressure through pipe *A*, and the return flow is through pipe *B*. The work-table is operated by piston-rod *D*, and the flow of oil to and from cylinder *C* is regulated by control valve *E* in conjunction with pilot valve *F*. The illustration has been made partly diagrammatic to show the arrangement more clearly.

When valve *E* is in the position shown, the oil from the pump enters through port *G*, which is connected with pipe *A*, and passes through port *H*, forcing the piston to the right. Oil from pipe *A* passes through port *J* and through the hollow pilot valve *F* and port *K*, thus exerting pressure against valve *E* which causes it to shift to the left-hand position shown. The arrows indicate the direction of flow. During this movement of valve *E* to the left, oil which previously entered the left-hand end of the chamber containing valve *E* is exhausted through ports *L*, *M*, and the main exhaust port *N*, as indicated by the arrows.

Now when a stop on the reciprocating part engages collar *P* on the pilot valve rod and moves the pilot valve to the right, port *K* is opened to the exhaust ports *M* and *N*, and oil under pressure flowing through port *J* passes through port *L* and shifts the control valve *E* to its right-hand position. The main inlet port *G* and port *Q* are now connected, so that the piston begins its movements to the left, and oil in the left-hand end of the cylinder is exhausted through ports *R* and then to the right through the interior of valve *E* and out through ports now opposite the main exhaust port *N*. The pilot valve *F* requires a movement of only 3/8 inch, and valve *E* is shifted quickly so that full port opening is obtained without delay and the flow of oil is not restricted.

CHAPTER X

SPEED-CHANGING MECHANISMS

Many different types of mechanisms for obtaining speed variations have been designed. These mechanisms, as applied to machines used in connection with manufacturing processes, permit the speed of cutters or other tools to be regulated to suit different materials or operating conditions, as in the case of machine tools. Numerous designs have also been developed for changing the speeds of moving vehicles. In all of these applications, the general object is to vary the speed of a driven member by mechanical means and independently of the driving engine or motor. Chapter XI of Volume I (pages 310 to 362) deals with different types of speed-changing mechanisms. The additional designs which follow are more or less special and embody interesting principles relating to the design of mechanisms of this general class.

Speed-Reducing Gearing for Operating Press Fixture.— An automatic fixture for a small punch press required a cam to operate it and it was necessary for the cam to make one revolution to seven revolutions of the punch press shaft. The compact mechanism for obtaining this speed reduction is sometimes known as "wobble gearing," owing to the eccentric motion imparted to one of the gears.

The punch press shaft *A* (see Fig. 1) has keyed to it an eccentric *B* (see also detailed view). This eccentric rotates within and transmits an eccentric motion to arm *C* to which is attached an internal or wobble gear *D*. Gear *D* meshes with and drives pinion *E* to which is attached cam *F*. This cam has two working edges for operating followers *G* and *H*, as these followers require different motions. At the lower end of arm C there is a stud J. One end of this stud is fixed to the press frame and the other end engages an elongated slot in arm C, thus preventing the latter from rotating about its axis, but permitting the axis to rotate around a circle equal in diameter to the throw of the eccentric.

The action of the mechanism is as follows: When the

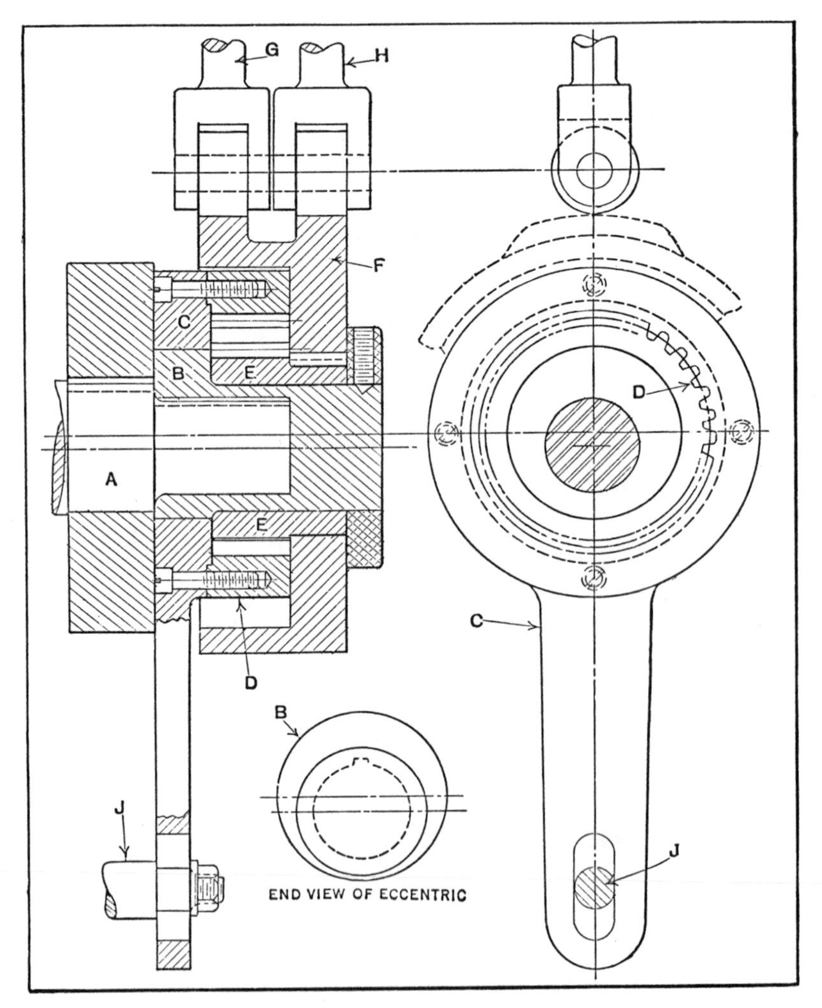

Fig. 1. Compact Speed-reducing Gearing for Operating Press Fixture

press shaft *A* rotates in a right-hand direction, the driven pinion *E* revolves in a left-hand direction. In one revolution of shaft *A*, the rotation of pinion *E* is equivalent to four teeth, this being the difference between the numbers of teeth in internal gear *D* and pinion *E*. Gear *D* has thirty-two teeth and pinion *E* twenty-eight teeth and $32 - 28 = 4$; therefore, pinion *E* will make one revolution for every seven revolutions of the punch press shaft, which is the reduction required. The gears are of 8 diametral pitch and the eccentric radius is 1/4 inch, giving 1/2 inch throw. The gear teeth are modified somewhat to provide clearance for the eccentric movement. All parts are made of machine steel. A 1/8-inch air hole (not shown) is drilled through part *B* opposite the end of shaft *A* to permit the air to escape when assembling *A* and *B*. The bearing surfaces also have suitable oil holes, which are not shown.

Nine-Speed Gear-Box with Single-Lever Control.—The single-lever control mechanism shown in Fig. 2 permits the operator to obtain instantly any one of nine different speeds. For instance, with a driving motor running at 960 revolutions per minute, the gear-box gives nine speeds ranging from 10 to 50 revolutions per minute. These speeds are obtainable, in the usual manner, with three pairs of sliding gears on two parallel shafts, by sliding each group of three gears. The gears and the gear-box of conventional design are not shown, but the yokes that control the sliding gears are indicated at *J* and *K*.

The interesting feature of this design is the arrangement for obtaining the nine changes of speed by means of the single lever *A*. A horizontal movement of lever *A* serves to rotate the segment gear *F*, which is in mesh with the rack teeth on the gear-shifting slide *J*, causing the slide to move endwise. A vertical or up and down movement of lever *A* imparts a similar sliding movement to the gear-shifting slide *K*, which has rack teeth meshing with the segment gear *H*. The range of movements imparted to the slides J and K by the horizontal and vertical movements of lever A permits any one of nine speeds to be selected.

When the handle A is moved horizontally, the housing B, in which the lever is pivoted at C, turns with the lever and the rod D. The teeth on gear E, secured to rod D, are a

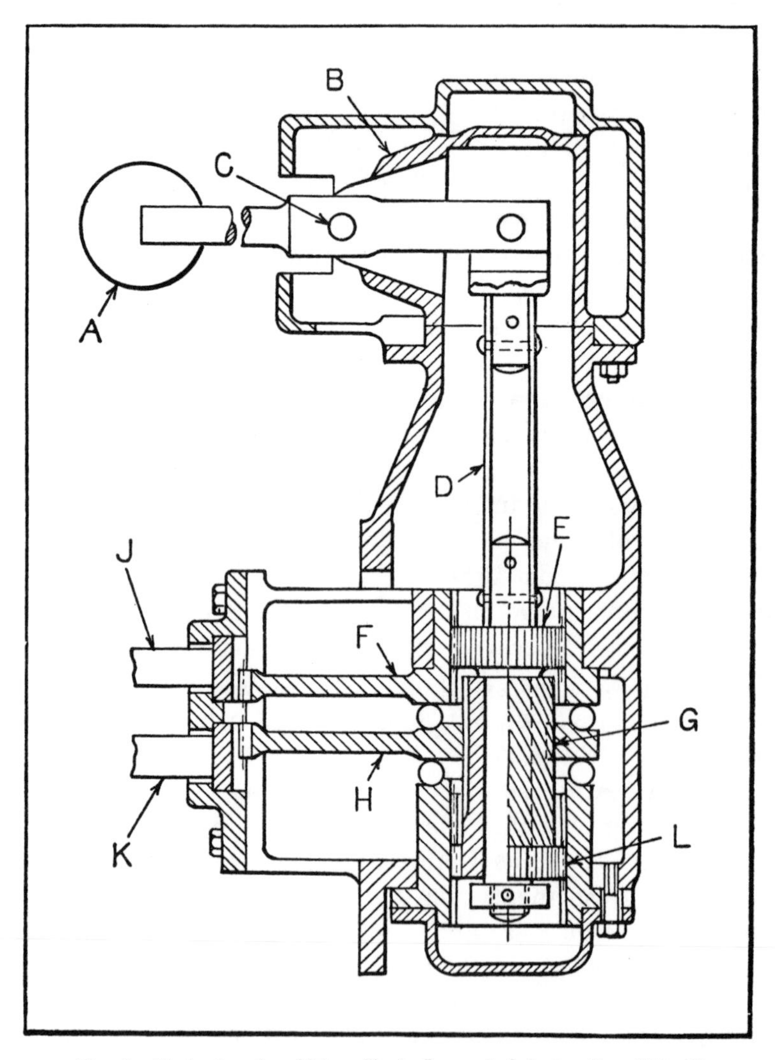

Fig. 2. Mechanism in which a Single Lever A Actuates Two Slides J and K that Control a Nine-speed Gear-box

sliding fit in internal teeth in the hub of the segment gear *F*. Thus a horizontal movement of lever *A* transmits a sliding movement to slide *J* through the rod *D*, gear *E*, and segment *F*. A vertical movement of lever *A* causes it to pivot on stud *C*, resulting in transmitting a vertical movement to rod *D*. The downward projecting end of gear *E* is a turning fit in the helical gear *G;* gear *G* has straight spur gear teeth at *L* meshing with teeth in the housing that prevent rotation of the helical gear. The vertical movement of lever *A* causes gear *G* to move vertically and impart a rotating motion to the segment *H*, which has internal helical teeth that are a sliding fit over the helical teeth on gear *G*.

Combination Differential and Speed Reducer.— The truck and tractor differential, Fig. 3, not only compensates for the variable speed of the rear wheels but has a second differential motion for obtaining two speeds at a ratio of 2 to 1; thus the pulling power of the car is increased at the low speed approximately 100 per cent. Both of these movements are enclosed in one casing.

In the sectional view of the rear axle assembly, the gears are shown enclosed in the housings *A* and *B*, which are mounted in ball bearings. A worm-gear *C*, clamped securely between these housings and driven from the engine by the worm, serves to drive the mechanism. The bevel gear *E* is locked by a pin to the housing *A*, while the gear *F* is free to turn in housing *B* and has a projecting hub on which the clutch *G* is held by a feather key.

By sliding this clutch to the right into engagement with the stationary member *H*, the gear *F* is prevented from rotating; and when the gear *E* is rotated, the pinions *I*, which are in constant mesh with bevel gears *E* and *F*, are advanced in the same direction as gear *E*. However, as these pinions are mounted on the four-armed spider *J*, their rolling action on the gears *E* and *F* will cause the spider to revolve at one-half the speed of the gear *E*.

The movement of the spider also carries the inner pin-

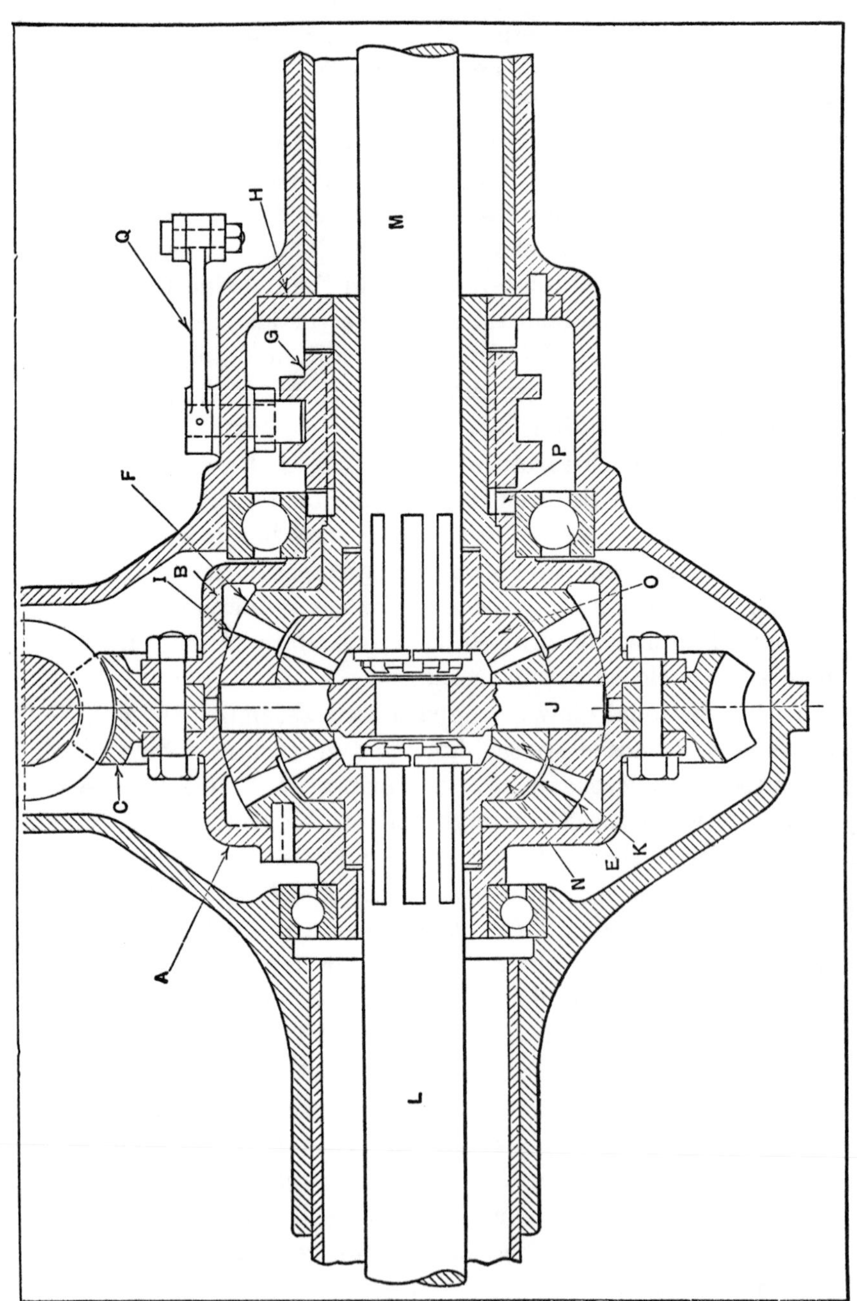

Fig. 3. Automobile Differential from which Two Speeds May be Obtained

ions *K* and the gears *N* and *O* around, driving the axles *L* and *M*. It is obvious that the gears *N*, *K*, and *O* form a differential similar to that found on most cars, permitting either axle to lag. When clutch *G* is moved to the left into engagement with the jaws *P* on the casing *B*, the entire assembly is keyed together and its action is the same as in an ordinary differential case. The spider *J* is free to float in the pinions *K* and *I*. The thrust of these pinions as they roll on the bevel gears is taken by the casings *A* and *B*, the spherical thrust seat being similar to that employed in a standard differential. The clutch *G* is operated by means of the lever *Q*, connected to a shifting lever near the driver's seat.

Compound Planetary Speed Reducer.—A compound planetary speed reducer, designed to give a large reduction in a unit of small size, is shown in Fig. 4. The gearing consists of a stationary gear *F*, doweled to housing *D*; a low-speed gear *R*, fastened to the low-speed shaft *E*; planetary pinions *G* and *H*; planetary gears *L* and *M*; and a high-speed pinion *P*, fastened to high-speed shaft *Q*.

Pinions *G* and gears *M* are carried on shafts *J*, while gears *L* and pinions *H* are carried on the sleeves *K*, the two assemblies being held on the planetary arm *N*. Pinions *G* and *H*, and gears *F* and *R*, have the same number of teeth; but gears *L* and *M*, although they are of the same pitch diameter, do not have the same number of teeth. In this case, *L* is a normal gear, while *M* has one tooth more than *L*, but is cut on the same pitch diameter.

The fact that gears *M* and *L* are both driven by the single pinion *P* makes possible the large speed ratio between the driver and driven shaft. It may be well to point out here that the ratio will be greatly decreased if pinion *P* is made with two sets of teeth in which one set is larger by a tooth than the other, and the gears *L* and *M* are made normal—that is, *M* is made to correspond with the lower half of the divided pinion *P*.

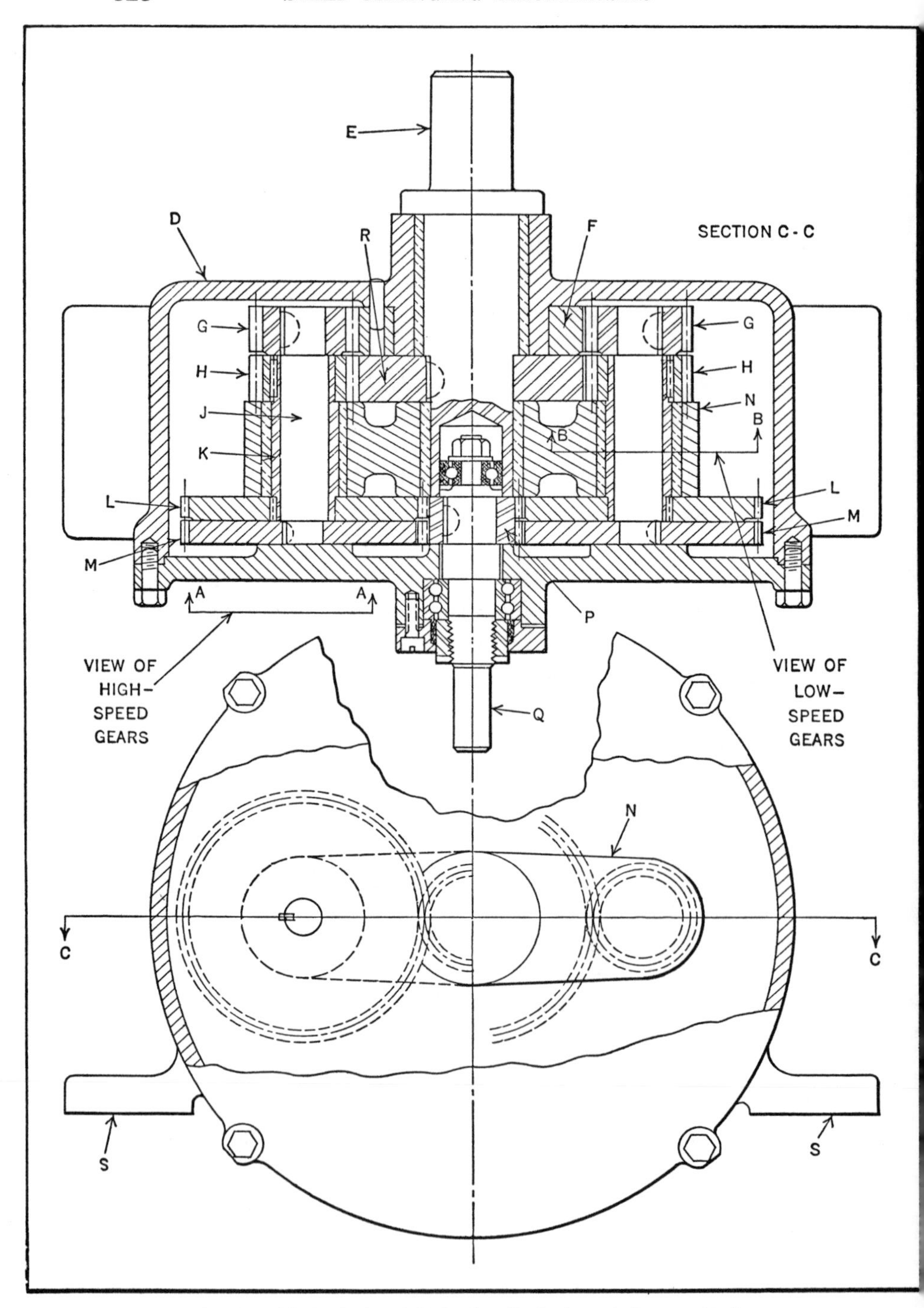

Fig. 4. High Ratio Speed Reduction Mechanism of Compact Design

Calculating the Speed Ratio.—The method of calculating the ratio of this kind of reducer is not very different from any other, although the fact that the movement goes through two sets of gears may make it confusing. An example will make the method clear.

Referring to the illustration, the number of teeth in the different gears of the mechanism are as follows: $F = 50$, $G = 20$, $H = 20$, $R = 50$, $L = 80$, $M = 81$, and $P = 32$.

First, assume that all the gears are locked tight and that the entire mechanism, case and all, is given one revolution. Thus, both shafts E and Q are given one revolution. Next, assume that arm N is held stationary, and that the case D is turned back one revolution. As this one revolution is made, we analyze the rotation of the various gears, noting first what happens to shaft Q and then to shaft E. The movement is added to or subtracted from the first revolution in each case, and the two results are set up as the ratio.

Equations can now be written from this information. Assuming that the first revolution was made in a clockwise direction, the second revolution is made in a counter-clockwise direction. It will be noted that when we turn the case back one revolution, while holding the arm still, the shafts revolve in a counter-clockwise direction; so we must subtract the calculated movement from the first revolution of both the driving and the driven shafts. Thus we have,

Movement of driver Q equals

$$1 - \frac{\overset{F}{50} \times \overset{M}{81}}{\underset{G}{20} \times \underset{P}{32}} = 1 - \frac{405}{64} = \frac{64 - 405}{64} = -\frac{341}{64}$$

and movement of driven shaft E equals

$$1 - \frac{\overset{F}{50} \times \overset{M}{81} \times \overset{P}{32} \times \overset{H}{20}}{\underset{G}{20} \times \underset{P}{32} \times \underset{L}{80} \times \underset{R}{50}} = 1 - \frac{81}{80} = \frac{80 - 81}{80} = -\frac{1}{80}$$

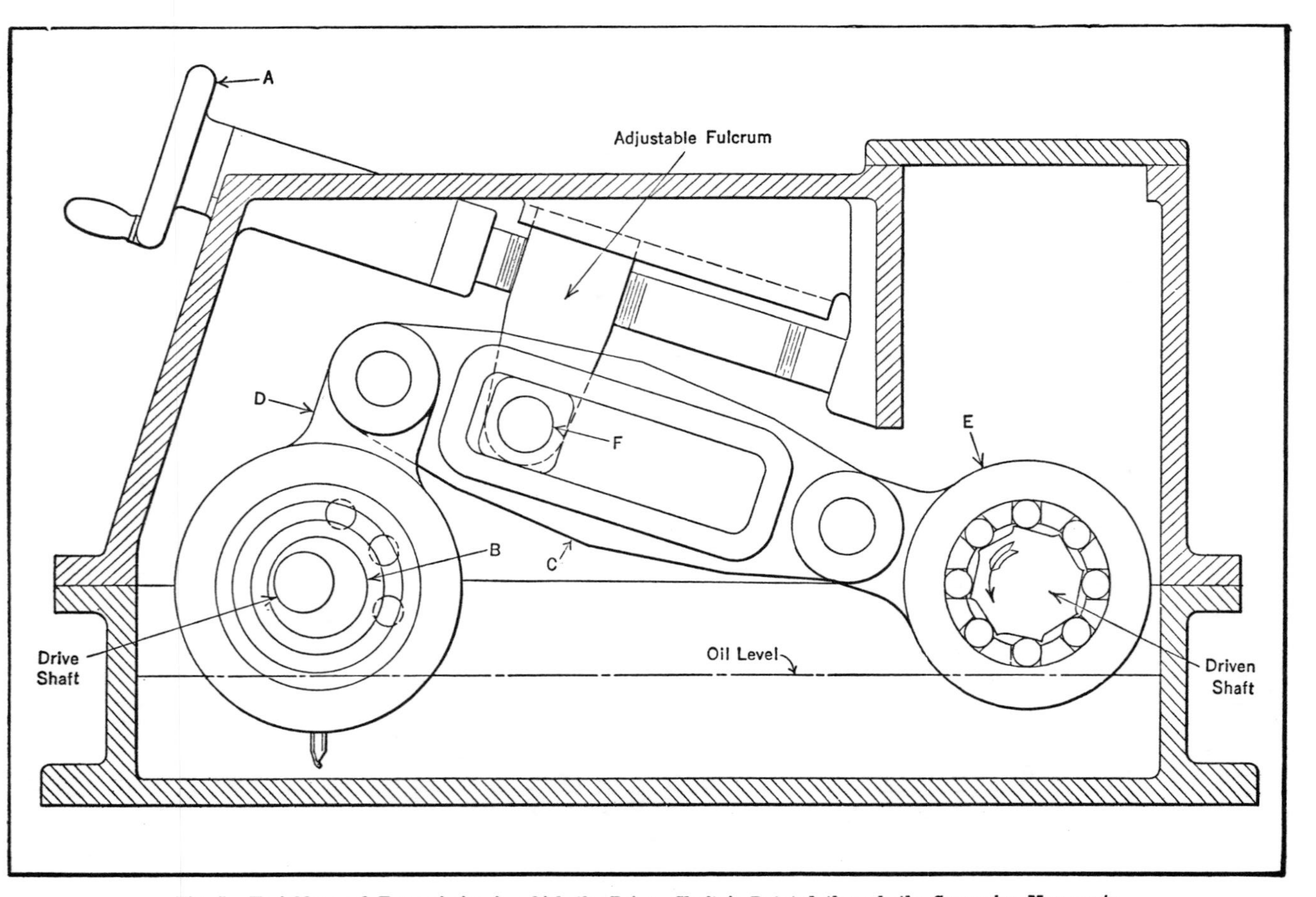

Fig. 5. Variable-speed Transmission in which the Driven Shaft is Rotated through the Successive Movements of a Series of Roller Clutches

Thus we have the ratio between the driving and driven shafts equals

$$\frac{341}{64} \div \frac{1}{80} = \frac{341 \times 80}{64} = \frac{426}{1}$$

It should be noticed that the equations for both Q and E are negative; hence their quotient is positive, which means that the driven shaft E runs in the same direction as the driver Q. If the positions of gears L and M are reversed, the same ratio is obtained, but with a negative sign, as will be found by working out the example as in the first case.

The effect obtained by making P in two pieces, with 31 teeth for the lower half to match the 81-tooth gear M, and 32 teeth in the upper half to match L with 80 teeth, will be to greatly reduce the ratio. Working out this example by the same method as was used for the first example shows the ratio to be 121.5 to 1.

The general construction of the speed reducer is shown quite clearly in the illustration. The low-speed shaft E runs in a bronze-bushed bearing, and the drive can be taken off to one side—that is, with a chain or gearing, if necessary. The high-speed shaft Q, in this case, was designed to be coupled directly to a motor. If it were necessary to drive shaft Q with a chain or gears, the double-row ball bearing would probably have to be split up into two bearings, one being arranged as in this design, and the other located close to the end of the shaft to take the radial load of the chain or gear. The method of mounting will naturally depend on conditions. In the application described, the case D was fastened to two channels running parallel with the shafts, legs S being provided for that purpose.

Gearless Variable-Speed Transmission.—In many drives, it is desirable to be able to shift from one speed to another without stopping the machine and also to be able to obtain any speed between the maximum and minimum. A design which meets these requirements is shown in Fig. 5. It is compact, and instead of using gears, chains, or belts for transmitting the rotary movement to the driven shaft, levers and roller clutches are employed.

In this particular model (built and patented by the Lenney Machine & Mfg. Co., Warren, Ohio), any speed between 15 and 150 revolutions per minute can be obtained with the motor running at 1750 revolutions per minute. The speeds are instantly changed by turning the handwheel indicated at *A*. On the drive shaft is mounted a series of eccentrics *B*. These eccentrics are connected to levers *C* by yokes *D*. Roller clutches on the driven shaft are connected to the levers by yokes *E*.

As the drive shaft rotates, the eccentrics impart an oscillating movement to the left-hand ends of the levers *C;* and as these levers are pivoted at *F*, their other ends will also oscillate and impart a rotary reciprocating movement to the roller clutches within the yokes *E*. Each reciprocating movement of the clutches will cause the driven shaft to rotate a fraction of a revolution; and as the eccentrics are spaced uniformly about the drive shaft, the impulses given the driven shaft will be successive and overlapping. In this way, a uniform rotary movement of the driven shaft is obtained.

The oscillating movement of the right-hand end of the links *C* determines the amount the driven shaft turns during each impulse, and this oscillating movement depends upon the position of the fulcrum *F* along the slot in the levers. For example, if the fulcrum is moved down toward the right by handwheel *A*, the reciprocating movement of the clutch will be shorter, and a longer time will be required to rotate the driven shaft. Obviously, the entire range of speeds is covered smoothly, enabling the mechanism to glide from one speed to another. Although not indicated, forward and reverse rotation of the driven shaft can be obtained by merely shifting a lever. This lever may also be shifted to neutral to stop the driven shaft.

Mechanism that Insures Changing Speeds According to Successive Gear Ratios.—An ingenious application of intermittent gears is incorporated in the mechanism shown in Fig. 6. This mechanism is designed for shifting change-gears axially in a metal-spinning machine. Provision is made for obtaining three different speeds and for shifting

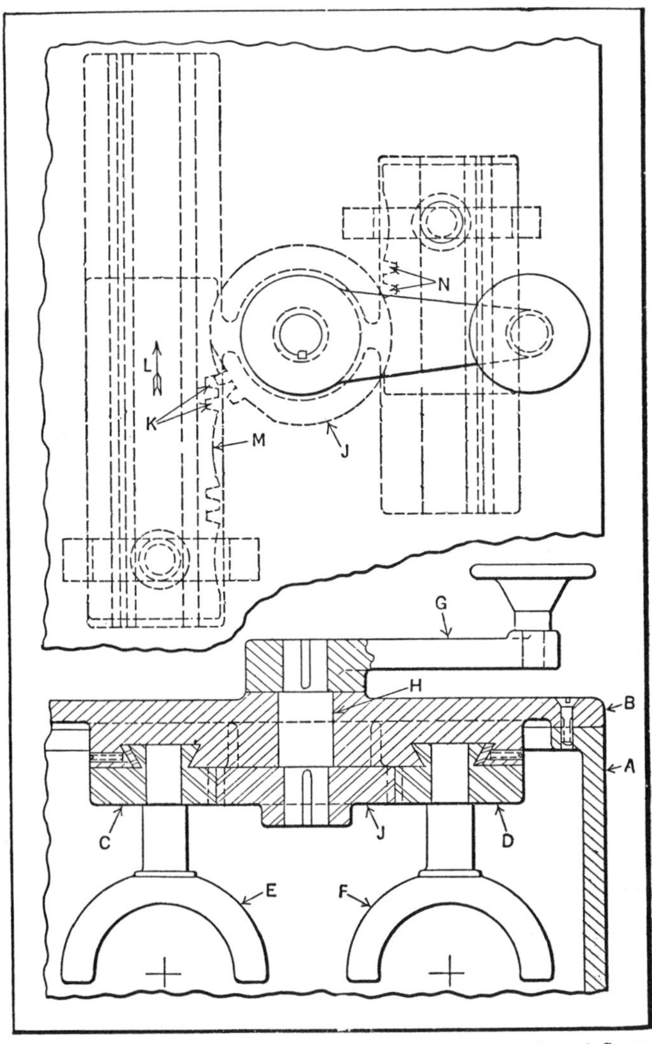

Fig. 6. Gear-shifting Mechanism that Insures Easy Changing of Gears in Order of Speed Ratios

the gears into the neutral position in the order of their ascending and descending ratios. Only one hand-lever is employed for manipulating the gears. The speeds are arranged in geometric progression and the provision for changing them in accordance with their ratios was made to avoid clashing of the gear teeth when changing speeds while the machine is in operation. With this arrangement, the pitch-line velocities of the gears to be engaged are so nearly the same that the teeth slide readily into mesh.

The gear-box is indicated at *A* and its cover at *B*. On the cover is mounted the entire gear-shifting mechanism, which consists essentially of the two slides *C* and *D* to which are attached the gear-shifting forks *E* and *F;* the hand-lever *G* keyed to shaft *H;* and the intermittent gear *J*, which is keyed to the shaft and engages the slides *C* and *D*.

With the hand-lever in the position shown, the change-gears (not shown) are in neutral. By rotating this lever in a clockwise direction, the two teeth in pinion *J* engage the tooth spaces *K* in slide *C*, causing the latter to move in the direction of arrow *L*. The movement of slide *C* continues until fork *E* slides the corresponding change-gear into mesh for imparting the lowest speed to the machine spindle. At this point, the cylindrical part of the pinion engages the corresponding depression *M* in the slide, locking slide *C* in a stationary position. To obtain the next higher spindle speed, the rotary movement of the lever is continued until the teeth in gear *J* engage the tooth spaces *N* in slide *D*. Up to this point, this slide has been locked in a stationary position by the cylindrical part of the pinion *J*.

As the lever continues to rotate, slide *D* is moved in a direction opposite that indicated by arrow *L*, causing the fork *F* to shift another gear into mesh and thus obtain the second speed. The next two speeds are obtained in like manner—that is, by continuing the rotary movement of lever *G* in a clockwise direction. Graduation marks on the gear-box cover and an arrow attached to the lever hub indicate the positions of the lever for the various speeds, as well as the position when the gears are in neutral. In order to shift the gear to neutral when the lever is in the "high speed" position, the lever must be swung through an angle of approximately 450 degrees. However, owing to the successive arrangement of the gears, their action in shifting is so smooth that the lever can be shifted very rapidly between these two points.

Speed-Changing Transmission of Hydraulic Type.— Transmissions for cars driven by internal combustion engines must be designed to provide the required speed changes, a reverse or backward motion, and a neutral position to permit of stopping the car while the motor continues to run. The transmission to be described provides unlimited speeds from zero to the high or direct drive, and all speed changes, as well as the neutral position and the reverse, are controlled by a single foot-pedal, there being no gear-shifting lever.

This transmission has been tested under road conditions as explained later. Fig. 7 shows a cross-sectional view, and Fig. 8 an end view of the "fluid clutch" or hydraulic part of the mechanism.

The two opposed cylinders *A* are attached to the engine flywheel, and another pair of opposed cylinders *B* is attached to shaft *D*, which is offset, or eccentrically located, relative to the main center line *x-x*. The four pistons *C*, Fig. 8, located in the four cylinders, are in the form of a one-piece cross with arms of equal length, located 90 degrees apart; consequently, the two pairs of cylinders always rotate together and one pair is held at right angles to the other.

Rotation of the cylinders and pistons may or may not be accompanied by a reciprocating motion of the pistons in the cylinders. Such a motion will occur, due to the offset position of cylinders *B* and shaft *D*, unless the pistons are

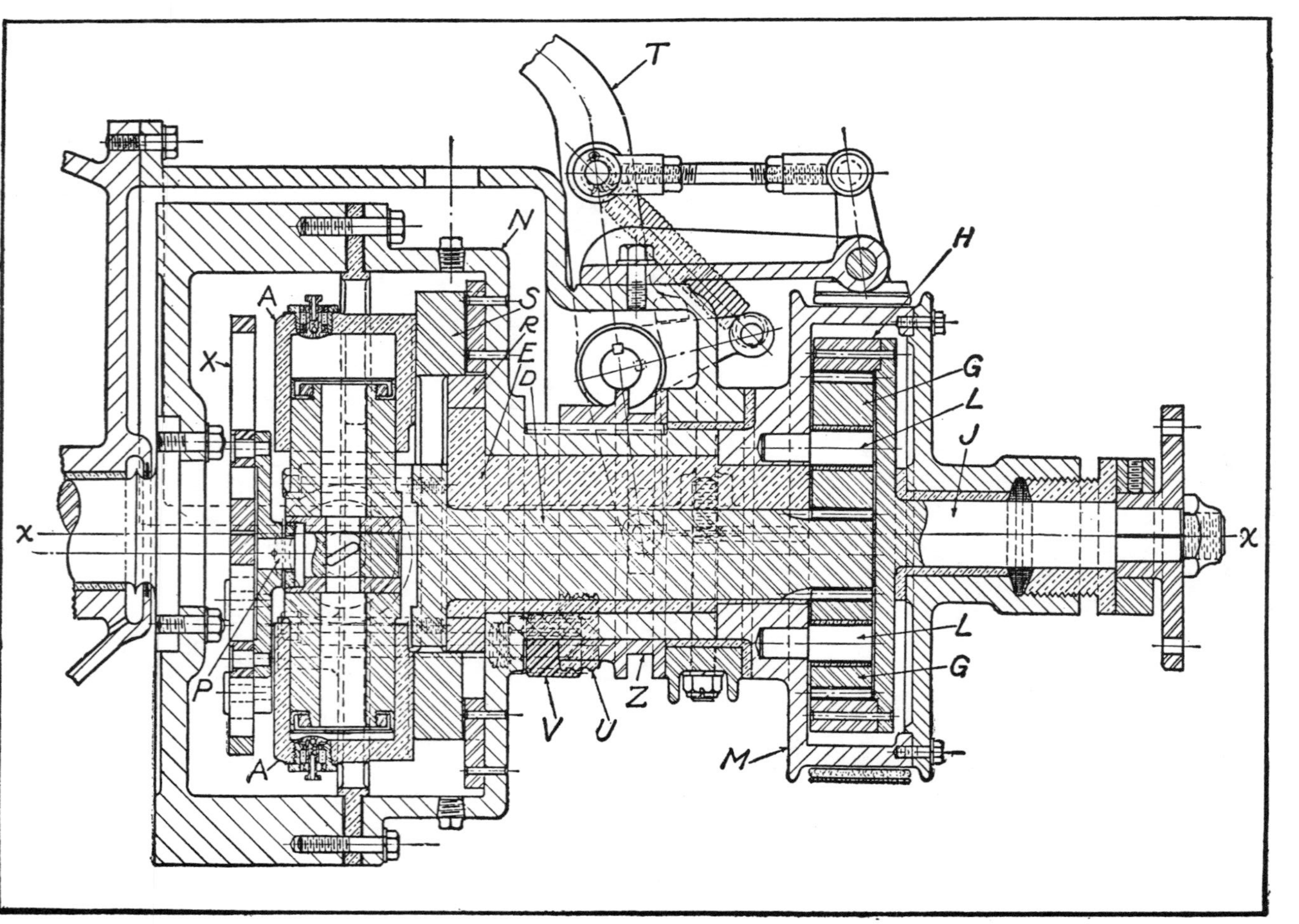

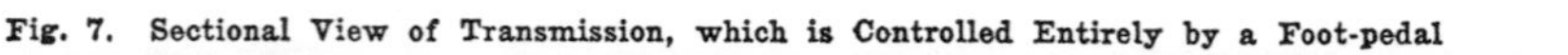

Fig. 7. Sectional View of Transmission, which is Controlled Entirely by a Foot-pedal

locked hydraulically so that movement is impossible. When the pistons are locked relative to the cylinders, the direct or high-speed drive is obtained.

How Direct Drive is Obtained.—The clearance spaces in all cylinders and the holes through the piston arms are always filled with lubricating oil. If the four-way plug valve *P*, Fig. 7, located at the intersection of the holes in the pistons, is closed so that oil is trapped in each cylinder, any movement of the pistons relative to the cylinders is prevented. (The control of this valve by foot-lever *T* will be explained later.)

When the pistons are thus locked, the various parts of the transmission from the motor shaft to the rear transmission shaft *J* rotate as a unit. These parts include, in addition to the cylinders, pistons, and eccentric shaft *D*, the eccentric bushing *E* (which is keyed to casing *M* and is free to turn with it in the main casing *N*), pinions *G* in casing *M* (which carries the pinion studs), internal gear *H*, and shaft *J* to which it is connected.

During this direct drive, the axis of eccentric shaft *D* rotates around the common axis *x–x* of the transmission, but shaft *D* does not turn about its own axis. The eccentric bushing *E*, in which shaft *D* is free to revolve, is forced to rotate about axis *x–x* by shaft *D*. While driving direct, pinions *G* cannot rotate about their own axes, because on one side they are in mesh with the driving pinion of *D*, which is locked against rotation about its axis, and on the other side they mesh with internal gear *H*, which offers resistance to rotation due to the fact that it is coupled indirectly to the wheels of the car. The result is that pinions *G* merely act as locking members or driving keys between the pinion of shaft *D* and internal gear *H*.

Reversing the Direction of Rotation.—The action of the mechanism during the intermediate speed changes and when in neutral will perhaps be easier to understand when the movement during the reverse drive is described. Valve

P, which was tightly closed for the direct drive, is wide open for reversal, thus allowing the oil to flow freely and permitting the pistons to reciprocate. The result is that eccentric shaft *D* now rotates about its own axis, but it also has a planetary movement about axis x–x, because when valve *P* is either partly or wide open, the eccentric bushing *E*, being keyed to casing *M*, is free to rotate when pinion gears *G* revolve inside internal gear *H*. When foot-pedal *T* is pushed down to the reverse position, it first opens valve *P* and then locks gear-case *M* and eccentric bushing *E* against rotation by gripping the gear-case with an external brake-band. The latter action is so timed that it does not occur until valve *P* is wide open and foot-pedal *T* controls the action of the brake, as well as that of the valve. With casing *M* and eccentric bushing *E* held stationary, motion from the motor is transmitted through cylinders *A*, the pistons, shaft *D*, pinions *G* (which now rotate about pins fixed in casing *M*), reversing internal gear *H* and shaft *J*.

Transmission in the Neutral Position.—If foot-pedal *T*, Fig. 7, is allowed to rise from the reverse to the neutral position, casing *M* and eccentric bushing *E* will remain released from the brake-band and valve *P* will be wide open. The revolving cylinders *A* then rotate cylinders *B* and eccentric shaft *D*, which merely turns about its own axis and axis x–x. The rotation of shaft *D*, however, is not transmitted to internal gear *H* and shaft *J*, because gear casing *M* and eccentric bushing *E* now are free to turn. The result is that pinions *G* merely rotate planetary fashion inside of gear *H*, because casing *M* and eccentric bushing *E* offer only slight frictional resistance to rotation, whereas internal gear *H* is coupled indirectly to the rear wheels of the car.

Action of Mechanism During Intermediate Speed Changes.—As foot-pedal *T* is allowed to rise from the neutral position toward the high-speed position, valve *P* is gradually closed. As it closes, there is a proportionate increase in the resistance to the flow of oil between the four cylinders; moreover, this resistance to the oil flow causes a corresponding increase in the resistance to the rotation of *D* about its axis until, finally, when valve *P* is completely closed, there is no such rotation, shaft *D* merely turning about axis *x–x*. However, when valve *P* is partially opened,

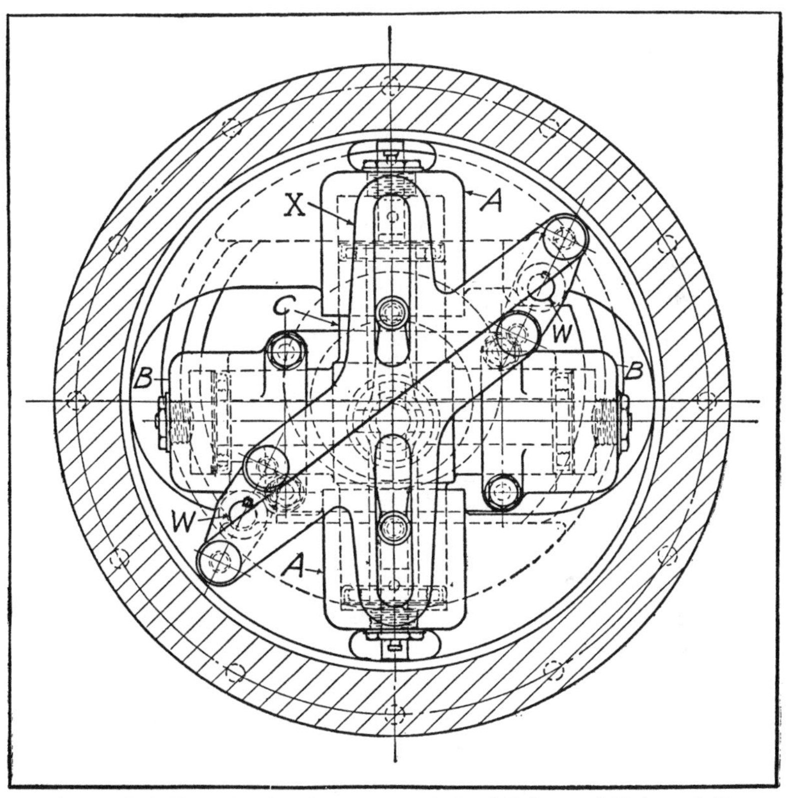

Fig. 8. End View, Showing Arrangement of the Cylinders and Pistons of the "Fluid Clutch"

shaft *D* has a planetary movement, there being rotation about its own axis and about axis *x–x*. The movement about axis *x–x* is, of course, accompanied by rotation of eccentric bushing *E*, which is forced to rotate by the studs in casing *M* when its pinions *G* revolve idly around *H* in the neutral position.

Now when shaft D begins to turn about axis x–x at an increased rate due to increasing the resistance of the oil, rotary motion will be transmitted to gear H at a rate depending upon the planetary movement of shaft D and its rotation about its own axis. The planetary movement increases and the rotation about the axis diminishes as valve P is closing. Finally, when the valve is entirely closed, thus preventing all rotation of shaft D about its own axis, pinions G act something like fixed keys that connect the pinion of shaft D with internal gear H, as previously mentioned.

The movement of foot-pedal T is transmitted to a sliding collar Z, which, in turn, slides a helical gear segment U that is continually in mesh with two helical pinions V keyed on short shafts W (Fig. 8). A quarter turn of these shafts can be obtained easily, as the gear segment slides in an axial direction while guided by a key to prevent rotation. This rotary motion of the pinion shafts is transmitted to the inside of the casing through oil-tight bearings. The elongated slots in the connecting link X allow space for the valve lever rolls to operate in when the pistons are reciprocating in the cylinders.

The car responds to the slightest touch of the foot-pedal without any jarring action or shocks. No trouble has been experienced from excessive heat generated by the compression of the fluid, as there is sufficient radiating surface; this has been proved by tests on the road. The main casing N of the transmission is made oil-tight and should be kept entirely filled with some good quality lubricating oil.

Safety valves are provided at the head of each cylinder, and they are set to blow at 1500 pounds per square inch. In conjunction with each safety valve, there is a sensitive one-way automatic check valve opening toward the inside of the cylinders. If a vacuum is created in the cylinders by the pistons due to insufficient oil as the result of leakage, the proper amount of oil will automatically be restored through the check valves. This is an important provision, since the cylinders must always be completely filled to obtain a smooth, even starting torque.

It is necessary to have two counterbalancing plates *R* and *S*, Fig. 7. One is used to counterbalance the two offset cylinders *B*, and the other to counterbalance the pistons. The plate *R* is forced to move in opposition to the cylinders by eccentric bushing *E*, and the other plate *S* is driven by plate *R*, but it is keyed so as to slide in direct opposition to the pistons.

This transmission has been applied to a car and subjected to various driving conditions during 1000 miles of road tests. While there is doubtless considerable sliding friction in the design shown, no difficulties have been experienced from overheating, although continuous runs up to forty miles per hour have been maintained for three hours. Nevertheless, the mechanical efficiency can be increased by the use of ball or roller bearings, especially between the eccentric bushing and the main casing. Other changes may also be made subsequently in the construction of this transmission.

Automatic Speed Compensating Mechanism.—In recording sound on a sensitized motion picture film, it is necessary that the film have an absolutely uniform linear movement through a microscopic light beam which photographs on it electrical vibrations coming from a microphone. This film is pierced at the time it is manufactured with two rows of very accurately spaced sprocket holes. The pitch of this spacing changes from film shrinkage—when the film is aged by being stored, or otherwise—often as much as one-half per cent; that is, five feet per thousand feet of film. When new film is used for recording sound, sometimes there is no shrinkage at all. Most frequently, however, it will be found that the shrinkage amounts to about one-fourth of one per cent.

Since an absolutely uniform motion of the film is neces-

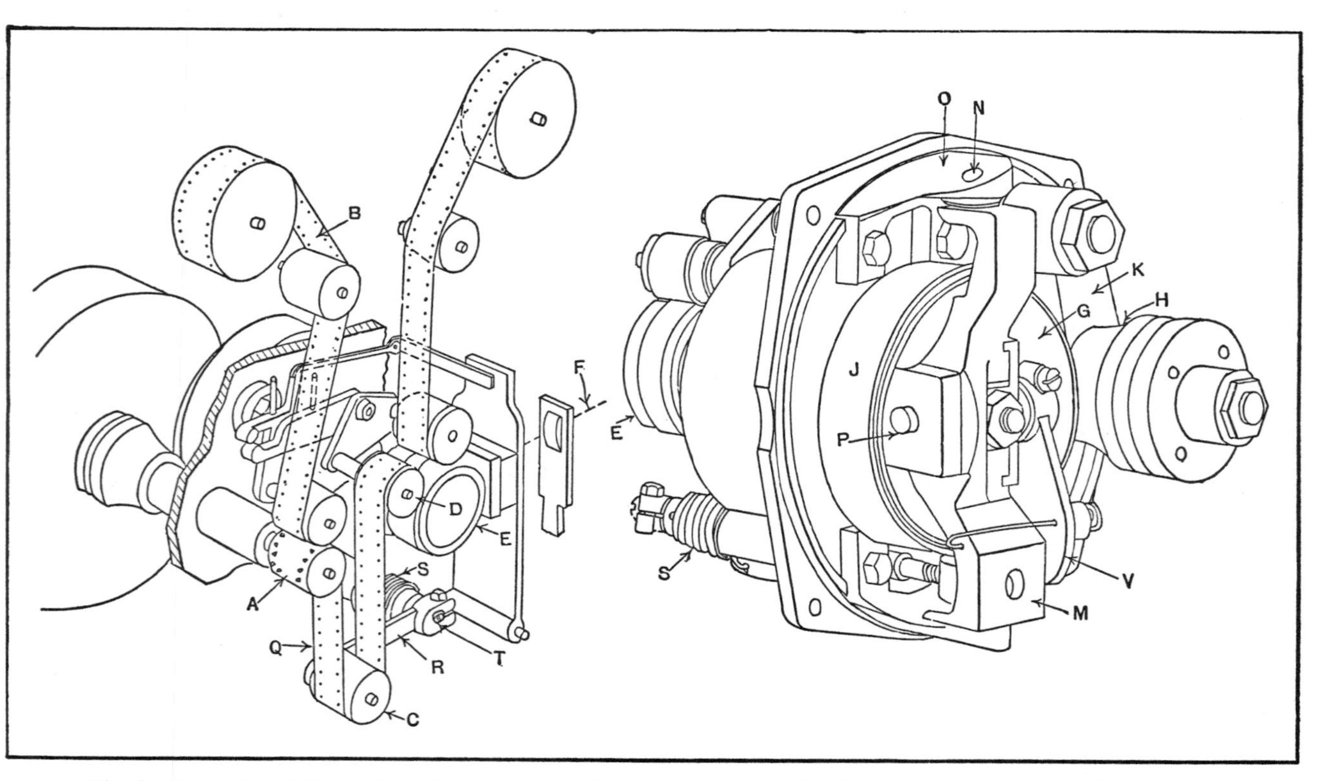

Fig. 9. Automatic and Precise Speed Compensating Mechanism for Maintaining Synchronism between Sound Record and Moving Picture Film Movement

sary for perfect sound recording, means must be provided to compensate for this shrinkage and thereby maintain synchronism between the sound record and sprocket holes, because the sound track record must synchronize with the pictures being taken by one or more cameras operated in synchronism with one or more sound recorders.

Referring to Fig. 9 (view at the left), the sprocket *A* causes the film *B* to travel in synchronism with a film passing through a corresponding camera, because the camera and sound recorder are actuated by synchronized electric motors. The film, after passing under the control roller *C* and over the roller *D*, snugly engages the drum *E* and is thereby caused to pass, with an absolutely uniform motion, through a microscopic oscillating light beam coming from a galvanometer, as indicated by the dotted line *F*.

Evidently, if the film has shrunk considerably, the periphery speed of the drum *E* must be correspondingly reduced in such a precise manner as to maintain the required uniformity in rate of film travel through the light beam. That is, the speed compensating mechanism about to be described must automatically select and maintain a speed ratio between the sprocket *A* and the drum *E* with a high degree of precision for the purpose of maintaining this uniform speed.

The perspective view at the right in Fig. 9 shows the mechanism as seen from the side opposite to that shown by the view at the left. Sectional views are shown in Fig. 10. By referring to these illustrations, it will be seen that control wheel *G* transmits motion from the driver *H* to the driven wheel *J* and drum *E*. The frictional contact between these wheels is maintained by a spring (not shown), which presses the driver *H* and its hinged bearing arm *K* toward the control wheel *G*.

The control wheel is journaled in a gimbal mounting formed by the members *L, M, N, O,* and *P*. The position of the control wheel is governed by the film loop *Q* through

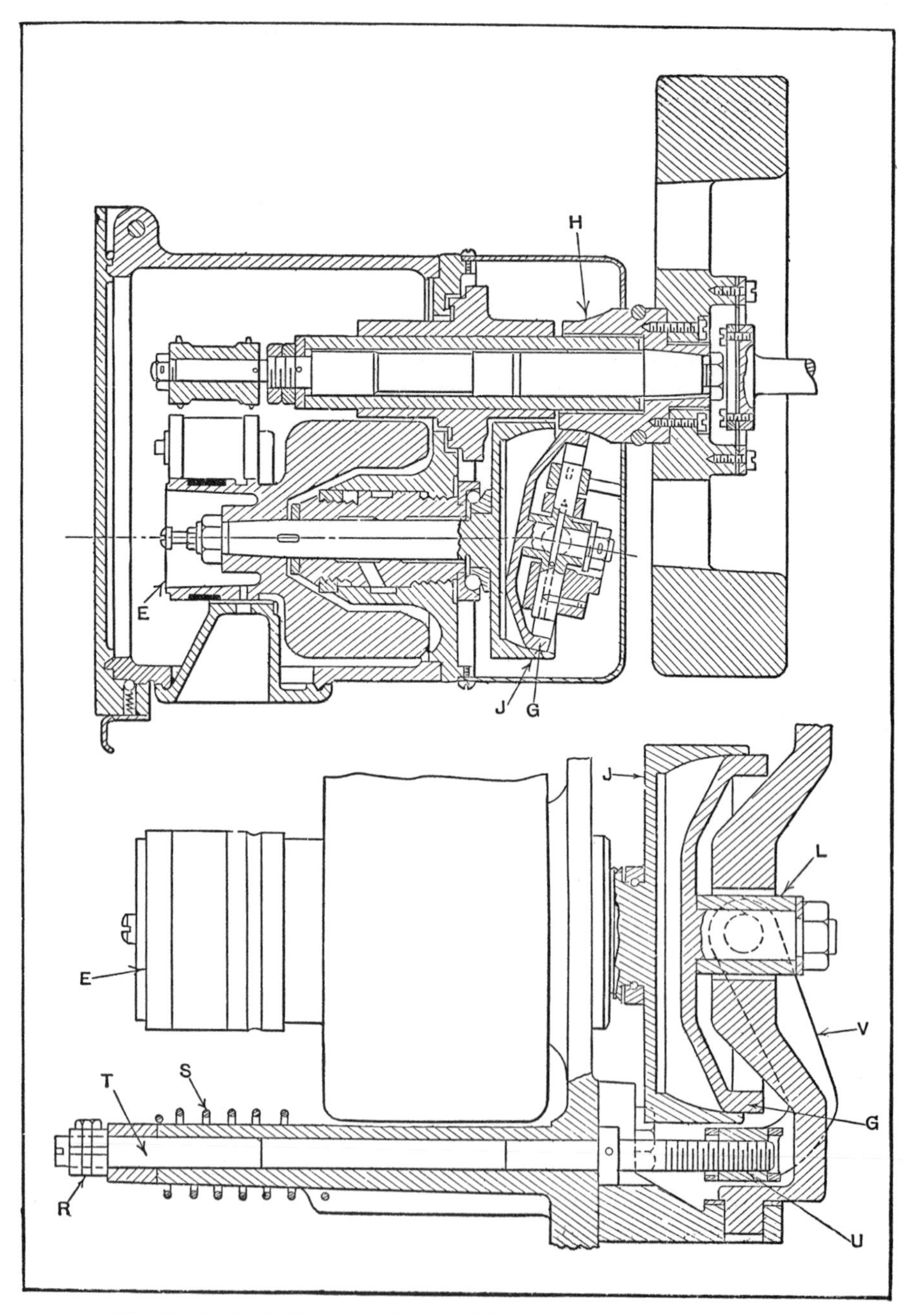

Fig. 10. Sectional Views, Showing how Minute Speed Variations are Obtained

the control roller *C*, lever arm *R*, spring *S*, control screw shaft *T*, nut *U*, and a swivel connecting with lever *V*. This lever is keyed to the trunnion *P* formed on the journal bearing *L* that supports the journal which is part of the control wheel *G*.

Owing to this construction, a very slight change (1/64 inch) in the position of the control roller *C* causes a corresponding change in the position of the control wheel *G* with respect to the driver *H* and driven member *J*. The friction contact surfaces between *G* and *H* are ground spherical, with great precision. The periphery of *G* is given a slight crowning to provide an actual contact friction driving surface of approximately 1/8 inch width. The slightest movement of the lever *V* tilts the control wheel *G* about its contact points with the wheels *H* and *J*, so that it automatically engages a new friction path that gives a corresponding change in speed ratio. Any slight wear or lost motion that may occur is taken up by means of springs.

In the apparatus, the arm *R* is 2 inches long, and a 1/64 inch movement of the control roller *C* gives approximately 1/800 turn of the screw shaft *T*, which has a double thread of 1/7 inch lead; consequently, the axial movement of the nut *U* is approximately 0.0002 inch. This slight movement of the nut automatically causes a new selection of speed ratio, thereby giving an extremely high degree of refinement in speed control. A very short time after starting under normal operating conditions, when running uniformly shrunk film, there is no appreciable change in position of the control roller *C* and its related parts.

Many sound recorders embodying the control mechanism described in the foregoing have been in successful operation for some time. The original model was developed in the Schenectady engineering laboratory of the General Electric Co. and operated daily over a period of two years, continually making perfect sound records without the least trouble.

CHAPTER XI

SPECIAL TRANSMISSIONS AND OVER-RUNNING CLUTCHES

The transmissions described in the preceding chapter are designed to permit changing the speed of the driven member relative to the speed of the driver. Even when speed changes are not required, some special type of transmission or connection between the driving and driven members may be necessary because of their respective positions or to provide for some other operating requirements. Interesting examples of these special transmissions will be found in this chapter.

One-Way Rotation with Reversing Driver.— The purpose of the mechanism shown in Fig. 1 is to obtain a one-way rotation for the driven shaft regardless of the direction of rotation of the driver. In other words, the driver *B* may at any time, at the will of the operator, reverse its rotation without changing the rotation of the driven sprocket *H*, and this result has been accomplished by a very simple mechanism consisting of few parts, as the illustrations show.

Gear *C* and sprockets *B* and *F* are mounted on and keyed to the sleeve, which is bushed and revolves freely about stud shaft *A*. Sprocket *B* is driven by a reversing motor. Spur gear *D* is driven by spur gear *C*. Gear *D*, sprocket *H* and sprocket *G* all revolve freely on stud *E*. Sprocket *F* drives sprocket *G*. Since sprockets *F* and *G* are connected by chain, they rotate, of course, in the same direction. Gear *D* always revolves in an opposite direction to that of sprocket *G*.

In the illustration the clutch teeth of sprocket *H* are

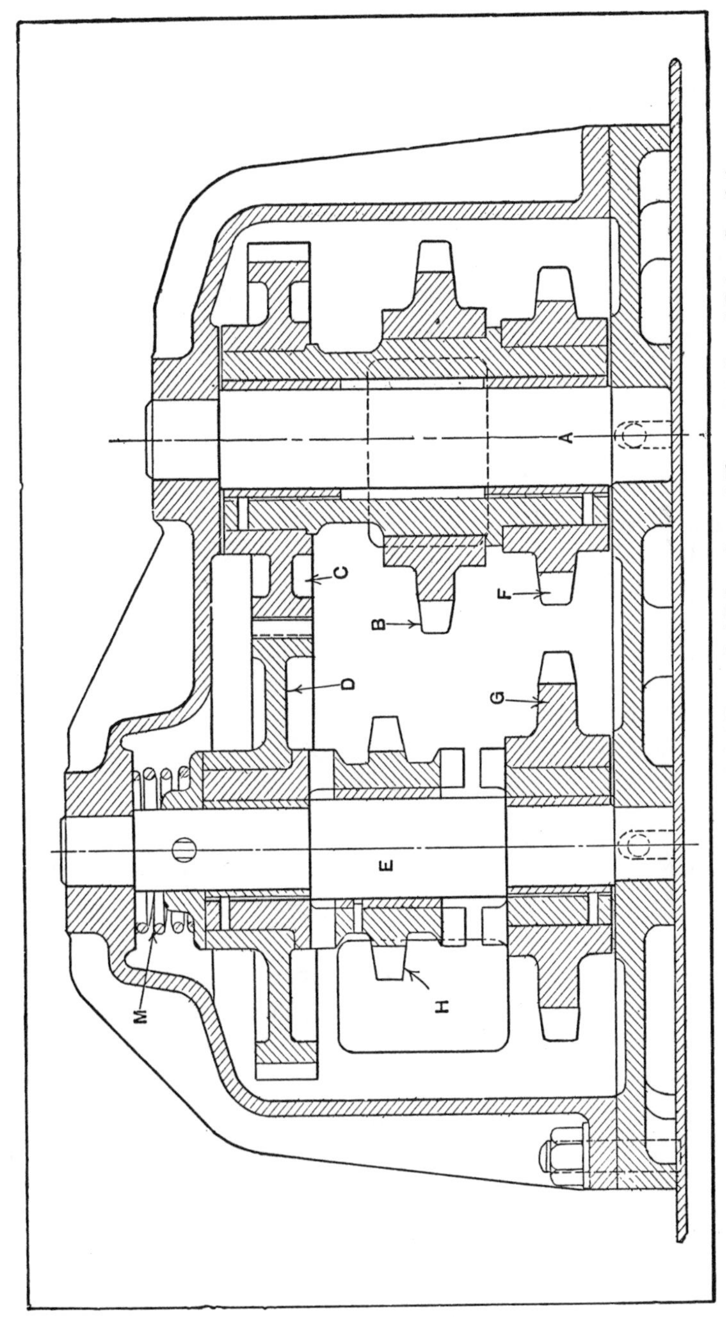

Fig. 1. One-way Transmission so Arranged that Reversal of Driver does not Change Rotation of Driven Sprocket

shown in mesh with the clutch teeth of gear *D*. Assume that gear *D* and sprocket *H* are revolving in a clockwise direction and that sprocket *H* is to continue rotating in that direction. Suppose now that the rotation of sprocket *B* is suddenly reversed to a clockwise direction, thus causing gear *D* to revolve counter-clockwise. This change of motion, owing to the shape of the clutch teeth, will throw sprocket *H* over into engagement with sprocket *G*, and the latter, which, of course, reversed its direction at the same

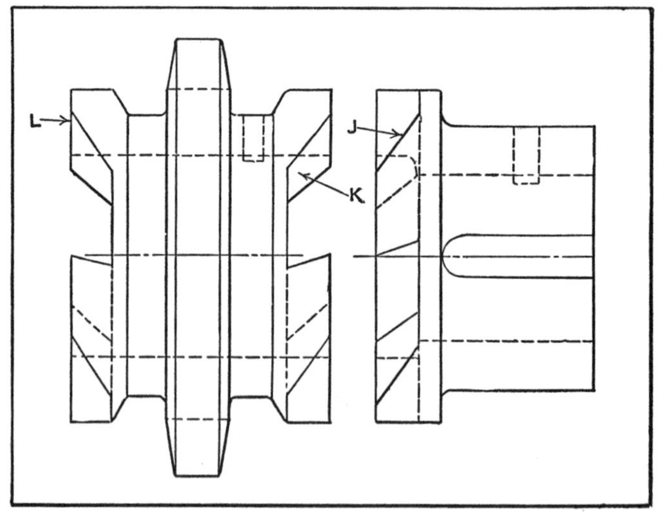

Fig. 2. Driven Sprocket of One-way Transmission and Form of Clutch Teeth Used

time as shaft *B* will now be rotating clockwise, which is the direction desired for sprocket *H*.

Why a reversal of motion causes sprocket *H* to shift from engagement with the gear over into engagement with the sprocket will be apparent by referring to the detail view Fig. 2, which shows this sprocket and the gear clutch. When this clutch (which is a duplicate of the sprocket clutch) is driving the sprocket, the under-cut surfaces of the clutch teeth are in engagement, but when the rotation of the driving clutch member is reversed, the tapering surfaces *J* act against the corresponding tapers *K* on the sprocket, thus exerting a wedging or cam action which thrusts the clutch over into engagement with the opposite side.

When this shifting of sprocket *H* occurs, it is evident that the tops or lands *L* of the clutch teeth might strike the tops of the teeth, say, on sprocket *G* instead of entering the spaces between the teeth. If this should occur, there would be a serious wedging action between gear *D* and sprocket *G,* because the width of the clutch part of sprocket *H* is somewhat greater than the clearance space between the clutches on gear *D* and sprocket *G.* If such wedging action should occur, however, it would be relieved instantly by the lateral movement of gear *D,* which is free to shift against the action of spring *M.* This relieving movement would be followed quickly by the return of *D* to its normal position as the clutch on sprocket *H* snaps into place. The machine on which this device is used consists of two conveyors driven by one reversing motor. One conveyor is reversed at the will of the operator, and the other must travel in one direction only. This mechanism has been fully covered by U. S. letters patent.

Gearless Transmission for Angular Drives.—An unusual form of transmission for shafts located at an angle is shown by the diagram, Fig. 3, which includes a side view and an end view. Motion is transmitted from the driving to the driven shaft through rods which are bent to conform to the angle between the shafts. These rods are located in holes equally spaced around a circle, and they are free to slide in and out as the shafts revolve. This type of drive is especially suitable where quiet operation at high speeds is essential, but it is only recommended for light duty.

The operation of this transmission will be apparent by following the action of one rod during a revolution. If we assume that driving shaft *A* is revolving as indicated by the arrow, then driven shaft *B* will rotate counter-clockwise. As shaft *A* turns one-half revolution, rod *C*, shown in the inner and most effective driving position, slides out of both shafts *A* and *B* during the first half revolution, and rod *C* will then be at the top; then during the remaining half, this rod *C* slides inward until it again reaches the innermost position shown in the illustration. In the meantime, the other rods have, of course, passed through the same cycle of movements, all rods successively sliding inward and outward.

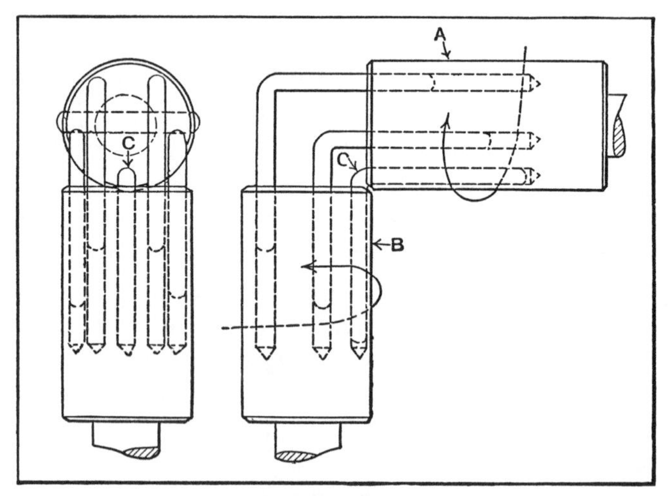

Fig. 3. Gearless Transmission Consisting of Shafts Connected by Rods which Slide in and out as the Shafts Revolve

Although this transmission is an old one, many mechanics are skeptical about its operation; however, it is not only practicable, but has proved satisfactory for various applications, when the drive is for shafts located permanently at a given angle. Although the illustration shows a right-angle transmission, this drive can be applied also to shafts located at any intermediate angle between 0 and 90 degrees.

One application that proved successful was on a special multiple-spindle drilling machine for drilling meter cases. This machine had between thirty and forty spindles equipped with small drills which revolved at 1500 to 1800 revolutions per minute. This transmission was used to replace universal joints consisting of forked ends, each of which was pivoted by means of screws to a connecting block. These universal joints rapidly deteriorated, but the sliding rod transmission proved durable and quiet.

In making this transmission, it is essential to have the holes for a given rod located accurately in the same relative positions in each shaft; all holes must be equally spaced both in radial and circumferential directions. The holes in each shaft must also be parallel to each other, and each rod should be bent to the angle at which the shafts are to be located. If the holes drilled in the ends of the shafts have "blind" or closed ends there ought to be a small vent hole at the bottom of each rod hole for the escape of air compressed by the pumping action of the rods. These holes are also useful for oiling. To avoid "blind" holes, the shafts may have enlarged ends with holes extending clear through the enlarged part or shoulder. This transmission may be provided with a central rod, located in line with the axis of each shaft and provided with a circular groove at each end for a cross pin to permit rotation of the shaft about the rod, the central rod simply acting as a retaining device for shipping or handling purposes.

Changing Relative Positions of Two Revolving Shafts.—The relative positions of two revolving shafts may be varied while in operation by the mechanism shown in Fig. 4. The speed ratio between shafts *A* and *B* is constant, but the relative rotative positions of the two shafts can be varied by means of the worm *H* and the worm-gear *M*. This mechanism is used in a wire-forming machine on which the timing of one section of the machine must be changed relative to the timing of another section.

The driving shaft *B*, supported by bearings *D* and *E*, carries the gear *J*, which is keyed to it. Bracket *F* is free on shaft *B*, and carries the stud *N*, which supports gears *I* and *K*. The two latter gears are free to rotate as a unit. Gear *L* is keyed to shaft *A*, which is supported by bearing *C*. Each of the gears *K* and *L* have 18 teeth, while gears *I* and *J* have 24 and 12 teeth, respectively.

Power is transmitted from shaft *B* through gears *J*, *I*, *K*, and *L* to shaft *A* at a 2 to 1 ratio in the same direction. Worm *H* is keyed to shaft *G*, which is supported on the upper end of bearing *E*. Shaft *G* carries a handwheel (not

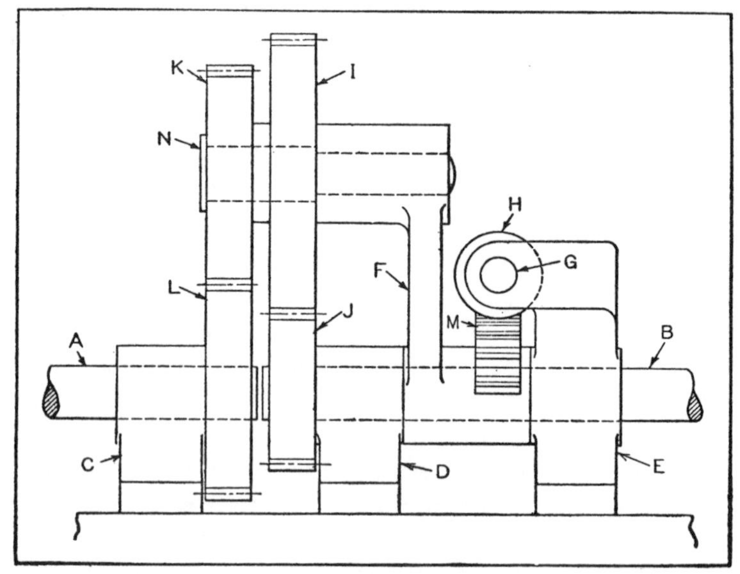

Fig. 4. Mechanism for Varying Relative Rotative Positions of Driving and Driven Shafts

shown) on one end, by means of which the worm is rotated. Worm *H* meshes with the worm-gear *M* on bracket *F*, causing the latter to rotate on shaft *B* when the handwheel on shaft *G* is turned. In Fig. 5, the dotted outlines of bracket *F* and gear *I* show the rotative movement around shaft *B* produced by the action of worm *H* and worm-gear *M*.

A better understanding of how the change of timing between shafts *A* and *B* is accomplished may be had by assuming these shafts to be stationary. Then, as bracket *F* is caused to rotate on shaft *B* as an axis, gear *I*, meshing with gear *J*, will be rotated on stud *N* as an axis in the ratio of 2 to 1. As the axes of shafts *A* and *B* coincide, gear *K* rotates around shaft *A*.

If gear *K* were independent of gear *I* it would revolve on stud *N* as an axis in the ratio of 1 to 1 with shaft *A*, but as gears *I* and *K* are fixed together, and must revolve as a unit at a ratio of 2 to 1 with shaft *B*, gear *L* will be revolved in the reverse direction. The ratio of rotation between

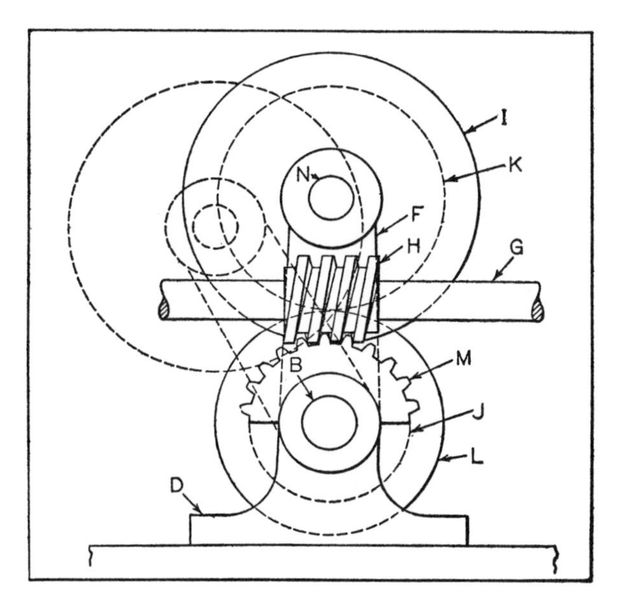

Fig. 5. End View of Timing Mechanism Illustrated in Fig. 4

gears *K* and *L* will still be 1 to 1, but each of the gears will revolve one-half turn in opposite directions. As gear *L* is keyed to shaft *A*, the latter must revolve with it. Under actual operating conditions, the effect is merely that of a gear train until the handwheel is turned, at which time shaft *A* is advanced or retarded as the bracket *F* is rotated forward or backward.

Changing Angular Velocity of Driven Member Twice During Each Revolution.—The velocity-changing mechanism to be described is incorporated in a hat-finishing machine. The finishing of the hat is done by a pad faced with sandpaper. This pad automatically travels from the top or center of the hat to the band while the hat is being revolved. An automatically controlled oscillating movement is also imparted to the finishing pad. The oval rotary motion mechanism referred to is required in order to keep the work in proper contact with the finishing pad.

Now, in order to finish the hat evenly all over, it is necessary to vary the angular velocity of the chuck on which the hat is mounted. If the spindle were revolved at a uniform speed, the front and back portions of the crown would be over-finished and the sides under-finished as a result of the difference in surface speed caused by the oval shape of the work. It is the purpose of the mechanism shown in Fig. 6 to increase the speed of rotation of the spindle as the front and back surfaces pass under the finishing pad.

As the oval head revolves, the quarter sections between the side and end portions must lift the working pad 3/4 inch in one-quarter revolution, thus increasing the pressure on the pad. As it passes from the end portion to the lower side, the pad drops 3/4 inch in one-quarter revolution, and this falling action gives less pressure on that portion. The oval head mechanism so acts that all portions of the section of the body with which the pad is in contact are raised to the same plane.

The same oval shape condition makes necessary the use of the differential angular velocity mechanism. The curvature of the end portions is sharper and the contact area of the pad on the hat is, therefore, less on these portions and the intensity of pressure greater. The side portions, being flatter, are subjected to less pressure by the pad. This condition would result in cutting the ends faster than the sides if the differential angular mechanism did not move the ends more quickly under the pad, the change in velocity being adjustable to suit the shape of hat.

The block that carries the hat is revolved by the spindle

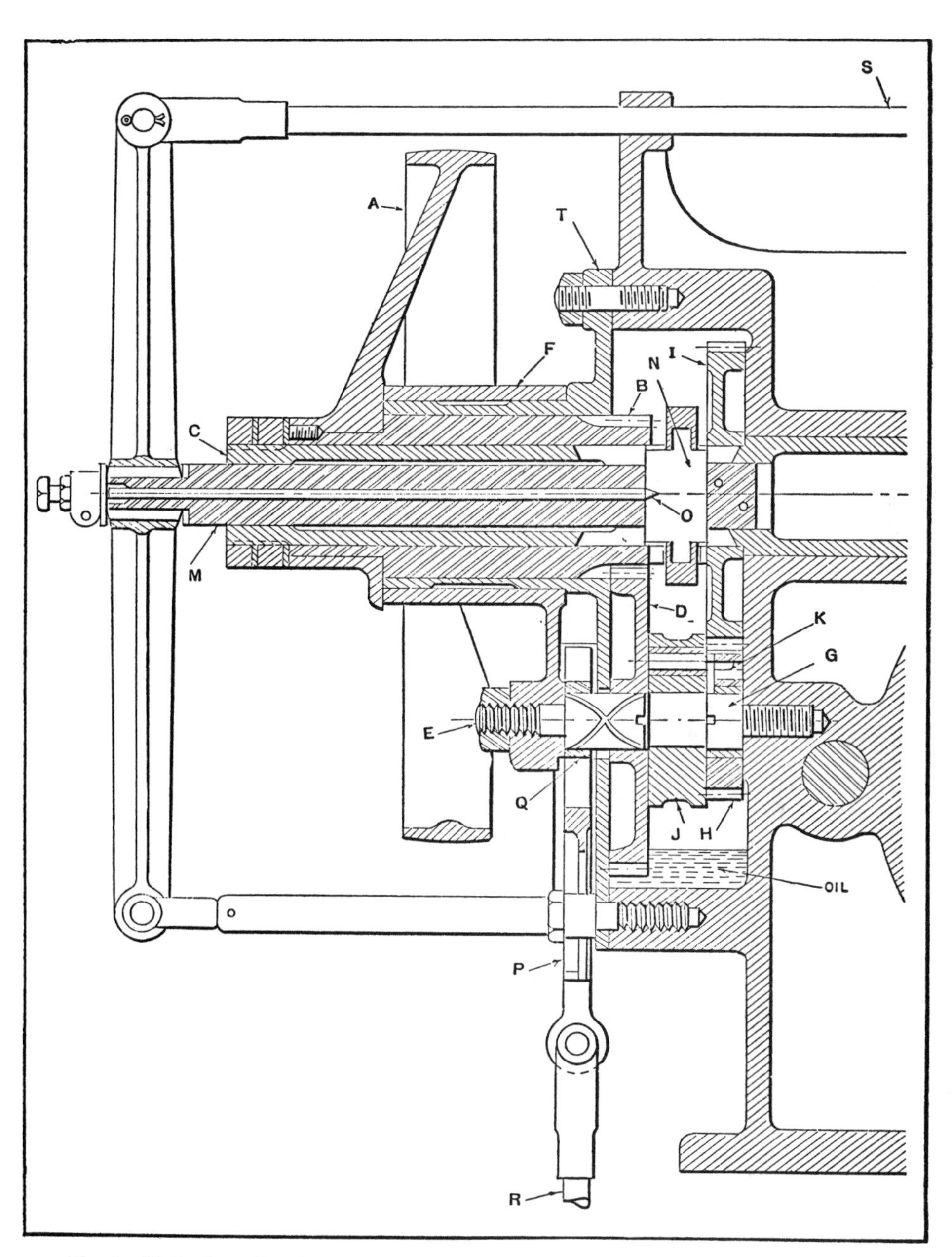

Fig. 6. Mechanism which Increases and Decreases Angular Velocity of a Shaft Twice During Each Revolution

C which extends through the machine frame at the right. By means of the control lever S, the spindle C can be connected directly with the driving pulley A so that it will run at the same speed, which might be required in finishing the band of a cylindrical or perfectly round hat. By moving the control rod S in the opposite direction, the variable velocity mechanism is brought into action. This mechanism causes spindle C to make one complete revolution to every four revolutions of the driving pulley A.

By means of the velocity-changing mechanism, the angular velocity of spindle C can be increased and decreased twice during one complete revolution. This change in velocity is synchronized with the oval motion mechanism, so that a uniform finishing effect is obtained. The amount of variation in the velocity is controlled by the cam-operated plate P which is actuated by the same cam that controls the oval motion mechanism.

Operation of Mechanism for Changing Angular Velocity.— The operation of the mechanism may be described by following the drive through from the pulley A to the spindle C, Fig. 6. Passing through a slot or opening in the push-rod M is a clutch finger N which is securely locked in the push-rod by the pointed rod O. Corresponding with the clutch finger opening in the push-rod are similar openings in the spindle C through which the clutch finger N freely passes.

When the push-rod M is moved backward by the control rod S, the clutch finger N engages the clutch face of the pinion B. The spindle C, through this contact, is driven at the same speed and in the same direction as the pinion B and its driving pulley A. When operating in this manner, the gear D, pinion H, and gear I all rotate on their free bearings and do not have any effect on spindle C.

When the push-rod M is pulled forward by rod S, the clutch finger N is disengaged from pinion B and engages the clutch face of gear *I* which then drives the shaft *C*. The drive is now transmitted through the pinion *B* to gear *D*, through link *J* to pinion *H* and thence to gear *I* at a reduction in speed of four to one as compared with the direct drive from the pulley *A*.

The gear *D* is mounted on the stud *E* secured to the sleeve member *F*, which can be swung about the hub bearing on the cover plate *T*. When the stud *E* is in axial alignment with stud *G*, gears *H* and *D* have the same speed and this speed is constant. The slide *P* has an angular slot in it which fits over the slide shoe *Q* on stud *E* and serves to move stud *E* of gear *D* out of alignment with the stud *G* in accordance with the movement imparted to rod *R* by the cam that also controls the oval mechanism.

In Fig. 7, angle x and dimension y indicate the amount

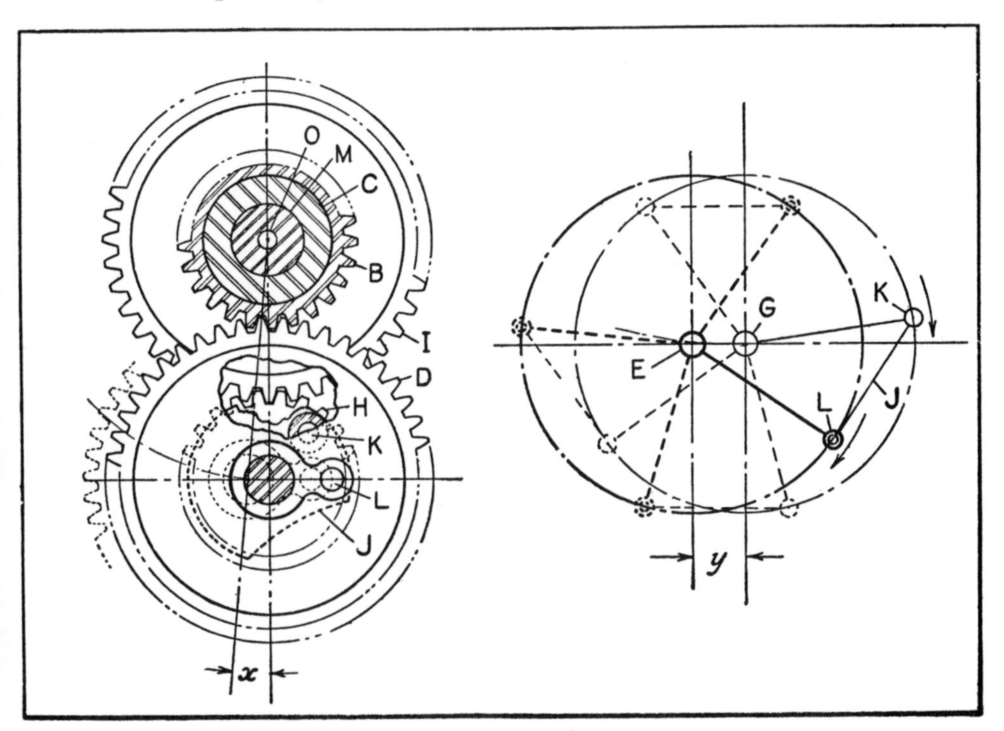

Fig. 7. Diagrams Used to Illustrate Operation of Velocity-changing Mechanism

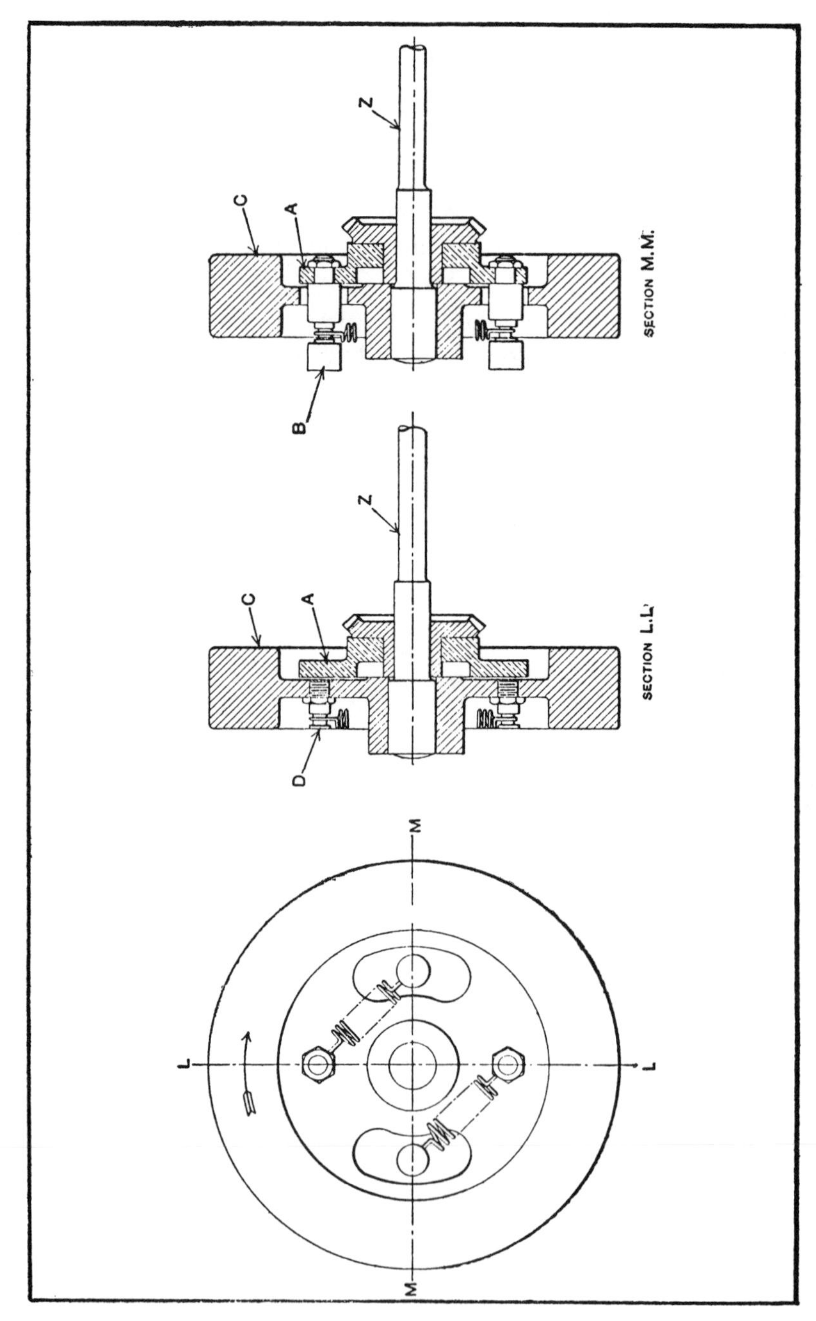

Fig. 8. Device for Converting Intermittent Motion to a Constant Drive

of "out of line" adjustment between the studs *E* and *G*. This adjustment is varied automatically from zero, when the finishing pad is at the top or center of the hat crown, to the maximum amount, when the finishing pad reaches the band of the hat. It is this changing of the axial relation of gear *D* and pinion *H* that controls the differential velocity. The diagram to the right will help make clear how the changing of the relative angular positions of the studs *L* and *K* and their connecting link *J* produces a variation in the velocity of the driven pinion *H*.

The change in angular velocity occurs once in each revolution of pinion *H*, but since pinion *H* makes two revolutions for each one of driven gear *I*, the change in the angular velocity of the latter gear and its spindle *C* occurs twice during each revolution. By changing the linkage connections between the cam and plate *P*, Fig. 6, it is possible to control the amount of variation in angular velocity. A segmental circular opening is provided in the lower portion of the cover plate *T* to allow sleeve *F* to swing on its hub bearing. The link *J* has no center bearing, and is held in position by the two drive pins *K* and *L*. The pin *L* projects from the inner face of gear *D* into a hole in link *J*. The pin *K* is made eccentric for convenience in assembling, but it would be possible to use a straight pin, as far as the operation of the mechanism is concerned.

Intermittent Motion Converted to a Constant Drive.—All moving projectors are equipped with some sort of intermittent device to cause the film strip to dwell at every picture. This is necessary, because running the film through the projector at a constant speed would result in a blur on the screen. With the advent of the "talkies," however, this motion had to be reconverted to a steady drive for the film strip while it passes through the sound-producing attachment, as otherwise the latter would not function correctly. This was done in one instance with the arrangement shown in Fig. 8.

The intermittent motion is transmitted to disk *A* through bevel gearing. This disk is equipped with two studs *B* which pass through elongated slots in the web of the flywheel *C* and are connected to two studs in the flywheel web by coil springs. The shaft *Z* is keyed to the flywheel. The intermittent motion transferred to disk *A* will be absorbed through the combination of the coil springs and the fly-

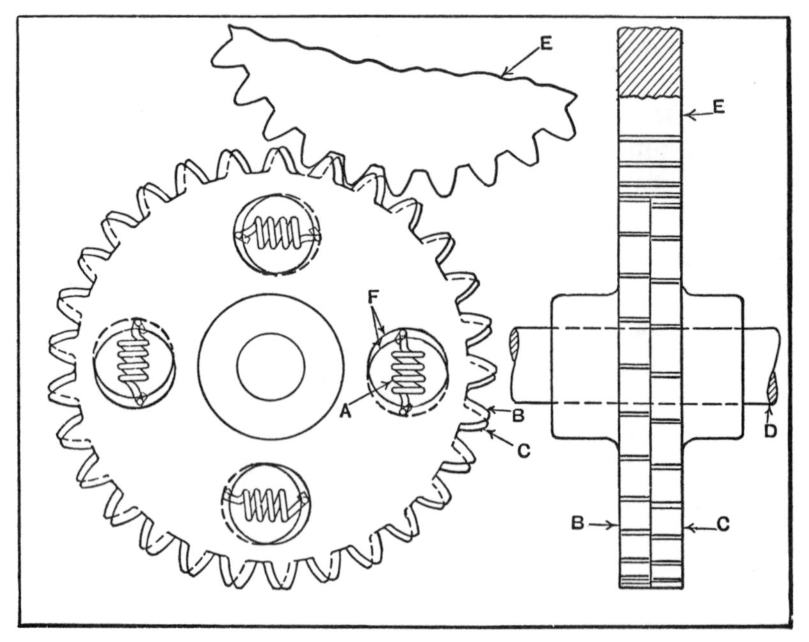

Fig. 9. Device for Eliminating Lost Motion in Gear Teeth

wheel, resulting in a constant speed for the shaft *Z*, which is the drive for the sound-producing attachment.

Double Gear for Eliminating Lost Motion in Gear Teeth.— In devices where gears are used, such as those operating graduated dials, it is often desirable to eliminate all lost motion resulting from wear occurring between the gear teeth. The gears shown in Fig. 9, which are only recommended for light transmissions, impart motion to a dial indicator that must register in both a clockwise and a counter-clockwise direction. It is obvious that any play in the teeth of the gears would produce inaccuracies in the dial readings.

The driving gear *E* meshes with both the gears *B* and *C*, the latter being fastened to the shaft *D*. In both gears *B* and *C* are drilled holes *F* for the springs *A*. One end of each spring is secured to gear *B* and the other end to gear *C*. The thickness of the teeth in both these gears is less than that of the driving gear *E*, so that normally, there would be considerable play between the meshing teeth. However, owing to the tension of the springs *A* the teeth in gear *C* are advanced ahead of those in gear *B* and serve to fill the tooth spaces in the driving gear. In this way, as wear occurs, it is obvious that all lost motion in the gear transmission is eliminated, and that no matter in which direction the gears are run, there will be no play between the teeth nor inaccuracy in the dial readings.

Over-Run Pawl Clutch to Permit Accelerating Driven-Shaft Speed.—The pawl clutch shown in Fig. 10 has been used very successfully in machines having camshafts or feed-screws which must be driven at accelerated or high speeds a part of the time. The low-speed shaft *S* drives the high-speed shaft *H* under normal operating conditions through the ratchet wheel *A* and pawls *G* of the collar *K*, which is keyed to the high-speed shaft. When the speed of shaft *H* is to be accelerated, a clutch mechanism, not shown, is engaged, and this drives the high-speed shaft at the accelerated speed through gearing connected with shaft *H* by gear *J*. The arrangement of the over-run clutch is such that it permits shaft *H* to be operated at the higher rate of speed without affecting the speed of shaft *S*.

Referring to the construction of the clutch, ratchet wheel *A* is keyed to shaft *S*. The pressure disk *B* is a sliding fit on shaft *S*, and is driven by ratchet *A* through pins *C*. The tension springs *D* tend to force the pressure disk *B* inward toward the ratchet wheel. Between disk *B* and the ratchet

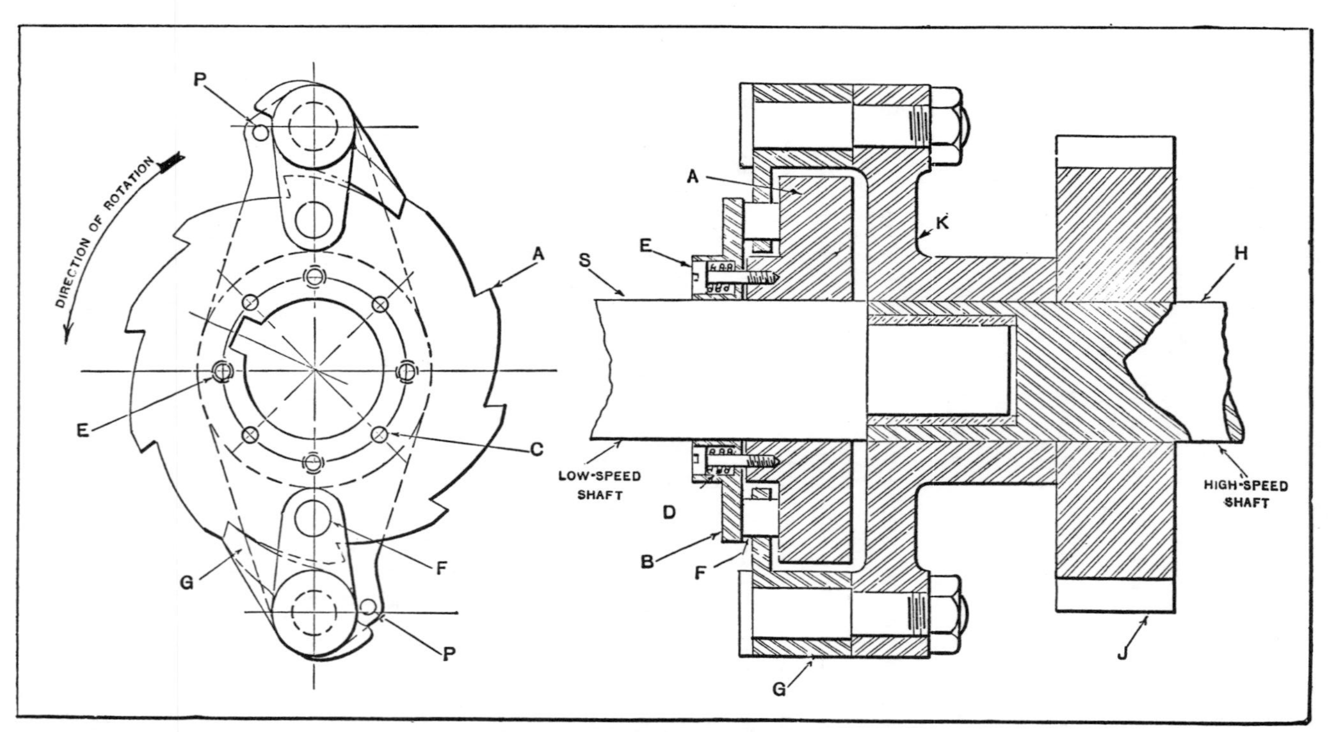

Fig. 10. Over-run Clutch which Permits Speed of Driven Shaft to be Accelerated without Affecting Speed of Driving Shaft

wheel are fiber drag-plugs *F*, which are set in holes in the ways of the pawls *G*. Behind the pawls are stop-pins *P* which keep the drag-plugs from swinging through too large an arc and thereby becoming entirely disengaged from the disk and the ratchet wheel. When the high-speed clutch is tripped and the gear *J* is driven at a higher speed than the ratchet wheel *A*, the pawls *G* are disengaged, and the plugs *F*, which then drag between the pressure disk and the ratchet wheel, cause the pawls to swing clear of the teeth in the ratchet member. The pawls remain in this disen-

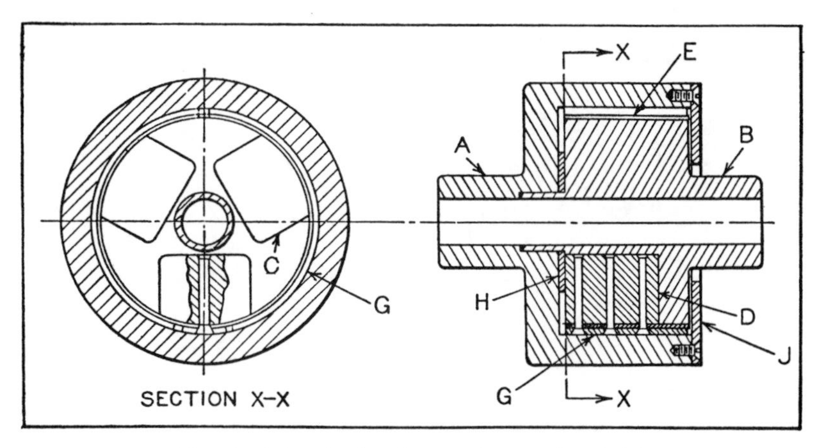

Fig. 11. Automatic Starting and Over-running Clutch

gaged position until the high-speed clutch is disengaged, at which time the drag on the plugs *F* is reversed, causing the pawls to be drawn down into engagement with the teeth and the ratchet wheel again.

Centrifugally Operated Starting and Over-Running Clutch.— In Fig. 11 is shown an automatic starting and over-running clutch of the centrifugal type. It consists of the driven housing *A*, in which rotates the driving clutch member *B*. Three cavities *C* are milled in member *B* to accommodate the three sliding weights *D*. To one of these weights is riveted the spring steel band *E*. Over the steel band and riveted to it is the brake lining material *G*. At *H* is a steel thrust washer, while at *J* is a retainer plate for keeping both halves of the clutch assembled.

The operation of the clutch is as follows: As the clutch is rotated by the driving member *B*, centrifugal force causes the weights *D* to move outward. This, in turn, expands band *E*, forcing the brake lining into contact with the inner surface of housing *A*. As long as the speed of driver *B* is maintained or increased, the gripping force is also maintained or increased, but the moment it is reduced, either from the slowing up of the driving force or because of resistance set up in the driven member, the sliding

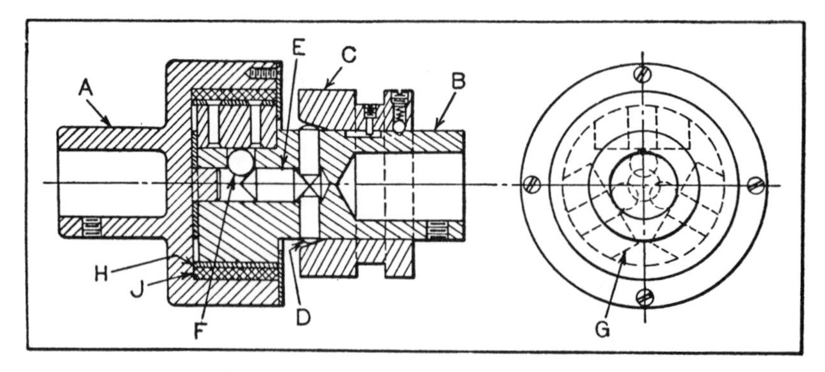

Fig. 12. Modified Design of Clutch Shown in Fig. 11

weights are forced inward by the spring band *E*, immediately disengaging the clutch and allowing half of the clutch to slip upon the other half. If the driving force is entirely cut off, the clutch disengages, allowing the driven half to over-run until it comes to rest.

Simplicity of construction and a large contact area are advantages of this clutch. It should be noted that the effective gripping power can be increased without increasing the diameter by merely lengthening the clutch. It should also be noted that, for a given velocity, the force exerted is proportional to the mass of the weights *D*, so that the larger the housing of the clutch. the greater will be the gripping force. This type of clutch can also be made in the form of two halves of a brake-shoe, a centrifugal weight being used to force the two halves apart. However, this type of construction would involve a problem of dynamic balancing.

Fig. 12 shows the same type clutch employed in the reverse manner. Here the housing *A* serves as the driver, the driven member being the housing *B*, which is caused to rotate when the clutch collar *C* is forced over the pins *D*. These pins, of which there are three, force the connector pin *E* against the three steel balls *F*. Pins can be used in

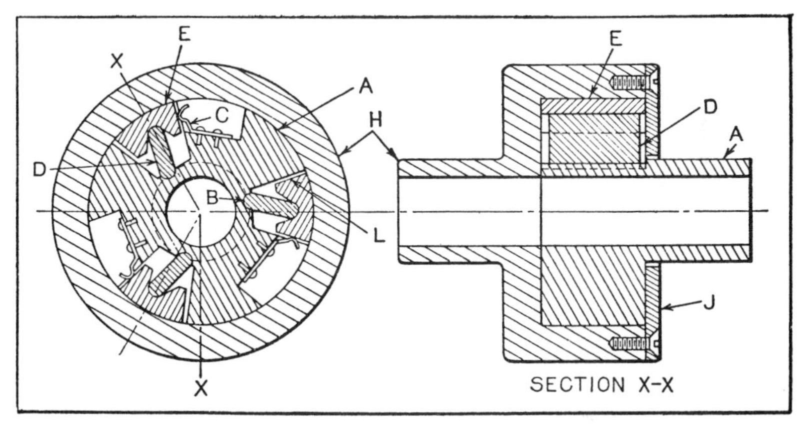

Fig. 13, Over-running Clutch with Toggle-actuated Shoes

place of the balls, if desired. The balls *F* force the three equally spaced sliding blocks *G* outward, causing the spring steel band *H* to expand, so that the brake lining material *J* will grip the outer housing, thus connecting the driving and driven members. The screw-pin is for the purpose of connecting the shaft collar *C* to the housing *B*, while the ball spring arrangement is provided to hold the shaft collar in either the open or the closed position.

Toggle Type of Over-Running Clutch.— An over-running clutch employing the toggle joint principle for obtaining the required locking action is shown in Fig. 13. It consists of the spider *A* having three milled slots spaced

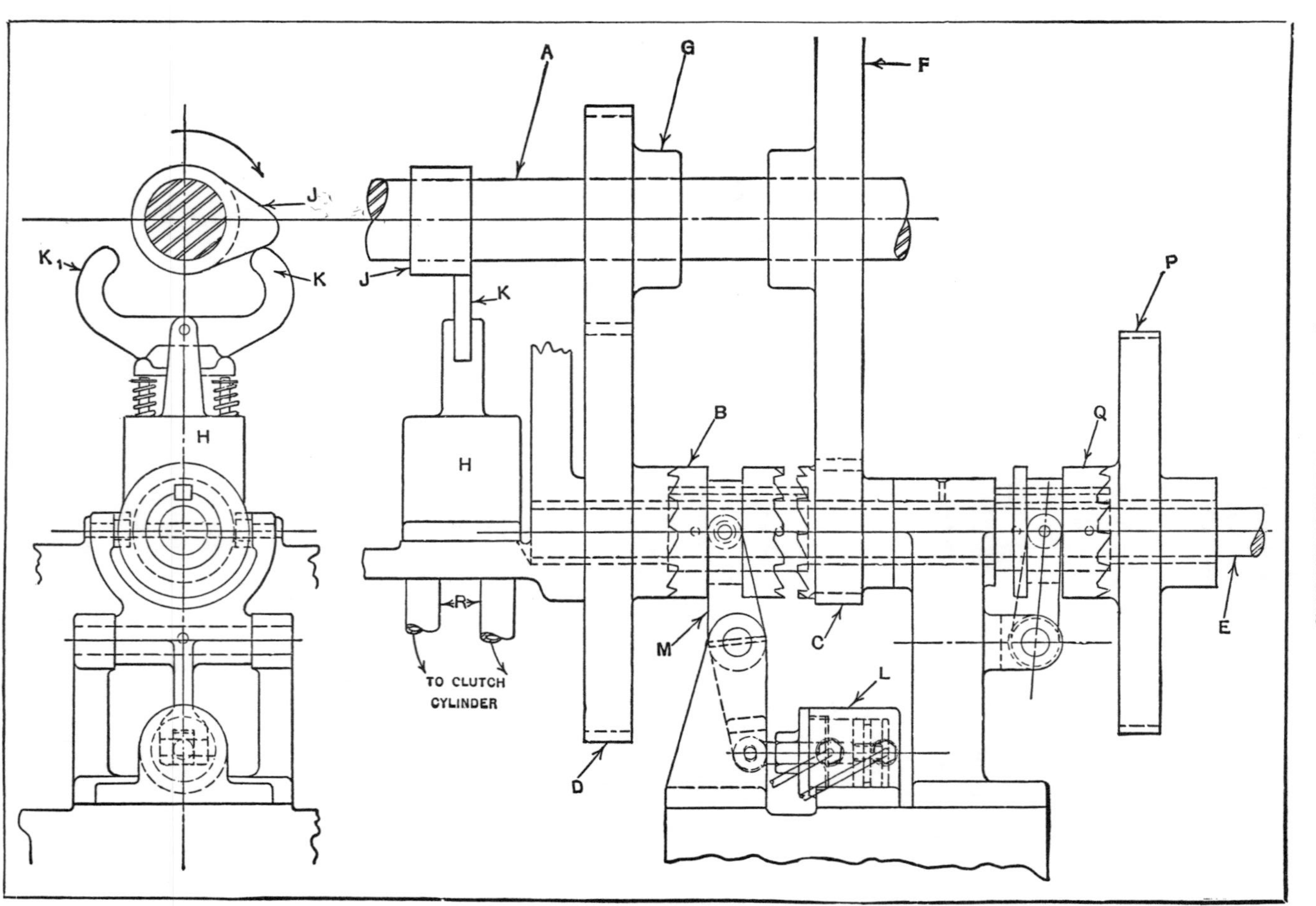

Fig. 14. Automatically Controlled Air-operated Clutch for Alternately Engaging High- and Low-ratio Gearing

120 degrees apart, the outer casing *H*, and the toggle levers *D*. At points *B*, which are slightly offset from the center lines of the slots, are milled semicircular recesses which act as seats for the toggle levers *D*. The shoes *E* are made slightly smaller than the width of the slots and have an outside diameter equal to the diameter of the spider *A*. Seats are milled in the rear sides of the shoes for the toggle levers *D*. The flat springs *C*, fastened in the slots in the spider, tend to keep the shoes *E* in contact with the inside of the outer casing *H*. A retainer plate *J*, held in place by screws, keeps the members of the clutch assembled.

The operation of the clutch is similar to that of any free-wheeling clutch. The toggle levers *D* are set at as slight an angle as possible, making due allowance for wear on the shoes *E*. After the shoes become worn, the faces *L* are machined or cut back. When the shoes have been cut back so that the toggles *D* become ineffective they are replaced.

Air-Operated Clutch for Two-Speed Drive.—The speed of a shaft that drives the feeding and indexing mechanism of a multiple-spindle drilling machine must be increased from 2.13 to 15 R.P.M. to permit indexing in 1 2/3 seconds. This speed change is controlled by a cam-operated, four-way air valve and an air-operated clutch which alternately engages the high- and low-ratio gearing.

Starting and stopping of the machine is controlled by hand-operated clutch *Q* (Fig. 14) which connects or disconnects gear *P* with shaft *E*. The main shaft *A* drives the feeding and indexing mechanism only. When the machine is drilling and the feeding mechanism is in operation, motion is transmitted from shaft *E* to *A* through the low-speed gearing *C* and *F*, as a result of the engagement of clutch *B* with gear *C*. When an indexing movement is required, clutch *B* is automatically shifted into engagement with gear *D*, thus driving shaft *A* through gears *D* and *G* and increasing the speed to 15 revolutions per minute, so that the indexing will be completed in the allotted time;

then the clutch is shifted back automatically to the feeding position.

Clutch *B* is shifted by means of compressed air acting against a piston which is within cylinder *L* and is connected to clutch yoke *M*. The admission and exhaust of the air to and from cylinder *L* is controlled by a cam *J* acting in conjunction with a four-way valve *H*. This valve connects with the main air line, and there are two 3/8-inch pipes *R* leading from it, provided with reducing bushings to fit the small tubes connecting with each end of the air cylinder *L*. When cam *J*, which is attached to shaft *A*, moves in the direction of the arrow, it comes into contact with end *K* of the air valve operating lever, thus admitting air to the left-hand end of the cylinder and exhausting it from the opposite end. The result is that clutch *B* is thrown into engagement with gear *D*, as the illustration shows. This high-speed drive of 15 revolutions per minute continues until cam *J* engages end K_1 of the lever, thus admitting air to the right-hand end of the cylinder and exhausting it from the left, which throws the clutch into engagement with gear *C*. An air pressure of 80 pounds per square inch is carried in the main line, and the total pressure exerted against the piston in cylinder *L* is about 98 pounds.

CHAPTER XII

SELF-CENTERING PIVOTED LEVERS AND SLIDING MEMBERS

It is sometimes necessary or desirable to provide certain machine elements with mountings that will permit them to be deflected from or forced out of their normal positions by other parts of the machine. Usually such elements must be so designed or equipped that they will return automatically to their normal positions when the deflecting forces are removed. The term "self-centering" is used in reference to the devices shown in Figs. 1 and 2, because they serve to return the elements to their central or normal positions when the deflecting force is removed. Probably the most familiar example of a self-centering device is the spring-actuated control rod employed to operate a dog-tooth clutch. With this type of clutch, the two members may not be in the correct angular relation to permit them to engage when the control lever is thrown over. Under these conditions, it is the function of the self-centering spring-actuated control rod to yield and permit the clutch lever to be thrown over without unduly straining the mechanism, causing the two parts of the clutch to engage as soon as they are in the proper positions.

Another application of self-centering devices is to the control levers of power-operated machines. For such applications, the lever is normally held in the neutral or central position by the self-centering device, and is returned to this position automatically as soon as it is released by the operator. Manual operation of the lever in one direction or the other places the machine either in forward or reverse motion, and its release causes the machine to stop.

Self-Centering Devices for Angular Movement.—The attachment of a weight, pendulum fashion, as shown at *A*, Fig. 1, is one of the simplest examples of a self-centering device having an angular movement. A device of this kind may be used for a lever having a range of movement of 20 degrees each side of the vertical center line. This type of self-centering device, however, may be objectionable on account of the inertia introduced by the weight. Another objection is that the weight offers little resistance to angular movements of small amount, as the self-centering force is zero at the central position. Still another objection is the tendency of the weight to oscillate after displacement.

The device shown at *B* is similar to the one at *A* except that spring tension is substituted for the weight. This type of self-centering device is often used and is fairly effective for some purposes. The self-centering effect, however, is zero at the mid-position of the lever, and effective centering forces are not developed until a considerable angular movement of the lever has taken place. The lever also has a tendency to vibrate after the decentralizing forces begin to act. In practice, short stiff springs are usually employed, which cause a heavy pressure to be exerted on the bearings or pivots of the lever.

The self-centering device shown at *C* typifies approved practice. Two clip levers pivoted about the boss of a lever are employed in this device. A pin carried by the lever is gripped between the two clip levers with a force depending upon the initial tension of the spring connecting the ends of the clip levers. Another pin of the same width as that on the lever is fixed to the stationary framework of the machine and interposed between the two clip levers. The result is that no movement of the lever in either direction can take place without forcing the clip levers apart against the resistance of the spring. The majority of self-centering problems involving angular movement can be solved by the application of the principles embodied in this device.

Self-Centering Device Applied to Electric Switch.—The self-centering device employed on a rotary snap-action electrical switch is shown at *D*, Fig. 1. The principle on which this centering device operates is similar to that of the device shown at *C*. In this case, a C-shaped spring is used to force the clip levers together, as the angular movement is only a few degrees. The centering forces are equal in both directions of angular movement in the devices shown at *C* and *D*.

It is sometimes desirable, however, that the decentering forces be more strongly resisted in one direction than in the other. A modification of the design shown at *C* to meet this requirement is illustrated at *E*. The unequal tension is obtained by employing two separate springs of unequal strength for the two clips. As the springs are anchored to the machine framework, it is a simple matter to provide them with means for adjusting the tension.

The designs shown at *C* and *E* have proved highly satisfactory in general practice, but when absolute precision in the self-centering action is necessary, the type shown at *F* is preferable. With the designs shown at *C* and *E*, any difference in the size of the two pins located between the clip levers will result in lost motion, whereas the design shown at *F* is free from this possibility.

In this design, one of the clip levers is dispensed with, and a spring is used to pull the main lever and the clip lever together. This provides a definite self-centering action without the possibility of the slightest lost motion, and has been successfully employed in the design of a shock absorber in a train of gearing connecting a gun to its elevation indicator. Under the shock of the recoil of the gun, the device yields, but immediately regains its proper position with respect to the indicating pointer. In this application, any lost motion would destroy the accuracy of the elevation indicator.

Roller and Spiral Cam-Operated Centering Device.—In the case of the self-centering device shown at *G*, **Fig. 1,**

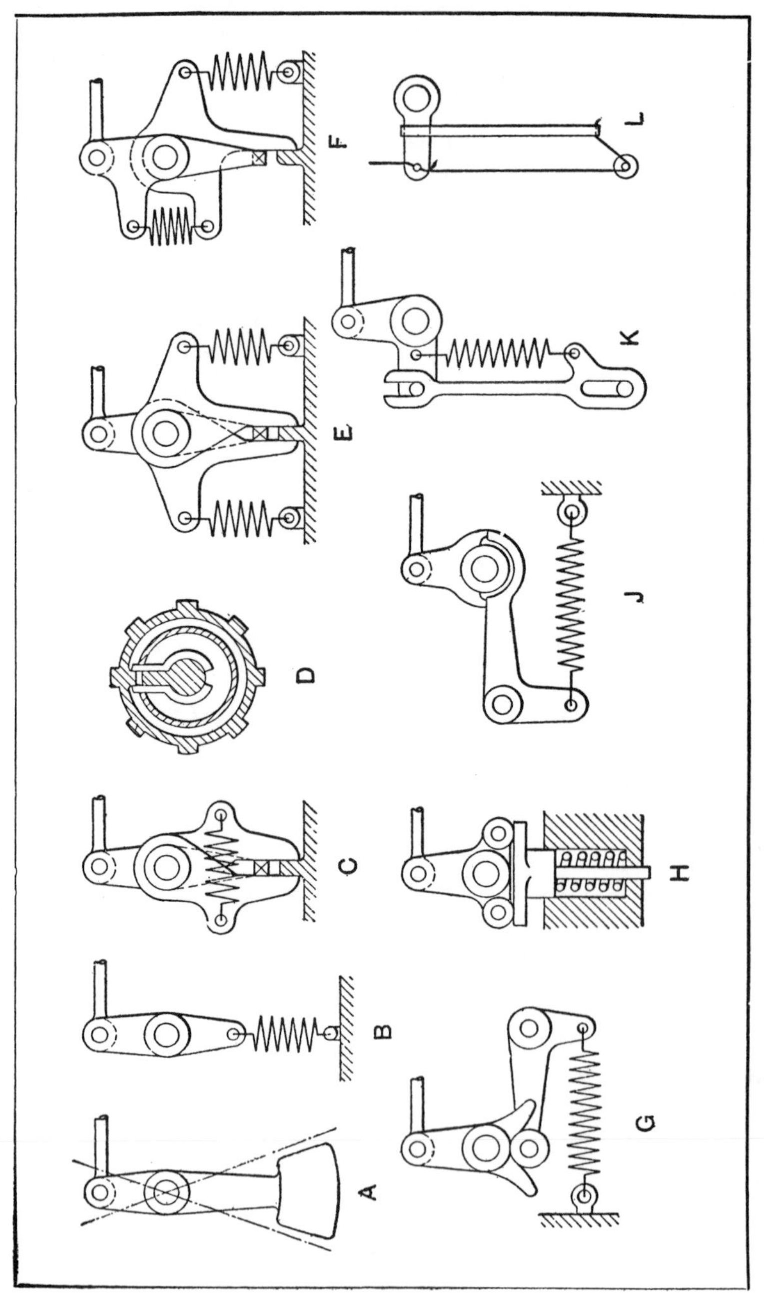

Fig. 1. Examples of Various Types of Self-centering Levers

the actuating forces are applied to the lever by a roller which is pressed into engagement with a spirally shaped cam through the action of a coil spring. The arrangement is similar to that of the "zeroizing" cam lever used in stop watches. By shaping the cam curves properly, any desired variation in the centering force can be obtained from the spring that presses the roller radially inward. A heavy pressure on the roller and the lever bearings would result if the pitch of the spiral cam were made small in order to secure a wide range of angular movement, and in such a case, it would be advisable to use ball bearings for the roller and for the lever bearings.

Another design, consisting of two arms extending in opposite directions and carrying rollers that are subjected to spring pressure from a radially sliding T-shaped piece, is shown at *H*. When the main lever is in the central position, the T-shaped piece exerts an equal pressure on both rollers. Angular movement of the lever in either direction results in one or the other of the rollers depressing the T-shaped piece against the action of the spring.

At *J* is shown a design similar to that at *H*, except that the rollers are omitted and a pivoted lever is used in place of the T-shaped piece. When there is a limited amount of space directly below the lever, this design may be used to advantage. The devices shown at *G*, *H*, and *J* possess one feature that is often desirable, namely, they permit the self-centering action to be rendered inoperative when desired by forcing the centering T-shaped piece or lever out of contact with the main lever.

At *K* is shown an interesting, effective and extremely simple self-centering device. There is no lost motion in this type, and it can be arranged to give equal or unequal resistance to movement in either direction. It requires but one actuating spring, which can be readily provided with a tension adjustment.

The device is simply a flat connecting-rod slotted at each

end, one slot being engaged by a pin on the lever, and the other slot by a pin secured to the machine framework in a fixed position. The spring is fastened to the connecting-rod and to the lever. If the upper end of the spring is connected at a point midway between the fulcrum of the lever and the connecting-rod, the self-centering forces will be equal in both directions. If the upper end of the spring is moved nearer the lever fulcrum, the self-centering force opposing clockwise rotation of the lever will be reduced, while the force opposing anti-clockwise rotation becomes increased. The action of this design is illustrated by the model shown at *L*. Three pins, a cardboard lever, a piece of wire, and an elastic band suffice to provide this working model of the device.

Self-Centering Devices of the Sliding Type.—One of the earliest types of sliding self-centering devices is shown at *A*, Fig. 2. The centering forces are supplied by two springs acting in opposite directions on the sliding member. When flexible springs are used, the centering or positioning action is rather indefinite with this type of device, especially if the springs are flexible. The position of the sliding member depends upon the value of the force opposing the self-centering springs.

At *B* is shown what may be considered the usual or standard practice in the design of self-centering sliding members. In this case, the springs apply the centering pressure to washers placed on each side of a collar located on the sliding spindle. The inner sides of the washers engage the sides of a member that projects from the machine frame. Any vertical movement of the sliding member is resisted by the projecting member on which one of the two springs acts.

The springs may exert equal or unequal pressure on the collars, as desired. When an equal resistance in either direction is desired, the design is frequently modified, as shown at *C*, so that only one spring is required.

Methods of Eliminating Lost Motion.—Unless the distance between the faces of the washers employed in the sliding self-centering device is equal to the corresponding distance between the faces of the projecting members, lost motion will occur. Where the possibility of the slightest lost

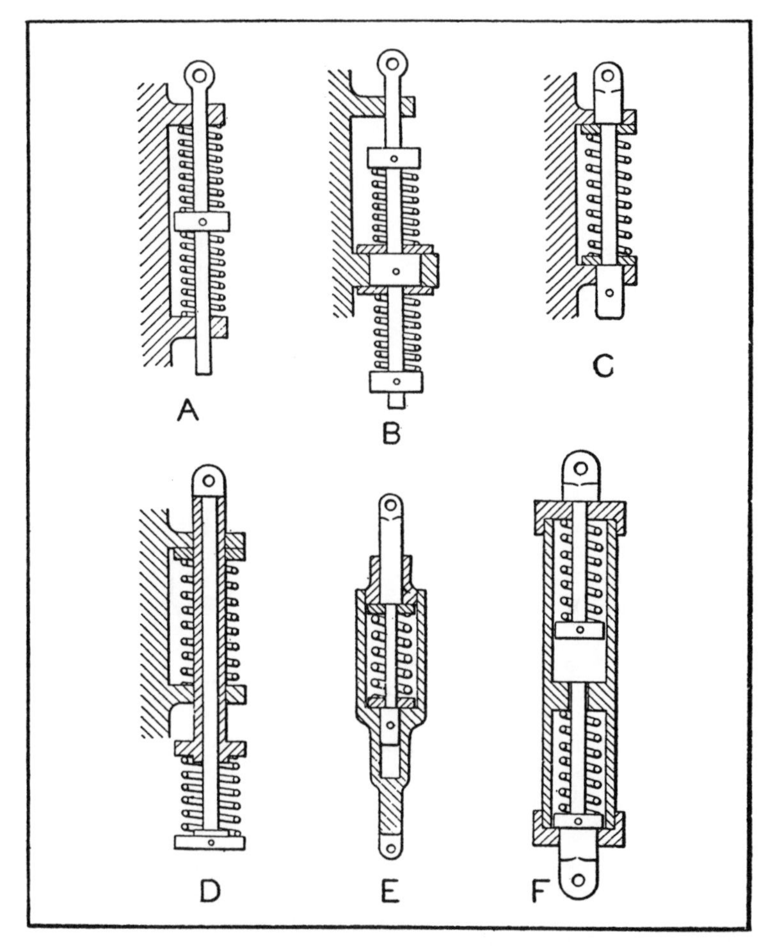

Fig. 2. Self-centering Devices of the Sliding Type

motion must be eliminated, it is better to use a device of the type shown at *D*, Fig. 2, in which the upper collar is pinned to the sliding member. In this design, the sliding member is mounted within a sleeve which is pressed upward against a projecting member on the machine frame by a spring. A depressing movement of the sliding member carries the sleeve with it. On the upward movement of this member, the sleeve cannot follow, and the sliding member rises out of the sleeve against the pressure of a second spring that acts between the two parts. With this arrangement, no

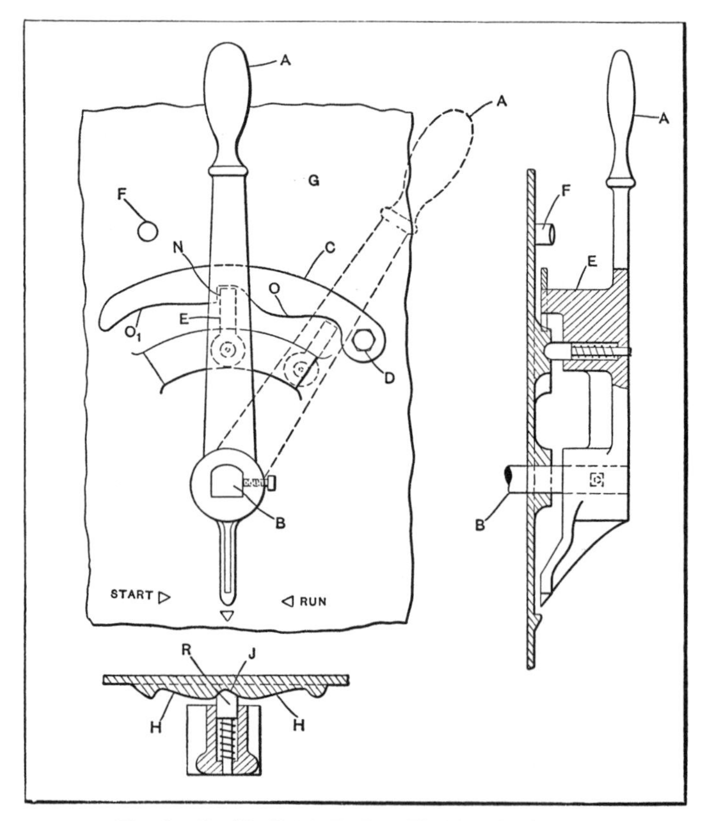

Fig. 3. Gherkin Latch Used on Electrical Equipment

lost motion is possible and the centering action is precise.

The device shown at *C* is commonly employed in the design of spring-actuated connecting-rods used in dog-clutch control link work and for similar purposes. At *E* is shown a typical connecting-rod of this class.

At *F* is shown a design, similar to that illustrated at *D*, which is intended for use where the possibility of the slightest lost motion must be eliminated. In this design, both springs are of the same size and provide equal resistance to either compression or extension of the connecting-rod.

Gherkin's Latch.—Fig. 3 shows a mechanism which is known as a "Gherkin's latch" which is used in conjunction with certain types of centering devices. This latch consists of a handle *A* which is mounted on shaft *B*. The handle has a projecting boss *E* which engages with the latch *C* that is pivoted on the pin *D*. The pin *D* is carried by the case *G* and the movement of the latch is limited by a second pin *F*. This latch is so constructed that, when the handle is moved slowly to the left, the latch will prevent the handle moving beyond the notch *N;* but if the handle is moved over to the extreme right—as shown dotted in the illustration—and then thrown quickly in the opposite direction, the boss *E* on the handle will leave the top of the incline *O* on the latch with sufficient speed to enable it to jump across the notch *N*, in which case the handle will come to rest in the position marked O_1 which is the running position. The pin *R* riding on the surfaces *H* will lock the handle in the cavity *J* when the action of the centering spring becomes effective. In this way, the handle is securely held in the central or off position.

Applications of Gherkin's Latch.— This latch may be used in connection with either of the centering devices shown in Figs. 4 and 5. The centering spring has one end attached to the shaft *B* of Fig. 3 and the opposite end to the case *G*. Referring to the illustration of the centering spring, or so-called "cross-legged spring" shown in Fig. 4, it will be seen that a coiled spring *A* is employed to connect the shaft *B* with the case. The effectiveness of this spring is dependent on the following conditions: first, that the ends cross each other in a line which is perpendicular to the horizontal axis of the spring; second, that the ends of the spring extend far enough above the outside of the coils to enable a pin or bar *C* and a fixed boss *D* to be inserted between these two ends. The pin or bar *C* is carried by the crank *E*, which, in turn, is carried by the shaft *B* of the mechanism shown in Fig. 3. An arm *G* or some other means provides for transmitting the motion so that when the shaft *B* is moved in either direction the cross-legged spring will be put in tension by having one end turned through the action of the pin *C*. This spring tension provides for returning the arm *G* and the shaft to which it is

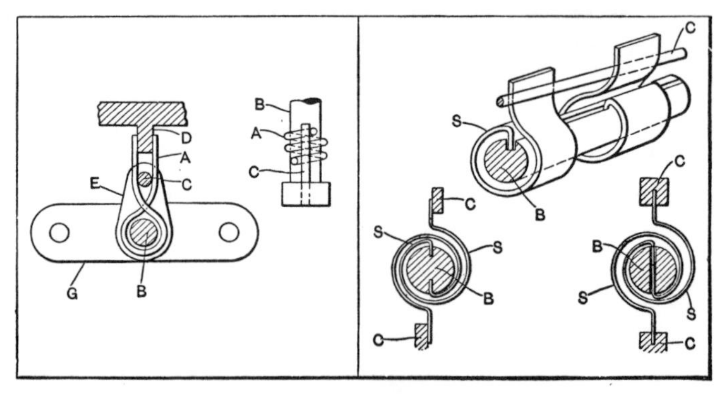

Figs. 4 and 5. Two Forms of Centering Devices that may be Used with a Gherkin Latch

secured to the central position when the action of the spring becomes effective. A very simple centering device is illustrated in Fig. 5, the only difference from the preceding type being that two separate springs are employed in this case. One end of each of these springs *S* is secured in a slot in shaft *B*, while the other end is held by the fixed pin *C*. When shaft *B* is turned in either direction, one of these springs is put under tension and this spring tension provides for returning the shaft to the central position when the action of the spring becomes effective. Two modifications of this design are shown in the lower corners of this illustration.

Miscellaneous Types of Centering Devices.— Another form of centering device is illustrated in Fig. 6. Referring to this illustration, it will be seen that there is a bracket *A* which supports a pivot *B* and a pin *C*. Two arms *D* and *E* are mounted on the pivot *B*, on which they are free to swing. On the ends of the arms are projections *F* and *G* which are secured to opposite ends of the spring *S*. These projections *F* and *G* also engage with bosses *J* and *K* on the moving member *H*. This moving member is carried by the pin *C*, upon which it is free to swing. Assuming that the moving member *H* is rotated in a clockwise direction, the centering device will assume the position shown. The arm *E* is held against the pin *C*, while the boss *G* on the arm *D* engages the boss *K* on the movable member *H*. Further rota-

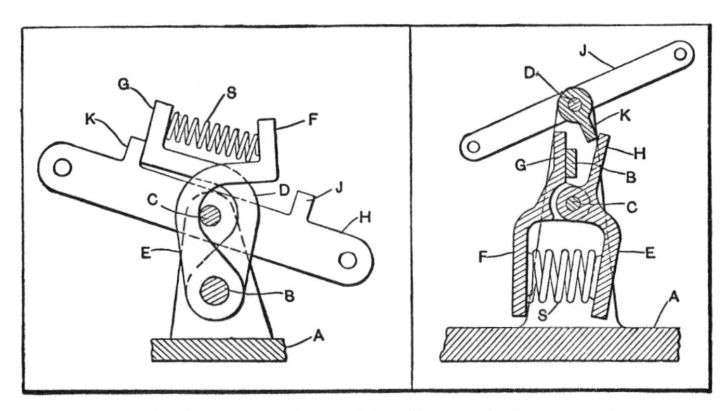

Figs. 6 and 7. Two Types of Double-arm Centering Devices

tion of the member *H* will compress the spring *S* and this will return the movable member to the central position when the action of the spring becomes effective. If the movable member *H* is rotated in a counter-clockwise direction, the action of the centering device will be exactly reversed, the arm *D* being held by the pin *C* while the boss *F* on the arm *E* engages the boss *J* on the movable member *H*, thus causing the compression of the spring *S*.

In the centering device shown in Fig. 7, the frame *A* is provided with a stop *B* and two pivots *C* and *D*. Two arms *E* and *F* are mounted on the pivot *C*, these arms being free to swing and provided with projections *G* and *H* at their upper ends. When the centering device is in its normal position, both of the projections *G* and *H* are in contact with the stop *B*, owing to the action of the compression spring *S* which forces the lower ends of the arms in opposite directions. It will be seen that the movable member *J* is mounted on a pivot *D* and provided with a projecting lug *K* on its lower side. This lug engages with the projections *G* and *H*, and when the member *J* is rotated in either direction, it swings one of the arms about the pivot *C* and compresses the spring *S*. When the action of this spring

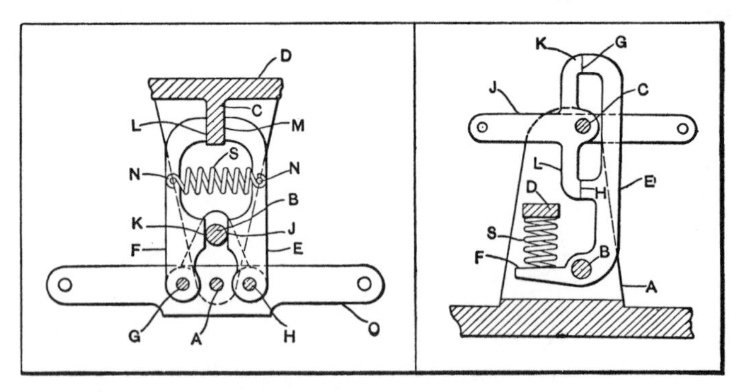

Figs. 8 and 9. Two Types of Centering Devices in which Sliding Contact of the Control Arms is Employed

becomes effective, it returns the arm to the normal position; and the projection at the upper end of the arm acts on the lug *K*, causing the movable member *J* to be returned to the starting point.

The centering device illustrated in Fig. 8 works on a somewhat different principle from that of the preceding types. The frame *D* carries a stop *C* and a pivot *A*. The movable member *O* is carried by this pivot *A*, and, in turn, carries two arms *E* and *F*. The arm *E* is carried by the pivot *H* and provided with two sliding surfaces *J* and *M* which work in contact with the pin *B* and the boss *C*, respectively. The arm *F* is supported on the pivot *G* and has two sliding surfaces *K* and *L* which work in contact with the pin *B* and the boss *C*, respectively. The tension spring *S* is secured to the arms *E* and *F* at the points *N*. Assuming that the movable member *O* is to be rotated in a clockwise direction, the arm *E* will be pulled down by the supporting pivot *H* and in so doing, the surfaces *M* and *J* will slide on the boss *C* and pin *B*. For the same reason, arm *F* will be pushed up, the surface *L* sliding on the boss *C* and the surface *K* sliding on the pin *B*. This action increases the distance between the pins *N* to which the spring is secured, thus putting the spring *S* under tension; and when the action of this spring becomes effective, it will return the movable member *O* to the starting point.

In the device shown in Fig. 9, the frame *A* carries two pivots *B* and *C*, and a boss *D* to which one end of the spring *S* is secured. The arm *E* is pivoted at *B* and has a projection *F* which engages the opposite end of the spring *S*; this arm also has two sliding surfaces *G* and *H*, which work in contact with the projections *K* and *L* on the movable member *J*. This movable member is carried on pivot *C*. When the member *J* is rotated in either the clockwise or counter-clockwise direction, one of its projections causes the arm *E* to be rotated in a clockwise direction about the pivot *B*. When this rotation takes place, the projection *F* on the arm *E* compresses the spring *S*. When the movable member *J* is released, the action of the spring *S* becomes effective and returns it to the central position.

Device for Small Angular Movements.—Fig. 10 shows a form of centering device which is limited in its application, but has been found particularly effective in those cases where it can be used. It is only applicable for returning mechanisms which have a relatively small angular movement. Referring to the illustration, it will be seen that the shaft *A* has a hub *B* secured to it. An arm or other means of transmitting the motion may be secured to the shaft *A*. The hub *B* has two recessed surfaces *H* and *J* that receive the ends *K* and *L* of the yoke *D*, which is held in position by means of a bolt *E*. The opposite end of this bolt is carried by a lug *C*, and the bolt carries a spring *S* and washer *F* which supports the pressure of the upper end of the spring. When the shaft *A* is rotated in either direction, the hub *B* raises the yoke *D*, thus compressing the spring *S*, and when the action of this spring becomes effective, it pushes down the yoke and returns the shaft *A* to the starting point.

Cam, Bellcrank, and Spring Combination.—One of the simplest forms of centering devices that can be used consists of the combination of a cam, bellcrank, and spring, illustrated in Fig. 11. Referring to this illustration it will be seen that the shaft *A* has the two-lobed cam *B* secured

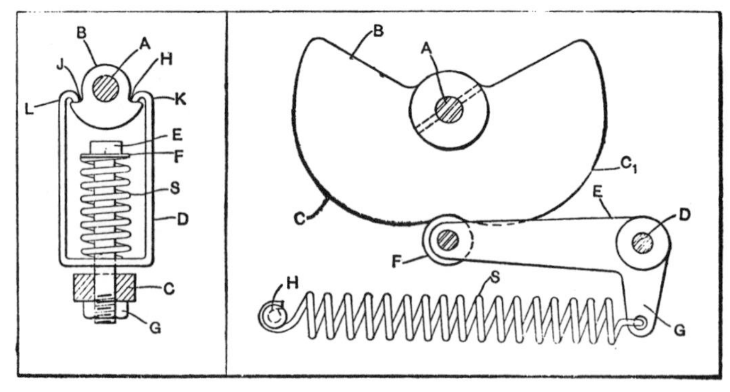

Figs. 10 and 11. Yoke and Cam Types of Centering Devices

to it. The bellcrank *E* is mounted on the pivot *D;* and this bellcrank has the cam roller *F* mounted at one end and the other end secured to the spring *S*. The opposite end of the spring is secured at the point *H*. When the shaft *A* is rotated in either direction, the lobe *C* or C_1 of the cam pushes down the end of the bellcrank; and this causes the crank to rotate about the pivot *D*. The result is that the opposite end *G* of the bellcrank swings out and places the spring *S* under tension. When the action of this spring becomes effective, it rotates the bellcrank in the opposite direction and returns the shaft *A* to the starting point through the action of the cam roller *F* on the cam.

Oil-Switch Control Centering Device.—A type of centering device which finds wide application in the control of oil switches is illustrated in Fig. 12. In this illustration, *A* represents a shaft to which a hub *C* is secured. This hub carries a projecting lug *D* which extends between the two arms *E* and *F* of the centering device. These arms are free to swing about the shaft *A*. The arm *E* is provided with an extension *K* to which one end of the spring *S* is secured, and a second extension *G* which limits the movement of the arm *E* in a clockwise direction through contact with the stop *B*. The arm *F* has an extension *J*, to which the opposite end of the spring *S* is secured, and a second extension

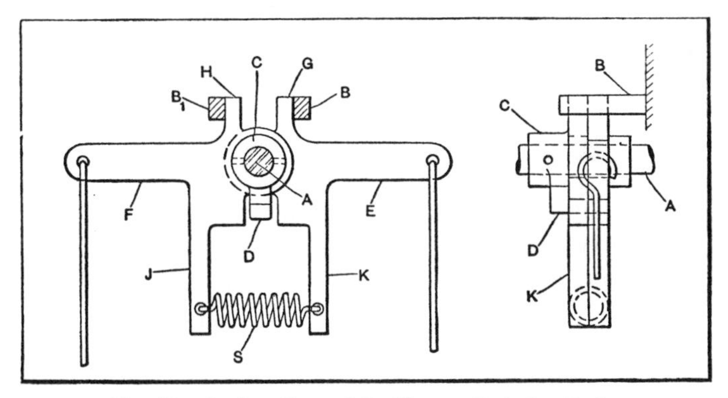

Fig. 12. Another Form of Double-arm Centering Device

H which limits the rotation of the arm *F* in a counter-clockwise direction through contact with the stop B_1. The tension of the spring *S* holds the arms *E* and *F* against their respective stops *B* and B_1. When shaft *A* is rotated, the hub *C* and its lug *D* are moved in either a clockwise or counter-clockwise direction as the case may be. This rotation causes either the arm *E* or *F* to be rotated against the tension of the spring *S*, and when this spring tension becomes effective it causes the shaft *A* to be returned to the starting point.

Lever-Returning Devices.— Figs. 13 and 14 show two simple devices for returning two levers to their normal positions. In the device shown in Fig. 13, the lever *C* is carried by the pivot *A* and the lever *D* is carried by the pivot *B*. An arm *G* is pivoted between these levers and connected to them by means of two links *E* and *F*. When arm *G* is rotated in a clockwise direction, it causes levers *C* and *D* to swing in about pivots *A* and *B*, thus placing the spring *S* under compression. When the force tending to bring the levers *C* and *D* together is released, the compression of the

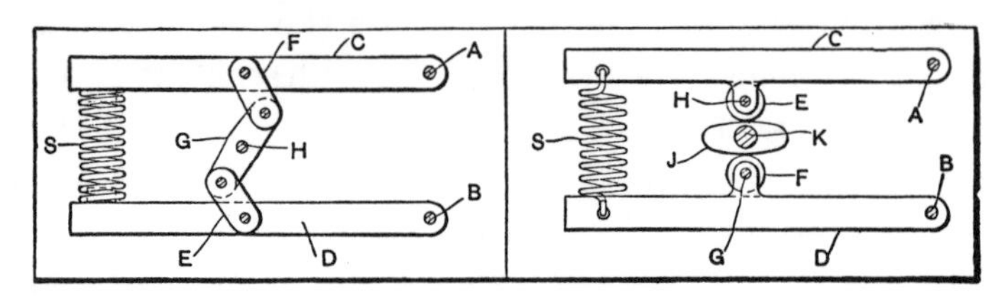

Figs. 13 and 14. Centering Devices for Returning Two Arms Simultaneously

spring *S* will return the levers to their normal positions. The arrangement of the mechanism shown in Fig. 14 is quite similar to that of the preceding illustration, except that the levers *C* and *D* are provided with cam rolls *E* and *F* which are mounted on pivots *H* and *G*. When the cam *J* is rotated in either direction, by turning the shaft *K*, it causes the levers *C* and *D* to swing outward about the pivots *A* and *B*, thus placing the spring *S* under tension. When the torque tending to rotate the shaft *K* is released, the tension of the spring *S* returns the levers *C* and *D* to the starting point.

CHAPTER XIII

MULTIPLE-LEVER MECHANISMS WITH DWELLING OR IDLE PERIODS AND OTHER SPECIAL LEVER COMBINATIONS

Mechanical movements may in many cases be derived in the simplest manner by the use of properly proportioned levers or combinations of levers and connecting links. Several lever combinations which are arranged to provide a period of dwell during the cycle of movements will first be described and these designs will be followed by certain special lever applications or other devices allied in some way to mechanisms of the lever type.

Multiple-Lever Mechanisms Designed to Obtain Dwells in Lever Movements.— Levers, in combination with links, can be used as a means for obtaining a dwell or idle period during the cycle of movements imparted to the lever of a driven shaft. The cam and follower-roll mechanism is perhaps the only simple one in which complete elimination of motion is obtained during the dwell period, but it cannot always be applied conveniently; moreover, it is often difficult to obtain sufficient movement by means of cams.

When a close approximation to complete elimination of motion during the dwell period will meet requirements, levers and links provide a simple solution of the transmission problem, particularly when the driving and driven units are not located too close together. Fig. 1 illustrates a case of this kind, in which a hand-lever (not shown) mounted on the driving shaft operates two distinct mechanisms near the ends of its stroke, while the middle section of the stroke operates the driven shaft, which receives no noticeable movement when the other two mechanisms are being actuated.

Advantages of the Lever-Type Dwell Mechanism.— The lever type mechanism has many advantages. For example, with the lever and link mechanism, it is a simple matter to increase or decrease the movement imparted to the driven member and change the direction of movement.

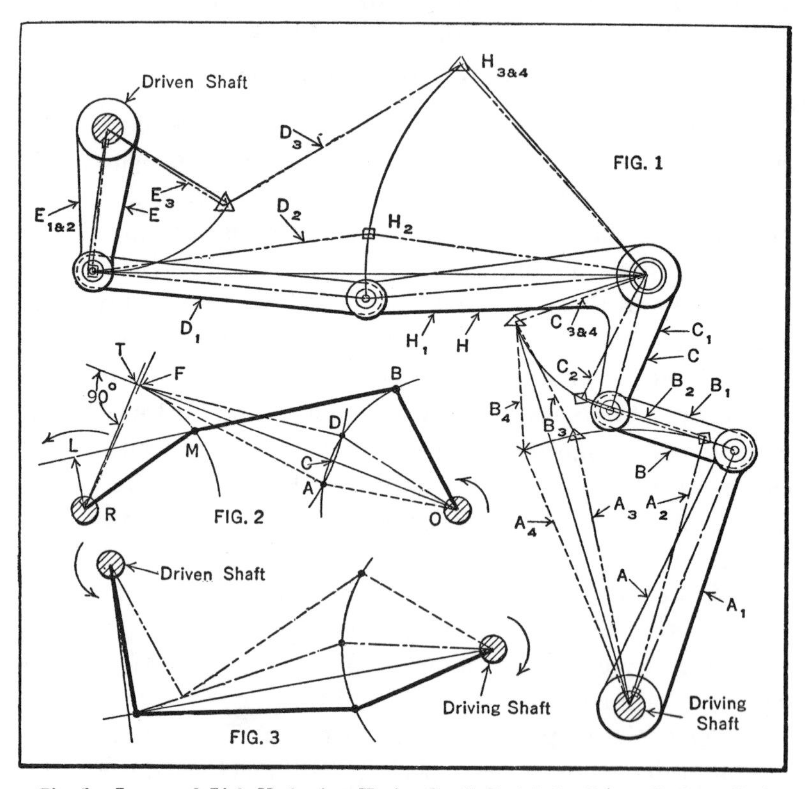

Fig. 1. Lever and Link Mechanism Having Dwell Period for Driven Shaft at Each End of Oscillating Movement of Driving Shaft. Fig. 2. Diagram Used to Illustrate Method of Laying out Lever and Link Mechanisms. Fig. 3. Arrangement for Obtaining Dwell at Beginning of Driving Lever Stroke with Shafts Turning in Opposite Directions

Such mechanisms can also be arranged easily to avoid obstacles. Only simple parts which can be made easily in any machine shop are required for the lever mechanisms. These mechanisms operate smoothly and quietly without requiring any attention other than an occasional oiling. All the levers shown in Fig. 1, together with some that are not shown in the illustration, were required for the transmission of motion from the front of a machine to a higher position at the rear. This is accomplished, with the additional feature of a dwell period of the driven shaft near each end of the stroke of the driving shaft.

Operation of a Lever-Type Dwell Mechanism.—In operation, the lever A (Fig. 1) on the driving shaft moves from position A_1 through A_2 and A_3 to A_4, serving to operate two mechanisms (not shown) from levers also mounted on the driving shaft. The movement from A_2 to A_3 is transmitted to the lever on the driven shaft, causing it to move from E_1 to E_3. Lever E on the driven shaft dwells while the lever on the driving shaft is moving from A_1 to A_2 and from A_3 to A_4. The first dwell of the lever on the driven shaft, as lever A moves toward A_4, occurs while the lever on the driving shaft moves from position A_1 to A_2, and while the longer or driving arm of the bellcrank H swings from H_1 through the neutral point to H_2, causing lever E to move only a very short distance, as indicated by the full line and the dot-and-dash line. The actual over-travel transmitted to lever E at this time amounts to an angular movement of only one minute. Even this small movement can be reduced by incorporating a similar arrangement at another point in the transmission system. As a matter of fact, the motion of lever E during the dwell period does not exceed 8 per cent of the total movement.

The movement of lever A from A_3 to A_4 results in transmitting practically no movement to link B and lever C. This movement simply causes positions C_3 and C_4 to become merged at one point, giving corresponding positions H_3 and H_4 to lever H. It is generally necessary to increase the length of the driving levers or decrease the length of the driven levers to compensate for the decreasing effects, on the driven levers, of the angular motions of the drivers as they approach the position of dwell. This can be done conveniently, but the designer must be careful not to reduce too greatly the moment arms of the driving forces in the links.

Incidentally, it may be mentioned that link and lever arrangements of this kind are self-locking against the reversal of the lever moments when the driving lever is at each end of its stroke. In the case illustrated, there was no necessity to utilize the toggle action of the levers for the production of heavy pressure or for locking purposes, but this feature might be useful in some cases.

Laying Out a Lever-Type Dwell Mechanism.—The general method of laying out a mechanism of this kind is illustrated in Fig. 2. This lay-out can be varied considerably without seriously affecting the results. First, the approximate or definite locations for the driving and driven shafts are laid out and the directions of the motions are determined, so that intermediate levers and links can be sketched in for a preliminary trial. Isolating one unit, as in Fig. 2, arcs representing the swing of the link-pins are drawn, and from the center of the driving shaft is drawn line *OCT* tangent to the arc of the driven link-pin.

It is not absolutely necessary to have line *OCT* in the tangential position, but this position ordinarily gives the best results. From the center of the driven shaft, line *RT* is drawn at right angles to line *OCT*. Now, *CT* represents the length of the connecting link. From *C* mark off at *D* and *A* arcs representing one-half the angle through which it is desired to eliminate the transmission of movement to *M*, and with length *CT* and centers *A* and *D* draw arcs which cut *OCT* at *F*. This point *F* is the final position of the driven link-pin. From *OA* lay off the point *B*, so that arc *AB* subtends the whole angle of rotation of the driving lever. Then point *B* is the starting position of the driving link-pin.

From point *B*, with a radius equal to *CT*, draw an arc cutting the arc of the driven link-pin at *M*. Now, *RM* indicates the initial position of the driven lever, while *BM* indicates that of the link. Thus, it will be seen that while the driving lever is given a continuous forward motion, the driven lever moves forward up to its final position and then has a slight additional forward motion, after which it returns to its final position. This last reciprocating movement between F and T is usually so slight as to be negligible. This slight movement also acts on the driven lever in such

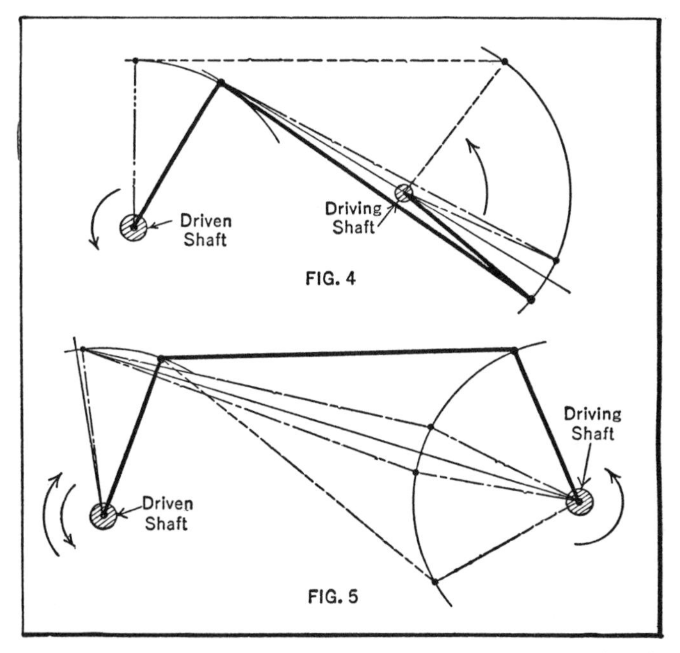

Fig. 4. Levers Arranged to Have Driven Shaft Dwell at Beginning of Driving Shaft Stroke with Both Shafts Turning Counter-clockwise. Fig. 5. Arrangement for Dwell at Mid-point of Driver Oscillation

a direction as to produce the minimum amount of angular movement.

In order to eliminate motion at the beginning of the stroke instead of at the end as just described, and at the same time retain the same direction of rotation for both levers, it will be necessary to carry the lever-pin to the right-hand side of the driving shaft. This necessitates using an overhanging transmitting lever and link, in order to allow the link to pass across the center of the shaft—see Fig. 4. If it is possible to reverse the direction of the driving lever, elimination of the motion at the beginning of the stroke can be achieved by the arrangement shown in Fig. 3.

Fig. 5 shows another arrangement in which the dwell occurs at the mid-point of the swing of the driving lever, so that the driven lever is given an oscillating motion with a dwell at the extreme left-hand position.

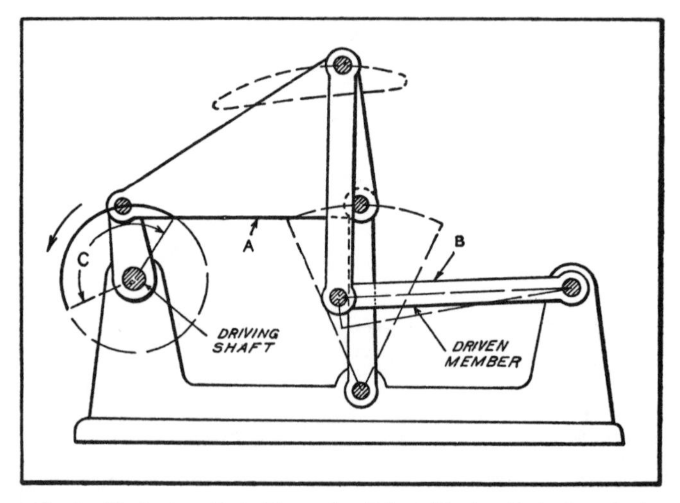

Fig. 6. Mechanism that Allows the Driven Member B to Remain in Position Shown while Crankpin of Driving Arm Passes through Arc C

Driven Lever which Dwells while Driving Crank Turns Part of a Revolution.— A link mechanism which allows the driven member to dwell for a relatively long interval is shown by the diagram Fig. 6. Any point in the flat triangular plate which constitutes the rod *A* can be used for a link connection. For the designing of such mechanisms, it is necessary to study the different forms of curves described by the various points on rod *A*. The forms of the curves traced by the different points on this rod vary distinctly according to their relative positions. To obtain long dwell periods, only the curves that have large circular sections on a part of their outline are used. Geometrical methods have been evolved to determine such curves, but they are rather complicated. However, some very remarkable improvements in dwell movements have been made by applying such formulas.

There are curves that show a near relationship to circular forms and there are others in which the radii of cur-

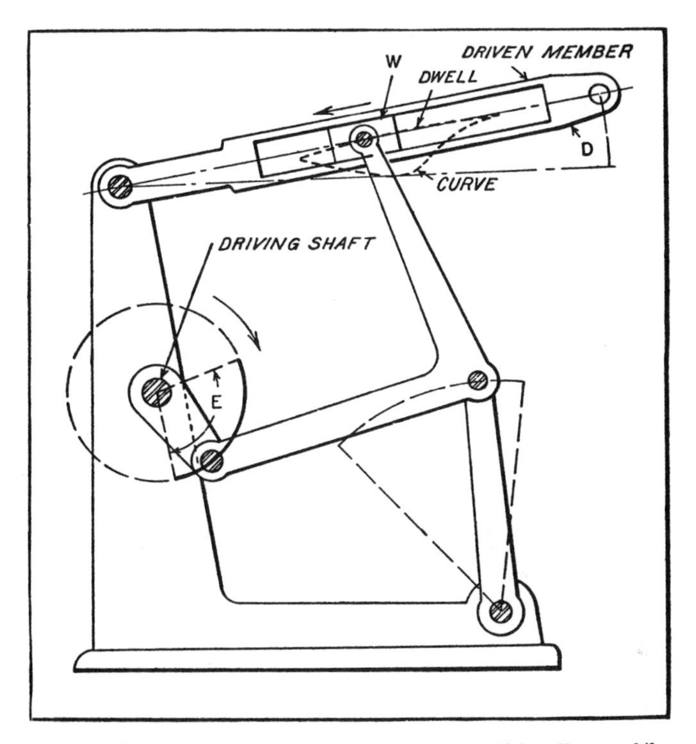

Fig. 7. The Driven Member D Remains in the Position Shown while the Crankpin of the Driving Arm Passes through Arc E

vature will increase to such an extent that the curves practically resemble a straight line. Mechanisms with dwells can be designed, in which the driven member connected with the base of the mechanism has a sliding way, such as shown at *W*, Fig. 7. A member with two vertical slide ways at 90 degrees or any other angle can be used in place of the swinging lever.

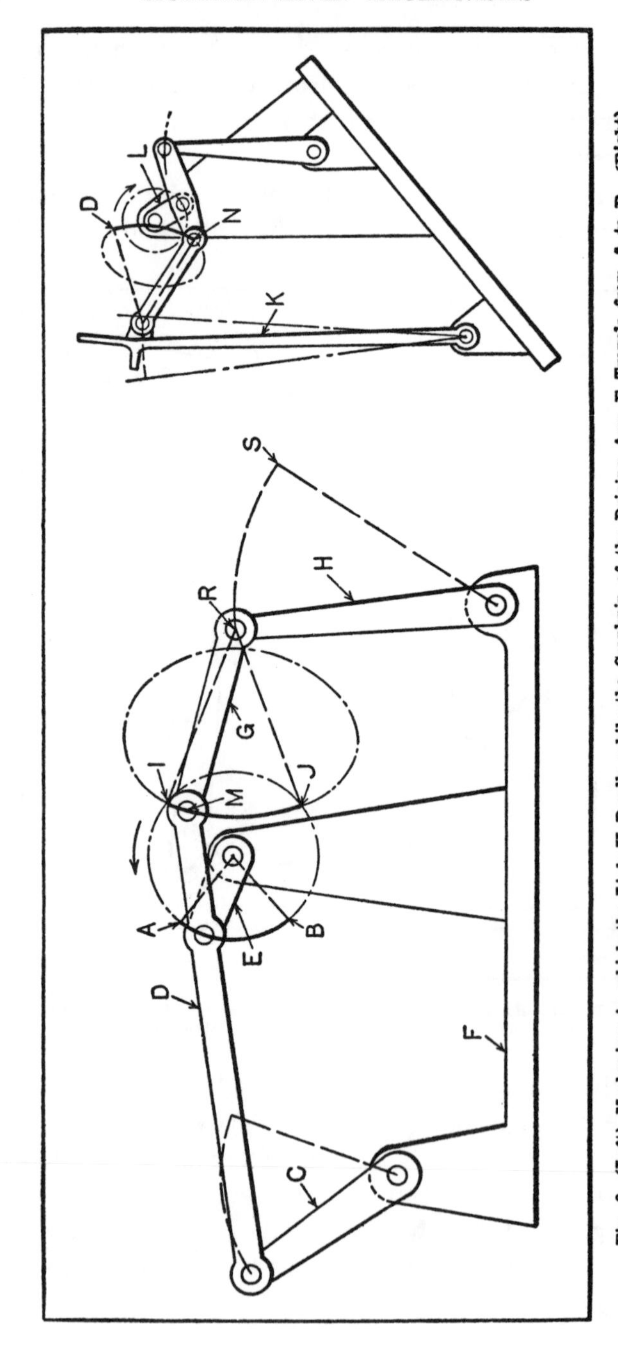

Fig. 8. (Left) Mechanism in which the Link H Dwells while the Crankpin of the Driving Arm E Travels from A to B. (Right) Application of Mechanism Shown at Left to a Textile Machine

Link Mechanism on Textile Machine for Obtaining Dwell in Lever Movement.— Crank-driven mechanisms which also provide lever movements having a dwell period are shown in Fig. 8. The view at the right shows a design applied to a textile machine which provides a dwell period for the lever *K* equal to nearly one-third the period required for a complete revolution of the crank-arm *L*. This mechanism is used on a weaving loom and has proved very successful. The pause or dwell obtained with this arrangement is of sufficient duration to permit the shuttle to pass from one side of the machine to the other.

Referring to the view at the left, which shows the principle of operation, it will be noted that the driving arm *E*, which revolves continuously in the direction indicated by the arrow, is connected to rod or link *D* at a point approximately one-fourth its length from the end connected to link *G* by the stud *M*. Link *G*, in turn, is connected with the upper end of the driven lever *H*, which oscillates through the arc *RS*, dwelling at *R* while the stud *M* travels from *J* to *I*. The lever or arm *C*, connected to the left-hand end of link *D*, also oscillates through an arc as indicated. The amount of dwell and the length of the arc through which the driven lever oscillates depends, of course, on the positioning and the lengths of the links and levers. The use of slide ways for varying the relative positions of the links may be advantageous in some of the many applications for which a mechanism of this kind is adapted.

High-Speed Oscillating Motion with Dwell at Each End.— It is possible, by means of a mechanism consisting of links and gears, but no guides or cams, to obtain an oscillating motion with a dwell at each end of the oscillation. A mechanism of this kind is shown in Fig. 9. This mechanism has two gears, the driving gear *A* and the driven gear *D*. Gear *D* is only half the diameter of gear *A*. It will be noted that the gear shafts are mounted in the fixed base *E* of the mechanism.

The lengths of crank-arms *F* and *G*, or the distance of the crankpins of the push-rods *H* and *I* from the centers of their respective gear shafts, as well as the lengths of the push-rods, can be of any suitable dimensions. The crank *F* works in a different phase from crank *G*. The push-rods *H* and *I* are connected by the joint *J*. If crank *F* makes one revolution, and crank *G*, by means of the spur gears,

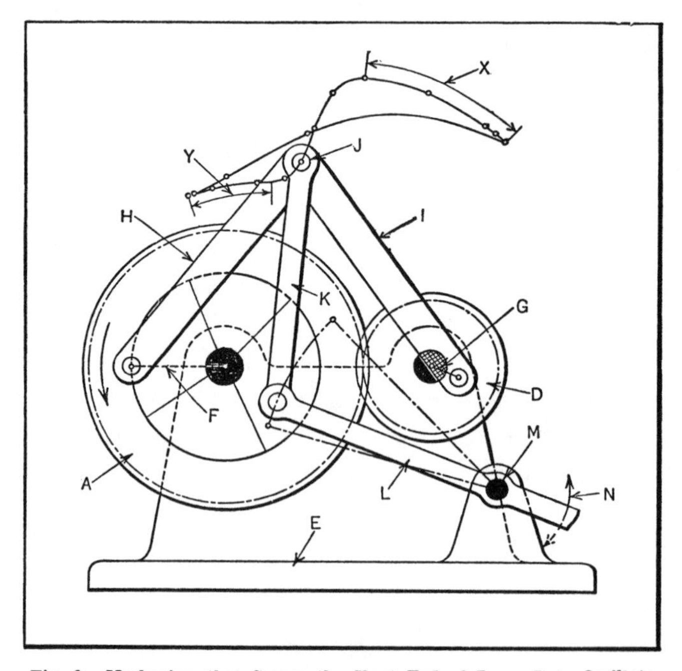

Fig. 9. Mechanism that Causes the Short End of Lever L to Oscillate through Arc N with Dwell at Each End of Arc

makes two revolutions in a reverse direction, joint *J* will describe a curve having the unusual form indicated.

The curvature is so selected from the possible forms that it has the same radius at the maxima and minima points. If this radius nearly corresponds to the length of member *K*, which is connected to joint *J* at one end and to swinging member *L* at the other, member *L* will be oscillated about a fulcrum bearing *M* in the base *E* and will pause or dwell while the joint J describes the arcs X and Y. Member L is thus given an oscillating movement indicated by arc N, with a dwell at each end of the arc. The members H and I in the design shown are of identical size. A mechanism of this kind can be used for high speeds.

Straight-Line Motion for Oil Circuit-Breaker.—Straight-line motions are not used extensively, since they are only adapted to certain special conditions, and they have some inherent disadvantages. The particular straight-line motion illustrated by the diagrams Fig. 10 (which represent side and end elevations of the links) produces an approximate straight-line between the two points A and B. This mechanism is part of a heavy-duty oil circuit-breaker. Straight-line and toggle linkages have had considerable application to this type of electrical apparatus.

The pivots of what might be called the main lever are located at A, C, and E. The fixed pivots or hinge points are at D and F. The links CD and EF are free to rotate around their respective fixed pivots or bearings. The line EG indicates a link connecting with a crank or other form of driving member. A force indicated by the arrow at A acts along the path AB, and the triangular lever ACE moves from the "closed position" shown in full lines to the "open position" indicated at B, C_1, and E_1. This latter position represents approximately the lower limits of the main lever movement. The load from A to B varies in a manner characteristic of circuit-breaker operation, and the velocity in each direction also varies from zero to as high as 12 to 15 feet per second in order to operate the brushes fast enough to open or close in from 1/3 to 3/4 second. The load at A may vary from 150 to 3000 pounds during normal operation and may be higher under certain conditions. The line AB is vertical in the apparatus, and the line FP at right angles to AB.

In considering the advantages of this design, first note that the main lever ACE is a triangle with the load at one

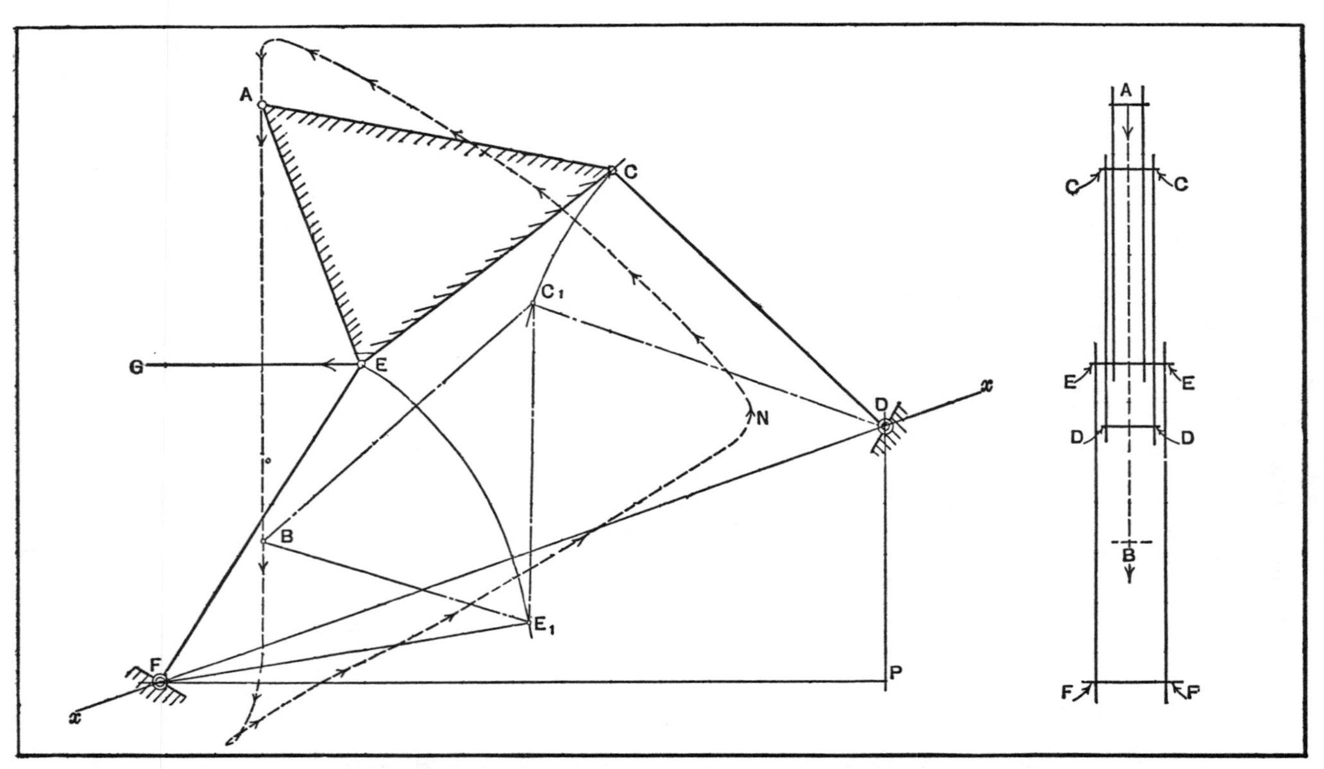

Fig. 10. Diagrams Representing Side and End Elevations of Straight-line Motion for Heavy-duty Oil Circuit-breaker

corner *A*. This form provides maximum strength and rigidity for a given amount of metal, and few straight-line motions, except the more complicated ones, have this feature. Second, the members *AE* and *EF* are struts well located with reference to the load at *A*. These struts are disposed to form the familiar toggle joint. Third, the rod from *A* to *B*, carrying the load, passes about midway between the points *E* and *F*, allowing proper clearances. Morever, the rod is in the center of the double link *EF*, there being a link on each side, as shown by the end elevation; this is also the case with link *CD*. As the result of this construction the pins are in double shear and with practically no bending due to overhang. The line from *A* to *B* deviates only slightly from a true one and is accurate enough for the purpose mentioned. This motion has the advantage that the pivot points and links can be varied at one place and compensated for somewhere else to an extent not possible with a number of other types. All pivot points, however, must be in the proper relation to obtain the most accurate line, although this does not necessarily require the particular arrangement shown.

The real geometrical reference axis of this linkage is indicated by the line x–x. Note that the dotted line and small arrows starting from *A* extend through *B* downward curving to the left and then, after making a small loop, extend upward through *N* and back to the starting point *A*. This line indicates the path which point *A* would follow if the motion were continued beyond point *B* and through a complete cycle.

The four-sided linkage *CDFE* is similar in principle to the skeleton for most of the straight-line and parallel motions from Watts down to the Roberts type. The radius arms *CD* and *EF* are of equal length, but would not need to be, if the main lever *ACE* were changed to an isosceles form of suitable length and the center *D* were swung clear over the top until directly over pivot *F;* then the Roberts compensating motion would result. It will be noted that the path of point *A* crosses the axis *x–x* twice. If all the links in the linkage *CDFE* were of different lengths and if *CE* were longer than *DF*, and *CD* shorter than *EF*, then a point near the middle of the oscillating link *CE* would cross the axis six times. This will be recognized as the more general case of the irregular four-sided linkage. When crossed linkages are used, as in Watts and some other motions, the line is crossed only twice. From a practical standpoint this motion has a few decided and inherent advantages.

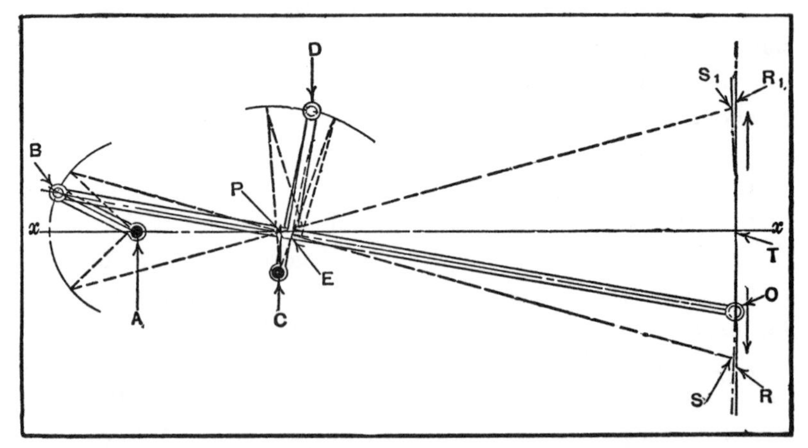

Fig. 11. Diagram of Straight-line Mechanism Used on Granite Gang Saw

Straight-Line Mechanism for Gang Saw.—The mechanical movement shown in the accompanying diagram (Fig. 11) is used in connection with a gang saw for sawing granite, to obtain an approximate straight-line motion with a combination of links. The bearings or pivots *A* and *C* are stationary. Link *AB* is free to turn about bearing *A*, and *CD* is free to turn about bearing *C*. The rigid bar or link *OB* has an extension *ED* at right angles to it, which is pivoted at *D* to the lever *CD*. As the end *O* is moved in the direction of the arrows, the pivot *B* swings about an arc having a radius *AB*, and the pivot *D* swings about an arc CD. The resultant movement of the point O is very nearly a straight line.

When this mechanism is applied to a granite gang saw, a slight rise at the ends of the stroke S and S_1 is required, so that links of special length are used. These lengths, in inches, are as follows: $AB = 8\ 1/2$; $OB = 66\ 1/2$; $EB = 22\ 13/16$; $AP = 13\ 1/2$; $CD = 16$; $ED = 12$; $RT = 12$; $R_1T = 12$. The rise at the end of the stroke is 1/4 inch, approximately.

Four mechanisms of this type are used on the granite gang saw. Each mechanism is so located that the center line x–x is vertical, the straight-line movement being horizontal. The machine is equipped with a rectangular "sash" in which there are numerous steel blades. Each corner of this sash is attached to one of these straight-line movements, and the sash is moved back and forth by a crank and connecting-rod. Steel-shot under the blades works against the stone and does the cutting at the rate of from 3 to 6 inches per hour. The object of the slight rise at the ends of the stroke is to allow the grains of shot to fall under the blades as the shot drops down from above. By shortening the dimensions AB and EB equally, the end O can be made to travel in an exact straight line for a certain distance.

Stroke Adjustment for Oscillating Lever.—The arrangement of levers shown in Fig. 12 provides a simple means of transmitting a variable oscillating movement to the lever D from the lever C. The oscillating or up and down angular movement of lever C about center H remains constant, while the oscillating or angular movement of lever D about center J can be varied to suit requirements. Adjustment of the angular movement of lever D is obtained by shifting the position of block A, which is in contact with the two levers. The levers are held in contact with block A by springs (not shown in the illustration). Lever C transmits the smallest angular movement to lever D when block A is in the position shown to the left. The largest angular movement of lever *D* is obtained with the block in the position shown by the dotted lines at *K*.

The track surfaces of the levers in contact with block *A* are set parallel with each other when the levers are at either the top or the bottom of their strokes, according to whether the motion is required to be more nearly constant in speed at the upper or the lower position. In Fig. 12 the levers are shown in their lowest positions. Contact block *A* can be adjusted by the machine operator to any position along the levers. The adjusting rod *B* is hinged to the control lever, so that it can swing in a vertical plane as lever *C* oscillates.

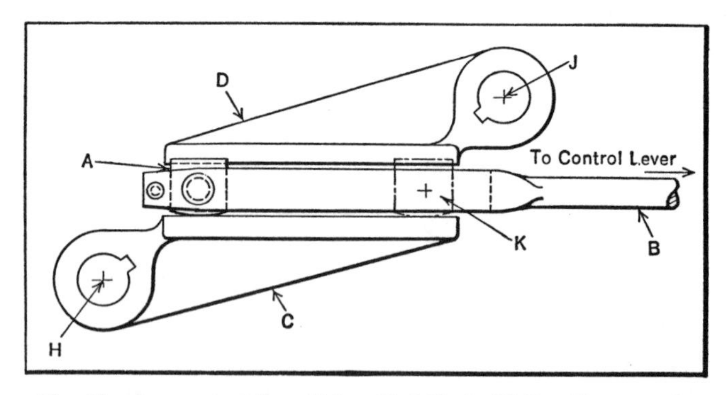

Fig. 12. Arrangement by which a Variable Oscillating Movement is Transmitted by Lever C to Lever D

If *A* is a single-piece block, it must either have a roller or else have a flat face contact with the track of the upper lever and a lower surface formed like the portion of a cylinder of relatively large diameter. The block and the tracks of the levers must be hardened. Grease lubrication is best where a flood of oil cannot be applied.

When the loads are heavy, sliding friction is likely to develop flats on block *A* and rough spots on the lever tracks. In such cases, the design shown in Fig. 13 is preferable. With this design, the sliding friction is reduced to a minimum by substituting rollers *F* and *G* for the plain bearing surfaces of block *A*, Fig. 12. On the upper side of the trunnioned separating block *E*, Fig. 13, is a single roller *F*, while on the lower side are two similar hardened rolls *G*. With this arrangement, only a slight sliding motion is possible between the rollers and the lever tracks, the movement of the rollers being limited by the end flanges of the block. The close fit between the side plates and the rollers, and the squaring effect of the end flanges of block *E*, tend to keep the rollers parallel.

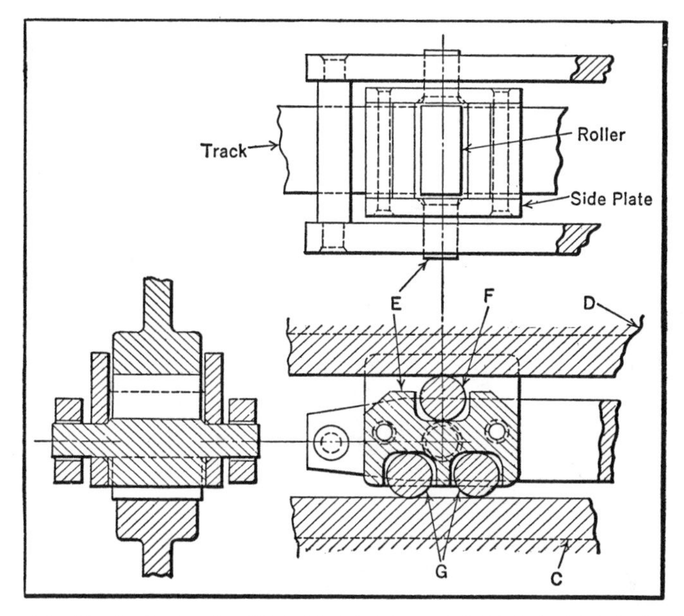

Fig. 13. Roller Bearing Block Used in Place of Solid Contact Block A, Fig. 12, when Lever D is Heavily Loaded

Cam and Rack Mechanism for Increasing the Movement of an Oscillating Lever.—When compactness is essential in a mechanism for producing a long movement with a short lever, the design shown in Fig. 14 may be used to advantage. The lever indicated at *G* is pivoted to a slide *D* confined in the guides of the machine. Cast on the lower end of the lever is a gear segment which meshes with a stationary rack *H*. Movement is imparted to the slide by the continually rotating cam *A*, which is mounted on a shaft in the stationary bracket *C* and engages roll *F* on the lever stud. Coil spring *E*, fastened to the slide and the machine frame, serves to hold the roll in engagement with the cam.

As the cam rotates in a clockwise direction from the position indicated, the slide *D* will move toward the right. The lever will, of course, travel with the slide, and owing to the engagement of the gear segment and the rack, will swing in a clockwise direction. The lever will continue to swing

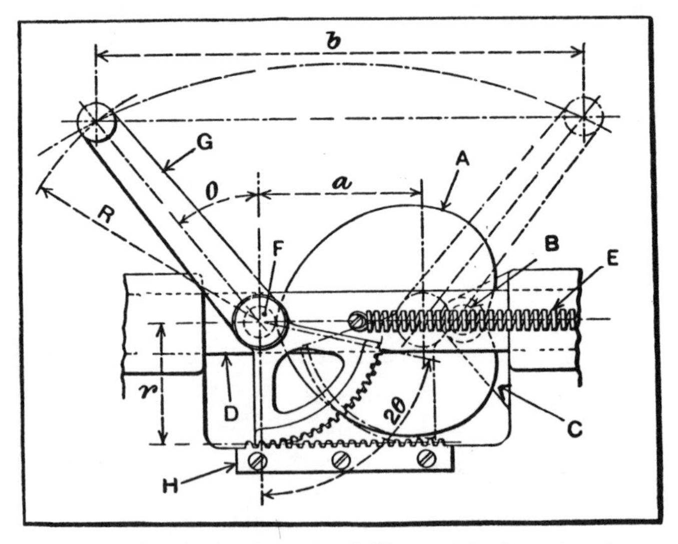

Fig. 14. Combination Cam-and-rack Movement for Increasing the Throw of a Lever

in this direction until the cam has carried the slide to its extreme position at the right. The position of the lever at this point is indicated by the dot-and-dash lines.

In determining the movement *b* of the upper end of the lever, it is only necessary to add the throw *a* of the cam to the normal angular movement of the lever, all measurements being taken horizontally. The movement of the lever can be varied by changing the throw of the cam, the pitch radius *r* of the segment gear, or the length *R* of the lever itself.

Push-Button Mechanism for Alternately Changing Position of Lever.—Fig. 15 shows a mechanism designed to change the position of lever *C* from the right to the left of the vertical center-line or vice versa each time the push-button *D* is operated. The cam-shoe *F* (left-hand view) which is an extension of push-button *D*, is in position for changing lever *C* to the position shown by the middle view, by coming in contact with the inclined face of pawl *G* when the button is pushed upward. The upward motion of cam-shoe *F* will cause pawl *G* to move in the direction of arrow *O*. When the cam-shoe reaches the end of the incline, the fulcrum of the connecting links *J* will rest against the flat face of the cam-shoe. Pawl *G* will then be lifted, causing lever *C* to swing and change from the right-hand to the left-hand position.

During this motion, the pressure of the shoe applied to the connecting links *J* automatically moves the link or pawl *H* toward the left. This obviously prevents pawl *G* from moving too far to the right and from becoming disengaged from the cam-shoe. The right-hand view shows the push-button *D* released and returned to its normal position by spring *K*. When the connecting links *J* are released from contact with the cam-shoe, the spring *L* causes pawls *H* and *G* to move toward each other until links *J* come in contact with pins *M* and *N*. Lever *C* is now ready to change from left to right when the push-button is again operated. The stem of the push-button *D* is square and is a sliding fit in the square hole in guide *E*.

Multiplying Action of Lever for Obtaining Quick-Acting Brake Movement.—Fig. 16 shows the construction of a mechanism designed to provide more than the customary amount of clearance for a brake-shoe without sacrificing any of the braking effect. This is accomplished by a system of levers that provide for a quick take-up of the clearance space, after which the brake movement is effected in the usual manner.

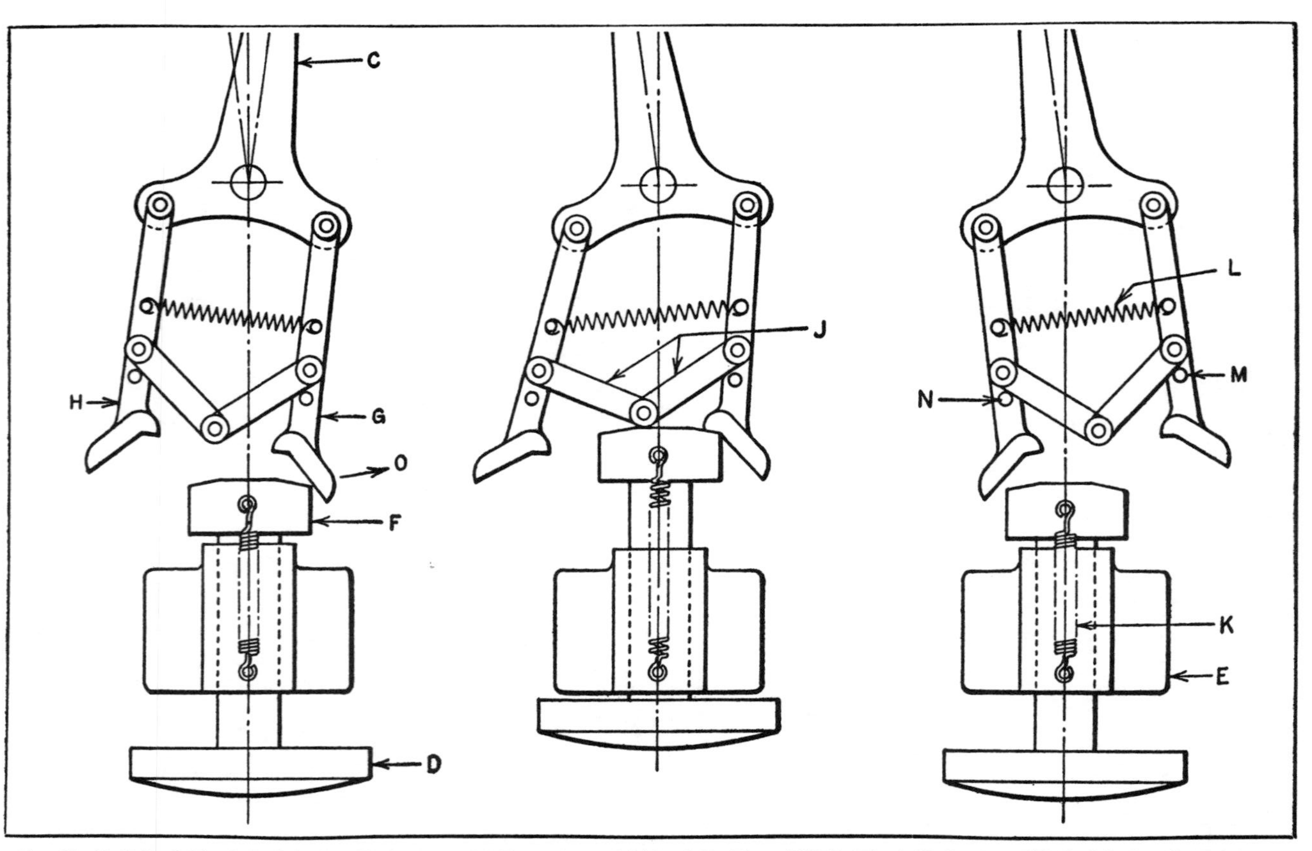

Fig. 15. (Left-hand View) Push-button Mechanism with Lever C in Right-hand Position. (Middle View) Mechanism with Push-button Pushed in to Transfer Lever to Left-hand Position. (Right-hand View) Push-button Released with Lever Remaining in Left-hand Position

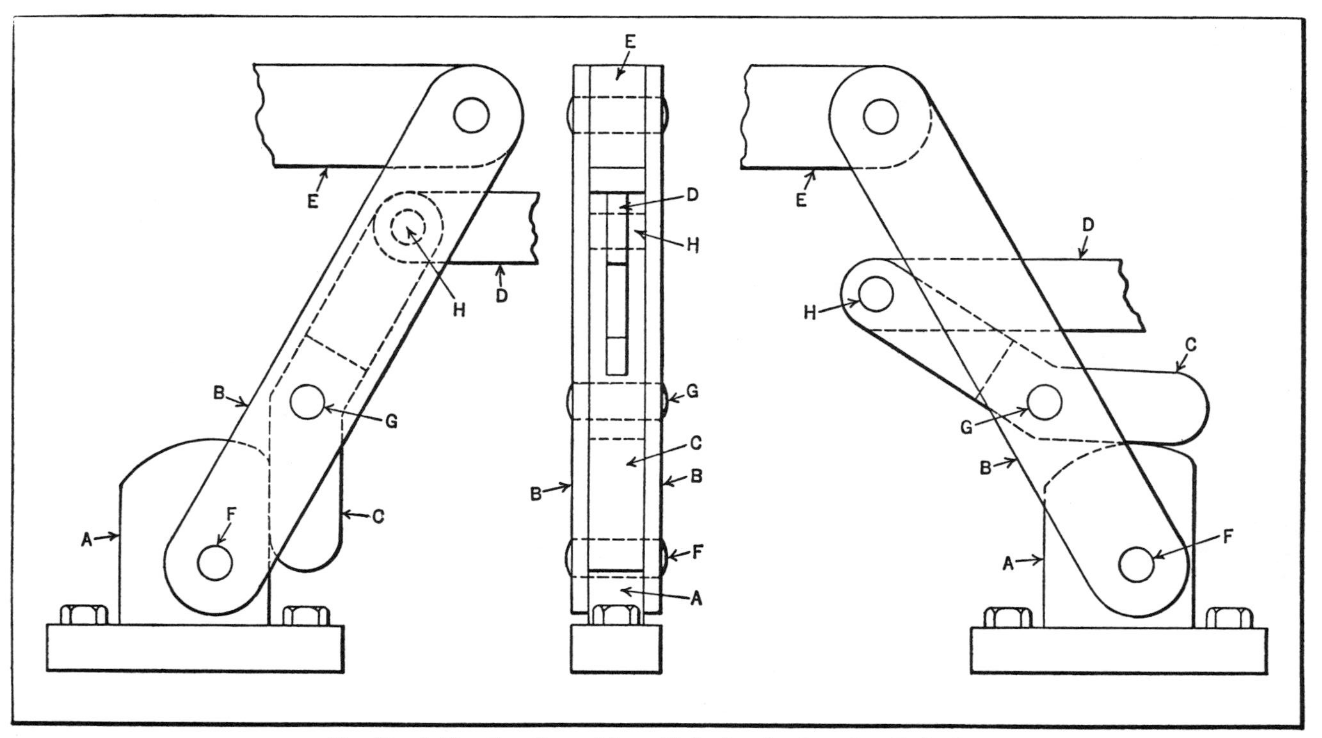

Fig. 16. Quick-acting Cam-and-lever Mechanism for Operating Brake-shoe

Part *A* is fastened to the stationary part of the machine and carries the pin *F* on which the double levers *B* pivot. The upper surface of part *A* is machined to conform with the arc of a circle of which pin *F* is the center. It will be noted that the pin *F* is located off center in part *A* and that the upper edge of part *A* terminates in a small arc-shaped surface on the right-hand end. Levers *B* carry between

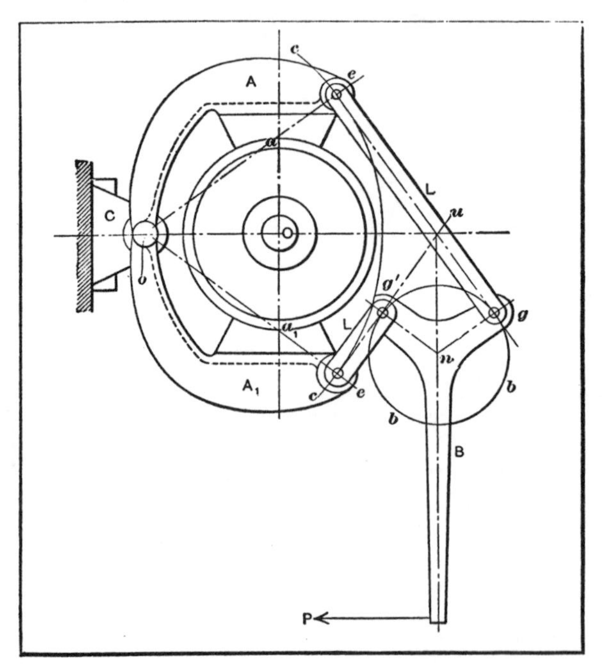

Fig. 17. Arrangement of Links and Levers for Operating the Clam-shell Type of Block Brake

them lever *C* which fulcrums on the pin *G* and carries at its upper end the bar *D* attached to it by the pin *H*. The opposite end of bar *D* is attached to the brake-shoe.

The side view at the left and the end view at the center show the arrangement with the brake-shoe in the released position. As the bar *E* is moved to the left, lever *B* fulcrums on pin *F*, and the upper right-hand corner of part *A* acts on lever *C*, which is caused to fulcrum on pin *G*. As the movement of the lower end of lever *C* is multiplied at the upper end, bar *D* is drawn forward quite rapidly in advance of levers *B*, thus quickly reducing the clearance space between the brake-shoe and the drum. As soon as the lower end of lever *C* has passed over the corner of part *A*, lever *C* ceases to act independently, and moves in unison with levers *B*.

Lever Mechanism for Block Brake.—The type of block brake known as the "clam shell" brake, Fig. 17, is often used in place of the band brake, over which it possesses the advantage of even wear on the blocks, and positive release, although not possessing as great a gripping power. The cast arms *A* and A_1 are pivoted at *o* to the frame of the machine, and carry blocks formed to grip the brake wheel. Links *L* connect these arms to the bellcrank *B*, having the floating center *n*. To lay out this brake to the best advantage, draw lines from *o* through the center points of contact *a* and a_1, on the rim of the wheel; also with *o* as a center, draw arc *cc*, intersecting these lines at points *e*. At these points, draw tangents to arc *cc*, intersecting at *u*, and draw *un*, bisecting angle *gug'*. Select a point *n* on *un* for the center of circle *b*, drawn tangent to *eg*, so that the required leverage will be obtained for the brake system.

When the brake is new, the exact nature of block contact is doubtful, and must be considered as only a line across the face of the blocks, but the wear on the blocks causes such a condition of pressures per unit area that the rate of wear is the same at all points of contact between the block and the drum. The point *o* may be placed below the wheel, making the axis aa_1 horizontal, the arms *A* and A_1 falling apart by gravity, when released. When the arms are not heavy enough to do this without one of them bearing against the wheel, while the other is free, light springs may be attached to points *e* to keep them apart, when released. The arms are sometimes extended so that points *e* may be connected by a spring which sets the brake, the release being made by toggles separating the arms when applied. The wheels of these brakes may be made V-shaped, the same as for band brakes. The blocks are often made to embrace a larger portion of the wheel than shown—sometimes nearly 180 degrees.

In Fig. 18 are shown two types of this brake, the fixed points being indicated by a dot within a circle, and the floating points by a plain dot. At *A* is shown a form of brake that is useful when there is no convenient way of pivoting the arms to the frame at *u*. The bellcranks *ace* and lever *mn* are pivoted as shown, but the point *u* is fixed in space only by its geometrical relations to points *a*. Since

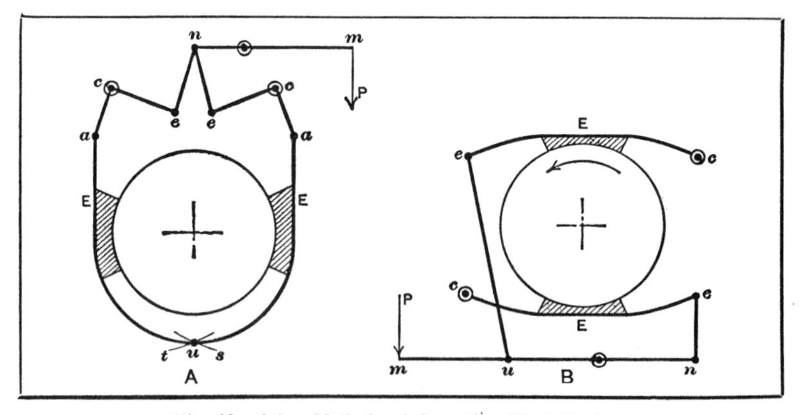

Fig. 18. Other Methods of Operating Block Brakes

the arcs *s* and *t*, struck from the points *a*, cross at *u*, it is evident that the point *u* becomes fixed in its relation to points *c* where the system is connected to the frame, and thus *u* is the fulcrum of the arms *E*, although not the point which receives the thrust of the brake arms, this being taken at points *c*. At *B* is shown a good type of brake in which both arms act as tension members in transferring the braking force to the fixed points *c*.

Safety Locking Device for Clutch Lever.— Safe operation of an extractor used for drying wiping cloths requires that the cover of the machine be tightly closed before the starting clutch is engaged. One device which complies with this requirement is shown in Fig. 19. The stationary casing *A* and the cover *B* enclose the rotating container (not shown) for the cloths. The cover is hinged at the left-hand side of the casing and can be secured in its closed posi-

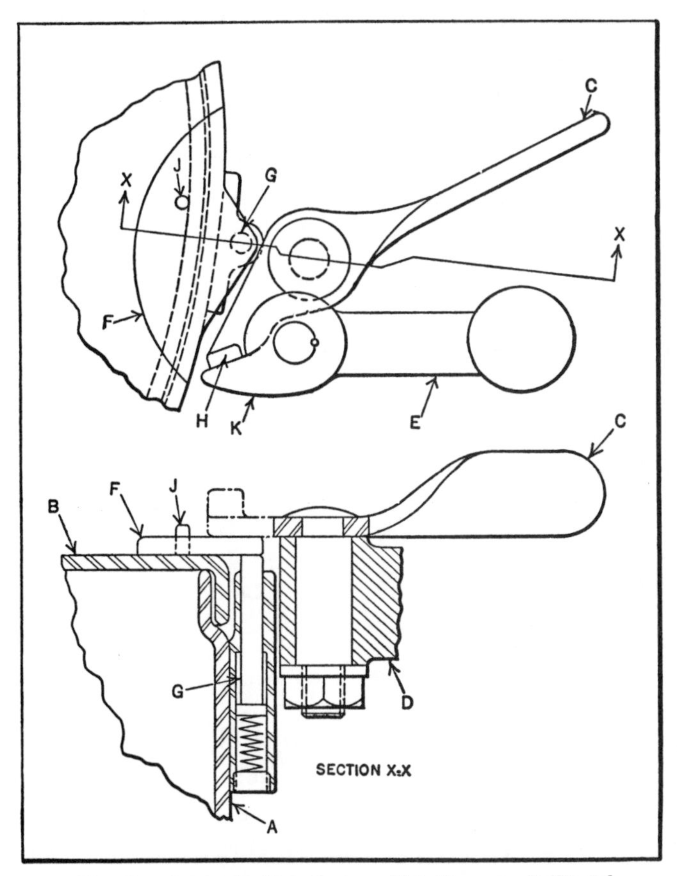

Fig. 19. Safety Locking Device which Prevents Accidental Shifting of the Clutch Lever

tion by the clamping lever *C*, pivoted on the stationary bracket *D*.

In the position shown, the cover can be swung upward on its hinge to allow loading or unloading of the container. While the cover is up, clutch lever *E* is prevented from being returned accidentally to its engaged position by the projection *H* on the lever *C*. Lever *C*, in turn, is prevented from returning to its clamping position by the spring-actuated plunger *G*, mounted on the casing. With this arrangement, both levers are automatically locked when the cover is up and are automatically released when the cover is in its closed position. As the cover descends, plate *F*, which is welded to the cover, depresses plunger *G* and allows lever *C* to be swung into its clamped position. At this time, finger *K* will clear the projection *H*, permitting lever *E* to be swung in position to engage the clutch. Plate *F* is provided with a stop-pin *J* to limit the clamping movement of lever *C*.

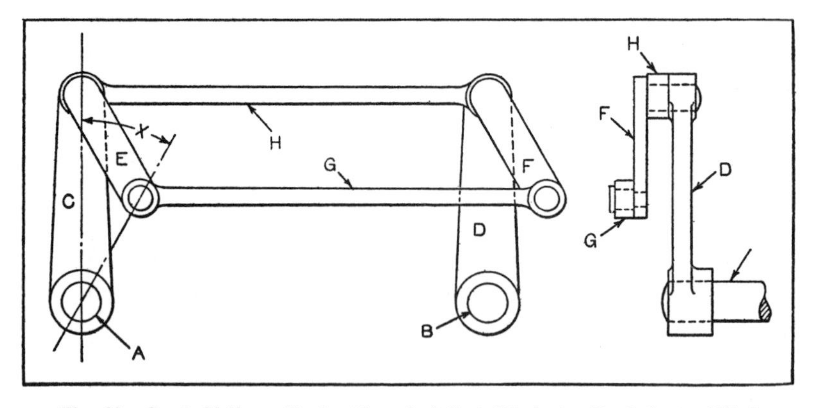

Fig. 20. Crank Motion with Auxiliary Rod that Eliminates Dead Center Effect

Crank Motion with Dead Center Eliminated.— When a rotary motion is transmitted from one shaft to another by means of cranks and a connecting-rod, the dead center positions may be avoided by the arrangement shown in Fig. 20. This mechanism is for a wire-forming machine and transmits power between two shafts located some distance apart. The purpose of the auxiliary rod *G* is to carry the driven shaft past the dead center positions. The question may be raised as to why a crank motion is used when a chain drive would produce the same effect. The reason is that on the machine in question, a reciprocating part of the machine passes into the space between the two shafts while the connecting-rods are "running over," or passing through the upper half of their cycle of rotation, withdrawing as the rods approach the center position. Obviously, this arrangement would be impossible with a chain or gear drive, which remains in the same position at all times.

Referring to the illustration, the shafts *A* and *B* carry the crank-arms *C* and *D*, respectively, which are connected by the rod *H* that runs free on its crankpins. In the actual installation, rods *G* and *H* were longer than shown. The length of these rods, however, does not affect the operation of the drive. The upper crankpins, which are keyed to the crank-arms *C* and *D*, carry the auxiliary arms *E* and *F*, which are set at an angle with arms *C* and *D*. Connecting-rod *G*, which is exactly the same length as rod *H*, connects arms *E* and *F*. Although this arrangement may be classed as being without a dead center, it really has two dead center positions, but there is a time element between the two which renders them both ineffective in arresting the driving motion. When one crankpin reaches dead center, the other is still approaching and is effective in forcing the first past the dead center.

It is essential that each pair of similar parts be of exactly the same length; otherwise, there will be a binding action. Although the length of arms *E* and *F* should be kept as short as possible for the sake of compactness, they should not be less than one-half the length of arms *C* and *D*. The movement will operate without any dead center effect over a wide range of positions for arms *E* and *F*, although the smoothest movement seems to be attained when the angle *X* is not less than 20 degrees.

CHAPTER XIV

FEEDING MECHANISMS AND AUXILIARY DEVICES

The expression "feeding mechanism" may indicate mechanical means of presenting parts successively for some manufacturing operation or this term may be applied to a mechanism for imparting a feeding movement to a metal-cutting or other tool. This chapter deals with feeding and allied mechanisms of various types and designed for miscellaneous application. It supplements the four chapters in Volume I which deal with this general subject (pages 447 to 519).

Hopper Feeding Mechanism Used in Soldering Fuse Plugs.— In manufacturing electrical fuse plugs, such as those used for house circuits, a thin fuse strip is soldered to a split rivet that has previously been assembled into the plug. The soldering is done on a special indexing dial machine in which the solder slugs are automatically dropped on the rivet on the inside of the plug. In another position of the dial, the plug dwells under a concentrated gas flame long enough to melt the solder slug. At the next station the operator places one end of a fuse strip in the molten solder. The dial is then indexed to another position where the plug is automatically ejected from the machine.

A detail view of the mechanism for automatically feeding the solder slugs is shown in Fig. 1. The dial indexes intermittently and is shown with a plug in position to receive two slugs which are held up in the end of the tube *B* by the stop *G* riveted to the arm *F*. This arm swings on the pin *I* and is normally held in position by the coil spring *C*.

The hopper *A* is fastened in the boss *D* of the machine by

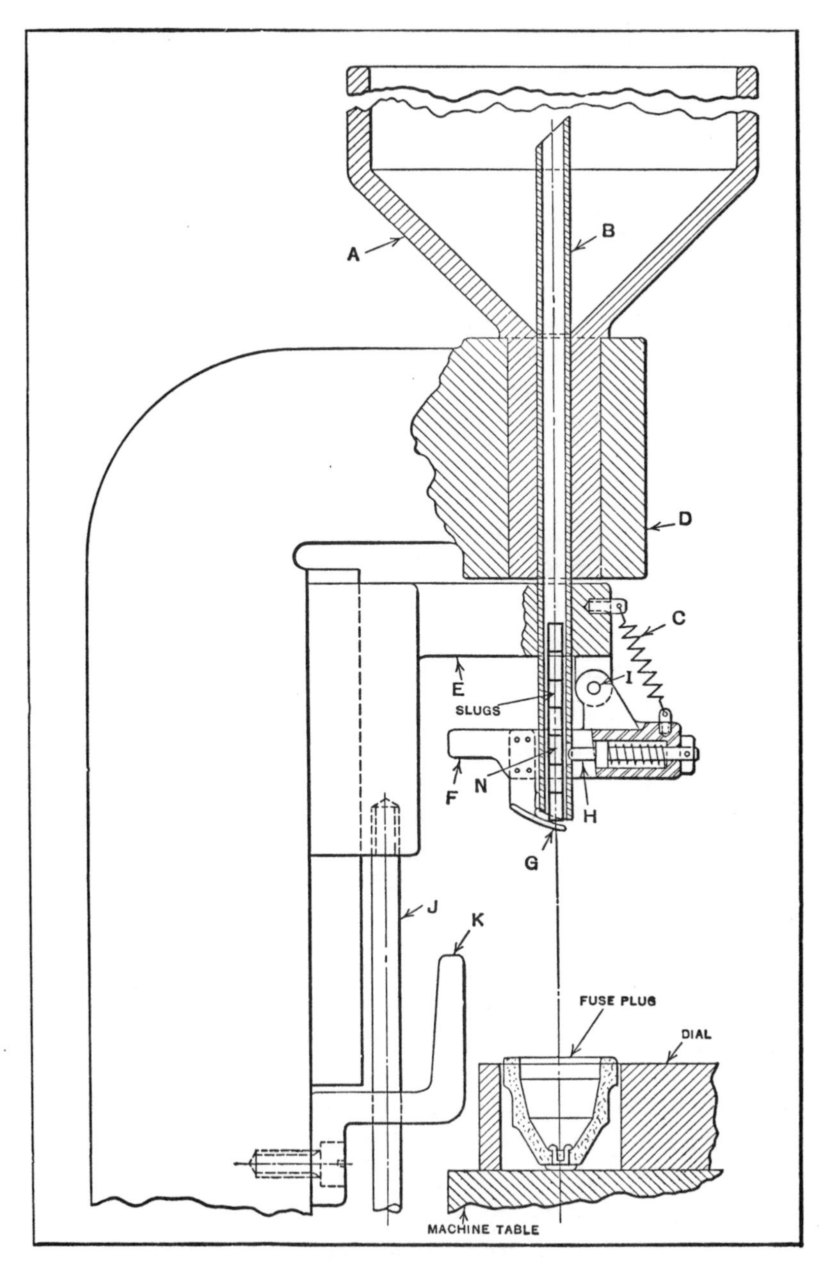

Fig. 1. Hopper which Delivers Two Solder Slugs to Each Fuse Plug in Soldering Machine Dial

a set-screw. Slide *E* carries arm *F* and tube *B*, and receives a vertical reciprocating movement through rod *J*, from a cam located on the machine. The slide completes one cycle during the indexing of each station. The top of tube *B* is cut at an angle to produce greater agitation of the slugs in the hopper. The slugs, collecting in the tube as the latter passes through the supply in the hopper, drop down on stop *G*.

At the end of the down stroke of slide *E*, the stop *K* engages projection *F*, swinging the latter up to the left. This motion withdraws stop *G* from the end of the tube and allows the two bottom slugs to drop out and into the fuse plug. Just before this happens, however, the spring-actuated plunger *H* forces the slug *N* against the side of the tube, holding back the flow of slugs until the return stroke of slide *E* disengages members *K* and *F*, permitting stop *G* and plunger *H* to return to the position shown in the illustration.

Agitating Device for a Pin Hopper.—Difficulty was experienced by a plant manufacturing electrical switches in maintaining a sufficient flow of switch pins from hopper to power press. The pins are made from brass rod and are about 3/16 inch in diameter by 3/8 inch long. The hopper is of simple design, consisting chiefly of a stationary conical shell in which the pins are placed. A length of tubing having an inside diameter slightly larger than the diameter of the pins is a slip fit in a vertical hole bored in the lower end of the hopper; this tubing is given a vertical movement, which is transmitted from the press crankshaft through a rack and pinion.

As the end of the tube passes through the pins, some of them drop into the tube and down to a connecting chute, which carries them to the press. It was found, however, that the pins had a tendency to collect around the tube horizontally instead of vertically, so that the number entering the tubing during each stroke was insufficient to supply the press. After some experimenting, this difficulty was overcome by the use of an agitating device, as shown in Fig. 2.

Bracket *A* is stationary, the upper end (not shown) supporting the conical shell with its tube while the lower part carries the agitating device. The tube is secured to the sliding bar *B*, both parts being raised to the upper position by means of the reciprocating arm *C* pivoted to the cross-head *D*. A double-end latch *E* is also pivoted to the cross-head, and engages teeth cut in bar *B*, as well as projections on the plate *F*, which is secured to bracket *A*.

In the position shown, the cross-head has already started its upward stroke, carrying with it bar *B* and the tube. Further movement of the cross-head causes the upper end of the latch to be forced to the left by the lower projection on plate *F*. At the same time, the lower end of the latch is forced out of engagement with the lower tooth in bar *B* and allows the bar to drop back by gravity, aided by the action of coil spring *G*. The lower end of the latch is held away from the bar momentarily only; thus, as the upper end passes the projection on plate *F*, its lower end at once enters the next tooth space in the bar *B*, which then continues its upward movement until the latch engages the next projection on plate *F*. At this time, and for each remaining projection, the action described is repeated.

At the end of the upward stroke, bar *B* is held suspended by the latch, but upon starting its downward stroke, the latch again comes into contact with the upper projection on plate *F* and is released from the bar, allowing the latter to drop to its lowest position. Here the slide is picked up once more by the latch after the cross-head returns to the bottom of its stroke. Incidentally, a bumper (not shown) is provided to take the shock upon the return of bar *B*. A series of reciprocating movements is imparted to the tube as it passes upward through the pins, agitating them sufficiently to cause a greater number to enter the tube. It should be mentioned that the success of a device of this

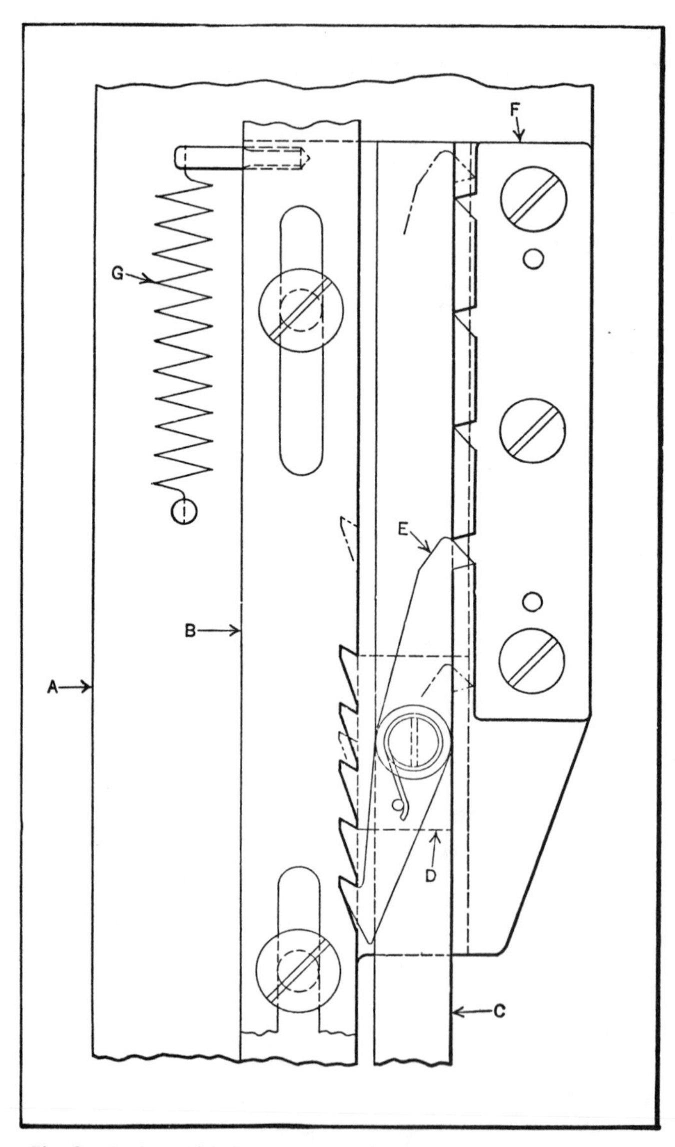

Fig. 2. Device which Increases the Flow of Work from a Hopper by Imparting Short Reciprocating Movements to the Feeding Tube as it Passes through the Parts

type depends upon the speed of the cross-head, because if the speed is too slow, the lower end of the latch will fail to catch the tooth in the sliding bar, thus rendering the device inoperative.

Mechanism for Feeding Granular Material Uniformly.—One of the methods used in a stamping mill for feeding

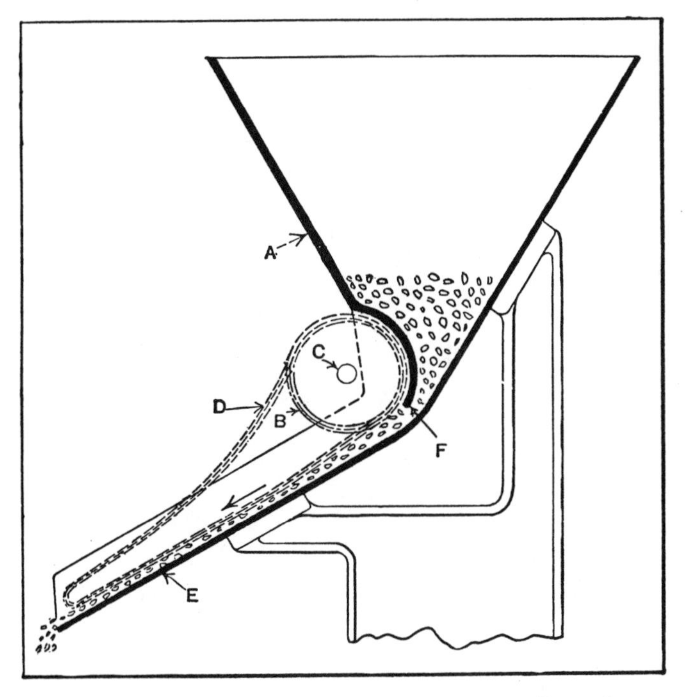

Fig. 3. Maintaining a Constant Flow of Ore from a Hopper by Means of Endless Chains

crushed ore at a uniform rate to a grading machine is shown in Fig. 3. The ore, which consists of pieces about the size of an egg, is dumped into the hopper *A* and passes through the opening *F*. As the ore must be delivered from the chute *E* at a uniform rate, some means must be provided for regulating its flow. This is accomplished in the following manner: A number of endless chains *D* passing over the sprockets *B* rest upon the ore as it flows down the chute.

If there is no movement of the chains, the friction resulting from the weight of the latter will prevent the ore from sliding down the chute. If motion is imparted to the chain so that it will travel at a constant speed in the direction of the arrow, the ore will be carried downward at a uniform rate and at approximately the same speed as that of the

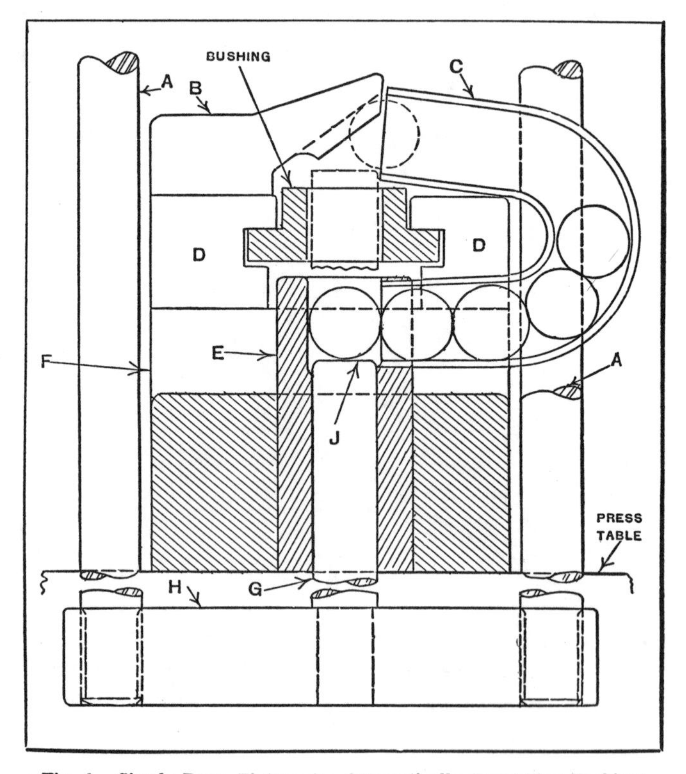

Fig. 4. Simple Press Fixture for Automatically Burnishing Bushings

chains. The movement of the chains is obtained by revolving the sprockets *B* on the shaft *C*, the latter being driven from the driving shaft of the grading machine. The nature of this ore is such that it will flow very freely, and little trouble is experienced from jamming at the mouth of the chute.

Automatic Ball-Feeding Attachment for Ball Burnishing.— Steel balls are frequently employed for burnishing holes when a very fine finish and an accurate job are required. With the automatic ball-feeding device shown in Fig. 4 the work can be burnished in a power press. The bushings to be burnished are fed into a chute, the end of which is shown at *D*, and carried down by gravity to a position directly over the reciprocating plunger *G*. When the plunger is at its lowest point, one of the balls in the return tube *C* rolls on the end *J*, directly under the hole to be burnished. As the plunger ascends, it pushes the ball up through the work. Continuation of this upward movement carries the ball against the angular surface of the block *B*, and into the return tube *C*, as indicated by the dotted outline of the plunger and ball when at their highest position.

The reciprocating motion of the plunger is derived from the ram of the press through the connecting posts *A* and the plate *H* in which the plunger is a drive fit. A clearance hole in the bolster plate is necessary to allow a through passage of the plate *H*. The bushing *E* is a drive fit in the fixture *F* and a slip fit for the plunger. One side of this bushing has an opening for the lower end of the return tube. This tube may be fastened by straps to the fixture, or it may be soldered to the bushing *E* and to the chute *D*.

Although not shown, the usual provision must be made for tripping the press clutch in case the work becomes jammed in the chute or fails to line up properly with the plunger *G*. One advantage of this fixture is that the wear incident to burnishing is distributed equally among a number of balls.

Chain Feed Mechanism with Periodically Accelerated Motion.— Fig. 5 shows a mechanism used for feeding tree trunks *T* to a sawing machine with a periodically accelerated motion that might also be applied to other machines. This motion is obtained by an interesting arrangement for simultaneously taking up slack in one side of the chain while giving out slack in the other side. The mechanism is driven by the spur gear *A* which drives gear *B* by means of an intermediate gear *C*, thus driving chain *D* at a uniform speed. The chain is carried over four pulleys in the base of the mechanism. The pulleys *E* and *F* are attached to a swinging arm *G*. This arm is oscillated by a crank and rod mechanism *H* and *J*, also connected to the driving gear *A*.

The gear *K*, driven by chain *D*, imparts the periodically varying feed motion to the rollers *M*, in contact with the tree trunk, by a second chain drive indicated by the light dot-and-dash lines at *L*. If pulleys *E* and *F* remained stationary, the feeding motion would be uniform. However, as the pulleys are on the periodically swinging lever *G*, the motion of chain *D* is changed in such a manner that it will remain stationary during one short period and will be accelerated or retarded in the other. The swinging lever *G* is slotted so that rod *J* can be adjusted to give the feeding motion required. A cam-actuated motion could be substituted for the crank motion obtained by crank *H* and rod *J*, the cam being given the profile necessary to obtain the desired motion.

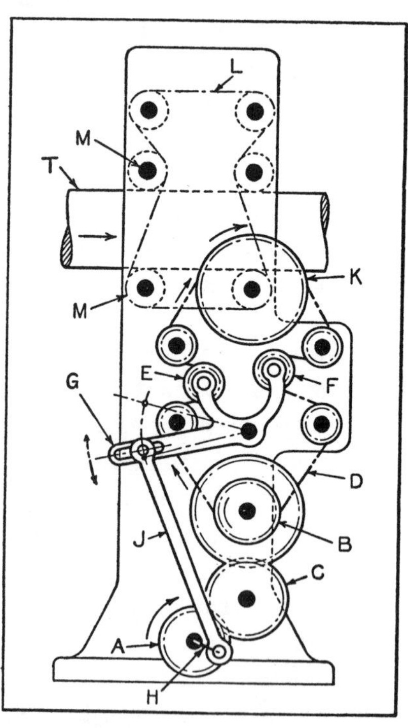

Fig. 5. Mechanism for Producing Periodically Accelerated Motion

Planetary or Differential Type of Feeding Mechanism for Internal Grinder.—The eccentric and differential type of feed controlling mechanism which is shown by a perspective view, Fig. 6, to illustrate the arrangement clearly, has been applied to internal grinders of planetary design, used for grinding holes in parts of such bulk that rotation is impracticable. The radius of the path of the grinding wheel, which has a planetary motion, is changed while the wheel is at work by an adjusting movement that is transmitted through differential gearing. The grinding wheel spindle 1 is located eccentrically in a cylindrical member 2, which is rotated to vary the radial position of the wheel. Center line *A* represents the axis of the main body 6 of the grinder head; center line *B* is the axis of cylindrical part 2; and *C* represents the axis of the grinding wheel spindle. The distance from *A* to *B* equals the distance from *B* to *C*, so that by turning part 2, axis *C* can be made to coincide with axis *A*, thus permitting the wheel to be located anywhere from a central position to its maximum position radially.

When the grinding wheel has been adjusted for a given cut, it has, in addition to rotation about its axis, a planetary movement about axis *A* of the grinding head. This planetary motion is obtained from the driving shaft which rotates head 6 through gears 9 and 10 at one end, and 5 and 8 at the other. The driving gears 8 and 9 are the same size, and the driven gears 5 and 10 are also equal in size; consequently, these two sets of gearing normally rotate at the same speed, but when a feeding movement of the wheel is required, gear 8, through an adjustment of the differential gearing located between gears 8 and 9, is caused either to lag behind or advance, thus shifting eccentric 2 through worm-gearing 3 and 4 and a screw gear which meshes with teeth on the inside of gear 5.

The action of the differential gearing will be explained in connection with Fig. 7 which shows a cross-sectional view. When the driving shaft *G* revolves, pinions *P* (14 in Fig. 6) which are mounted on studs fixed in a stationary housing of the differential, also revolve. This rotation of the pinions is transmitted to the internal gear *Q* (12 in

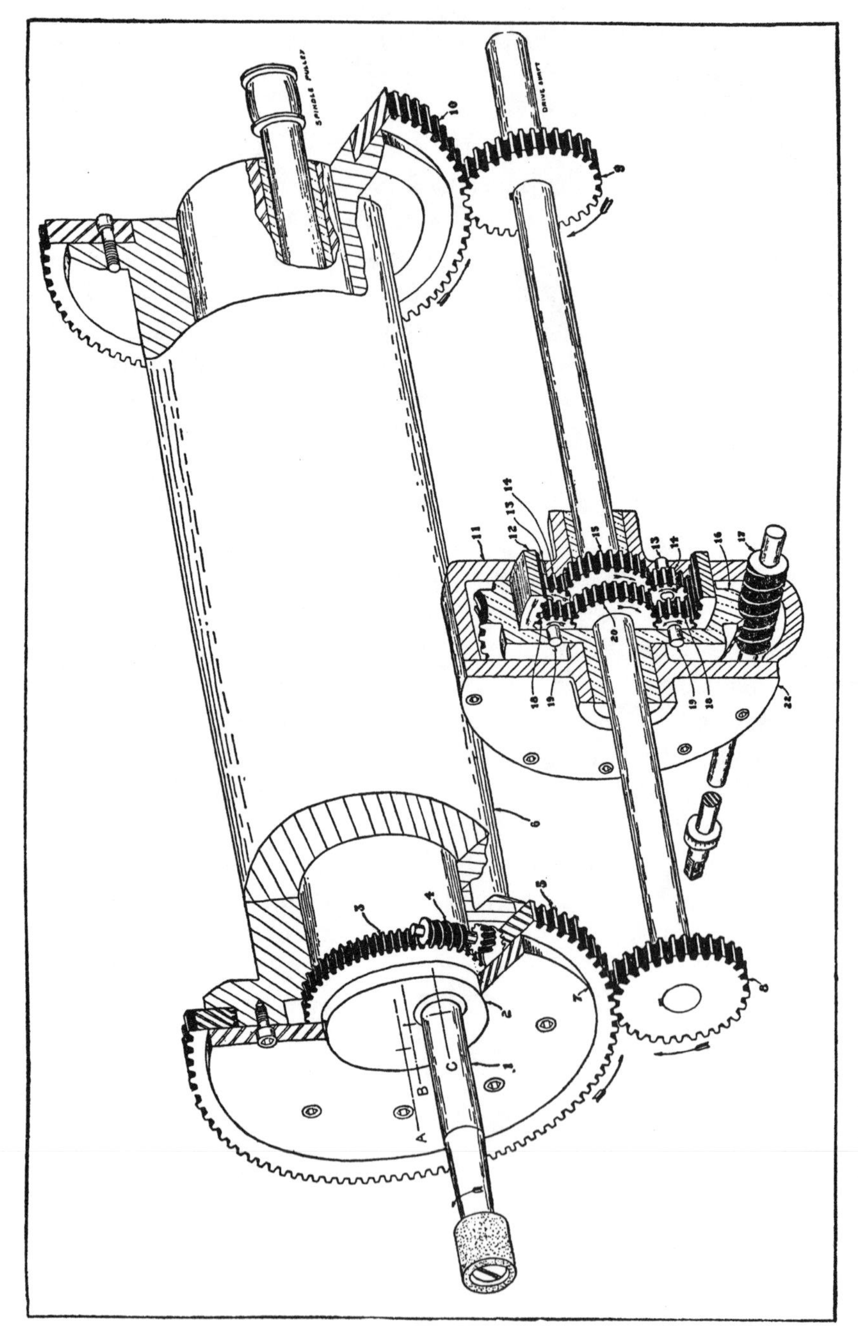

Fig. 6. Perspective View of Eccentric Feed Mechanism which is Operated by the Adjustment of Differential Gearing

Fig. 6) which is free to turn within worm-gear *R* (16 in Fig. 6). As internal gear *Q* revolves, it drives pinions *S* (18 in Fig. 6) which are mounted on pins fixed in worm-gear *R*. Pinions *S* rotate the left-hand section G_1 of the driving shaft at the same speed as the right-hand section, except when a feeding movement occurs.

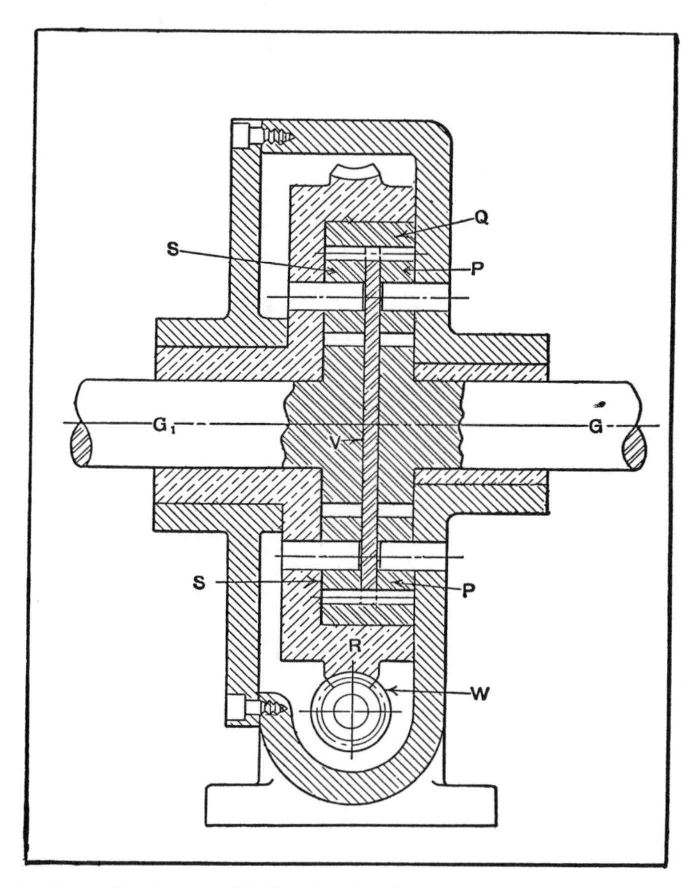

Fig. 7. Sectional View of Differential Gearing of Eccentric Feed Mechanism

For adjusting the grinding wheel in or out, worm *W* is turned by hand, thus turning worm-wheel *R*, which changes the position of pinions *S* relative to pinions *P*. If pinions *S* are advanced or moved in the direction of the rotation of

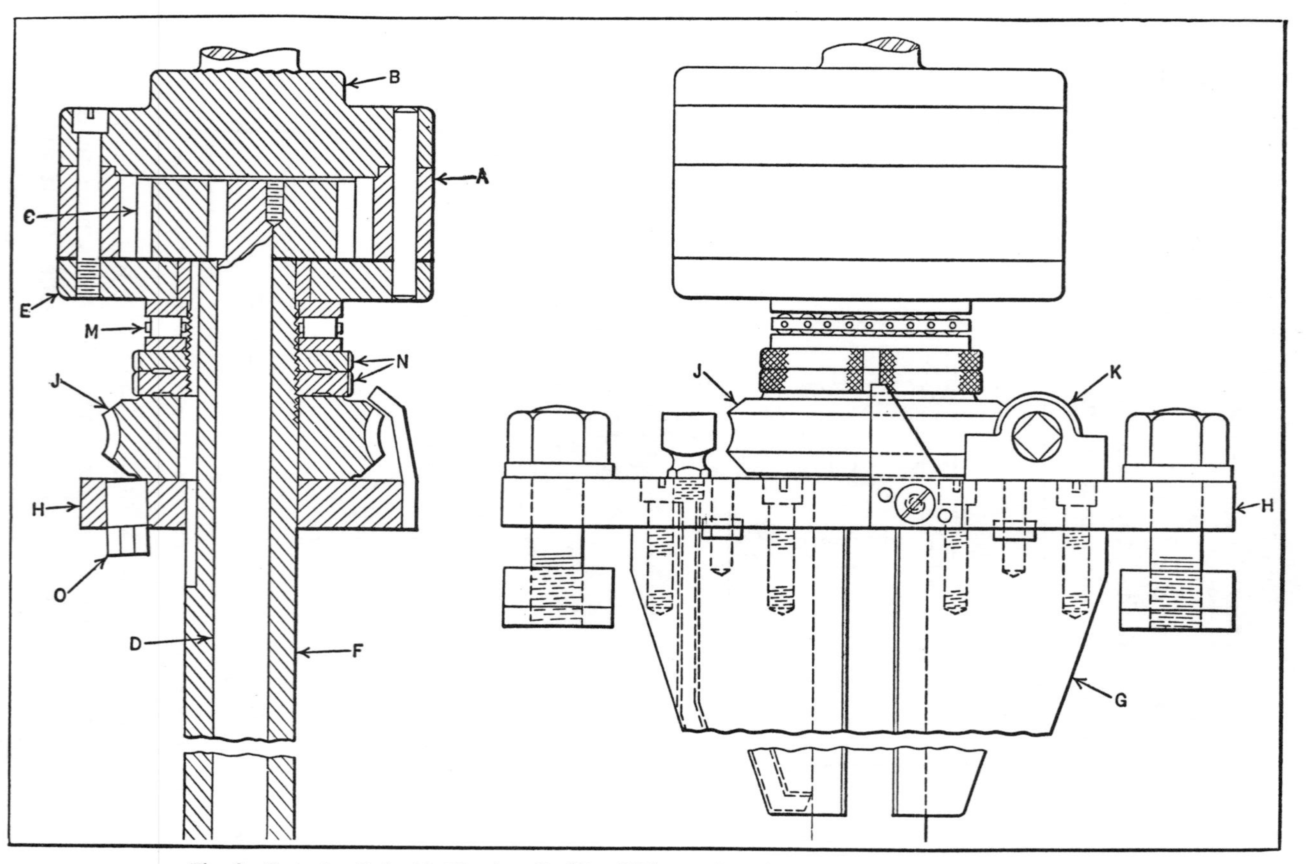

Fig. 8. Recessing Tool with Planetary Feed for Milling an Irregular Recess at a Depth of 8 1/2 Inches

internal gear Q, then during this period of adjustment, shaft G_1 will turn somewhat slower than G; consequently, there will be a movement of gear 5, Fig. 6, relative to the main head, thus causing worm-gearing 3 and 4 to rotate and changing the position of the grinding wheel. On the other hand, if the rotation of worm-wheel R is such as to move pinions S in the direction of the rotation of internal gear Q, the speed of shaft G_1 will be accelerated relative to G, thus adjusting the grinding wheel in the opposite direction. A spacer plate or disk V (Fig. 7) is located within the internal gear Q and between the two sets of gearing. The internal teeth of gear 5 (Fig. 6) were cut on a lathe, the indexing being done by disengaging the feed-screw

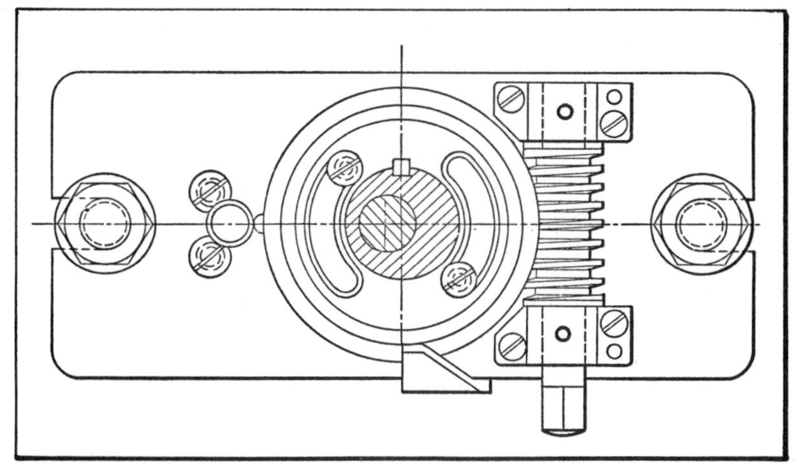

Fig. 9. Plan View of the Recessing Tool, Showing How the Cutter Shaft is Fed toward the Work by an Eccentric Sleeve

gears, and the teeth of the screw gear which engages the internal teeth of gear 5 were cut on a horizontal universal boring mill. The cutter used was a duplicate of the internal gear and formed the teeth by a generating action.

Planetary Adjustment for Feeding Deep-Hole Recessing Milling Cutter.— An ingenious application of planetary gearing to tool design is shown in Figs. 8 and 9. This tool is used in a radial drilling machine for milling, at a depth of 8 1/2 inches, an irregular recess in the ports of a steel forging, as indicated at *A* in Fig. 10. Internal gear *A* (Fig. 8) is fastened to the shank *B*, and meshing with this

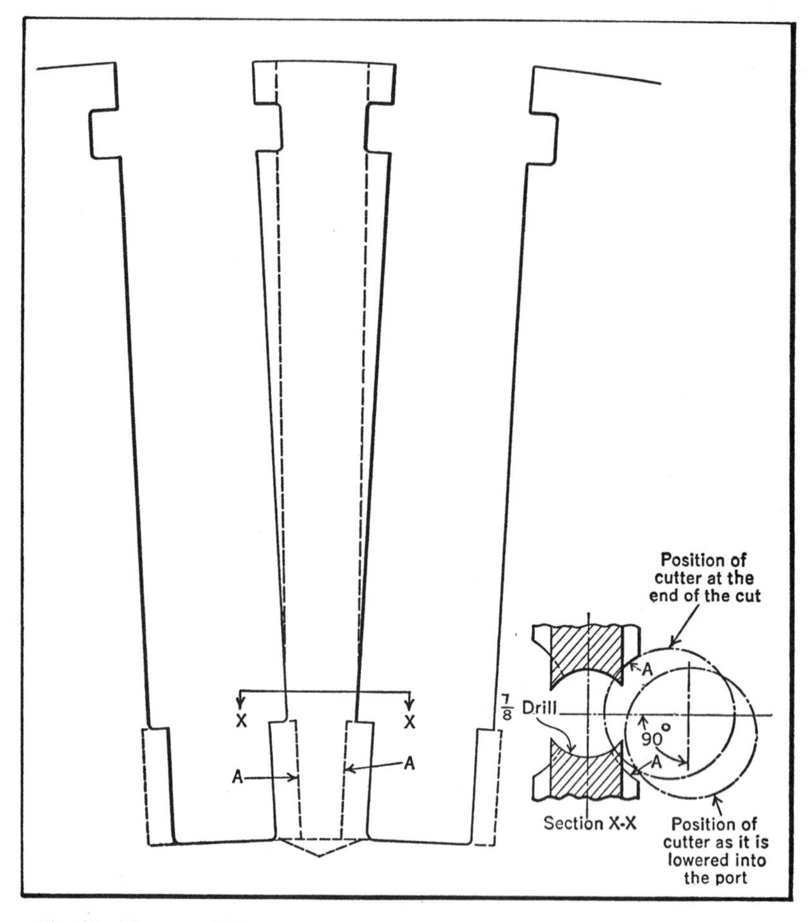

Fig. 10. Diagram of Work, Showing the Irregular Recess Cut by the Tool in Fig. 8

gear is the pinion *C*, secured to the end of the cutter shaft *D*. Endwise movement of both pinion and shaft is prevented by the retaining plate *E*.

The eccentric sleeve *F* provides a bearing for the cutter

shaft and is supported along its length by the two jaws *G*. These jaws are secured to the plate *H*, upon which is mounted the worm-wheel *J*, which is keyed to the eccentric sleeve. Meshing with this worm-wheel is the worm *K*, which serves to rotate the eccentric sleeve for feeding the cutter (attached to lower end of shaft *D*) into the work. The thrust of the cutter is taken by the roller bearing *M* through the check-nuts *N* on sleeve *F*.

To recess the port *A*, Fig. 10, the cutter is lowered to the bottom of the port, and the plate *H*, located by pin *O*, is fastened by T-bolts to the top of the forging. The jaws *G* are made slightly less in width than the port, so that they serve to centralize, as well as to support the eccentric sleeve. As the machine spindle rotates, the internal gear revolves about the pinion, rotating the latter with the shaft and cutter.

To start the cut, the eccentric is rotated by hand through the worm and worm-gear, a handwheel (not shown) being provided for turning the worm. As the eccentric sleeve rotates, the cutter is swung into the side of the port. The greatest depth of cut is reached when the worm-wheel has rotated 90 degrees, as indicated in Fig. 10 by the dot-and-dash circles representing the cutter. When the cut is completed, the worm-wheel is reversed to withdraw the cutter from the recess. The tool is then removed and set up in the next port, where the other recess is cut.

Friction-Grip Wire-Feeding Device.— The device shown in Fig. 11 can be used for feeding wire on any wire-forming or other machine requiring an accurate feed. The outer shell *S* is mounted on a slide or other reciprocating part of the machine which has a movement equal to the desired feed. Cage *C* is carried inside shell *S*, bearing with an easy sliding fit at both ends. This cage has two holes diametrically opposite each other into which balls *B* are placed. The holes are slightly larger than the balls, so that the balls move freely within them. The holes are not full size clear through to the central hole in part *C*, but end in a conical seat, thus allowing the balls to project through to the wire *W*, but preventing them from falling through when there is no wire there. Spring *P* pushes the cage so that the balls are carried into the taper portion of *S* and against the wire.

In action, the device is moved in the direction of the arrow shown on the diagram. As the balls are in contact with the wire and the shell, any resistance to the movement of the wire causes them to roll into the taper and grip the wire more tightly. This movement is very slight and does

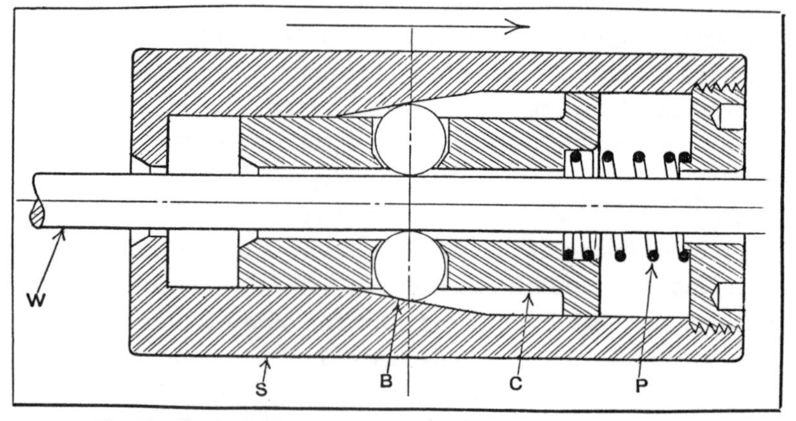

Fig. 11. Simple Design of Wire-feeding Device of Friction-grip Type

not affect the accuracy of the device for ordinary purposes. As the shell starts on its return stroke, the balls roll out of the taper, thereby releasing their grip on the wire. Theoretically, three balls should be used in this device, but it has been found that two balls are entirely practical. It will be noticed that several sizes of wire can be fed and that variations in the wire do not affect the accuracy of the feed.

Adjustable-Speed Wire-Feeding Unit for Wire-Cutting Machine.— Various lengths of wire for carrying electric current are used in the manufacture of a certain product. These wires are cut to length and the ends are stripped of insulation on a well-known make of wire-cutting machine.

The wire comes to the factory wound on heavy spools, a full spool weighing approximately 175 pounds. The inertia of such heavy spools caused an uneven feeding movement that resulted in a variation in the lengths of the wires. To overcome this trouble, an operator was employed to turn the spool, so that a small amount of wire would be kept slack at the feeding end of the machine. It became evident, however, that such an arrangement would be too expensive, as the job appeared to be one that would last for several years. The attachment illustrated in Figs. 12 and 13 was built to eliminate the necessity for hand-feeding.

The spool of wire is indicated at the left, Fig. 12. There are three large sheaves *C, D,* and *E,* and one idler *F.* Two of the main sheaves *E* and *D* are drivers, being geared together and driven by a sprocket and chain which is connected to a cross-shaft *H* at the end of the machine *A,* Fig. 13. In the driven sprocket is a free-wheeling clutch. The shaft to which this clutch is keyed is squared for a crank-handle. This permits the train of sheaves to be turned forward independently of the driving mechanism.

The original driving shaft of the machine was replaced by a longer one. To this longer shaft is secured a large friction disk *G.* Adjoining this disk and attached by suitable mountings to the end of the machine is the cross-shaft *H.* To the end of this shaft is secured the driving sprocket-wheel. Also keyed to this cross-shaft is a small friction wheel *I* which contacts with and is driven by the large friction disk. Set into a groove in the cross-shaft is a small screw *J* (see enlarged view). This screw is threaded through a half-nut which is secured to the side of the driven wheel. The screw is retained in the shaft by means of a plate which is fastened to the end of the shaft. By means of this screw, the driven wheel is adjusted across the face of the driving disk to obtain any desired ratio of speed. To facilitate setting the driven wheel, the cross-shaft is scored and numbered every half inch, the numbers corresponding

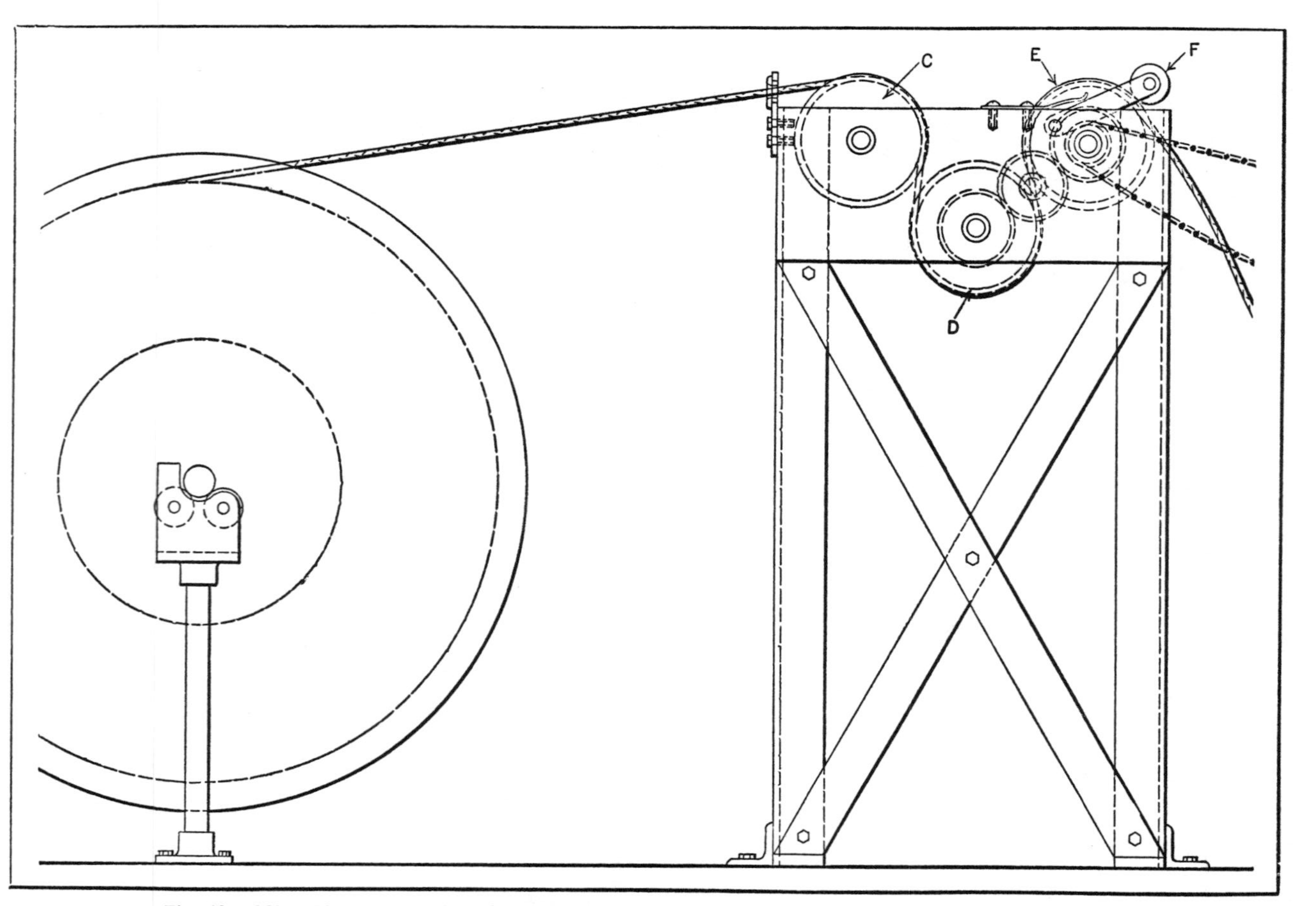

Fig. 12. Adjustable-speed Friction-driven Feeding Unit for Wire-cutting Machine—See Continuation, Fig. 13

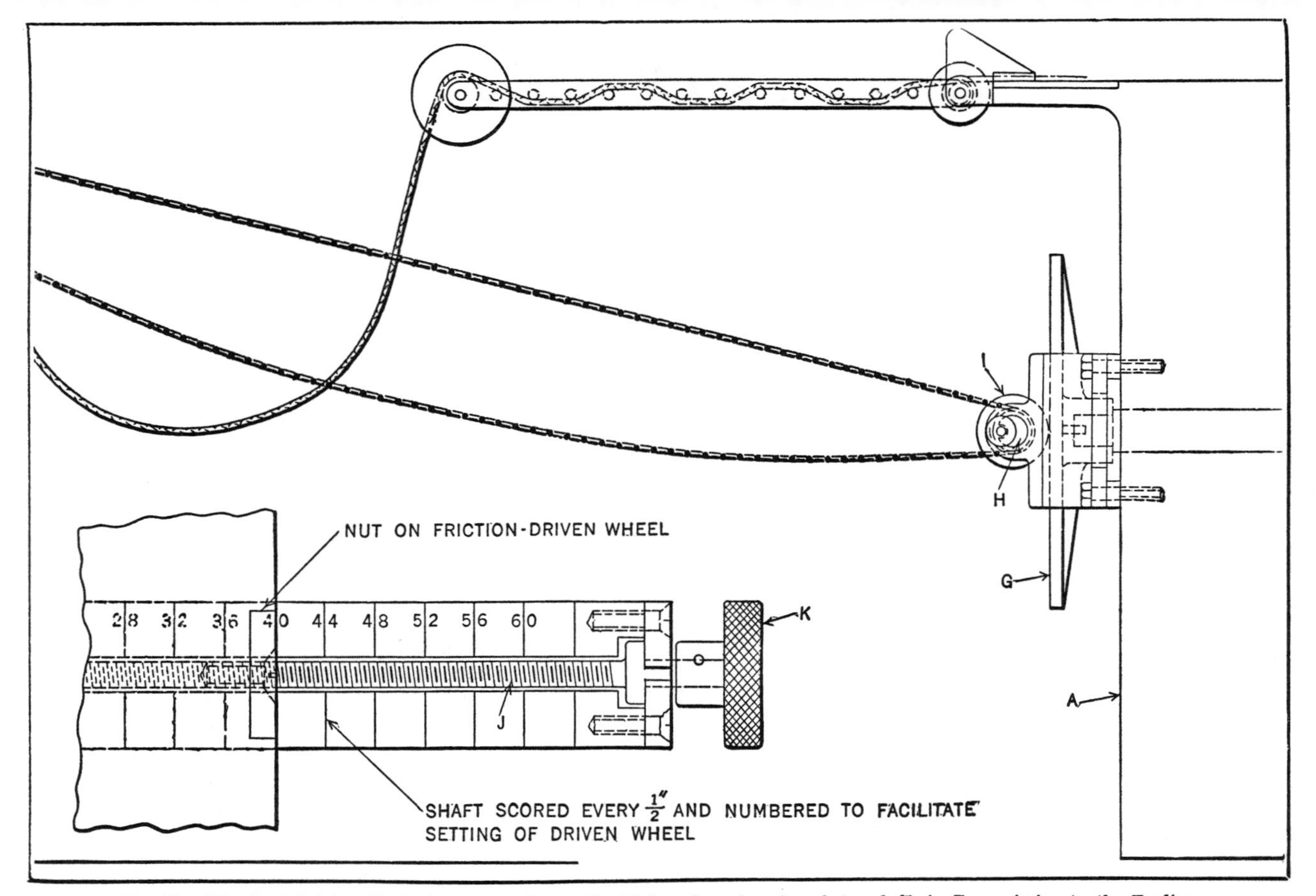

Fig. 13. Wire-feed Driving Mechanism which Transmits Motion through a Sprocket and Chain Transmission to the Feeding Sheaves, Fig. 12

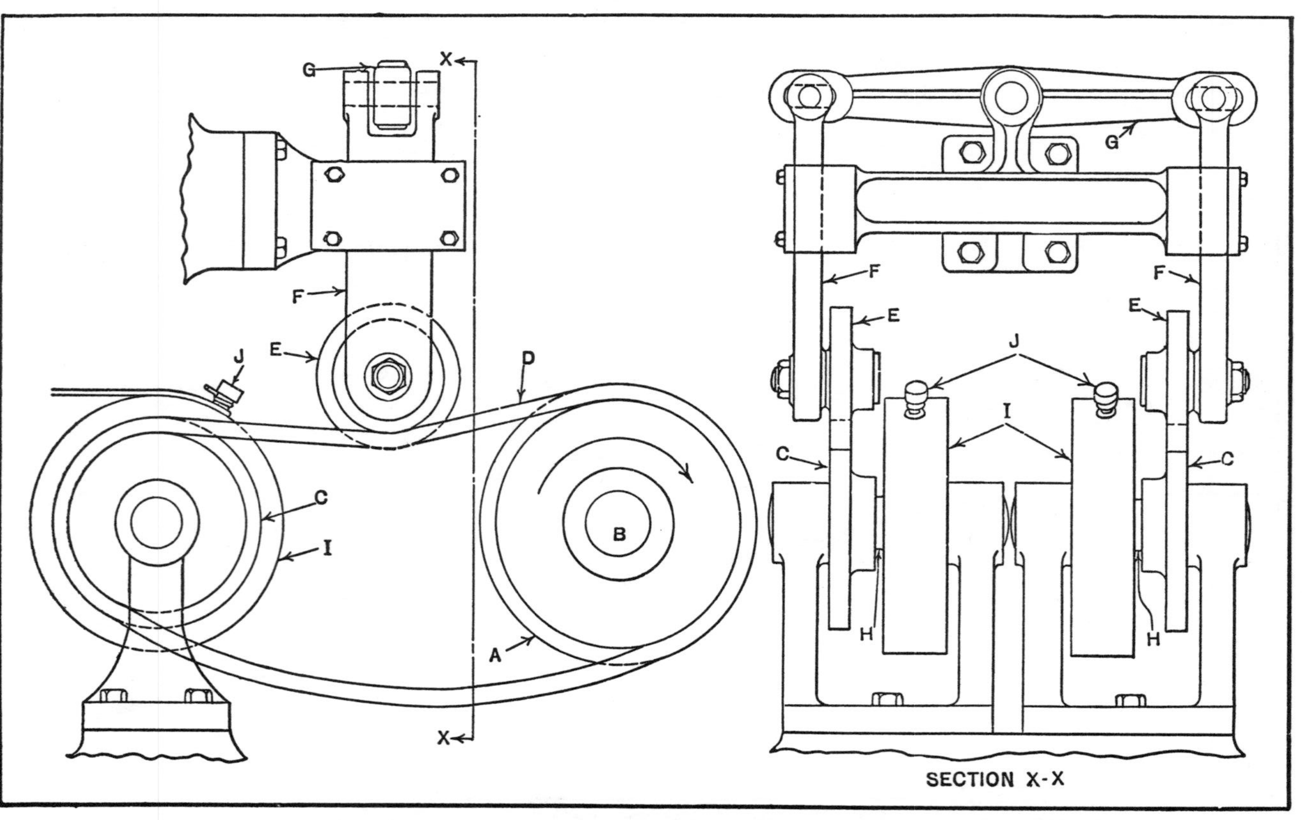

Fig. 14. Device in which the Slack in Two Sprocket Chains Equalizes the Tension in Two Wires Being Stretched

to the length of wire that will be fed by the rolls for that particular setting.

In operation, after the wire has been laced through the feed-rolls and into the machine, the operator places the crank-handle on the squared end of the shaft previously referred to, and turns the feed-rolls forward, feeding a small amount of slack wire ahead of the wire-cutting machine. The wire-cutting machine is then started, which also serves to start the feed-rolls. If the length of wire is one that has been cut previously, the setting, no doubt, will be fairly accurate and the same amount of slack wire will be maintained. However, should the length of wire to be cut be an odd size, an approximate setting is made. An occasional glance from the operator, while pursuing other duties, determines whether the feed-rolls are losing or gaining on the wire-cutting machine. In either case, a turn of knob *K* at the end of the screw in the cross-shaft readjusts the feed. This can be done while the machine is running.

Automatic Wire-Tension Equalizer.—On a special machine for producing a wire product, numerous strands of wire are woven or interlaced around two lengthwise strands. After the required number of interlacings have been made, the two lengthwise strands are pulled tightly and the whole locked together. It is essential, however, that both lengthwise strands have the same degree of tension during this locking action. Hand methods of tensioning had been used until the attachment shown in Fig. 14 was developed. This attachment automatically maintains an equal tension on the two wires.

The two sprockets *A* fastened on the driving shaft *B* drive the two sprockets *C* by means of the chains *D*, which have considerable slack. This slack, because of the direction of the drive, will be normally at the bottom. A uniform tension is maintained in the wires by means of the two idler sprockets *E* carried on slides *F*, which, in turn, are connected by the equalizing lever *G*. Sprockets *C* and

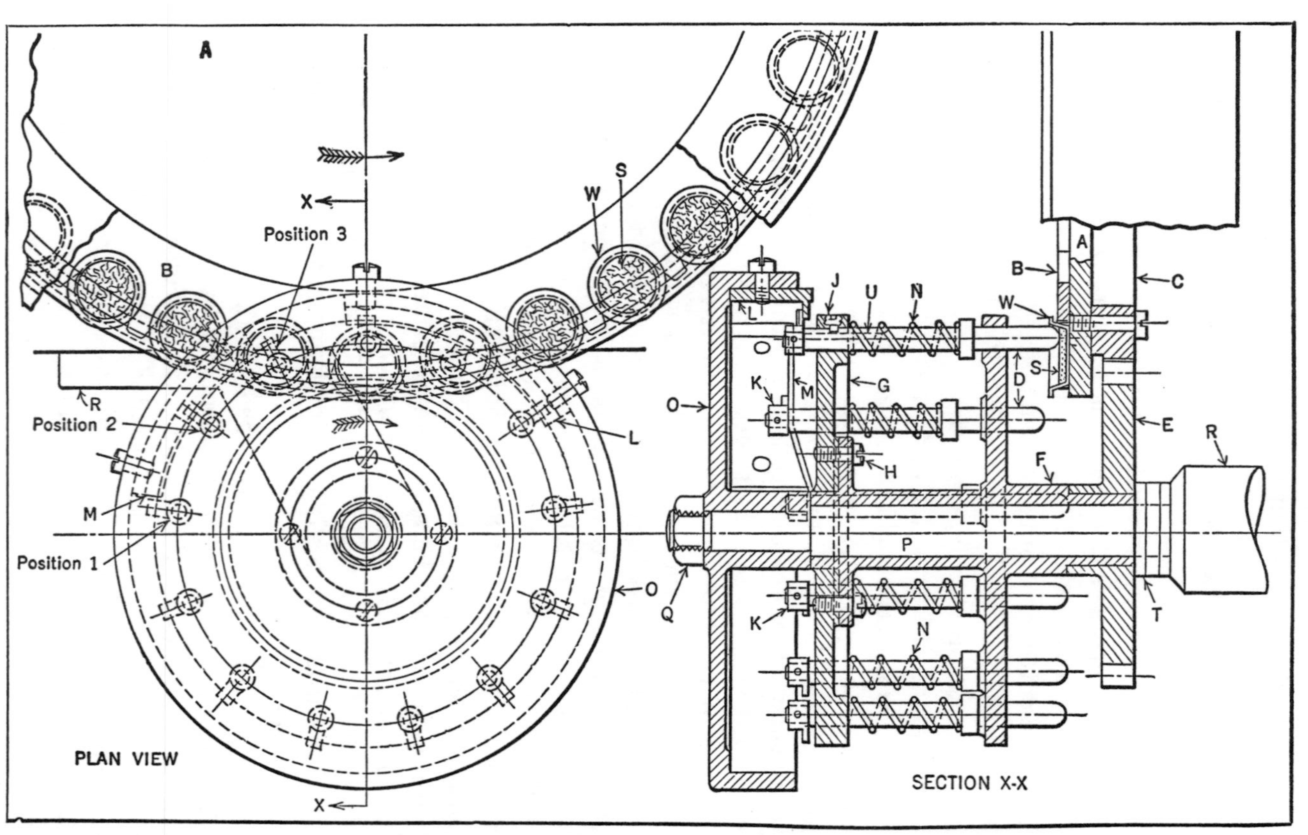

Fig. 15. Mechanism that Automatically Removes Caps W from Transport Wheel A when the Disk-feeder Fails to Supply the Caps with Cork Disks S

drums *I* are fastened on the short jack-shafts *H*. The lengthwise strands of wire are fastened to studs *J* on drums *I*.

When the wires are ready for tensioning, shaft *B* is given a slow rotary motion in the direction of the arrow. This motion is transmitted to the sprockets *A* and *C* and the drums *I* through the chains *D*. The idler sprockets *E* operate on the tight or load-carrying side of the chains *D*. Therefore, any increase in the tension of the wires will produce a corresponding increase in the tension of the chains *D*, and also in the pressure against the idler sprockets *E*.

As long as the tension of the two chains *D* remains equal, the lever *G* will be inactive. However, as soon as this tension becomes unequal, the sprocket *E* on the chain having the greater tension will be forced upward, causing the lever *G* to force the other sprocket *E* downward until the tension again becomes the same in both chains. Except at the very beginning of the tensioning operation, this attachment scarcely seems to operate. The slightest difference in the tension of the two wires is transmitted to the chains and idler sprockets, causing an almost imperceptible equalizing movement of the lever *G*.

Mechanism for Removing Incompletely Assembled Caps from Conveyor Wheel.— Cork disks *S*, Figs. 15 and 16, are assembled in caps *W* by automatic machinery having a conveyor or transport wheel *A*. Occasionally, the mechanism for feeding the cork disks into position for assembling becomes jammed, with the result that caps *W* pass to the ring *B* of the transport wheel *A* unprepared for the operations that are to follow.

The automatic mechanism shown in Fig. 15 for removing the caps that fail to receive cork disks was developed after some experimental work and attached to the transport wheel. This self-contained mechanism prevents the incompletely assembled caps from continuing along the line and thus causing unnecessary expense for useless work. The

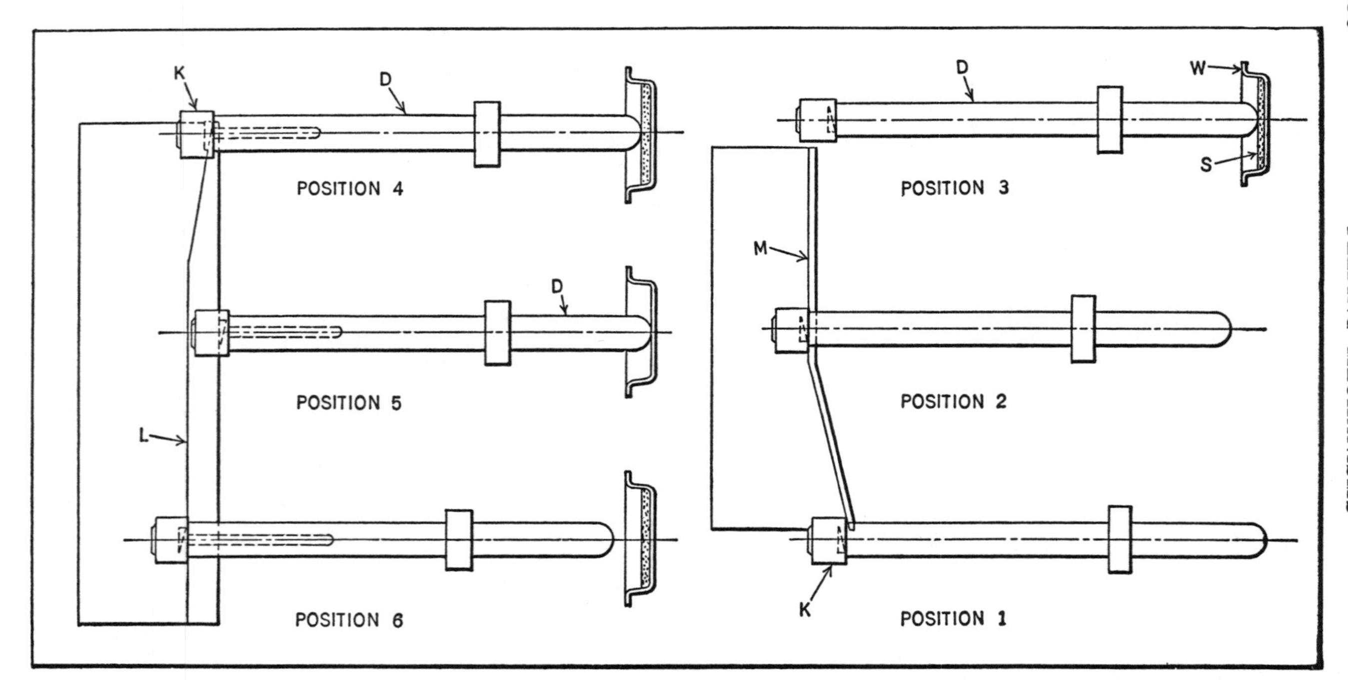

Fig. 16. Diagrams Illustrating Principle Employed in Mechanism for Removing Caps W from Machine when the Feeding Device Fails to Supply Them with Cork Disks S

assorting device operates satisfactorily and performs the desired sorting operation at a point that could not be reached readily by the operator's hand. The principle is simple and should be easily adaptable to other work of a similar nature.

Ring *B* is fastened to disk *A*, the caps *W* being carried in machined recesses provided in ring *B*. The ring-gear *C* is attached to the bottom of disk *A*, and through its connection with the gear *E*, serves to actuate the automatic assorter. Gear *E* is fastened to the flange member *F*, which is part of a spool consisting of two flange members *F* and *G*, held together by screws *H*. The spool rests on the thrust bearing *T* and revolves on the spindle *P*, which is part of the stationary support *R*.

Holes are drilled and reamed in the flanges *F* and *G* to permit a sliding fit for the plungers *D*. In order to prevent the plungers from twisting in their bearings, a slot *U* is provided in each one in which a set-screw *J* acts as a key. A light spring *N* keeps the plungers down. Each plunger is equipped with a lifter *K*, pinned in place at its upper end. The combined cover and cam-holder O is held in position on spindle *P* by the nut *Q*. Two cams *L* and *M* are fastened to the cam-holder *O*.

Operation of the Automatic Cap Assorter.— In operation, the movements of plungers *D* are controlled by cams *L* and *M*. The position of the first cam *M* is shown in the view to the left. In Position 1, Fig. 16, the lifter *K* is shown making its initial contact with the cam surface. Position 2 shows the plunger resting so as to clear the cap entirely. In Position 3 the plunger is shown released and dropped into the cap under the action of spring *N*, Fig. 15. Positions 1, 2, and 3, Fig. 16, correspond to the points marked 1, 2, and 3 in Fig. 15.

The function performed by cam *L* is illustrated in the views to the left, Fig. 16. In Position 4, plunger *D* is shown resting on top of a cork disk inserted in the cap. The lifter

K, in this case, is raised high enough to permit it to make contact with the cam surface. In Position 5, plunger *D* is shown resting on the bottom of a cap, with no cork disk in place. In this case, the lifter *K* is positioned too low to make contact with the cam surface, so that it passes under cam *L*. Since the plunger is not raised out of the cap in the latter case, it causes the cap to be removed from its position in the transport wheel by the plunger *D*. In Position 4, however, the cap moves on, because the plunger is raised sufficiently to allow the cap to retain its position in the transport wheel.

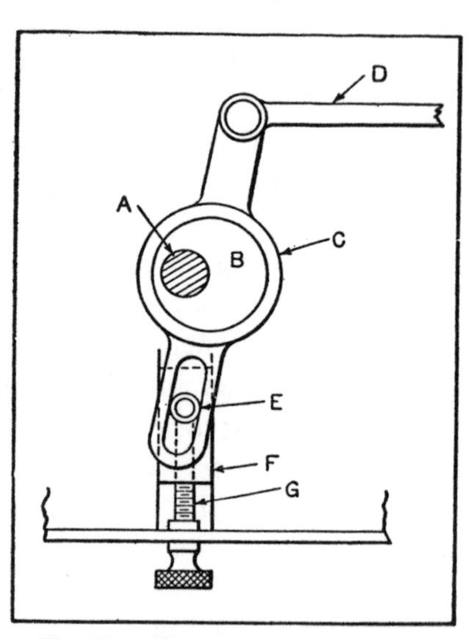

Fig. 17. Adjustable-stroke Mechanism

This device can be operated successfully at relatively high rates of production. In increasing the operating speed, it becomes necessary to place cams *L* and *M* farther apart, in order to give plunger *D* time to drop into the cap before the lifter *K* makes contact with cam *L*. The space between the plungers on the automatic assorter should be equal to the distance between the caps which are located on the transport wheel.

Adjusting Stroke without Stopping Machine.—The feeder slide on a wire machine is operated by an eccentric mechanism (Fig. 17), which has an arrangement for varying the length of stroke while in operation. The shaft *A* rotates the eccentric *B*, which carries strap *C*. The upper end of the strap carries an arm, the motion of which is transmitted to the slide through the rod *D*. The lower end of strap *C* has a slotted arm, which is positioned by the roller *E*. This roller is carried on block *F* which slides in a groove in the bed of the machine and is adjustable by means of the screw *G*.

The illustration shows the eccentric at its extreme right-hand position. The arms on strap *C* gradually assume an angular position after the eccentric has passed the top center. The effect is that of a lever with *E* acting as the fulcrum, so that rod *D* has a movement greater than would be obtained by direct connection with the eccentric strap *C*. If the roller *E* is moved toward or away from the shaft *A*, the effective length of the slotted arm is increased or decreased; and as the length of the upper arm remains constant, its movement is accordingly increased or decreased.

Adjustable Stroke-Feeding Mechanism for Sewing Machines.—The feeding mechanisms of sewing machines used for commercial production work must be designed to handle a great variety of fabrics. This requires a wide range of adjustment in the length of stitch. The parts of the mechanism must be so proportioned that they will be durable and require a minimum of power for their operation. While the method of adjusting the length of stitch should be simple and positive, it need not be of a character suitable for adjustment by the operators.

Referring to the feed-dog shown at *B*, Fig. 18, it is necessary that the path of travel of this part while above the throat plate *N* in the working part of its cycle of motion be approximately a straight line. It is also desirable that the working path of the feed-dog be capable of being tilted in either direction from a line parallel with the top of the throat plate. The mechanism shown accomplishes these several objects in the manner to be described.

The feed-dog *B* is attached at or near the front end of the feed-bar *A*. The rear end of this feed-bar is supported by rocker arm *C* by means of shaft *D*, about which it is free to pivot. In a similar manner, the front end of the feed-bar is carried by rocker arm *E*, which is free to pivot about the pin connection *F*. The rear rocker arm *C* is pivoted at its

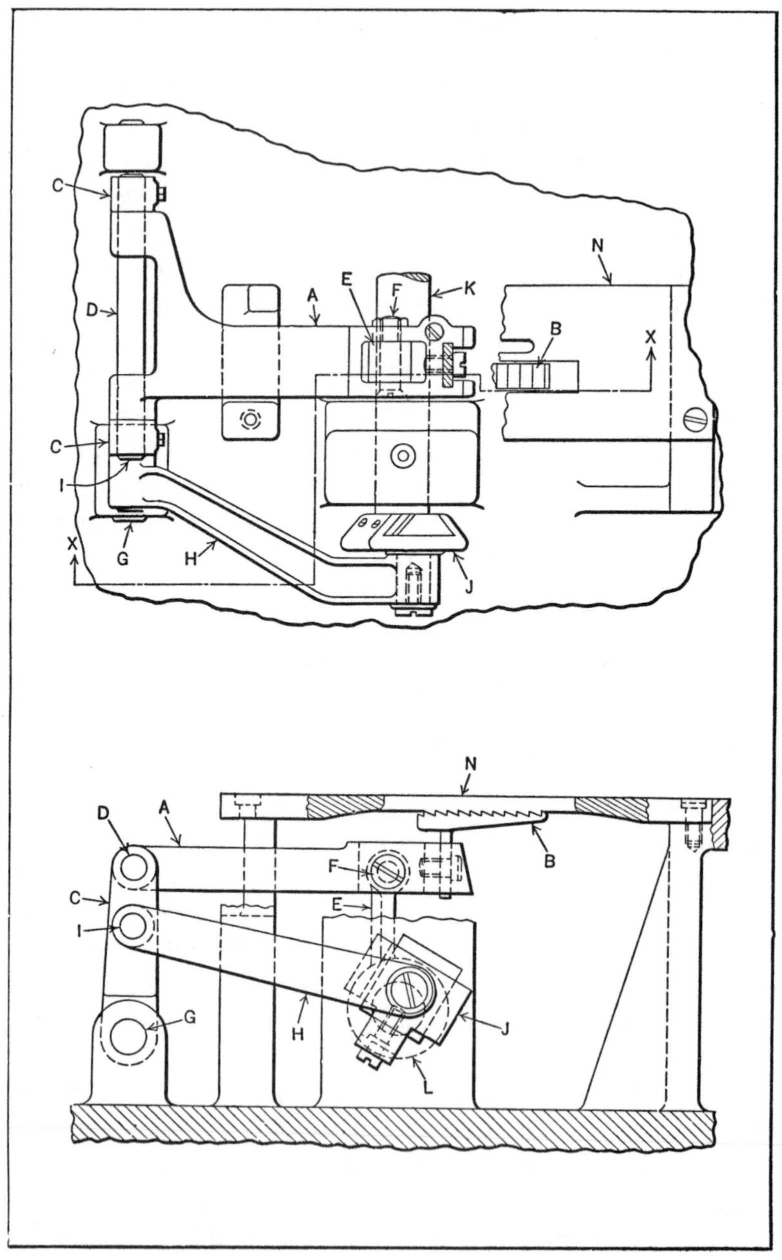

Fig. 18. Adjustable-stroke Feeding Mechanism for Sewing Machine

lower end to the frame of the machine at G and is driven by link H. Link H is pivoted to arm C at bearing I. The other end of link H is pivoted to the adjustable feed-crank J, carried on the end of main shaft K. The lower end of rocker arm E is in the form of an eccentric strap L, which engages an eccentric, driven by the main shaft K. This eccentric is termed the "feed-lift eccentric." Both the feed and lift motions are positive, and their combined action on the feed-dog results in a path of motion relative to the

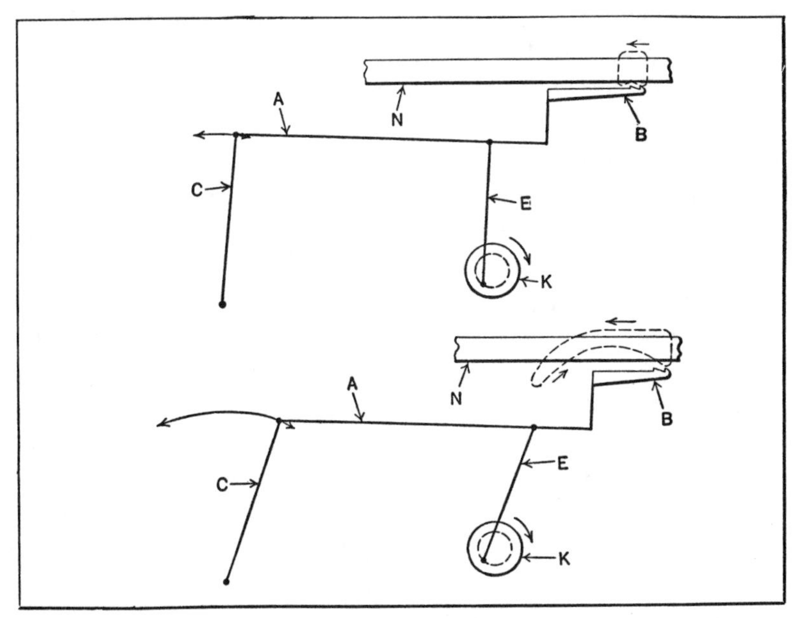

Fig. 19. Diagrams Showing Adjustments of Sewing Machine Feeding Mechanism for Both Short and Long Feeding Strokes

throat plate N such as is illustrated by the dotted lines of the two diagrams Fig. 19.

This arrangement causes the rear end of feed-bar A to rise and fall twice with each revolution of the shaft K. The amount of this rise and fall depends upon the length of the rocker arm and its angular displacement each side of the vertical position. In a like manner, the front end of the feed-bar will rise and fall, due to the relation between it and the front rocker arm. The front end of the feed-bar is also caused to rise and fall by the rotation of the lift-eccentric on the main shaft. It is evident, therefore, that the rise and fall of the front end of the feed-bar will be the result of these two actions. The rise and fall of the feed-dog will be similar to that of the front end of the feed-bar, but not exactly the same, depending upon its size and location relative to the front end of the feed-bar.

The upper diagram Fig. 19 shows the adjustment for a relatively short stitch, and the lower diagram, the adjustment for a relatively long stitch. These views indicate the relationship between the rocker arms, feed-bar, feed-dog, throat plate, and lift-eccentric. The dotted lines in these illustrations show roughly the path of the toe of the feed-dog. The path of the heel would be similar to that of the toe, but not exactly the same. By a suitable proportioning of the parts and adjustment of the angular relationship between the feed-crank and the lift-eccentric, the feed-dog may be caused to emerge through the throat plate parallel to the latter member and to travel a very nearly straight line parallel with the top of the throat plate.

By altering the angular relationship between the feed-crank and the lift-eccentric, the feed-dog may be caused to emerge from the throat plate toe first; that is, the feed-dog may be tilted backward at a slight angle. By changing this angular relationship in the opposite direction, the heel of the feed-dog may be caused to rise first. These various relations of feed-dog to throat plate are desirable because of the feeding requirements of different kinds of fabrics and the kind of seam required. In the design described, excessive wear and violent velocity changes have been avoided. This enables the mechanism to be operated at high speeds with relatively small wear and with a comparatively small consumption of power.

Mechanism for Operating Magazine Feed-Slide.—Electrical knife switches are automatically assembled on their slate bases in a machine equipped with a magazine that feeds the bases to a dial by means of a pusher-slide. An interesting feature of this magazine is that, although the stroke of the feed-slide is only 4 3/4 inches, this slide serves to transfer the base over a distance of 7 1/2 inches. Referring to the illustration (Fig. 20), the magazine is shown at *A* fastened to the machine frame *B*. The feed-slide is indicated at *D*. It rests on the top of the machine and operates between two guides, one of which is shown at *F*. The slide is reciprocated by the oscillating lever *G* through the

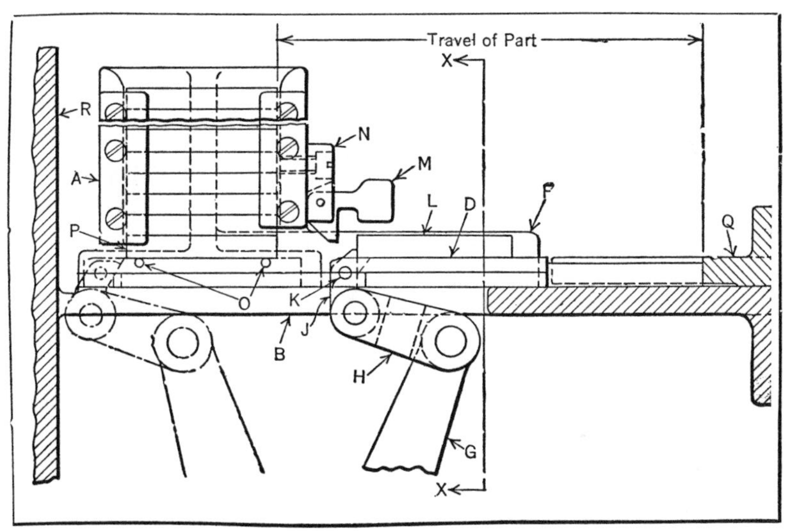

Fig. 20. Mechanism that Feeds Knife Switch Bases from Magazine A to Slots in the Dial Q

link *H* and the stop *J*, stop *J* being attached to the slide by the pivot pin *K*.

The lever *G* is shown in its farthest position to the right, where it has carried the base *L* out of the magazine. As the lever and slide move toward their left-hand position, base *L* is carried against the counterweighted pawl *M*, pivoted to the block *N* which is secured to the magazine. Continued movement of the lever toward the left causes the pawl to push the base off the slide, so that when the latter has reached its extreme left-hand position, the base drops on top of the machine frame *B* between the guides.

Since the bottom base in the stack rests on the pins *O*, the stop *J* must clear the bottom base as the slide moves to

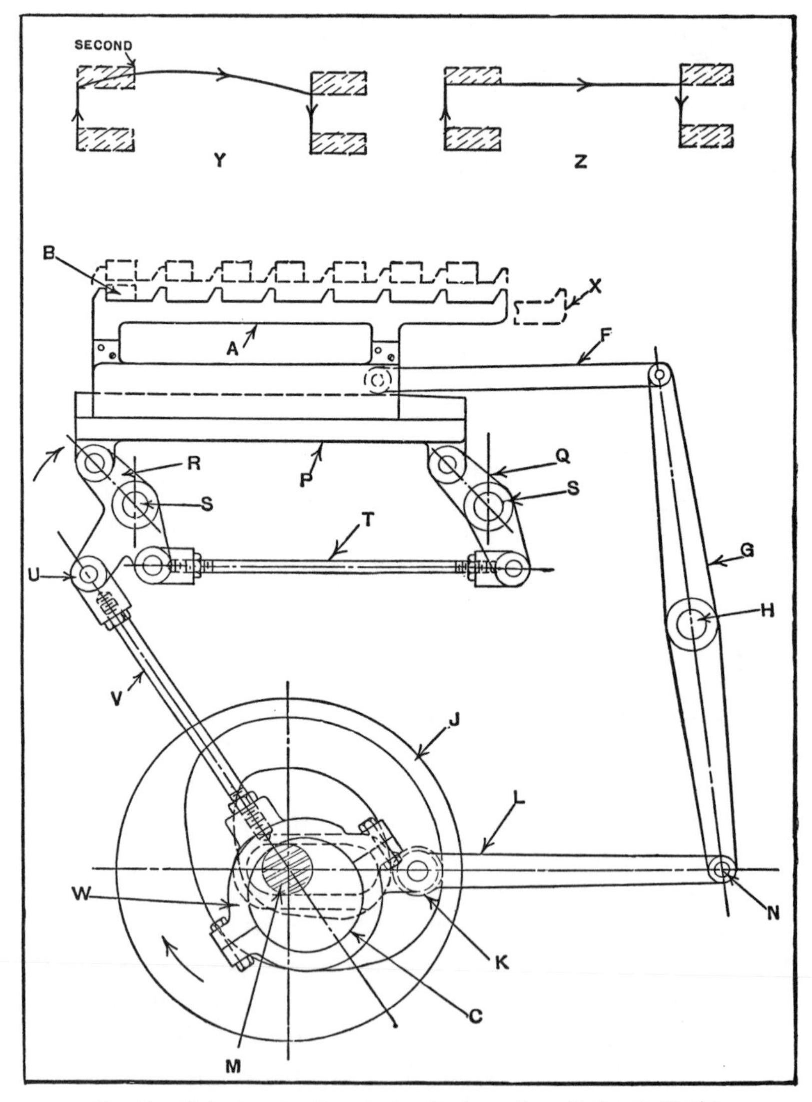

Fig. 21. Mechanism for Transferring Packages from Station to Station

its left-hand position. Obviously, as the lever swings toward the left, the stop will swing in a clockwise direction, so that its protruding point recedes below the top of the slide. The stop remains in this position until the lever and slide are at the left-hand end of their stroke, as indicated by the dot-and-dash outline. When the lever moves toward the right, the dog automatically swings back to its former position and pushes base *P* out of the magazine. In the meantime, base *L*, resting on the machine, is pushed by the front end of the slide into a slot in the dial *Q*. An added advantage of this mechanism is that it can be used effectively where space at the left of the magazine is limited, as it is in this case by the machine wall *R*. Ordinarily, a long slide would have been used which would have required considerable clearance at this point.

Transferring Parts from Station to Station.—In a machine for wrapping packages, the conveying mechanism shown in Fig. 21 is employed for transferring the packages to each successive station. In doing this, the transfer arms *A* must pick up the packages, carry them toward the right to the next stations, lower them into position, and then, after dropping enough to clear the bottom of the packages, return to their starting position. A package partly wrapped is deposited automatically on the carrier at *B* when the mechanism is moving up from the position shown.

The diagram *Y* indicates the path through which the packages theoretically are moved during one cycle, although they rest on bars in the upper position while the carrier drops below them when returning, the diagram representing the transfer arm travel. An eccentric in combination with a cam is used to obtain this movement, although two cams could be used that would cause the package to follow the path indicated at *Z;* or two eccentrics might be used if the motion imparted would be suitable.

The two carriers *A* support both ends of the packages, while the slide beneath supports the carriers and is connected by link *F* to the lever *G* pivoted on stud *H*. The slide is operated through this lever arrangement by the cam *J* which engages the roll *K* attached to the yoke *L*. Shaft *M*, which drives the entire mechanism, passes through the yoke, while a pin at *N* pivots the yoke to the lever. Supporting the slide that carries the transfer arms *A* is a bracket *P* to which the slide is dovetailed. This bracket is mounted on two levers shown at *Q* and *R*, which are free to pivot on the studs *S*. Connecting link *T* ties these levers together, and increases the strength of the assembly. Forming part of the lever *R* is an arm *U* to which is pivoted a connecting-rod *V* fastened to the eccentric strap *W*. Both the eccentric *C* and the cam *J* are pinned to the drive shaft *M*.

In operation, as the shaft *M* revolves, the movement of the eccentric causes levers *R* and *Q* to oscillate in the direction of the arrow, thus raising bracket *P*. In the meantime, a dwell on the cam prevents the slide from moving to the right or left. When the carrier reaches the position shown by the dotted outline, the cam operates lever *G*, so that the slide is moved to the right; the dwell on the cam then holds the lever stationary until the eccentric swings the lever *R* back so that the carrier will be in the position indicated at *X*. The cam then operates lever *G* to bring the slide back to its starting position. As the eccentric travels continuously through an arc, at no time will the slide be held stationary, the path of the carriers being curved; however, for practical purposes, this departure from a straight line movement may be disregarded. If it is desired to control this mechanism so that there is no up or down travel while the slide is traversing, a cam may be substituted for the eccentric. The dwells and rises on the cams may then be varied so that the path of travel will be as shown at *Z*.

Elevating Pile of Sheets to Keep Top Sheet in Alignment with Feed Rolls.—The top sheet in a pile is automatically kept in approximate alignment with a pair of feeding rolls by the mechanism shown in Fig. 22. It is used for feeding sheets to a paper-tube rolling machine and can be readily adapted to the feeding of metal sheets as well.

The rectangular sheets *A* are stacked on the vertical slide *B* mounted on the machine *C*. This slide is given a vertical

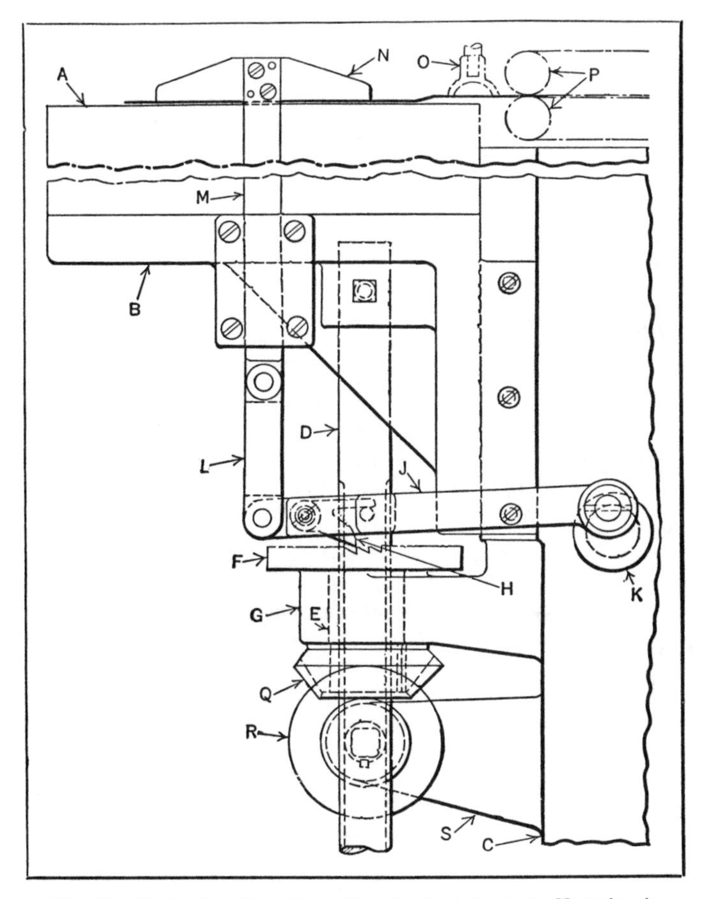

Fig. 22. Mechanism that Keeps Top Sheet of Stack in Magazine in Line with Feed-rolls

feeding movement by means of the screw *D* secured to the slide. The screw engages a nut *E* which is an integral part of the ratchet wheel *F*. Bearing *G*, cast on the machine frame, serves as a support for the nut and ratchet wheel.

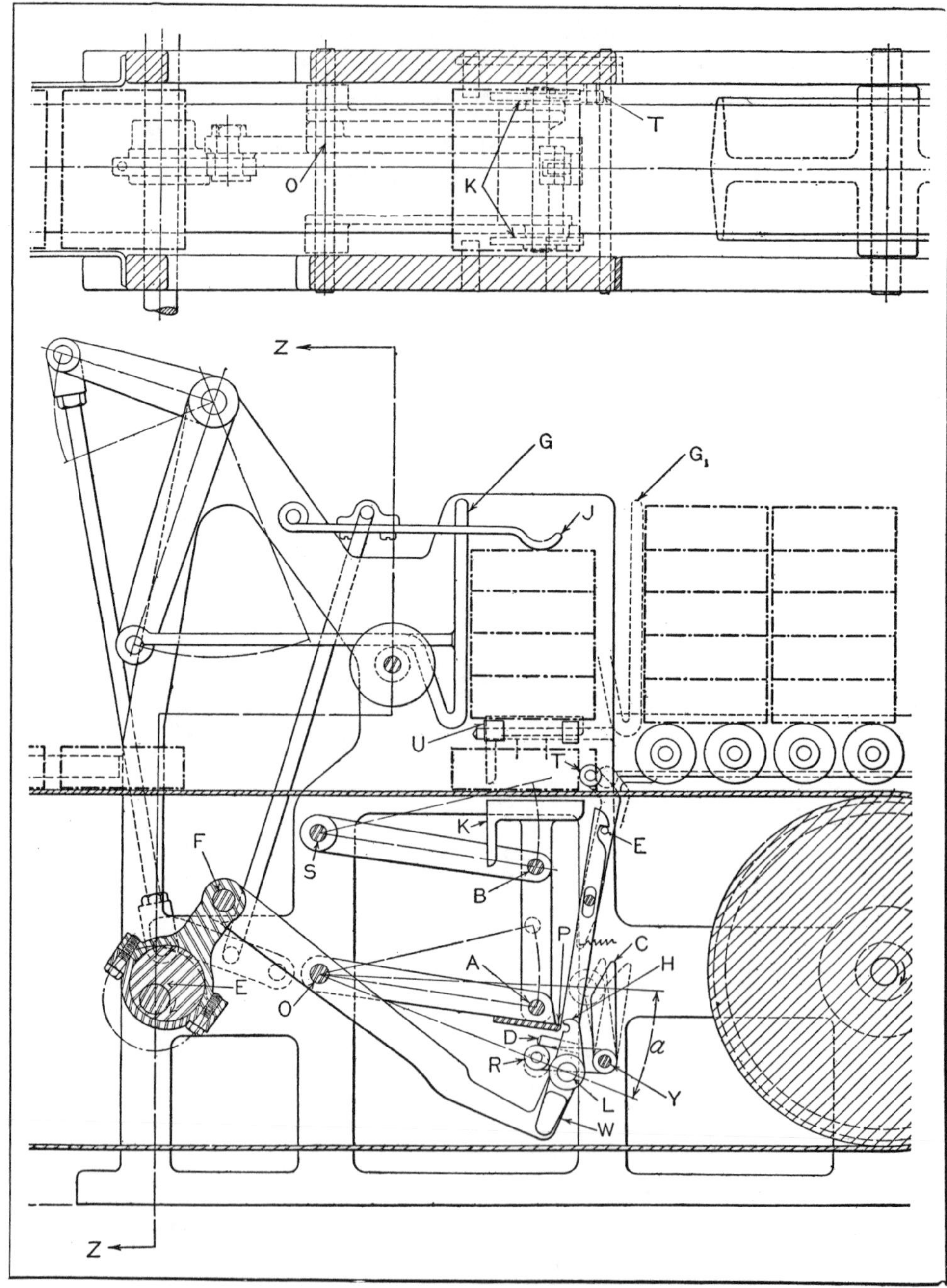

Fig. 23. Stacking Mechanism at End of Conveyor Belt which Raises Pieces from Belt, Arranges them in Stacks, and Transfers the Stacks to a Roller Carrier

The latter is rotated to feed the screw and slide *B* upward by means of the pawl *H* pivoted to the oscillating bar *J*. This bar receives its movement through the constantly rotating crank *K*, and at its left-hand end is connected to link *L*. Link *L*, in turn, is connected to the bar *M* which slides in a guide on the slide *B*. At the top of bar *M* is a cross-piece *N*, which rests on the top of the sheets.

As indicated, slide *B* is loaded with sheets and the suction cups *O* have raised the end of the top sheet preparatory to carrying it forward and between the moving belts on the rolls *P*. These belts then transfer the sheet to the rolling mechanism. This top sheet is under the cross-piece *N*. The pawl *H* remains out of engagement with the ratchet wheel and there is no upward feeding movement of slide *B* until *N*, together with bar *M*, link *L*, and arm *J*, drops down far enough to cause pawl *H* to engage the teeth in the ratchet wheel, rotating the latter and feeding screw *D*.

Now, as cross-piece *N* travels upward with the top sheet, arm *J* will once more lift pawl *H* out of engagement with the ratchet wheel and thus stop the feeding action of the screw and slide *B* when the top of the pile of sheets has been elevated to the required level.

To permit reloading of the magazine, bevel gears *Q* and *R* are provided. Gear *Q* is keyed to the nut, while gear *R* is keyed to a shaft which turns freely in bracket *S* cast on the machine. The shaft for gear *R* is square at its outer end to accommodate a hand-crank used for moving the slide manually to its loading position. At this time, pawl *H* is swung up out of engagement with the ratchet wheel.

Mechanism for Stacking Articles at the Delivery End of a Conveying Belt.—The basic mechanical motion used in the mechanism illustrated in Fig. 23 has various applications and should be of interest to machine designers. In this case, it is applied to the problem of stacking articles which are being carried along a conveyor belt. Its advantage for this work is that it will handle articles in varying quantities or singly, as the case may be. In this instance, the articles, shown by dot-and-dash lines, are stacked five high and are discharged on a roller conveyor as shown.

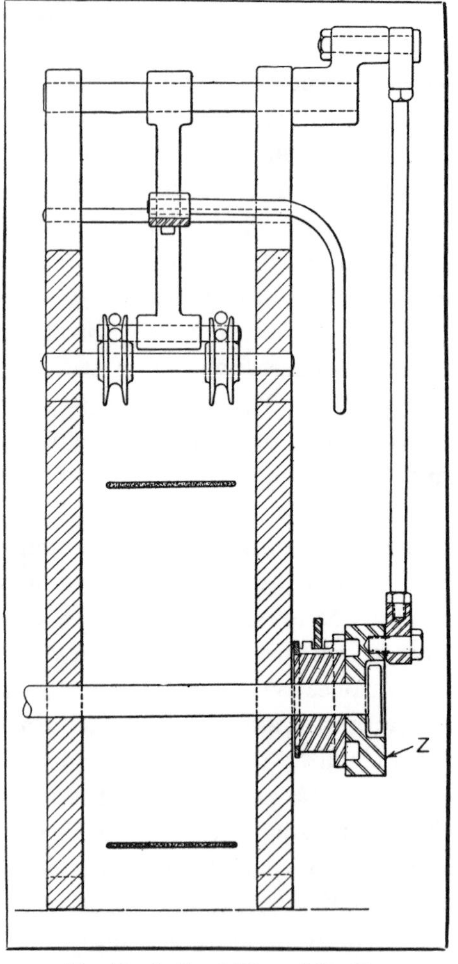

Fig. 24. Sectional View of Stacking Mechanism Shown in Fig. 23

The operation of the mechanism is based on the "firing" of a trigger T which is moved to the position shown by the dotted lines by a single article, which will then allow the conveyor belt to slip or pass beneath until the elevator K rises to the upper position, indicated by the dotted lines. This places the article at the bottom of the preceding articles which make up the stack. The article is prevented from dropping back by four latches U which are hinged in the side walls. Each article, of course, supports the one above it until the pile is complete, when the entire stack is moved to the right.

The action of the lower part of the mechanism is as follows: An oscillating movement is imparted to the rocker arm OL by the driver shaft which rotates continuously. This shaft carries an eccentric E and, in turn, is connected to arm *OF*. This arrangement causes arm *OL* to oscillate through the angle a. The secondary arm *HLW*, which is carried on arm *OL*, will pick up plate *P* under certain conditions, to be described, and swing arms *OA* and *SB* to their upper positions.

It is, of course, understood that these arms carry the elevators *K* and that the "pick up" action is accomplished as a result of the "firing" of the trigger *T* which allows the catch arm to drop off the small pin *E* and the lower end *R* to assume the position shown by the dotted lines. The cam arm *YC* and arm *YR* are tied together and turn or swing as one piece on the pin *Y* fixed in the frame. Part *HLW* also carries a small pin *D* which strikes roller *R*, and when in the lower position, locks into plate *P* on the upward part of the oscillating movement of arm *OL*. At the same time, cam *C* is moved to the position indicated by the dotted lines to the left. This action relatches the trigger *T* on pin *E* if no article is in position to be raised to the stack.

The discharge action may be arranged to take place on the return or downward part of the stroke. When the fifth article has raised arm *J*, the latter arm, which is connected to a bolt clutch finger of standard design, causes the part indicated at *Z* (Fig. 24) to rotate one complete revolution. Through suitable connections, which are clearly indicated, this action moves pusher *G* to the position indicated by the dotted lines at G_1 and returns it to its normal position. This movement transfers the stack of five pieces to the roller carrier and completes the cycle, after which the operations are repeated automatically.

Variable Rotary Movement for Operating Shell Hopper.— Brass shells are fed to a thread-rolling machine by means of a rotary hopper attached to the machine. After extensive experiments, it was found that a variable rotary movement of the hopper drum increased its efficiency; that is, more shells per minute could be fed by the drum when the pulsating movement was used. The mechanism shown

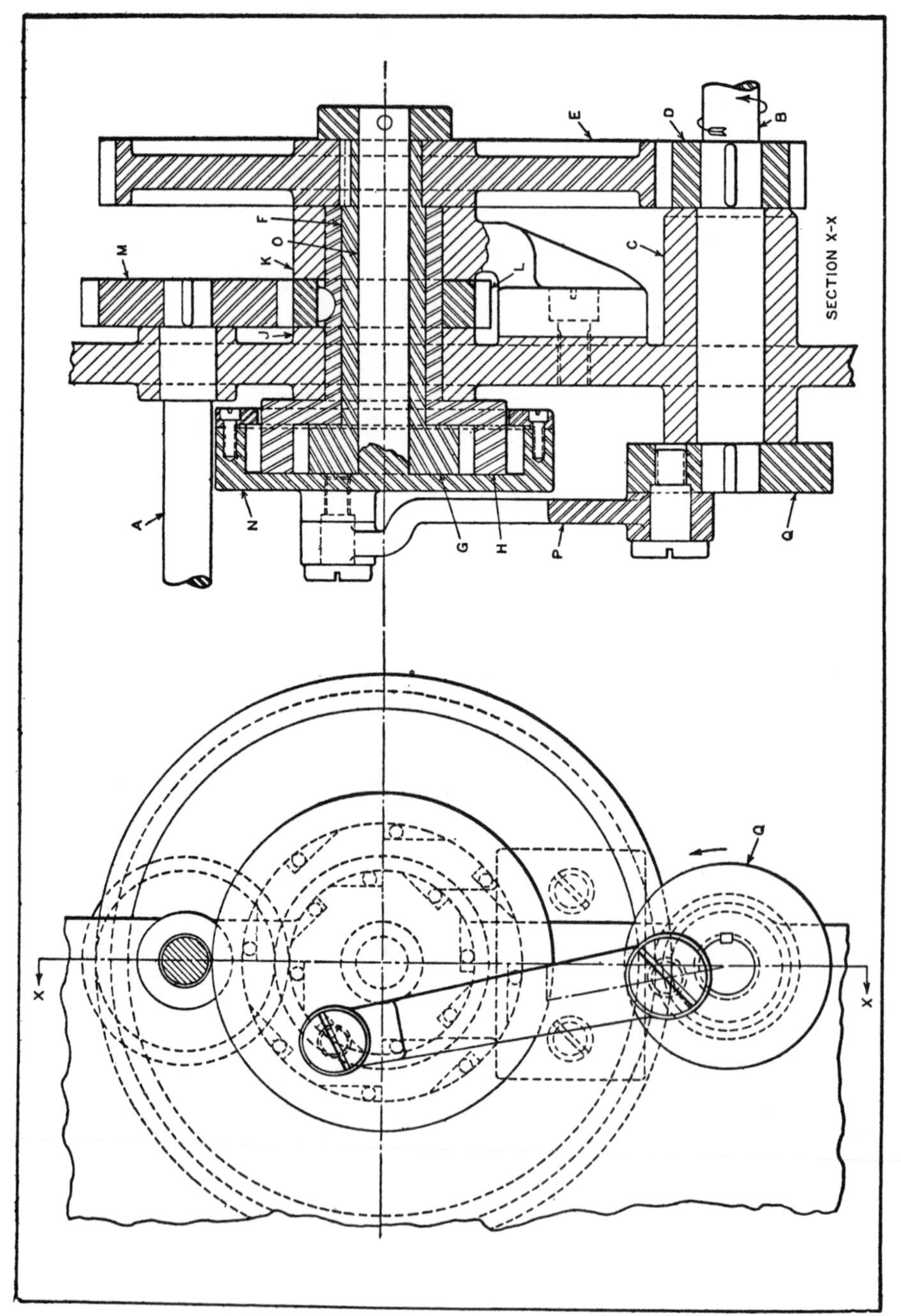

Fig. 25. Mechanism by which the Shaft B, Rotating at Constant Speed in One Direction, Imparts a Variable Rotary Movement to Shaft A

in Fig. 25 was designed to give the required pulsating movement.

With this arrangement, the drum is rotated a partial revolution at a slow velocity through a train of gears and a double roller clutch. The remaining part of the revolution is imparted by a crank which causes the roller clutch to over-run so that the drum rotates at a relatively rapid velocity.

The drive-shaft *B* is supported in the bearing *C*, which is an integral part of the machine. On this shaft is keyed the pinion *D*, which meshes with the gear *E*, keyed to sleeve *F*. The left-hand end *G* of this sleeve forms the core of a roller clutch; the outer ring for this core is indicated at *H*. The sleeve on ring *H* is supported in the bearings *J* and *K*, and its bore provides a bearing for the core sleeve *F*. Keyed to the ring sleeve is the gear *L*, which meshes with gear *M*, keyed to the drum shaft *A*.

It will be noted that ring *H* forms the core for the second or outer roller clutch, the ring for this clutch being indicated at *N*. The long shaft *O*, integral with this ring, is a free fit in the bore of core sleeve *F*, and a collar at its right-hand end serves to lock in position all the members supported in bearings *J* and *K*. On an offset boss on the side of ring *N* is pivoted the connecting-rod *P*, the lower end of which is connected to the crank disk *Q*, keyed to shaft *B*.

As shaft *B* rotates one-half revolution in the direction of the arrow, core *G* turns in a clockwise direction (see end view), rotating ring *H* and gear *L* with it. As a result, gear *M* and drum shaft *A* turn at a constant velocity in a counter-clockwise direction. In the meantime, crank *Q*, through rod *P*, rotates ring *N* in a counter-clockwise direction; but, as the clutch rolls between members *H* and *N* are free at this time, this movement does not affect the movement of shaft *A*. However, as soon as shaft *B* completes one-half revolution, the crank reverses the rotation of ring *N*. Now as this ring rotates at a much higher velocity than core *G*, the rolls between members *G* and *H* will be released, so that ring *H* will over-run and rotate gears *L* and *M* and drum shaft *A* at a high velocity. This high velocity of shaft *A* continues until shaft *B* has completed its second half revolution, after which the movement of ring *N* is again reversed, thus permitting the rolls to wedge between members *G* and *H*. This will cause member *H* to rotate the drum shaft at the slow velocity. There is practically no over-run of the drum when its velocity changes from high to low, owing to the frictional contact of the drum with the shells in the hopper. These slow and fast movements of shaft *A* are repeated alternately, imparting the required pulsating movement to the hopper drum.

CHAPTER XV

FEEDING AND EJECTING MECHANISMS FOR POWER PRESSES

Power presses and dies are utilized for such a large variety of manufacturing operations that many different types of feeding, ejecting, and other mechanisms have been designed. In fact, a large volume could be filled with mechanisms of this class alone; hence this chapter is not intended as a treatise on such mechanisms but it does contain illustrated descriptions of a number of feeding and ejecting mechanisms which incorporate in their design certain mechanical principles likely to be of value to users of a book on the general subject of mechanism.

Inverting Shells After they Leave the Hopper.— Some hoppers used for feeding shells to power presses are designed so that the closed end of the shell will enter the feed-tube first. To permit this type of hopper to be used for work in which the shells are required to enter the press dial with the closed ends at the top, some means must be provided for inverting the shells after they leave the hopper and before they enter the dials.

This may be done by employing the device shown in Fig. 1. Here it will be seen that the shells leave the hopper tube and drop into recesses in the disk *A*. These recesses are equally spaced and the disk is indexed one space for every cycle of the press. The indexing occurs during the upward stroke of the ram. Motion is transmitted to the disk for this purpose by means of the link *B* and the lever *C*. At one end of lever *C* is mounted a pawl which engages the ratchet wheel *D*. The ratchet turns freely on the shaft *E* and transmits the required rotary motion to the disk *A* by means of friction washers (not shown).

In the position shown, a shell has just entered the top depression in the disk, with its closed end at the bottom, while at the lower part of the disk another shell has dropped into the press dial with its closed end at the top. One-half revolution of the disk *A* is required to invert each shell.

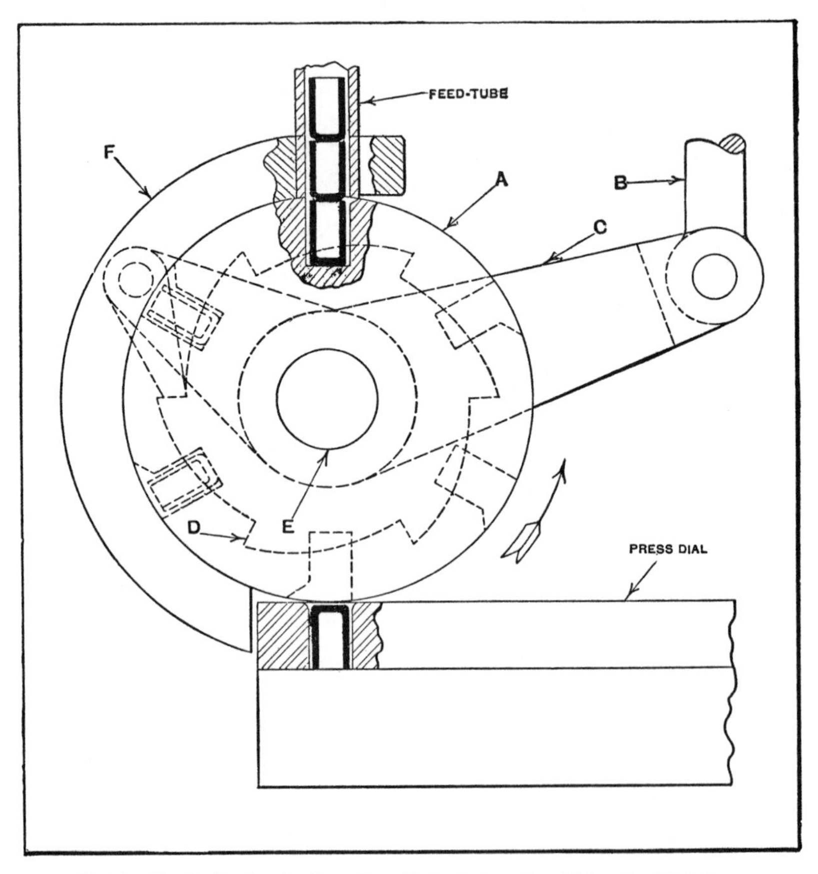

Fig. 1. Simple Device for Inverting Shells before they Enter the Dial Press

The stationary guard *F* provides for retaining the shells in the disk. It will also be noted that one corner of each of the impressions in the disks is beveled. This is done so that as this corner passes the hopper feed-tube it will not jam the shell in the end of the tube, but will force it upward into the tube. In case the stroke of the press is such as to cause the disk to be indexed more than one division, the link *B* can be equipped with a coil spring acting against the connecting member of the press ram, and a stop can be provided for lever *C*, so that the latter will oscillate only the required amount.

It may also be added here that the friction drive for the disk *A* provides a means for stopping the disk automatically in case of jamming when defective shells are fed through, in which case the guard *F* should be made removable, so that the shell can be extracted. After the shell is extracted, the disk must be rotated by hand until it assumes the correct position relative to the ratchet wheel. Corresponding lines scribed on both of these members may be employed for this purpose. It is evident that this arrangement may also be used for feeding shells into the dial with their closed ends at the bottom, provided, of course, that they leave the hopper tube with their closed ends at the top.

Hopper Attachment for Feeding all Shells with Their Closed Ends Up.— Regardless of whether shells are fed from a hopper with the closed end at the top or at the bottom, the attachment shown in Fig. 2 will deliver them to the press dial with the closed end at the top. This device greatly simplifies the design of the hopper, inasmuch as no attention need be paid to the position in which the shell leaves the hopper opening. By modifying the design of the attachment shown, the shells may also be delivered to the dial with the closed end at the bottom. Hence, by constructing two demountable attachments, shells may be made to enter the dial with the closed end at the top or at the bottom, only one hopper being employed in both cases.

The shells are fed from the hopper into the tube *A*. From the tube, they drop into openings in the annular ring *B*, which is given an intermittent rotary motion by means of the ratchet wheel *C* and ratchet lever *D*. This lever is given an oscillating movement through the link *E* which is carried

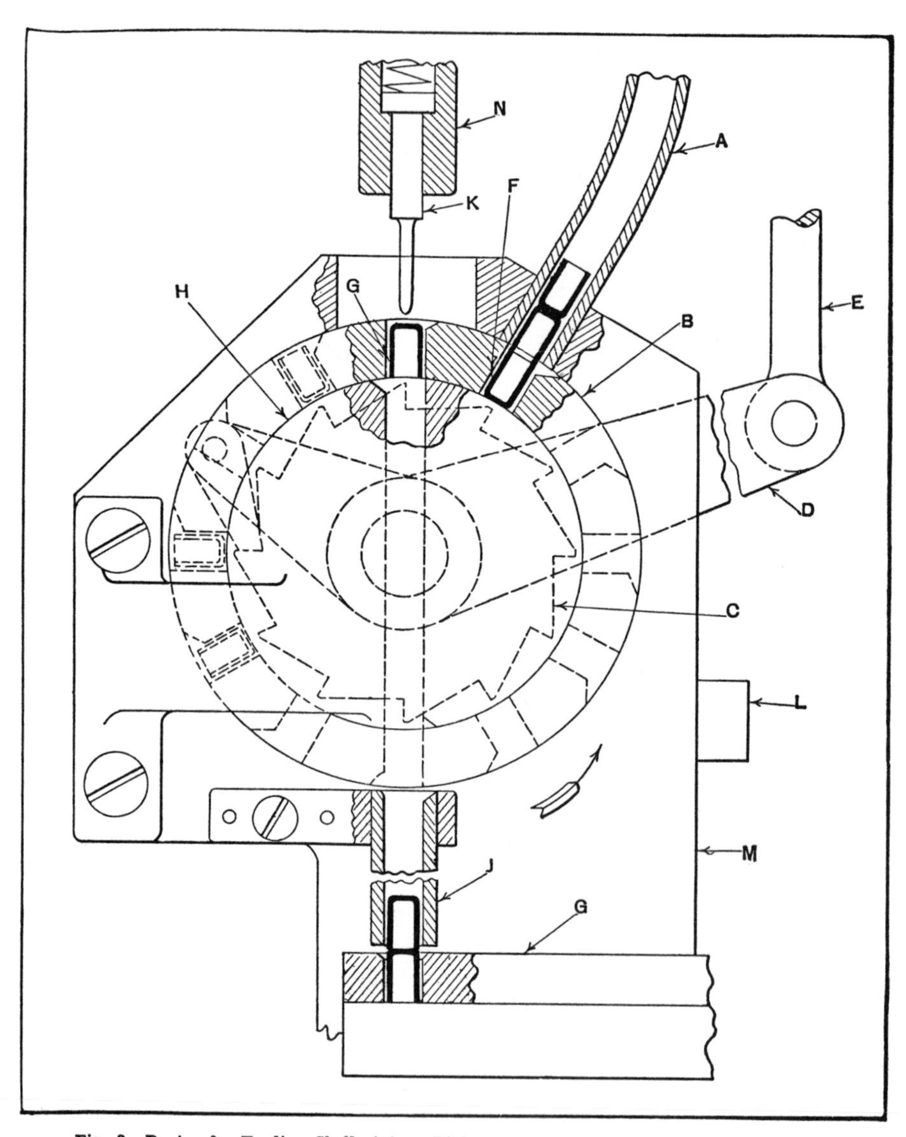

Fig. 2. Device for Feeding Shells into a Dial with their Closed Ends up Regardless of their Position as they Leave the Hopper

in the press ram; and for every cycle of the ram, the ring *B* is rotated one division. If all the shells were to drop into the ring in the position indicated at *F*, they would drop out of the lower end of this ring in the proper position to enter the press dial *G*. But when a hopper of the design mentioned is employed, the shells can take the position shown as they pass down into the tube *A*. Hence, provision must be made for delivering all the shells to the press dial with the closed end at the top. The shell *G* must be inverted before it reaches the dial. Instead of the shell being carried around with the ring, however, it will be forced down through a hole in the stationary core *H* by means of the plunger *K* secured to the press ram; and from there it will pass once more through the ring and into the tube *J*.

When a shell in the position indicated at *F* is indexed to the top of the ring, it will not pass down through the core *H*, as the plunger *K* will simply enter it without making contact. The shells are prevented from dropping by gravity at the top of the ring by a spring (not shown) which bears against the side of each shell as it passes this point. The tube *J* is made long enough to contain six shells, or half the number that can be held in the ring *B*. This length of tube was necessary, as, with a shorter tube, it is possible for the shells to pile up and be carried around once more past the tube *A*.

The ring is operated through a friction disk from the ratchet wheel *C*, and the position of the holes in the ring, relative to the hole in the core *H* and the tube *J*, is governed by the downward movement of the lever *D*, which is limited by the stop *L*, secured to the base *M* of the device. The base, in turn, is securely fastened to the press bed. Lever *D* strikes stop *L* before the downward stroke of the press has been completed, link *E* sliding, against the pressure of a coil spring, in a projection on the press ram. In case a damaged shell is indexed into position under plunger *K*, the plunger, instead of moving downward, will be held suspended, as it telescopes within the holder *N*, and the shell will be prevented from passing down into core *H* (see also simple device for feeding shells open end up, Volume 1, page 455).

Feeding Split Rivets to Power Press.—The automatic feeding mechanism, Fig. 3, is used for feeding split brass

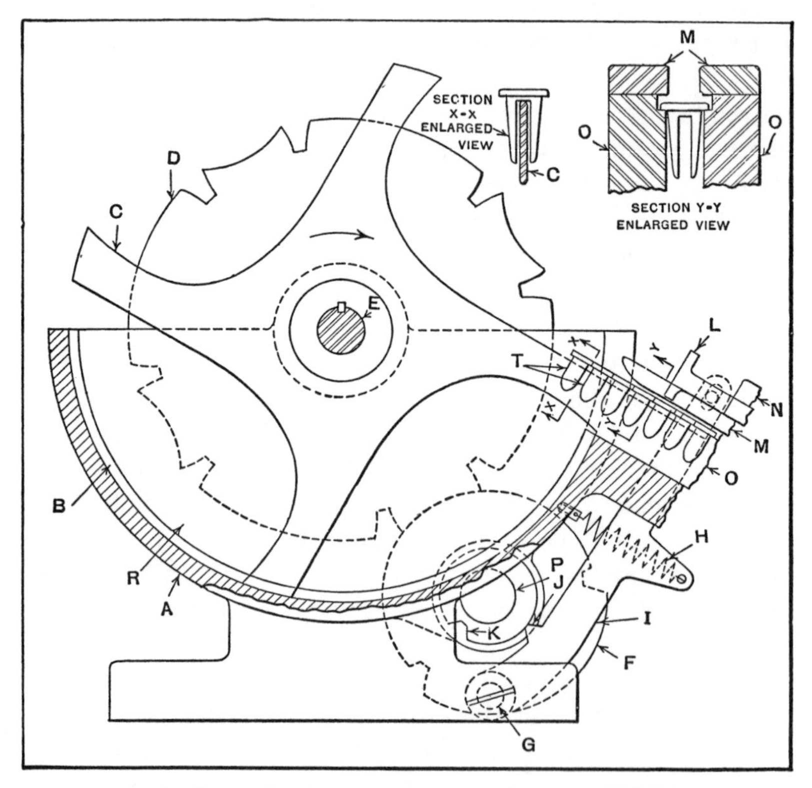

Fig. 3. Rotary Hopper for Feeding Split Rivets to a Dial Press

rivets to a dial on a power press, where they are assembled into porcelain fuse plugs. The hopper *A*, fastened to a bracket on the press in a position above the dial, is equipped with the four-bladed member *C*. For each revolution, member *C* is given an intermittent motion including four equally spaced dwells, through the action of the gears *D* and *F* keyed to shafts E and P, respectively. Shaft P is the driving member for the hopper, and receives its motion through a chain drive connected to the press crankshaft.

The rivets are placed in the reservoir R, and as member C revolves in the direction of the arrow, some of them will straddle the blades, as shown in the cross-sectional view x–x. As the member C continues to revolve, the rivets slide along the curved edge of the blade until they arrive at the position shown at T. Each blade dwells long enough at this point to permit the rivets to slide off and into the chute O. To facilitate the delivery of the rivets to the chute, the latter was constructed to support the rivets under their heads (see section y–y), the strips M acting as a guard to prevent them from piling up on each other on the incline.

The top of the chute is kept clear of incorrectly delivered rivets by the sliding finger L. This finger receives its motion from the lever I, pivoted at G, and is actuated by the revolving cam K which engages the lug J. The spring H carries the finger toward the hopper, while the outward movement is positive. The plate N serves as a gib for the sliding finger L.

Feeding Round Pins to a Dial Press.—A hopper for feeding round pins to a dial press is shown in Fig. 4. With this design, the open end K of the revolving conical coil of tubing E is continually passing through the mass of pins in the hopper reservoir L, some of which enter the tubing at K and slide, both by gravity and by the pushing action of the pins entering, down the incline to the center of the coil where they pass into a stationary tube F leading to the press.

The bracket C, fastened by screws to another bracket A on the press, has two bosses which serve as bearings for the turned shank M on the revolving member G. The latter, driven by the flanged pulley B, which is connected by a belt to a pulley on the press crankshaft, serves as a holder for the coil E. The lower end of this coil is straight and passes

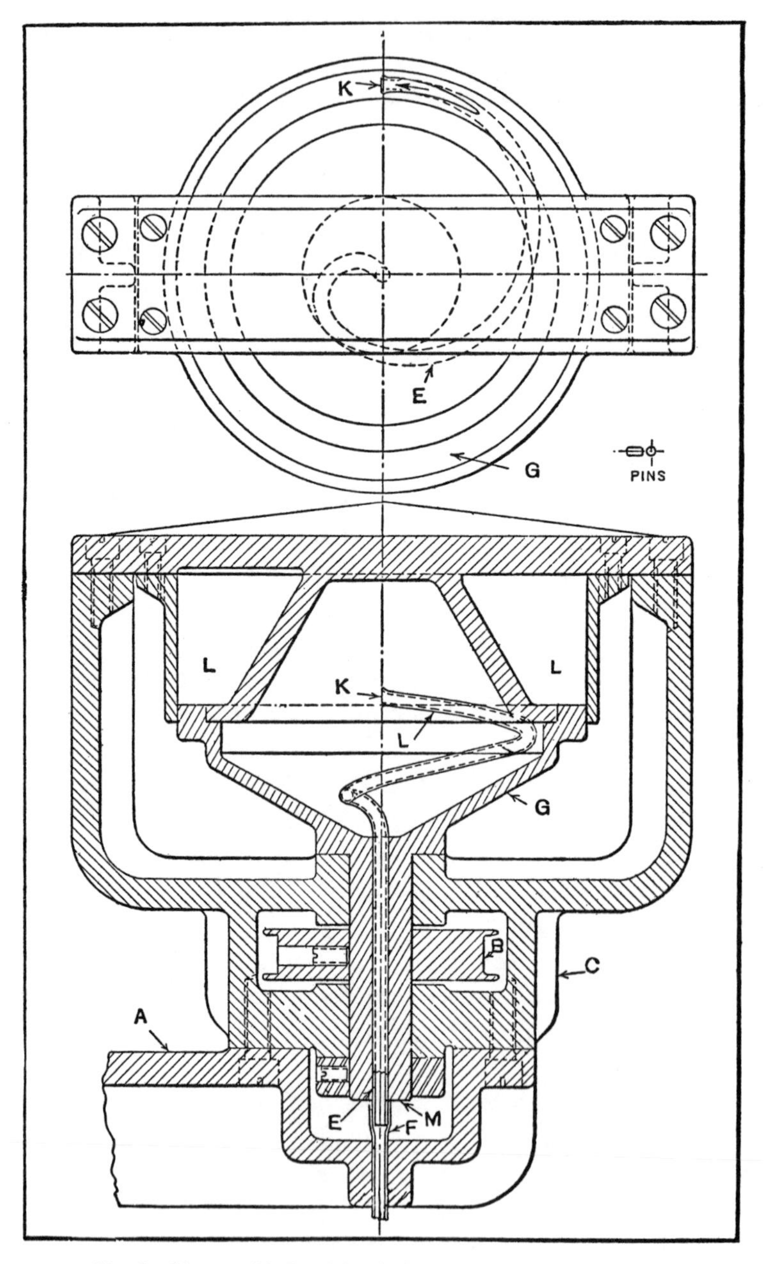

Fig. 4. Hopper with Revolving Coil of Tubing through which Pins are Fed to Dial

down through the shank *M* into the swaged end of the stationary tube *F*, while the upper end of the coil passes at an angle through the face of the outer flange on the member *G*, as illustrated at *N*. The hole for the tubing in member *G* at *N* is first cut through with an end-mill, and after the tube has been properly located, it is babbitted in place. The babbitt is next doweled to member *G* to complete the joint. With this type of hopper, the length of the pins to be fed governs the diameter of the coil. Such a hopper could not be used for very long pins, as the diameter of the coil would have to be so large as to be impractical.

Ejecting Shells that Enter Hopper Chute Wrong Side Up.— Among the many problems encountered in designing hoppers for feeding shells to dial presses is the delivering of each shell to the dial right side up. This is accomplished in one plant by means of the device shown in Fig. 5. The device is attached to the table of the press at the dial end of the chute, and is equipped with a vertical plunger for keeping in the chute those shells that have their open sides up. Another plunger ejects from the chute the shells that have their closed sides up.

A screw shell *A* is shown at the end of the chute *B*. This section of the chute is adjacent to the dial (not shown) and is secured in the bracket *C;* the bracket, in turn, is fastened to the side of the press table. Slide *D*, equipped with a spring-actuated ejecting plunger *E*, is mounted in bracket *C*, and is given a reciprocating movement by means of the bell-lever *F*. This lever is oscillated by a link on the arm *G*, which is bolted to the press ram.

A vertical sleeve *H*, also secured in this arm, supports the spring-actuated plunger *J*. This plunger serves to prevent those shells that have their open sides up from being ejected from the chute. For example, the shell *A* is shown with its open side up; now, as the press ram descends, plunger *J* enters and bottoms in the shell. As the ram continues to descend, the plunger remains stationary; in the meantime, the slide *D* has advanced toward the left, forcing plunger *E* against the side of the shell. Continued downward movement of the ram will merely result in both plungers *J* and *E* collapsing into their holding members. Thus, plunger *J* acts as a lock, preventing plunger *E* from forcing the shell out of the chute.

Just before slide *D* has completed its movement toward

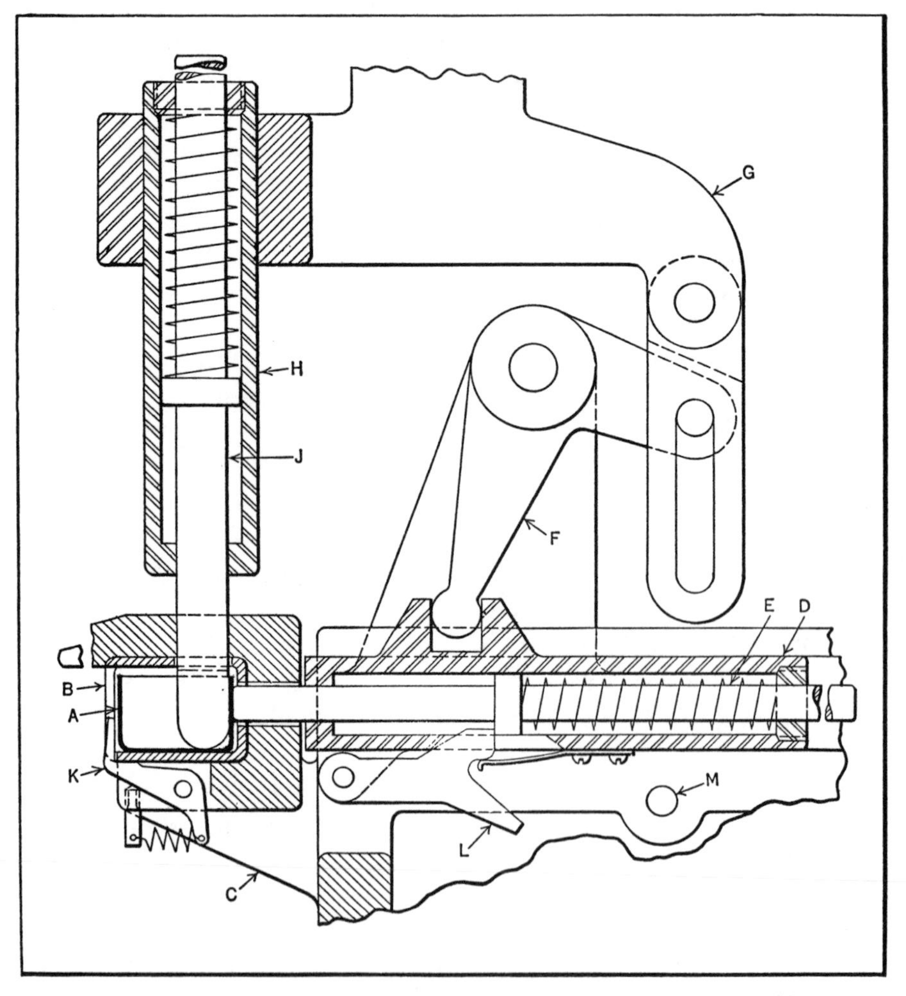

Fig. 5. Device for Ejecting Shells from Hopper Chute which Enter Chute Closed Side Up

the left, the latch *L* snaps into place behind the shoulder on plunger *E*, locking this plunger to the slide. Consequently, when the slide is returned toward the right, the pressure of plunger *E* on the shell will be released before the vertical plunger *J* leaves the shell. If this provision for locking plunger *E* were not made, the shell would be forced from the side of the chute. However, just before slide *D* has reached its extreme position at the right, latch *L* is disengaged from the plunger by coming in contact with the sta-

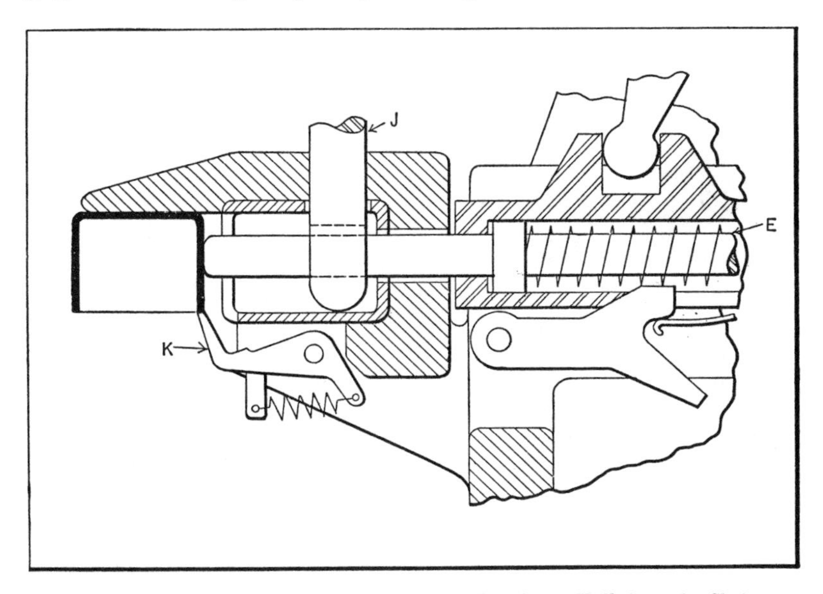

Fig. 6. Action of Device Shown in Fig. 5 in Ejecting a Shell from the Chute

tionary pin *M*. This allows plunger *E* to return to its ejecting position in the slide. At this time, however, both plungers are out of the chute; hence the shell is free to enter the dial.

When a shell enters the chute with its closed side up, it comes into position under plunger *J*, which, in this case, rests on top of the shell. Then, when plunger *E* moves toward the left, the shell will be pushed out of the chute, as shown in Fig. 6. During the ejecting process, the vertical

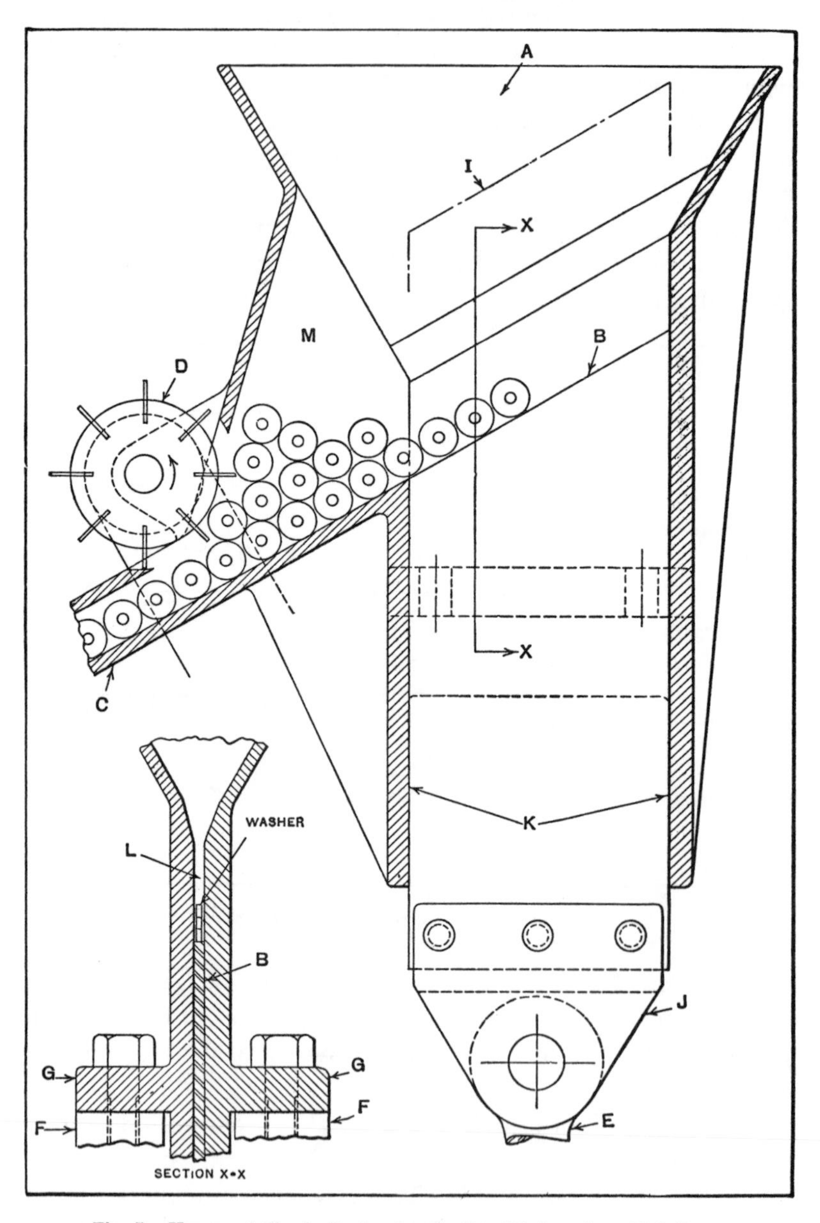

Fig. 7. Hopper of Simple Design for Feeding Washers to a Dial Press

plunger merely drops off the edge of the shell and to the bottom of the chute. This plunger is forked to allow it to straddle plunger *E*. It will also be noted that a guide rail *K* is provided to retain the shells in the chute while the ejecting device is inoperative. This rail is spring-actuated to allow it to return to its normal position. The ejected shells drop on a belt conveyor and are returned to the hopper.

Hopper for Feeding Washers to a Dial Press.—The hopper shown by the diagram Fig. 7 was designed for feeding special brass washers to a dial press, where they are assembled to electrical toggle switch levers. As slide *B* passes up and down through the mass of washers at *A*, some of the washers drop into spaces *L* or *M*, which are slightly wider than the washers, and roll down the incline at the top of the slide and thence into the chute *C*. The slide is shown in its lowest position, the highest position being indicated by the light dot-and-dash lines at *I*.

The reciprocating movement of slide *B* is obtained from a crank which is connected to the bracket *J* on the slide by the link *E*. The washers are prevented from jamming at the entrance to the chute by the wheel *D*. This wheel, revolving in the direction of the arrow, is driven by a sprocket and chain from the hopper crankshaft, and is equipped with eight flat spring projections which agitate the washers sufficiently to insure a uniform flow down the chute. The sides of the hopper are carried down at *K* to resist the side thrust of the slide set up by the hopper crank. Any dirt which may enter the hopper will pass out through the chute. The lugs *G*, cast integral with the hopper, provide a means for fastening the hopper to the bracket *F* on the press.

Automatic Ejector of Lift Type for Press Dial.—Fig. 8 shows a device for ejecting an assembled shell from the dial of an inclined press. The device is attached to the press ram by bracket *J*, which holds the post *B* with the pins *E* on which fingers *A* are pivoted. The operation is as follows:

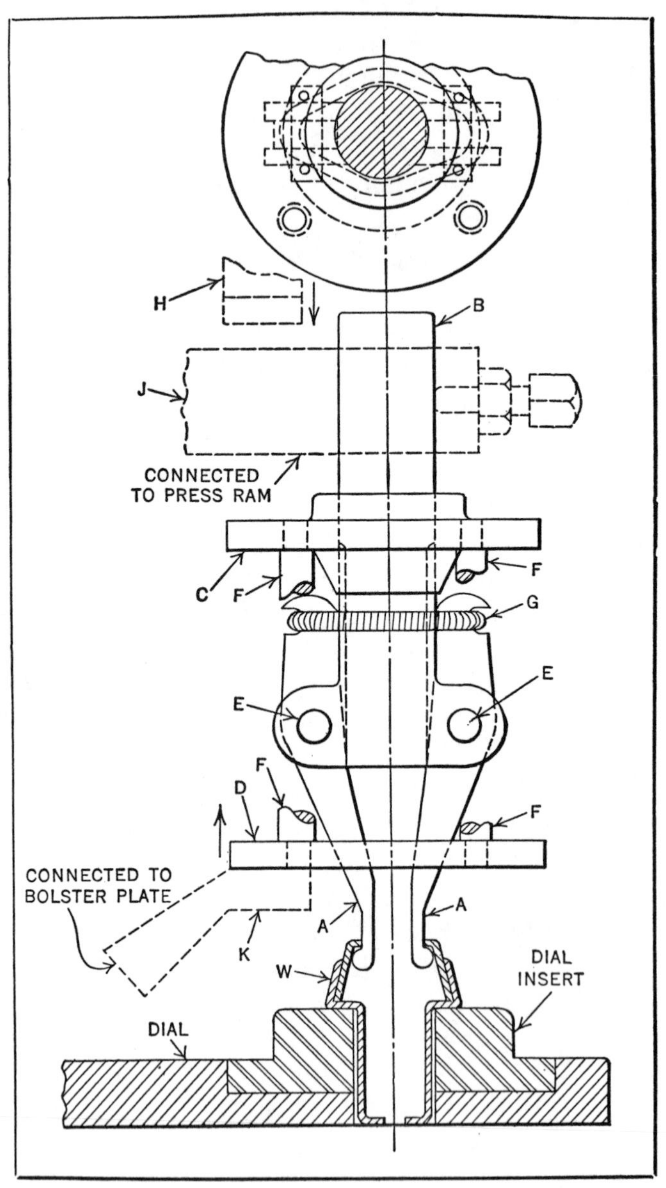

Fig. 8. Device for Ejecting Assembled Part from Dial of Inclined Press

When the ram starts to descend, the fingers *A* are closed at the lower tips, so that they will enter the hole in the work *W*, which consists of two assembled shells. Before the ram reaches the end of the down stroke, the stop *K* pushes the sliding plates *D* and *C* upward. These plates are connected by shoulder pins *F*. The illustration shows the beveled part on plate *C* leaving the fingers *A* so that the spring *G* is permitted to open the lower tips inside the work.

When the ram starts to ascend, the shell is gripped and lifted up out of the dial. Before the ram reaches the top of the upward stroke, the stop *H*, connected with the press frame, pushes plate *C* down. The beveled part of plate *C* then comes in contact with the upper end of fingers *A*, causing the tips to close and allow the work to drop down a chute and slide into a container, thus completing the cycle. The press is operated at a speed of about 75 revolutions per minute.

Device for Ejecting Fuse Plugs from Dial Press by Lifting Fingers.—A device used for ejecting porcelain fuse plugs from the dial of a press is shown in Fig. 9. In this press, the plugs are assembled with the metal caps that retain the isinglass covers. To eject the plugs, two spring fingers *A*, pivoted on the swinging arm *B*, are provided. The arm is mounted on a cam bushing *C*, the bore of which is a slide fit over the post *D* to allow vertical and rotary movement of the bushing. The post is stationary, and is secured to the press by means of the cast-iron bracket *E*. In the side of post *D* is secured the pin *F*, which engages the cam slot in the bushing and imparts the required oscillating motion to the arm *B* during the vertical movement of the bushing.

A flange at the top of this bushing engages the bracket *G*, which is secured to the press ram, so that as the ram reciprocates, the arm *B* will be given a combined vertical and oscillating movement.

As shown, the fingers *A* have gripped the plug preparatory to ejecting it from the dial. In this position, the ram is at its lowest point. As it ascends, the plug is lifted clear of the dial, and further upward motion of the ram causes the pin *F* to slide in the angular part of the cam groove and swing both bushing and arm enough to carry the plug over the edge of the dial. At this point the top ends of the two fingers *A* come in contact with the stationary stop *H*,

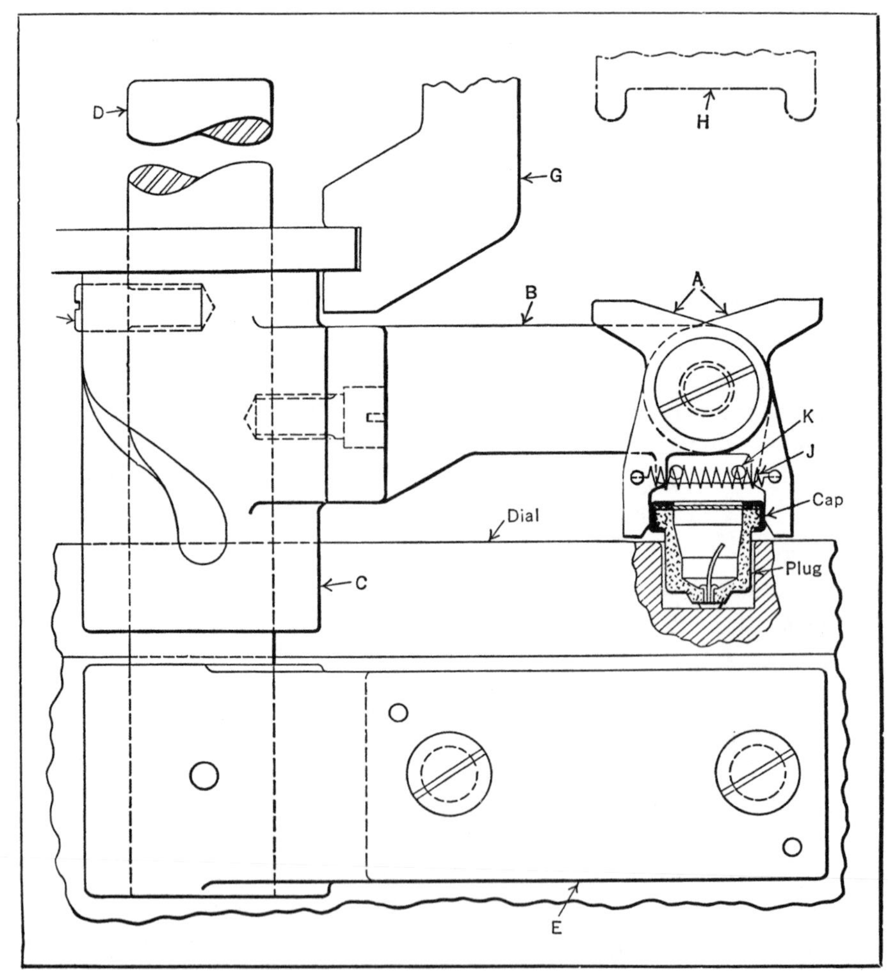

Fig. 9. Ejecting Device for a Dial Press which Lifts the Work from the Dial and Deposits it in a Container or Chute

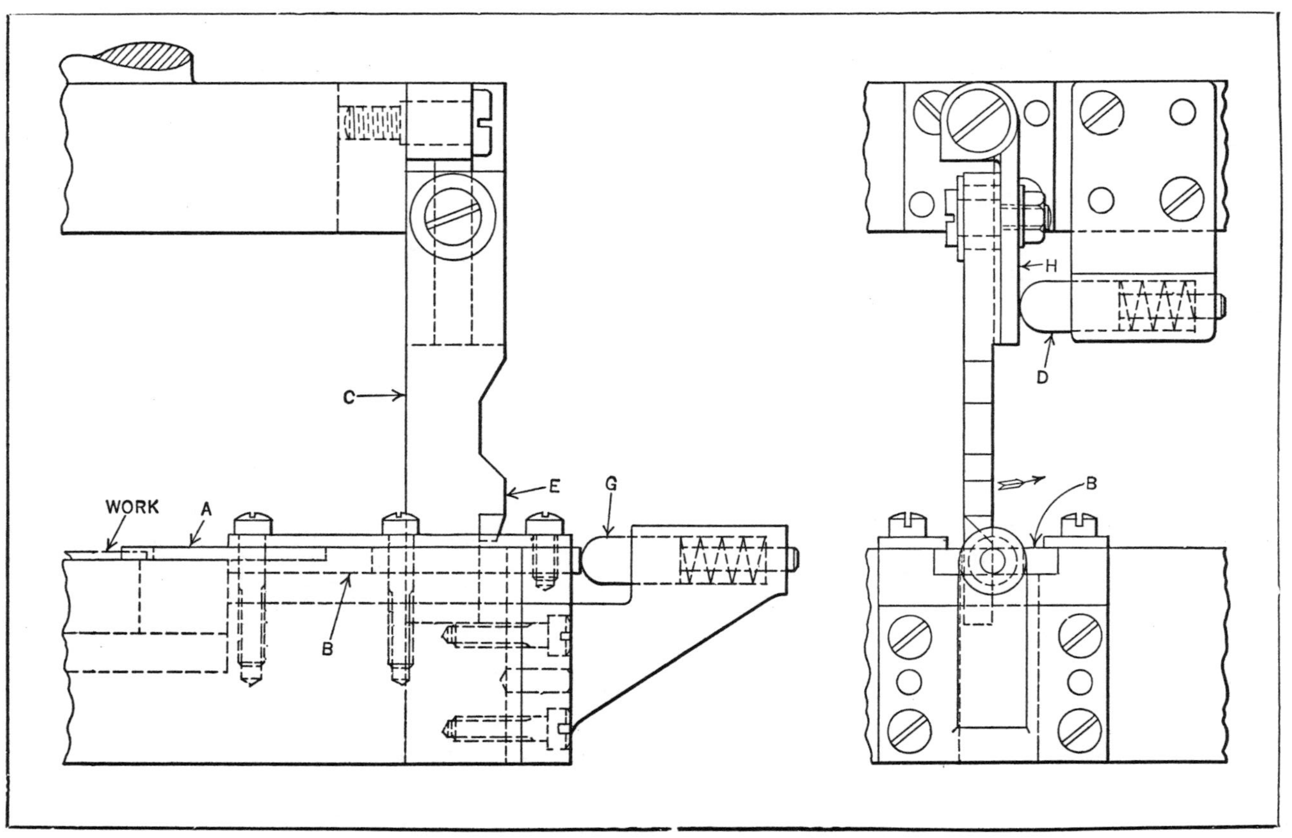

Fig. 10. Die Equipped with a Cam-arm that Imparts a Forward and Return Stroke to a Slide while the Ram Ascends. The Slide is Stationary while the Ram Descends

which opens the fingers, allowing the plug to drop into a chute at the side of the machine. From the chute, the plug slides into a container. A coil spring *J* is provided to give the required gripping pressure to the fingers. Also, in order to locate the fingers properly over the plug, centralizing pins *K* are provided. Although the stop *H* is shown in the illustration directly over the plug, this is not its actual position; it is placed at one side of the dial so that it will release the fingers only when they are above the chute.

Magazine Die with Stop which Shifts while Ejecting Work.— Several automatic magazine dies used for piercing

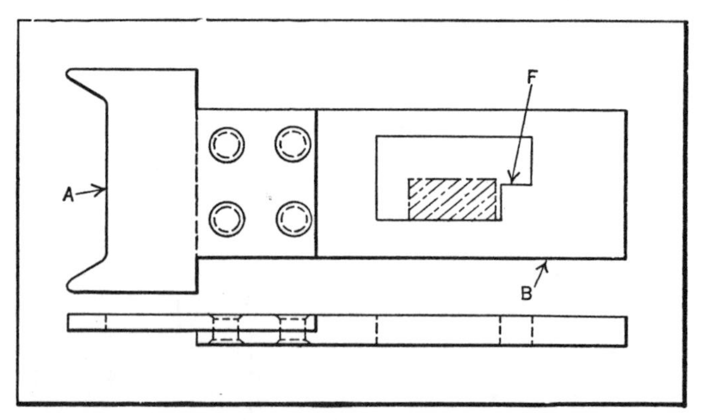

Fig. 11. Stop Operated by the Mechanism Illustrated in Fig. 10

and trimming operations on previously blanked parts are equipped with stop mechanisms like that shown in Fig. 10. During the upward stroke of the press ram, this mechanism serves to move the work locating stop *A* toward the right to permit the finished part to be ejected and a new blank to be put in place. Before the ram has reached the top of its stroke, the stop returns to its former position against the new blank. During the entire downward stroke of the ram, the stop is stationary.

As indicated in Fig. 11, stop *A* is riveted to the slide *B*, which is mounted in the die bolster. Cam-arm *C* (Fig. 10) is pivoted to the punch-holder and can be swung in the direction of the arrow. At the top of the stroke, the lower end of this arm assumes a position corresponding with the sectional area in Fig. 11, and is normally held there by the spring-actuated plunger *D*. As the press ram descends, the angular edge at the bottom of the projection *E* on the arm engages edge *F* on the slide, causing the arm to swing in the direction of the arrow. Upon the continued descent of the ram, the projection *E* passes the corner at *F*, allowing the arm to swing back to its normal position. During the entire downward stroke, slide *B* is held against the die by the spring-actuated plunger *G*, and therefore remains stationary.

As the ram begins to ascend, the angular edge at the top of projection *E* engages the under side of the corner at *F*, causing the slide to be moved toward the right. At this point, the finished part is ejected from the die and a new blank is slid into place by means of the magazine feed slide (not shown). As the ram continues its upward movement, the projection *E* leaves the slide and the slide is returned by plunger *G* to a position against the new blank. The blank is thus located centrally over the die in readiness for the next downward stroke of the punch. The action of the stop slide can be timed accurately by adjusting the arm to its correct position along the hinge *H*.

Oscillating Arm for Dislodging Pieces that Obstruct Hopper Feed Exit.—In using hopper feeds of the flat, revolving disk type shown in Fig. 12, there may be trouble due to jamming of the work on the aligning strip *A* at the point where the pieces leave the hopper. The purpose of strip *A* is to line up the work so that it will enter the chute opening in a predetermined position. The clogging of the hopper exit results in loss of production and unnecessary wear on the punch and die members. As an operator often runs three or four presses of this kind, jamming or clogging of the work in the manner referred to may not be noticed immediately.

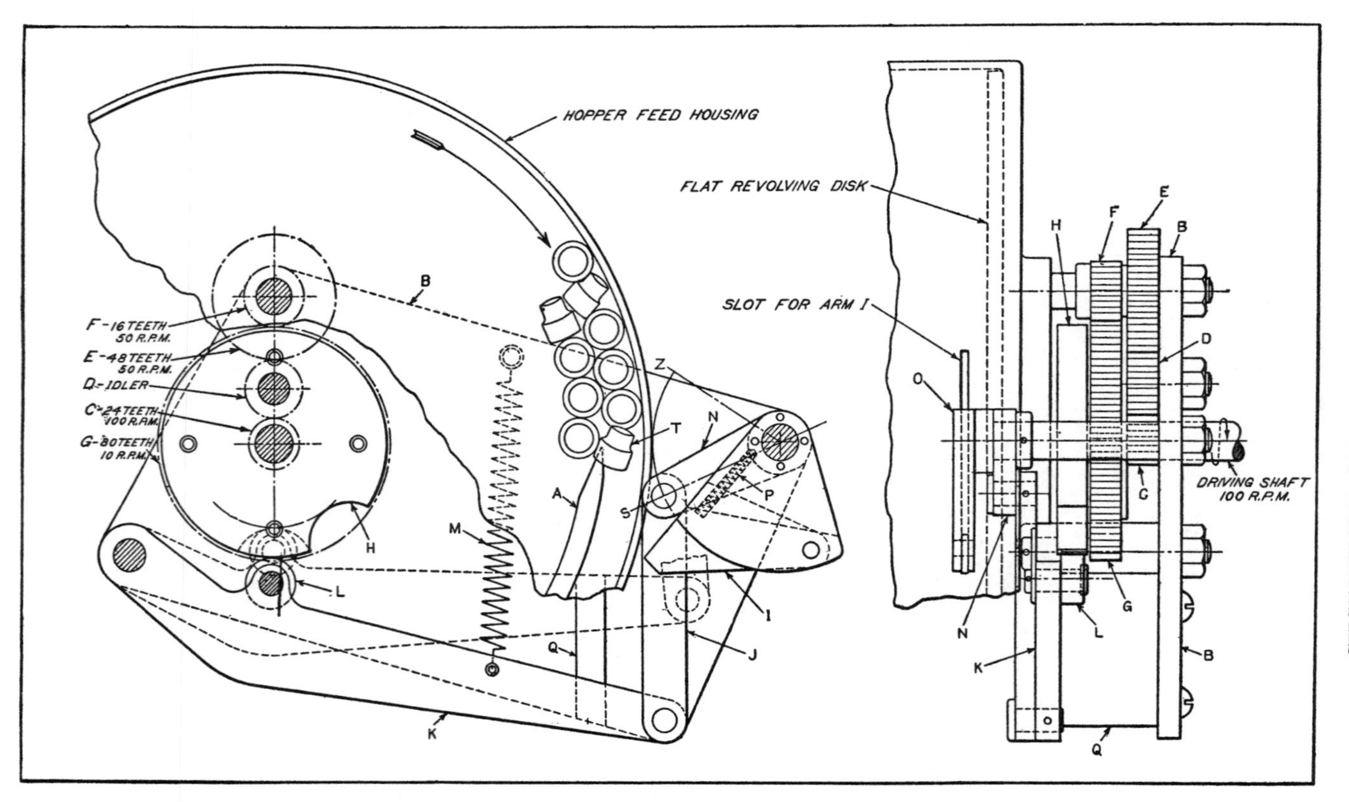

Fig. 12. Revolving Disk Type Hopper Feed with Mechanism for Dislodging Clogged Pieces

The entire device is mounted on the baseplate *B*. This plate is secured to the under side of the hopper feed housing. The gear *C* is an integral part of the driving shaft, which revolves at a speed of 100 revolutions per minute. The idler *D* transmits motion to the cluster gears *E* and *F*. Gear *F* meshes with gear *G*, which is free to revolve around the driving shaft and is riveted to cam *H*. Cam *H* revolves at 10 revolutions per minute.

The piece of work *T* which has obstructed the exit of the hopper is swept away or dislodged in the following manner: The circular cut-out on the cam *H* causes a sweeping motion of arm *I* from *S* to *Z* when the cam-roll *L* drops into the cut-out. The cam-roll is attached to lever *K*. Link *J* connects lever *K* to lever *N*. Lever *N* is riveted to the holder *O* of the arm *I*. Cam-roll *L* is kept in contact with the periphery of the cam by spring *M*. The small spring *P* provides flexibility for the arm *I* on the return movement. The shoe *Q* serves as a guide and steadyrest for lever *K*. The mechanism described can, of course, be applied to work of various shapes by making suitable alterations.

Transfer Mechanism for Stacking Parts on Rods as They Leave the Die.— In making parts such as shown in the lower right-hand corner of Fig. 13, it was necessary to stack them on rods with the irregular-shaped holes in correct alignment as they left the combination piercing and cut-off die. Stacking the parts in this manner facilitates subsequent operations. The transfer mechanism shown in Figs. 13, 14, and 15 provides an efficient means for stacking the parts. It is mounted at the right-hand end of the die and is operated by a cam attached to the punch-holder, the parts being stacked on rod *W*.

Referring to Figs. 13 and 14, the punch-holder *A* carries the piercing punch *B* and the cut-off punch *C*. On block *D*, which is secured to the die-bed, are mounted the slides *E* and *F*. Slide *E* carries the auxiliary cross-slide *G* which supports the shaft *H* at its left-hand end. The right-hand end of this shaft is supported by a double over-hanging bearing which is part of slide *E*. The left-hand end of shaft *H* is enlarged and recessed to slip over the end of the work shown at *J*.

Slide *F* is given a reciprocating movement by means of the cam *K* on the punch-holder through lever *L* and roller *M*. Bracket *X* supports lever *L* at its upper end. Slide *E*

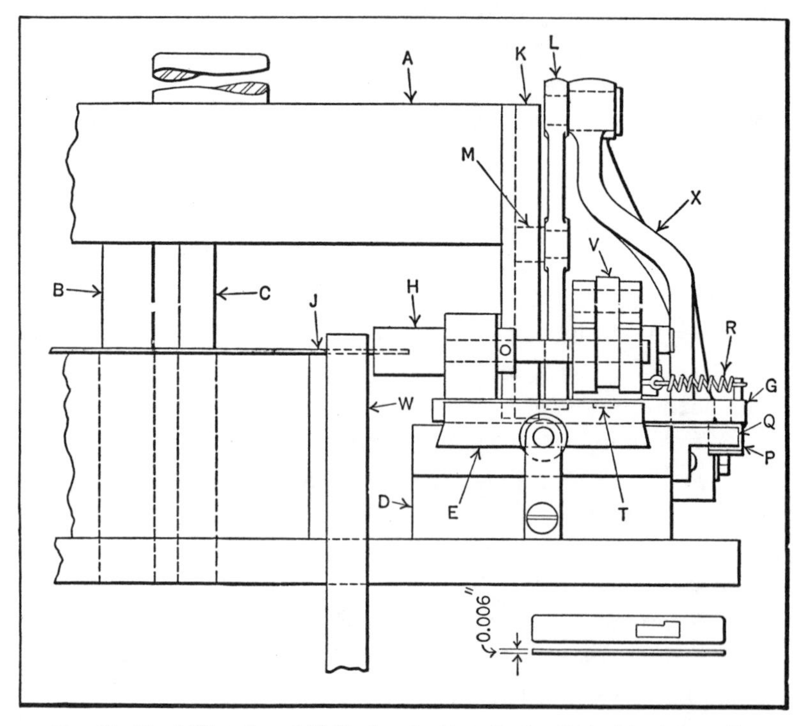

Fig. 13. Front Elevation of Mechanism for Transferring Parts Direct from the Die to the Stacking Rod

is backed up by the spring *N* on the flanged stud *O*. The flange of this stud serves as a stop for slide *E*. Slide *G* carries roller *P*, which is held in contact with cam *Q* by spring *R*. Slide *E* carries the spring-actuated latch *S* which engages slot *T* in slide *G*, and is connected to slide *F* by the slotted link *U*.

As indicated in all three views, the ram is in its lowest

position. The strip has been fed toward the right, its end entering the recess in shaft *H*. The part is then pierced and cut off by punch *C*. Slides *E* and *F* (see Fig. 14) are in their extreme right-hand positions. Latch *S* is held up out of engagement with slot *T* by link *U*, the screw at the rear end of which is in contact with the rear of slide *F*. Slide *G* is at its extreme left-hand position (see Fig. 13)

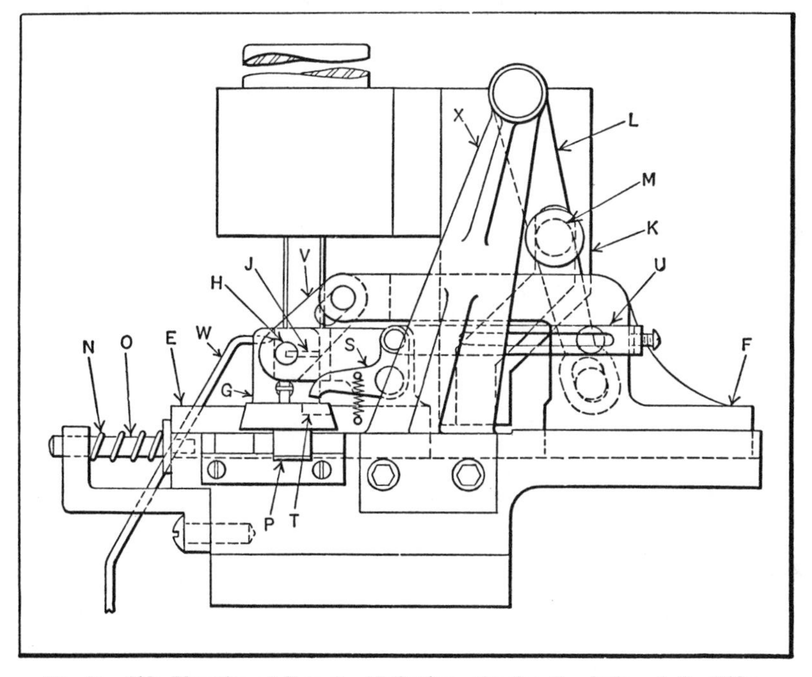

Fig. 14. Side Elevation of Transfer Mechanism, Showing the Action of the Slides Operating the Transfer Member

with the roller *P* in contact with the low part of cam *Q*. It will be noted in Figs. 14 and 15 that the front edge of work *J* is in alignment with the axis of the shaft *H*, so that when this shaft rotates, the work will rotate around its own front edge.

As the ram ascends, the cam *K*, acting on lever *L*, moves slide *F* toward the left until its inner end is in contact with the inner end of slide *E*, which, up to this point, has remained stationary. Prior to this, the motion of slide *F* has been transmitted to link *V*, which causes shaft *H*, to which the link is keyed, to revolve 90 degrees. At this point the work is standing on edge. Latch *S* has been permitted to swing downward; but as it is not in alignment with groove *T*, it merely rests on top of slide *G*.

Further upward movement of the ram causes slide *F* to push slide *E* toward the left, so that the work is placed over rod *W*. As the ram continues to move to the top of its

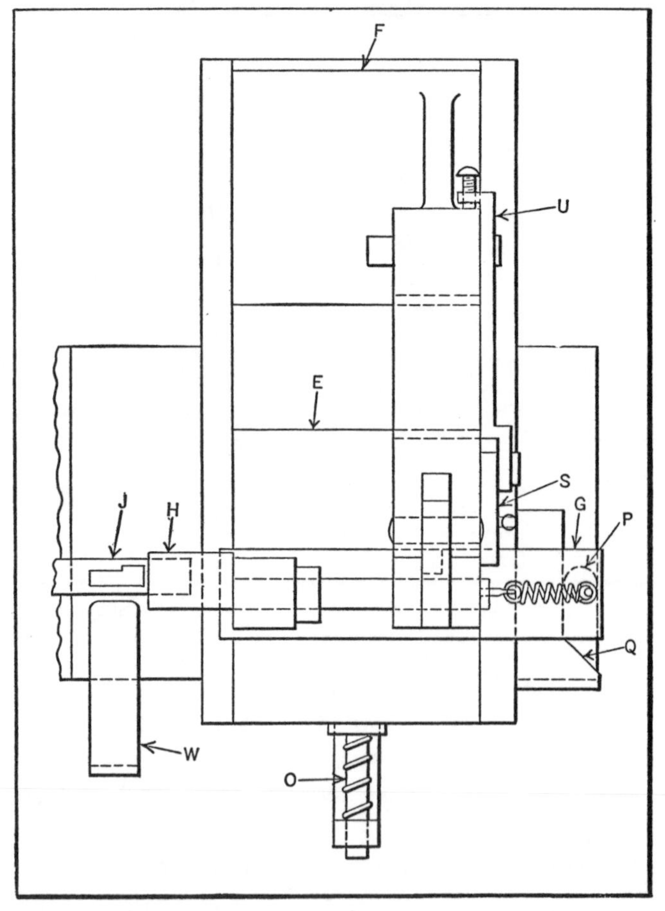

Fig. 15. Plan View of Transfer Mechanism, Showing the Position of the Part after Entering the Transfer Member

stroke slide *G* is drawn to the right by the action of roller *P* on cam *Q*. The work cannot follow this movement, being restrained by the rod *W* in the work-slot; hence shaft *H* is entirely removed from the work, allowing the latter to drop to the bottom of the rod. At this point in the cycle of operations, slide *G* is in its original position, at which time latch *S* is free to drop into the groove *T*, thus locking slide *G* to slide *E*.

As the ram descends, slides *E* and *F* move toward the right until the flange of stud *O* comes in contact with block *D*, which discontinues the movement of slide *E*. The movement of slide *F*, however, continues, and through link *V* revolves shaft *H* to its original position. As slide *F* approaches the end of its return movement, it engages the adjustable screw on link *U*, disengaging latch *S* from slide *G* and causing this slide to be drawn to the left (see front elevation, Fig. 13) by spring *R*. After slide *G* moves toward the left, the enlarged end of shaft *H* is in position to receive the end of a new part, thus commencing another cycle of movements. Rod *W* is about 4 inches high and will accommodate 500 pieces. It can easily be removed from the die when it has been completely filled, and replaced by another rod, after which the cycle of operations is repeated.

CHAPTER XVI

MISCELLANEOUS MECHANISMS OR MECHANICAL MOVEMENTS

Whenever mechanisms have a similar function or a common operating characteristic, they have been grouped together in chapters both in this volume and in Volume I to assist the user in finding whatever general type of mechanism may be wanted. In this chapter will be found mechanisms of such a miscellaneous character that they cannot be placed in any general classification.

Spherical-Elliptical Movement of Sewing Machine Double-Lock Stitch Mechanism.— Many of the sewing machines used in manufacturing clothing, bags, awnings, etc., use what is commonly known as a double lock stitch. These machines have two needles operating at right angles to each other. The lower needle operates beneath the throat plate that supports the goods being stitched. This needle is commonly termed a "looper," as it does not pierce the goods, but passes into and out of the loop of thread made by the other needle in its vertical motions. The loop of thread is formed at the desired location below the throat plate during the upward motion of the needle by permitting the thread to become slack at the proper time, thus causing it to buckle and form the loop. A mechanism used to impart the required motion to the lower needle, or looper, is shown in Fig. 1.

The looper *A* is required to pass very close to the needle in taking up the loop of thread. It must hold this thread loop and position itself on the opposite side of the needle's path by the time the needle has descended below the throat plate in forming the next stitch. From this it is apparent that the looper must have a back-and-forth motion at right angles to the needle and at right angles to the direction in which the goods travels. In addition, the looper must have a back-and-forth motion at right angles to the needle in the line of travel of the goods being stitched. This latter motion is commonly called the "avoider" motion, since its object is to avoid or prevent interference with the needle. The path followed by any point on the looper *A* consists of a closed curve that resembles an ellipse bent to fit the surface of a sphere.

The main shaft *B* carries the flat strap eccentric *C* and the ball-joint eccentric *D*, which drive the rocker arm *E*, mounted in the spherical bearing *F*. Arm *E* carries the looper *A*. Eccentric *C* imparts motion to arm *E* by means of pin *G*, which passes through the center of bearing *F* with which it is in sliding engagement. This gives the looper *A* an oscillating movement. Eccentric *D*, by means of its connection with the ball-ended bellcrank *H* and link *J*, causes arm *E* to oscillate about the center of pin *G*, so that *A* moves back and forth between the positions shown by the full lines and by the dotted lines at *M*.

It will be noted that eccentrics *C* and *D* impart motions to *A* which are approximately at right angles to each other, in producing the spherical-elliptical movement. Eccentric *C* gives the needle the "avoider" motion, while eccentric *D* imparts the motion that causes the looper to pass in and out of the thread loop formed by the needle that pierces the goods.

Considering the action of eccentric *C* in producing the "avoider" motion, it will be noted that only one component of the circular motion of this eccentric imparts motion to the looper, the other component resulting in pin *G* sliding through the ball joint in *F*. With reference to eccentric *D*, it will also be noted that only one component of this circular motion is imparted to looper *A* through its connection thereto by means of bellcrank *H*, link *J*, ball-pin *K*, and rocker arm *E*, whereas the other component of this motion causes

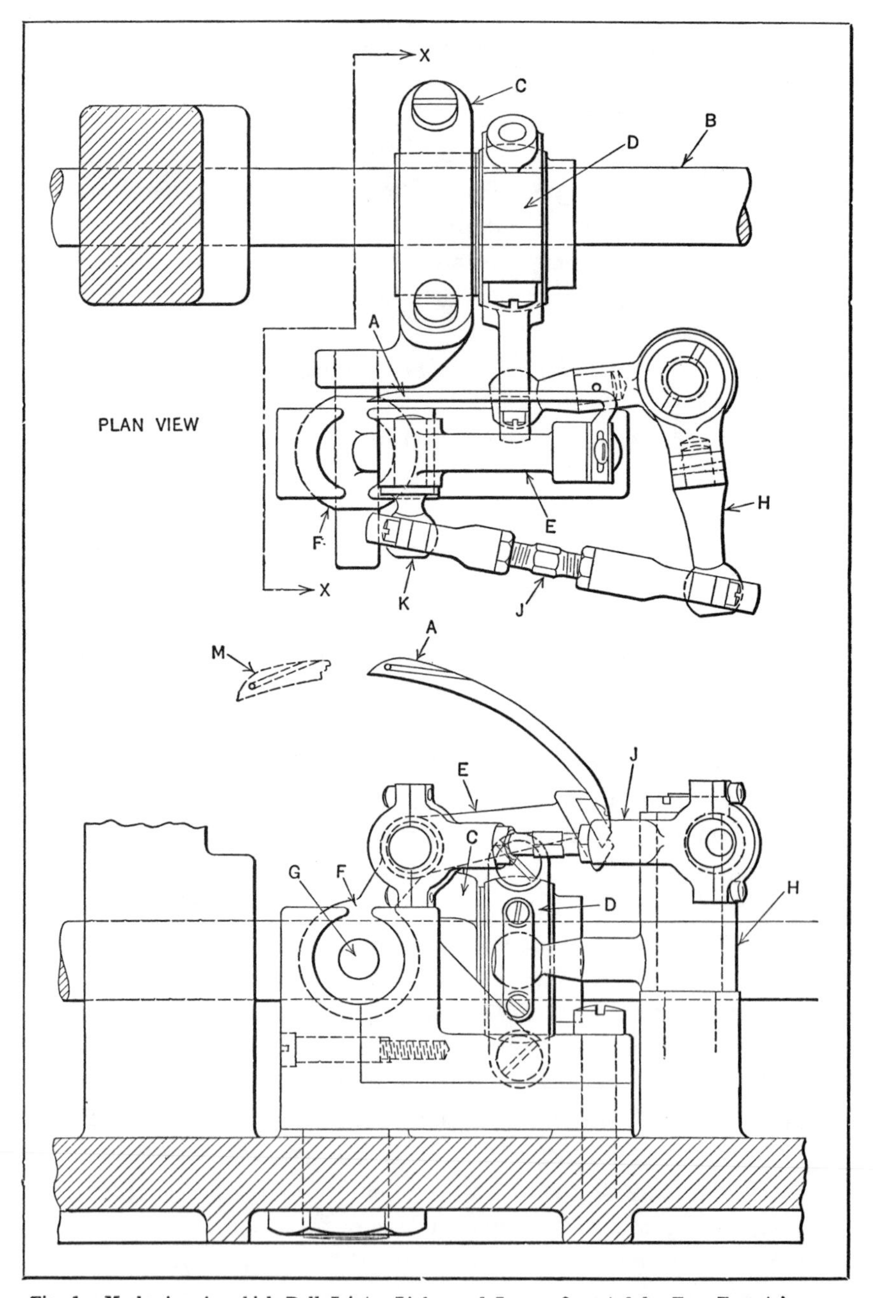

Fig. 1. Mechanism in which Ball Joints, Links, and Levers Operated by Two Eccentrics are Used to Obtain Spherical-elliptical Movement of Member A

rotation about its connection with the bellcrank *H*. Thus the motion imparted to the looper *A* may be considered as being the resultant of two circular motions at right angles to each other. The relative values of these two motions have been changed by means of levers, in order to give to each the desired amplitude.

The motion described could be obtained from a single eccentric, except for the mechanical difficulties encountered in obtaining suitable proportions for the components required for such motion. The motion produced by the mechanism illustrated is the resultant of two simple harmonic motions

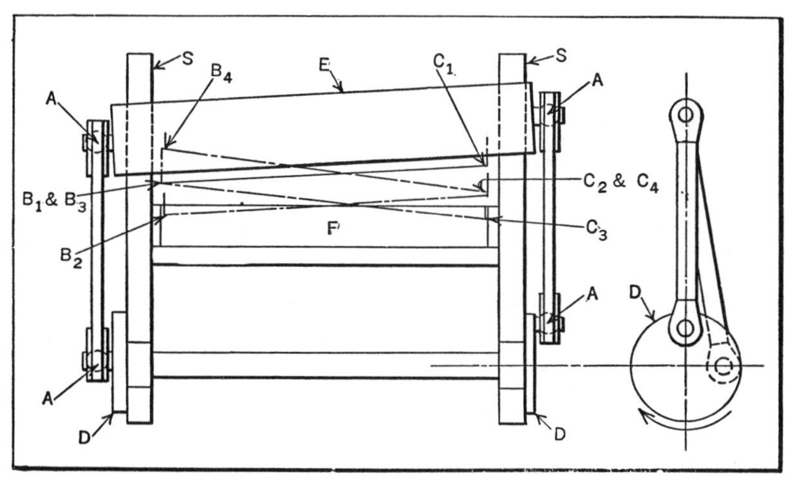

Fig. 2. Mechanism for Producing Shearing Motion

at right angles to each other, each being modified by the length and angularity of the connecting links. This form of looper drive is adapted to high speeds, since it is relatively simple, has few parts, and the few rapidly moving parts required can be made quite light. There are no violent changes in velocity, and the energy changes and friction losses are thus kept at a minimum.

Shearing Motion which Varies Angular Position of Blade.—The mechanism shown in Fig. 2 was applied to a shearing machine to obtain an action approximating that of hand-operated shears. The mechanism is arranged to make one complete cycle of movements automatically when the driving clutch is engaged. The sides *S* of the shearing machine serve as guides for the cross-bar *E* which supports the movable blade above the stationary blade *F*. As the crank disks *D* revolve, the ends of the movable blade are carried to different positions, as shown at points B_1, B_2, B_3, and B_4 on the left, and by C_1, C_2, C_3, and C_4 on the right. When the left-hand end of the blade is at B_1, the right-hand end is at C_1, and when the left-hand end is at B_2, the right-hand end is at C_2, and so on. This wabble or shearing action is obtained by locating one of the crankpins 90 degrees ahead of the other.

The successful operation of the mechanism is made possible by the spherical pin and connecting-rod bearings *A*. These bearings were easily produced from soft steel balls purchased from stock and bored out to a press fit on the bearing pins. The boxes at the ends of the connecting-rods were formed by pouring babbitt over the balls while the two parts were mounted on a surface plate in their proper relative positions. A few strokes with a soft-faced hammer served to loosen up the bearings sufficiently to permit them to operate satisfactorily.

Grease lubrication is provided by a hole drilled through the bearing after the babbitt was poured. This hole crossed another hole leading from the outer end of the pin. Side motion of the holder *E* is prevented by using curved guides having sufficient clearance to prevent binding. Machines equipped with this shearing mechanism are used in connection with the production of cotton batting and similar fibrous parts.

Transmitting Motion to Indexing Plunger by Steel Balls.— A rather unusual application of steel balls for transmitting motion to an indexing plunger on a dial press is shown in Fig. 3. The plunger *J* slides in the fixed bearing *I* and receives its motion from the lever *A* through the steel balls *H*. The lever oscillates about the fixed stud *B*, and its lower end engages a slot in the member *C*. The latter is a sliding fit in the stationary bearings *D* and *E*, and as the lever *A* oscillates, member *C* forces the balls up the tube, causing the plunger to move upward into the dial. The plunger is returned to the position shown by the coil spring *K* which also keeps the balls in contact with member *C*. It is obvious that with this simple device the tube containing the balls may be bent to almost any shape desired, permitting it to clear any member of the machine.

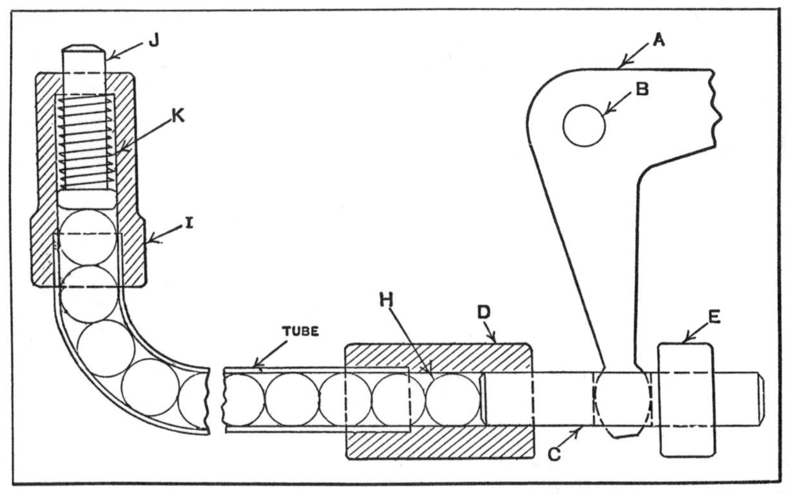

Fig. 3. Transmitting Motion to a Plunger with Steel Balls Confined in a Tube

Counter Used on Type-Setting Machine.— In setting type on a linotype machine (for lay-out pages), each line in the different sections is counted so that the type set will agree with the lay-out calculations. With the instrument shown in Fig. 4, however, no mental effort is required, as the actual count is registered automatically on the dial at *X*, which is returned to its zero position by one simple motion of the operator's hand.

The lower part of the counter is bolted to a stationary bracket on the frame of the machine. An adjustable finger

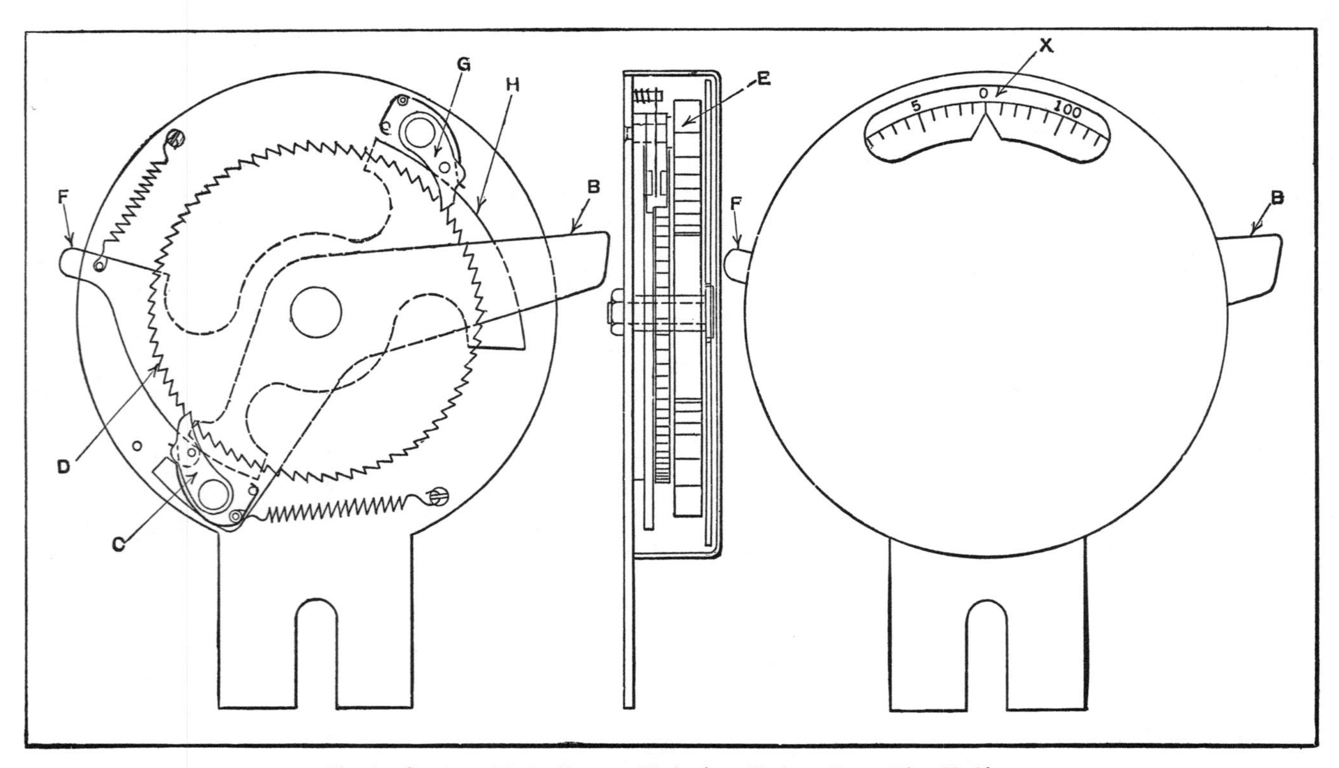

Fig. 4. Counter with Rapid-return Mechanism, Used on Type-setting Machine

attached to a reciprocating member on the machine strikes the end of the lever *B*, causing the pawl *C* to advance the ratchet wheel *D* one tooth, or one graduation on the dial. The dial continues to register until the required number of lines to be cast is indicated. The operator then depresses lever *F*, causing the dial to return once more to the zero position ready for the next group of lines. When the lever *F* is depressed, the cam-plate *H* forces the pins in pawls *C* and *G* outward, disengaging the pawls from the ratchet wheel and allowing the clock spring *E* attached to both cam-plate and dial to return the dial to the zero position. Stop-pins (not shown) are provided to limit the return movement of cam-plate *H* and lever *B*, as well as that of the dial on its return to the zero position. The counter is located on the machine so that the dial is at right angles to the line of vision, assuring easy and accurate reading.

Mechanism for Advancing and Lifting Parts to Clear Lugs of Conveyor Chain.—In a production line where a chain conveyor is used, it is sometimes necessary to lift the article conveyed clear of the lugs on the chain while the chain is in continuous motion.

The mechanism shown in Fig. 5 was developed to meet a requirement of this kind. The article conveyed in this case is required to be rapidly advanced ahead of the chain travel at the last station and brought into position where it can be elevated by an auxiliary mechanism before it is overtaken by the chain lugs.

The action of this entire unit may be summarized as follows: As the chain conveyor carrying the work travels along, arm *E* swings up and in back of the work, rapidly advancing it throughout its forward motion. After the work has been lifted clear, the arm swings in back of the next article.

In the side view the articles conveyed are indicated at *A* and *B*. There is a pair of chain conveyors at *C*, which travel in the direction indicated by the arrow *D*. The arm

E is in contact with the work and has advanced it ahead of the position it would normally have reached as a result of the conveyor movement. The mechanism is shown at the completion of the advance movement. This mechanism has a grooved cam *F* which revolves with shaft *G*. Over the block *H* is assembled a yoke *J*, which carries a roll *K* that rotates in the cam groove, causing yoke *J* to oscillate horizontally.

This action, through the medium of rod *L*, causes lever *M* to rock shaft *N* back and forth. A similar movement is imparted to a pair of levers *P* which, in turn, transmit a

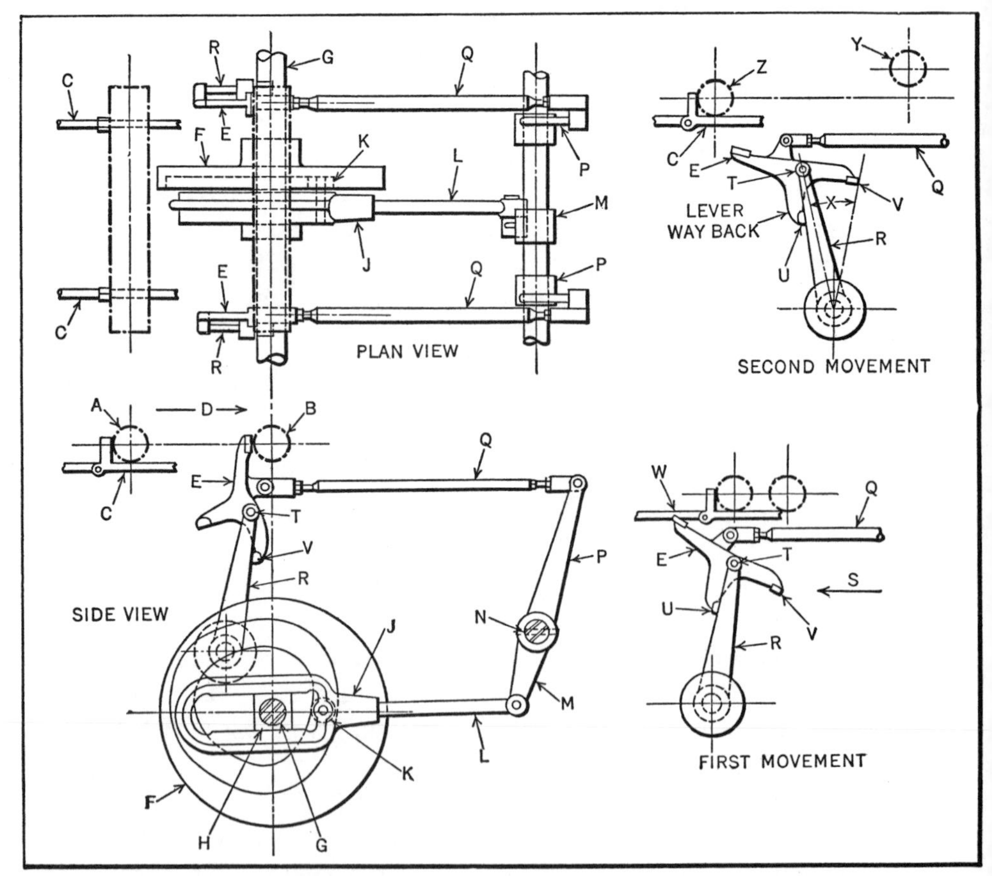

Fig. 5. Mechanism for Raising Work from Chain Conveyor and Advancing it to Operating Position

longitudinal reciprocating movement to a pair of rods *Q*. Rods *Q* transmit a rocking and transverse movement to the unit consisting of arms *E* and levers *R*, which results in advancing the work *A* and *B* and returning the unit to its starting point.

The side view shows the work *B* fully advanced, while the lower view at the right shows the position of the various members during the first portion of the return movement of arm *E*, as rod *Q* advances in the direction indicated by arrow *S*. This movement causes arm *E* to pivot about

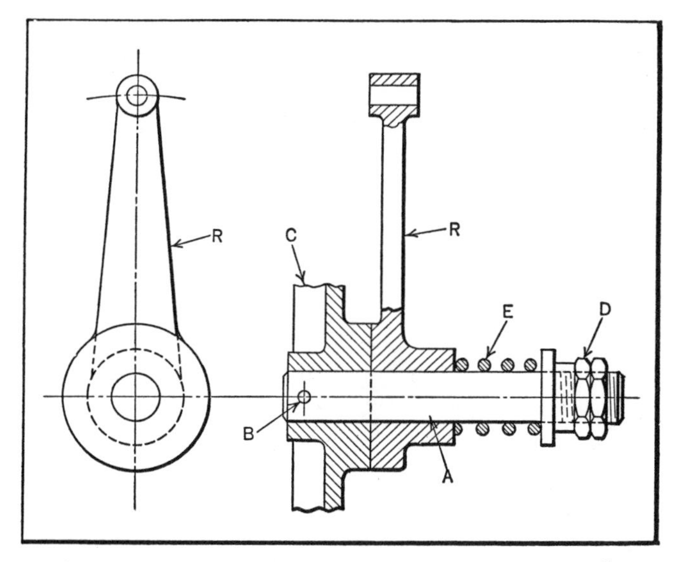

Fig. 6. Details of Friction Drag Applied to Lever R, Fig. 5

stud *T* as a center until lug *U* comes in contact with the side of the lever. It will be noted that, during the advance movement, the lug *V* was in contact with the other side of lever *R*. When the lug *U* is in the position shown in the view in the lower right-hand corner, the top of arm *E* is below the level of the work, as indicated at *W*, although lever *R* has not yet been moved. Continued movement of rod *Q* in the direction indicated by arrow *S* causes lever *R* to be pushed to the left until the entire mechanism assumes the position indicated in the upper view, where lug *U* is still in contact with the lever and the arc of travel is indicated by *X*.

Another article has now been brought forward, as shown at *Z*, and the position of arm *E* is such that rod *Q*, in pulling arm *E* to the right, will cause it to pivot about stud *T* until it swings behind the work *Z*, when lug *V* will be in contact with lever *R*.

Continued movement of rod *Q* will cause the article at *Z* to be rapidly advanced ahead of the chain travel, so that another mechanism (not shown) can pick up the work and raise it clear of the chain lugs into the position indicated at *Y*.

To insure that arm *E* will pivot about stud *T* during the first movement of the advance action and the first movement of the return action without imparting any movement to lever *R*, the latter lever is constructed as illustrated in Fig. 6. Lever *R* is shown as pivoting on the short shaft *A*, which is pinned at *B* to the side frame *C* of the machine. Two lock-nuts *D* permit the tension exerted by the spring *E* against the lever to be so adjusted that its movement is retarded until there is a definite pull in either direction that is sufficient to assure the correct functioning of the mechanism.

Air-Chuck Valve that Reduces Air Consumption Forty Per Cent.— In shops using a large number of air chucks for machining purposes, the air consumption may be reduced as much as 40 per cent by utilizing the type of pneumatic apparatus to be described. The full-line pressure is commonly used for both opening and closing the chuck. While the full-line pressure is required for closing the chuck, only a fraction of this pressure is needed to release the jaws.

By means of the valve shown in Fig. 7 the same air is employed for opening the jaws as is used for closing them. The principle can be more clearly explained by referring

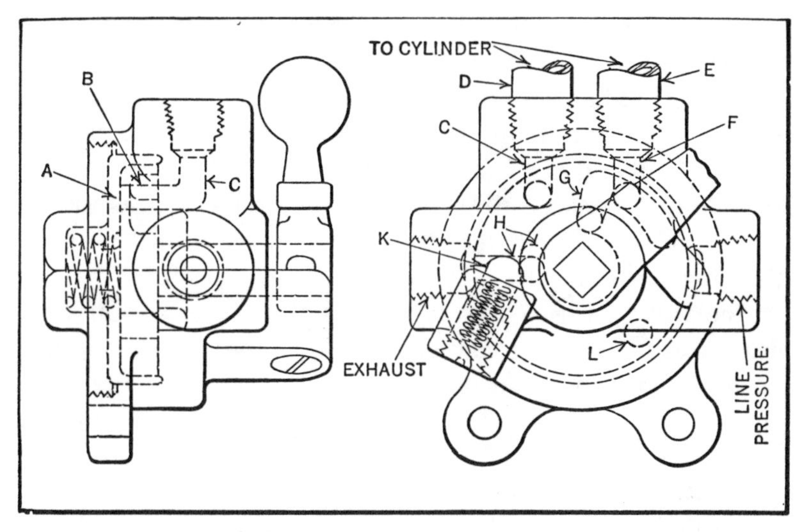

Fig. 7. Pneumatic Valve by Means of which the Same Air is Used for Opening and Closing a Chuck

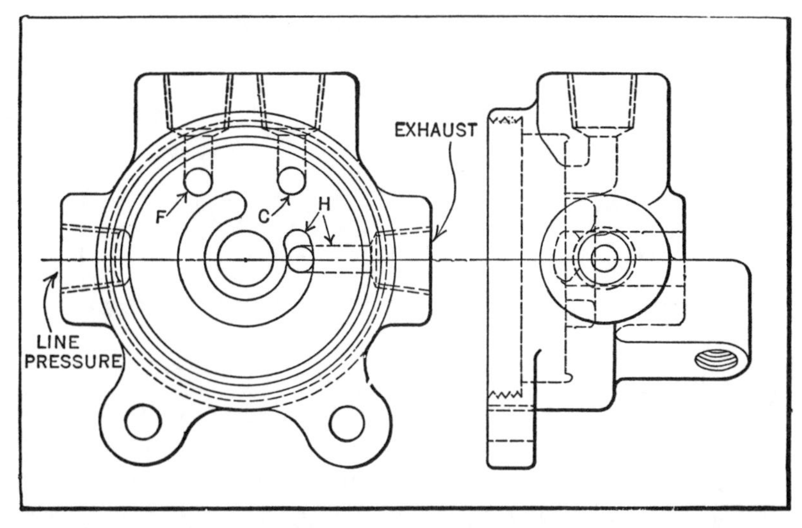

Fig. 8. Detail View of Body of Air Valve, Showing Positions of the Various Ports

to a diagram of the air cylinder (Fig. 10). This cylinder is designed so that when the chuck is closed, the space on both sides of the piston will be about the same, as indicated.

To release the jaws, the piston must be moved toward the

left. This is done by exhausting air from side *A* into side *B* until the pressure in both sides is equal. The air in side *A* is then exhausted into the atmosphere, leaving side *B* with about one-half of the line pressure—about 50 pounds —to expand and push the piston toward the left, in this way releasing the jaws. It is very seldom that a pressure of more than 45 pounds per square inch is required to release the chuck jaws.

The valve (Fig. 7) that controls the air in the manner described is of the rotary self-seating type, which requires no packing. The position of the various ports in the valve body and disk are shown clearly in the detail views, Figs. 8 and 9, the reference letters corresponding in all views.

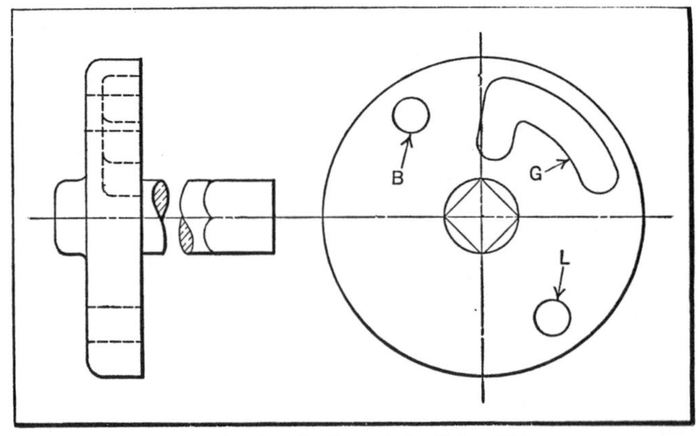

Fig. 9. Detail View of the Valve Disk

With the valve lever in the position indicated in Fig. 7, the air entering at line pressure passes over the disk at *A*, up through the hole *B* in the disk, out through port *C* in the body, and into the pipe *D* which leads into the left-hand side of the cylinder. The air entering the cylinder forces the piston toward the right and closes the chuck. During this movement, the air is exhausted from the right-hand side of the cylinder through pipe *E*, port *F* in the body, port *G* in the disk, port *H* in the body, and out through the exhaust opening to the atmosphere.

To release the jaws, the lever is swung toward the left until it comes into contact with the spring stop *K*. As the lever commences its movement toward the left, ports *G* and *H* are disconnected, thus closing the exhaust from the right-hand side of the cylinder. At this time, ports *C* and *B* are also disconnected, closing the inlet and confining the air at line pressure to the left-hand side of the cylinder. Continued movement of the lever causes port *G* to connect ports *C* and *F*, so that the air is by-passed from the left-hand side of the cylinder to the right-hand side until the pressure on both sides of the piston is equal. Further movement of the lever disconnects port *G* from port *F*, thus closing the latter and confining the air to the right of the cylinder.

As the lever comes in contact with the spring stop *K*, port *G* connects ports *C* and *H*, and the air on the left side of the cylinder is exhausted into the atmosphere. Thus, air at a pressure of approximately one-half the line pressure is left on the right of the cylinder. This air expands, pushing the piston toward the left and opening the chuck jaws. If for some reason this pressure is insufficient to release the jaws, the lever is forced farther toward the left, depressing the spring stop *K*. This additional movement of the lever causes port *L* to connect with port *F*, so that air at line pressure enters at the right of the piston and overcomes the resistance. When released, the lever will immediately spring back to the unchucking position, shutting off the line pressure.

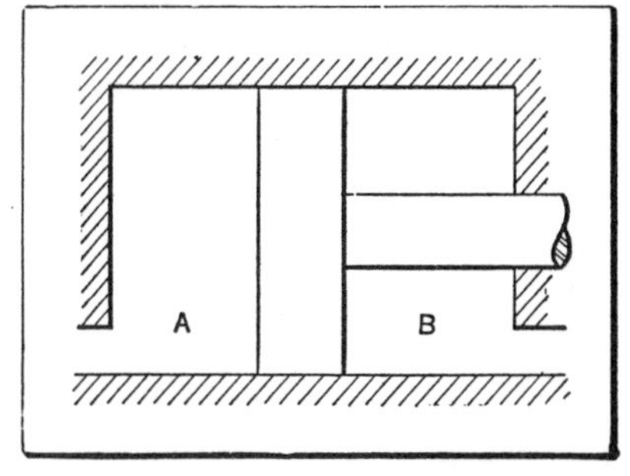

Fig. 10. Diagram of Air Cylinder with Piston in "Closed Chuck" Position

Mechanism for Winding Spherical Cores for Golf Balls.—Cores for golf balls are produced by winding a soft rubber band on a spherical rubber center. The thin rubber

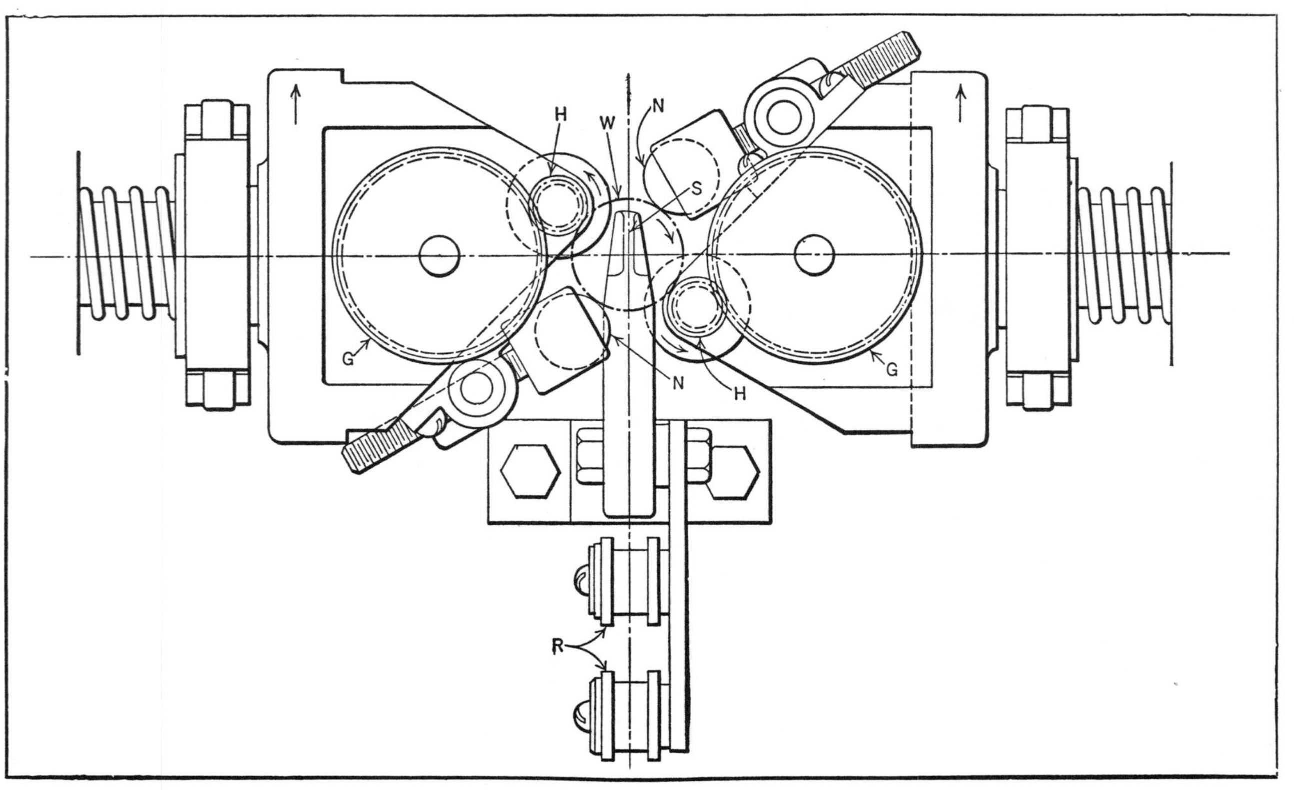

Fig. 11. Plan View of Winding Mechanism Shown in Fig. 12

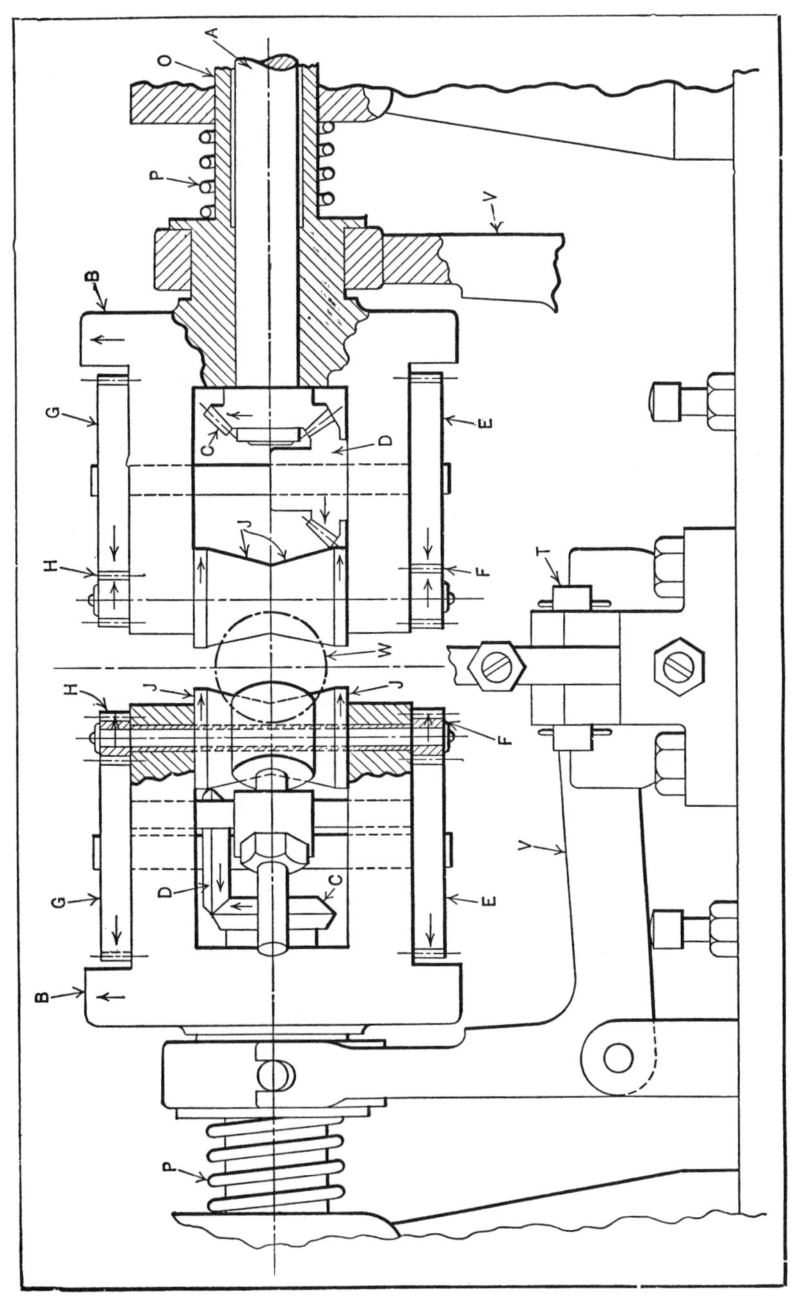

Fig. 12. Mechanism for Winding Thin Rubber Band Evenly over the Surface of a Small Spherical Center of Rubber

band is approximately 1/16 inch wide and must be wound evenly about the rubber center under uniform tension. The completed core must have a true spherical form. These requirements necessitate a somewhat complicated winding or rotational movement, in order to have the crossing points of the rubber band evenly distributed over the spherical surface of the core, which is constantly increasing in size during the winding operation.

The principal elements of the mechanism that provides the necessary winding movement are shown in Figs. 11 and 12. The work is started by placing the rubber center between the four truncated cones or rollers *J* (Fig. 12) mounted in the right- and left-hand winding heads *B*. The rollers are kept in contact with the core by the compressive action of the helical springs *P*, which allow the heads to recede equally as the diameter of the core increases. The bellcranks *V* equalize and centralize the outward movements of the winding heads so that the core is always kept in the central position. The bellcranks are connected with a treadle that permits the operator to withdraw the heads, so that the finished core *W* can be removed. The two ball casters indicated at *N* (Fig. 11) serve to guide or retain the core in its proper position between the rolls.

The two winding heads *B* (Fig. 12) have hollow shafts *O*, which are geared to a countershaft (not shown) at the rear of the mechanism. The countershaft is belt-driven and is provided with a handwheel at one end to allow the mechanism to be operated slowly by hand for making adjustments. The shafts *A*, which run inside shafts *O*, are also geared to the countershaft and run at a somewhat higher speed than shafts *O*. Shafts *A* provide for the secondary revolving movements imparted to the core by the four rollers through the bevel gears *C* and *D*, and the intermittent gears *E*, *F*, *G*, and *H*. The shafts of gears *H* drive the two upper rollers *J*, and the shafts of gears *F* drive the lower rollers *J*, mounted on the two opposed heads *B*. Thus different movements are imparted to the upper and lower rolls.

The four rollers have their conical surfaces spotted to give a better frictional grip on the core. The winding operation consists of feeding the rubber band by hand over the first and under the second tensioning rolls *R* (Fig. 11) and on through the guide slot *S* down on the soft rubber core which revolves rapidly in two directions simultaneously. The feed guide with the tensioning rolls attached swings on a fulcrum pin *T* (Fig. 12). A helical spring (not shown) attached to the guide provides the proper winding tension. The winding movements are obtained from the constant rotation of the heads *B* in the direction indicated by the arrows, combined with the motions imparted to the core by the four rollers *J* driven by the intermittent gears *E, F, G,* and *H*. These movements are so timed that the rubber band is wound on the rubber core evenly.

When the core reaches the finished size, the machine is stopped automatically through an electrical contact made by the spindle of one of the receding heads. While the approximate winding movements described appear quite simple, the exact path followed by any particular point on the spherical surface of the core is somewhat complicated, as will be apparent when the effect of the difference in speed of the shafts *A* and *O* is considered in conjunction with the rotational movements obtained by the four intermittently rotated rollers *J* and the constant rotational movement of the heads *B*.

Loading and Discharge Door Control for Enameling Oven.— An oven for baking the enamel on automobile wheels is so arranged that the wheels roll down gravity runways through the oven and are automatically discharged into runways which lead them either to the next operation or to the loading dock for shipment. The baking time for a

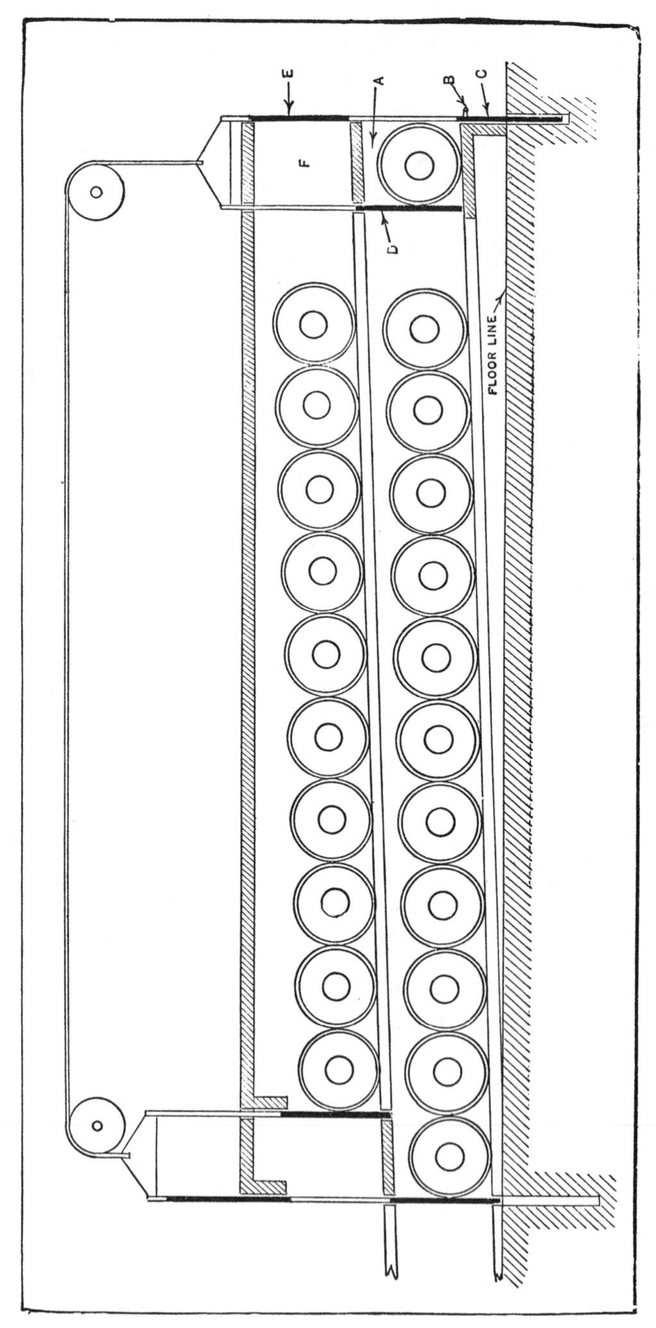

Fig. 13. Vertical Longitudinal Section of Enameling Oven which Automatically Discharges a Row of Baked Wheels when an Unbaked Row Enters the Oven at the Loading End

given length of oven depends entirely on the rate at which the wheels are put in.

In operation, the wheels are placed twelve abreast in compartment *A* (see vertical longitudinal section, Fig. 13). The operator, using handle *B*, then raises the door, which is made of three panels—*C*, *D*, and *E*—all mounted rigidly on a common frame and suspended from above. As the door is raised, panel *C* rises, closing compartment *A*, and panel *D* rises, letting the twelve wheels roll into the oven. Simultaneously panel *E* rises, opening compartment *F*, which has been closed at the back by panel *D*. Compartment *F* is now ready to be loaded with twelve wheels, after which the door is lowered again to reload compartment *A*.

The door at the discharge end is similarly constructed, and provides an automatic escapement for the wheels. The doors are connected by two cables which pass over sheaves above the oven and thus counterbalance each other. The two sheaves at each end are keyed to the shafts in order to keep the doors on an even keel and make them work freely in their guides. The two doors being connected, the discharge door is automatically lowered when the charging door is raised and vice versa, thus releasing a row of wheels every time a row is put in and keeping the oven full all the time. The man at the loading end sprays the wheels and places them in the runways, closing the door when he has inserted twelve wheels. Then, without further labor or attention on the part of the operator, they are baked the proper length of time and delivered to the next operation. With a given number of wheels per hour to be baked, any desired length of baking time may be obtained by making the oven of the proper length. Another point to be noted is that the doors are closed practically all the time, which cuts down the loss of heat to a minimum and also adds greatly to the comfort of the operator.

Short-Stroke Mechanism for Operating Fixture Lock-Pin.— A semi-automatic facing fixture used on a single-spindle drill press carries six castings that are equally spaced around the circular table of the fixture. As each casting is indexed around to the machining position, where it is faced by a profie cutter, the fixture table is locked in the proper position by a 3/8-inch tapered pin which enters a tapered hole. The mechanism to be described is for operating this stop-pin. Fig. 14 shows sectional and plan views of the stop-pin mechanism. The slide *A*, which is lo-

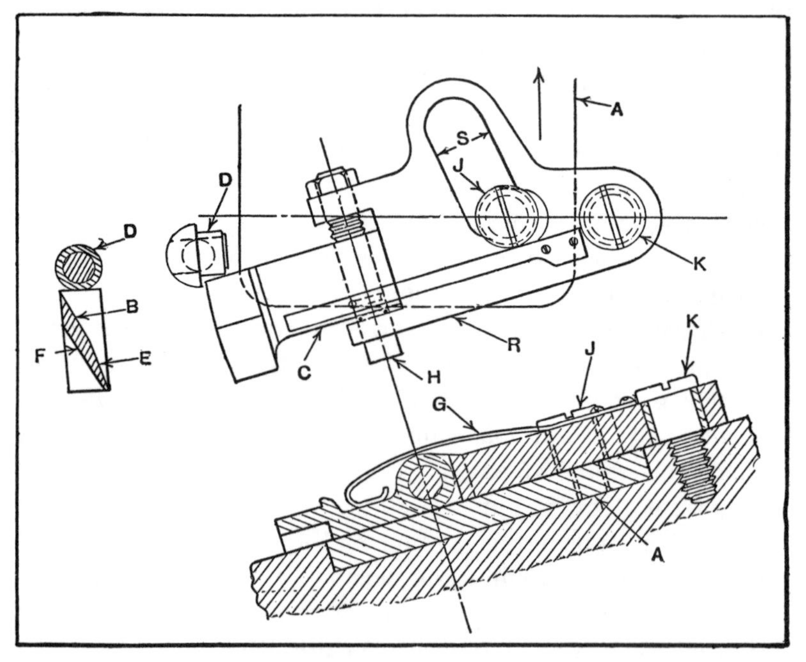

Fig. 14. Mechanism for Operating the Pin that Locks Table of Fixture During Machining Operation

cated on one side of the fixture, has a stroke of 1 5/8 inches. One and one-half inches of this stroke are utilized for indexing the fixture 60 degrees and the additional 1/8-inch movement is all that is necessary for withdrawing the stop-pin.

Attached to slide *A* is a pin *J*, which engages a slot *S* in lever *R*. The lever *R* is free to swing about the fixed pin

K, and carries an extension *C*, which is free to swing about pin *H*. The outer end of lever *C* has inclined surfaces *B*, *E*, and *F*, which come into contact with a roller *D* on the stop-pin when lever *C-R* is turned about pin *K*.

The plan view shows the mechanism in the position it occupies just before the stop-pin is withdrawn. When slide *A* moves in the direction of the arrow, pin *J*, acting against lever *R*, brings inclined surface *B* beneath roller *D*; 1/8-inch movement of slide *A* is sufficient to lift the stop-pin out of engagement. Before roller *D* has passed the end of surface *E*, the hole from which it was withdrawn has been moved from beneath the pin by the indexing movement of the fixture derived from slide *A* through a pawl (not shown), which engages one of six indexing pins. Soon after the end of surface *E* has passed roller *D*, the swinging movement of the hinged lever *C-R* discontinues, as slot *S* has moved around to a position parallel to the movement of slide *A*; consequently, slide *A* and pin *J* continue their movement without affecting the position of the slotted lever, and this additional movement of *A* is utilized to index the fixture 60 degrees.

During the return movement, slide *A* ejects the machined casting. When pin *J* engages the curved end of slot *S*, the lever *C-R* is forced back to the position shown, and the inclined surface *F* engages the opposite side of roller *D*, forcing it into the next stop-pin hole as lever *C* swings upward about the pin *H* against the action of spring *G*. As soon as the end of lever *C* has passed roller *D*, the lever is forced downward by spring *G* to the position shown in the sectional views. When in this position, lever *C* rests against slide *A*, as shown by the lower sectional view. The fixture is now in position for machining the next casting, after which the cycle of operations just described is repeated. The pin *H*, about which lever *C* swings, is made adjustable to compensate for wear or for variations caused by regrinding the chasers and the edges of the facing tool.

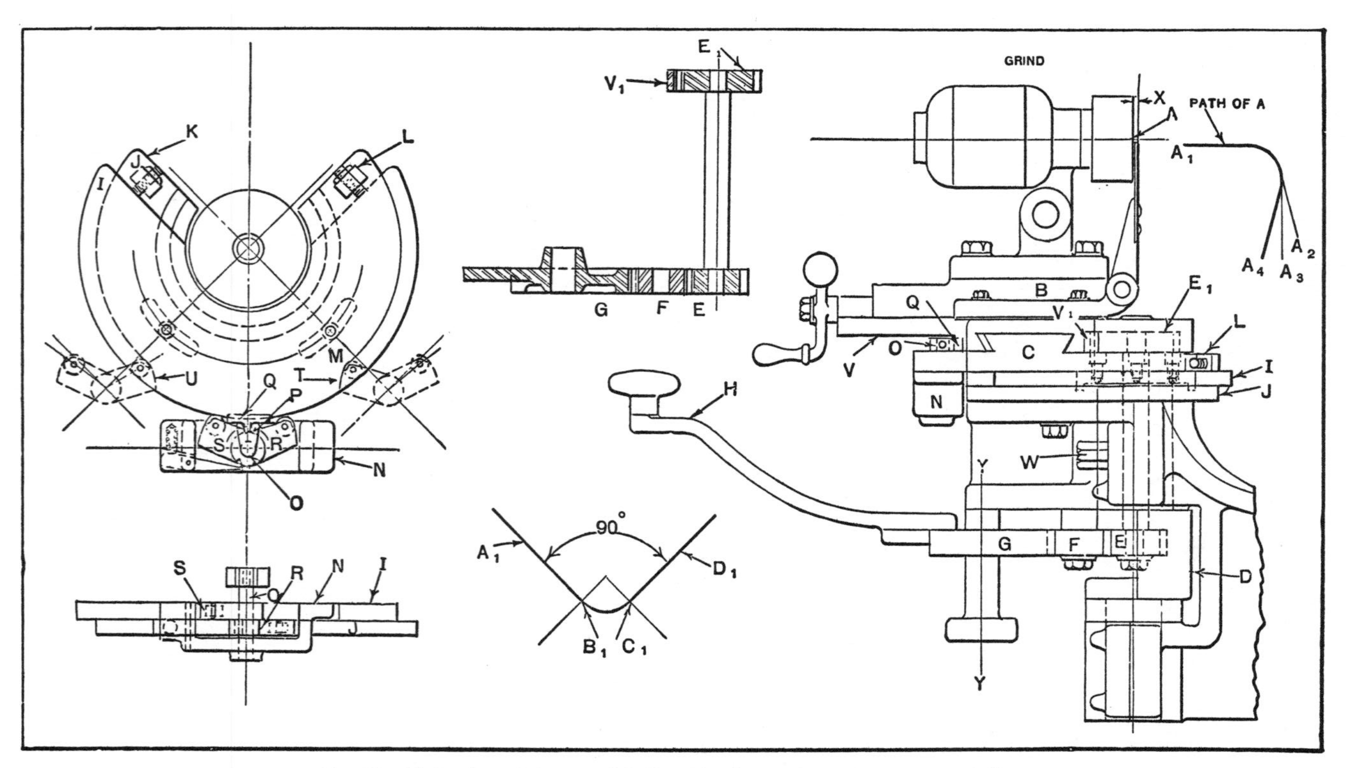

Fig. 15. Mechanism of Contour Grinding Machine Designed for Cutter Grinding

Mechanism of Contour Grinding Machine.—The mechanism shown in Fig. 15 controls the traversing movement of a grinding wheel in such a manner that hardened cutters can be quickly and accurately ground to contours such as are indicated by the three outlines (see upper right-hand corner). By simply moving the handle *H* from left to right, the grinding surface of the wheel at *A* will follow a path, such as is indicated by the line $A_1B_1C_1D_1$ (lower center diagram). In this case, the wheel grinds the straight surface from A_1 to B_1, and without pausing forms the radius from B_1 to C_1, continuing to the point D_1, the side C_1D_1 being ground at right angles to side A_1B_1 and the corner formed in one continuous pass.

By reversing the movement of handle *H*, the wheel is caused to travel back along the same path to the starting point A_1. Provision is made for setting the machine to grind the sides A_1B_1 and C_1D_1 to any included angle from 70 to 110 degrees instead of 90 degrees as shown, and with a radius of curvature at the corner of from 0 to 1 1/2 inches.

During the movement of the grinding wheel from A_1 to B_1 the handle *H* revolves about the center line *Y-Y*, transmitting the required longitudinal movement to carriage *V* which slides on *C*. The gear train that transmits motion from gear *G* to the rack V_1 secured to carriage *V* is shown in section. It will be noted that handle *H* is secured to gear *G*, which is actually a segment gear.

When the grinding surface of the wheel reaches point B_1, the revolving movement of handle *H* about axis *Y–Y* is automatically stopped and continued pressure on the handle causes the carriage *V* and slide *C* to revolve about the vertical axis of shaft *D* until the grinding surface of the wheel reaches the point C_1, when further movement about the axis of shaft *D* is stopped, and lever *H* again revolves about axis *Y–Y*, causing the wheel to travel from C_1 to D_1.

The grinding wheel is directly connected to the motor, which is mounted on the cross-slide *B* on carriage *V*. The distance *X* from the center line of shaft *D* to the grinding face of the wheel determines the radius of curvature of the corner. Carriage *V* is fitted to the slide *C*, which is secured to the top surface of shaft *D*. Shaft *D* thus supports the slide *C*, carriage *V*, cross-slide *B*, and the grinding wheel and its driving motor. These parts all swing together about the axis of shaft *D* during the corner-forming portion of the traversing movement.

Underneath slide *C* and fastened to the supporting frame or member are two plates *I* and *J*, also shown in the plan view (see left-hand illustration). These plates have stops *K* and *L* and cut-out portions on their peripheries which terminate in the cam surfaces at *T* and *U*. The spacing between *T* and *L* on plate *I* is definitely fixed, as is also the spacing or relationship of *K* and *U* on plate *J*. These plates are adjustable around the center of shaft *D*, and may be clamped in any desired position. The angle between *T* and *U* determines the angle of rotation of shaft *D* and slide *C* in forming the corner and, consequently, the included angle between the sides of the piece ground.

Midway of the length of slide *C* and in line with the center of the shaft *D* is a vertical shaft *O*, supported by bearings on slides *C* and *N*. On the upper end of shaft *O* is a two-tooth segment of a gear which meshes with a single rack tooth *Q*, attached to the mid point of the front face of carriage *V*. Also fixed to shaft *O* are two arms *S* and *R*, provided with rollers which make contact with the peripheries of plates *J* and *I*. The arms are offset vertically, so that *S* is in line with plate *I*, and *R* in line with plate *J*. The plan view shows the relative positions of the parts at the middle point of the corner-forming portion of the traverse movement. At this stage of the traverse movement of the grinding wheel, the shaft *D* is free to rotate in its bearings, and a movement of handle *H* to the right causes slide *C* and all parts attached to it to swing to the right about the vertical axis of shaft *D*. With the tooth *Q* in mesh with the two teeth of segment P, as shown, and the roller in arm S in contact with the periphery of plate I, movement of carriage V along slide C is positively prevented.

Rotation of slide C will continue until arm S drops into the cut away portion T, at which point slide C makes contact with stop L and further rotation about the axis of shaft D is positively prevented. When the roller of arm S drops into T and the parts are locked by stop L against further rotation about the axis of shaft D, the tooth Q is released from contact with the locking teeth of segment P by the rotation of shaft O. This leaves the carriage V free to slide on member C, so that resumption of the turning movement of handle H about axis Y–Y will complete the grinding movement along line C_1D_1.

When the movement of handle H is reversed, carriage V will slide along C until tooth Q enters the space between the two teeth of the segment P on shaft O. The continued motion of V and, consequently, the rotation of shaft O, causes the arms S and R to rotate until the roller in S is free of the opening T and the roller in arm R is in contact with the periphery of plate J.

When R makes contact with J, further motion of V on C is prevented, but rotation of C about D is permitted until R enters U and C makes contact with stop K, thus completing the corner-forming movement from C_1 to B_1. The carriage V can now slide on C and the revolving movement of handle H to the left about axis Y–Y completes the movement of the wheel from B_1 to A_1.

When the rotating movement required in forming the corner to a radius is taking place, all parts of the mechanism are locked against relative movement with each other, and the motion of handle H acting on gear E through gears G and F causes the complete assembly to revolve about the axis of shaft D. The combination of movements is caused by the continuous movement of H. There is no hesitation in the movement at the corners when rotation starts or

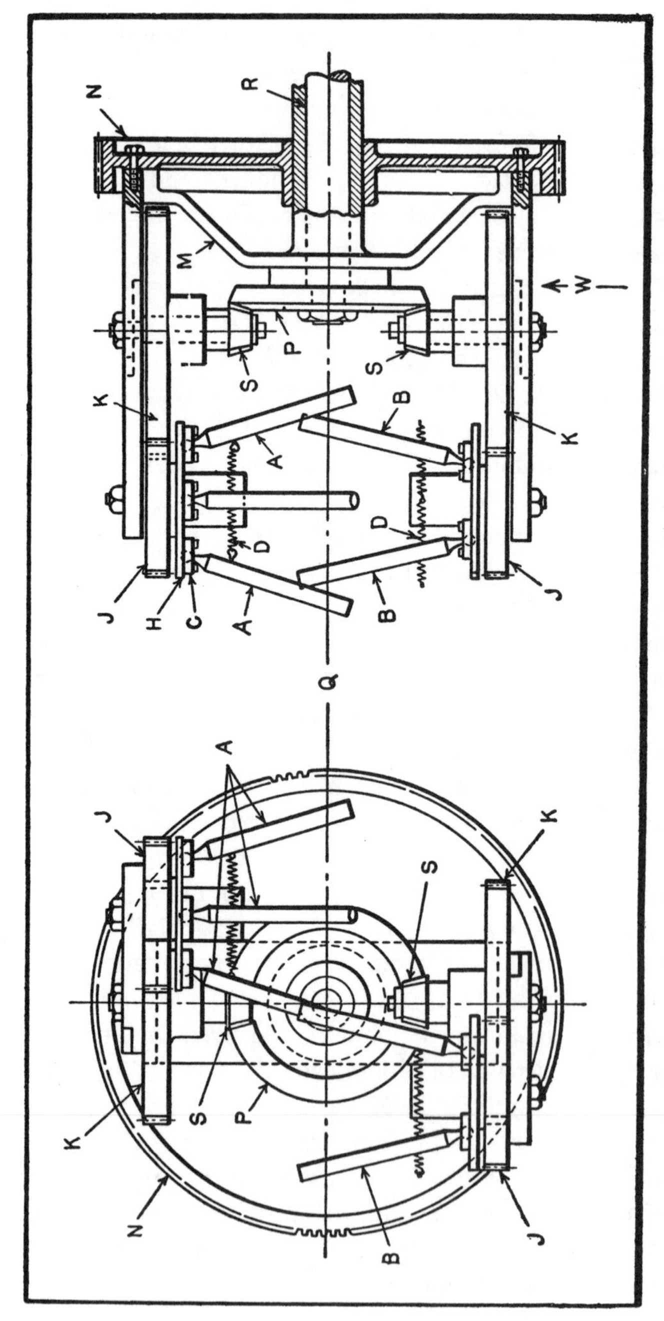

Fig. 16. Mechanism for Rotating Groups of Rods Arranged to Smooth Wrapper over Tapered Ends of Cigars

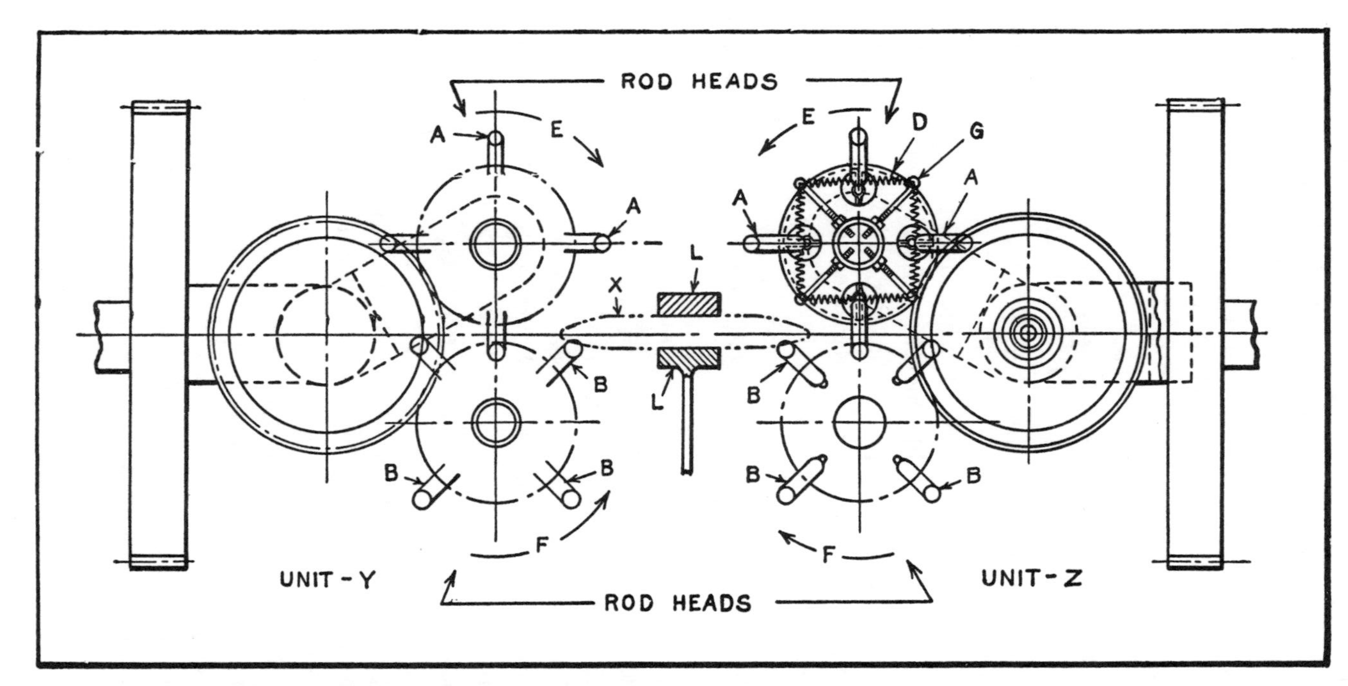

Fig. 17. Diagram Illustrating Operation of Wrapper-smoothing Mechanism Consisting of Two Units Like the One Shown in Fig. 16

stops, and no lost motion in the locking and unlocking action of the mechanism.

Mechanism for Smoothing Foil Wrapper Over Ends of Tapered Package.—Rods having a compound rotary motion are employed to smooth the creases of foil-wrapped cigars or other objects having tapering ends. Four round rods are arranged in a group at *A*, Fig. 16. Four rods are also arranged in a group at *B*. Each rod has a ball-shaped end mounted in a two-piece socket *C* in which the rod is free to pivot.

A group of springs *D*, of which eight are used in each unit, tie the rods flexibly together. In the center of Fig. 17 is indicated a tubular part *X* having an irregular shape with tapering ends, such as a foil-wrapped cigar. The cigar has been encased in its wrapper cylindrically and is to have the foil smoothed out and creased taperingly at the ends so that it folds smoothly over the irregularities and depressions of the rough-shaped ends.

To accomplish this, two entire mechanisms or units such as the one illustrated in Fig. 16 are employed, as shown in Fig. 17. The left-hand unit is shown at *Y*, and the right-hand unit at *Z*. The upper rods *A* are revolved in the direction indicated by the arrow *E* while the lower rods *B* are revolved in the direction indicated by the arrow *F*, the direction of rotation being determined by the bevel pinion drive. As each rod comes in contact with the wrapping on the work *X*, it smooths and creases it. The springs yield sufficiently to allow the rods to ride smoothly over the wrapping.

Each rod head, of which four are shown in Fig. 17, has eight springs *D* attached to four retaining screws *G*, two springs being attached to opposite sides of each of the rods *A* and *B*. The rod sockets are mounted on plates *H*, Fig. 16. Plates *H* are attached to gears *J*; thus the revolving of each large gear *K* causes the smoothing rods *A* and *B* to revolve continuously and stroke the work lengthwise while the foil-wrapped part is held stationary in gripping fingers *L*.

In addition, the entire rod-actuating arrangement is mounted on a bracket *M*, which, in turn, is attached to a large gear *N*. Gear *N* is caused to revolve, thereby carrying the entire arrangement about the axis *Q*. Thus the rods not only work lengthwise along the wrapping but radially as well. These combined movements serve to draw the wrapping tight. The bevel gear *P* is revolved by shaft *R* which causes bevel pinions *S* to drive gears *K* that operate the bar-holding members. The rod-actuating units are comparatively small, and form one section of an entirely automatic machine.

Mechanism for Preventing Creep of Wire-Mesh Conveyor Belt.— Wire-mesh belts are used in a certain plant for feeding lacquered parts through drying ovens. Because of unequal stretching of the wire, however, the belt had a tendency to creep to one end of the driving roll and against the machine frame. This often resulted in damage to the belt. To overcome this difficulty, an idler roll was incorporated, as indicated at *G* in the diagram Fig. 18.

This roll is pivoted at *C*. End *D* is automatically swung either to the right or left, according to the direction of belt creep. For example, if the belt *E* creeps toward the rear of the driving roll *F*, idler *G* immediately swings toward the right, causing the belt to return to its normal path. If, on the other hand, the belt creeps toward the front, the idler will swing toward the left and return the belt as before to its normal path.

The automatic movement of this idler is produced by means of the ratchet mechanism shown in Fig. 19. This mechanism is indicated at *A* in Fig. 18, and operates a screw represented by the dot-and-dash line *B*. This screw engages a nut on the swinging bearing *D*. Base *A*, Fig. 19, is stationary, and on it is mounted ratchet wheel *B* and the double pawl *C*. The ratchet wheel is pinned to shaft *D*

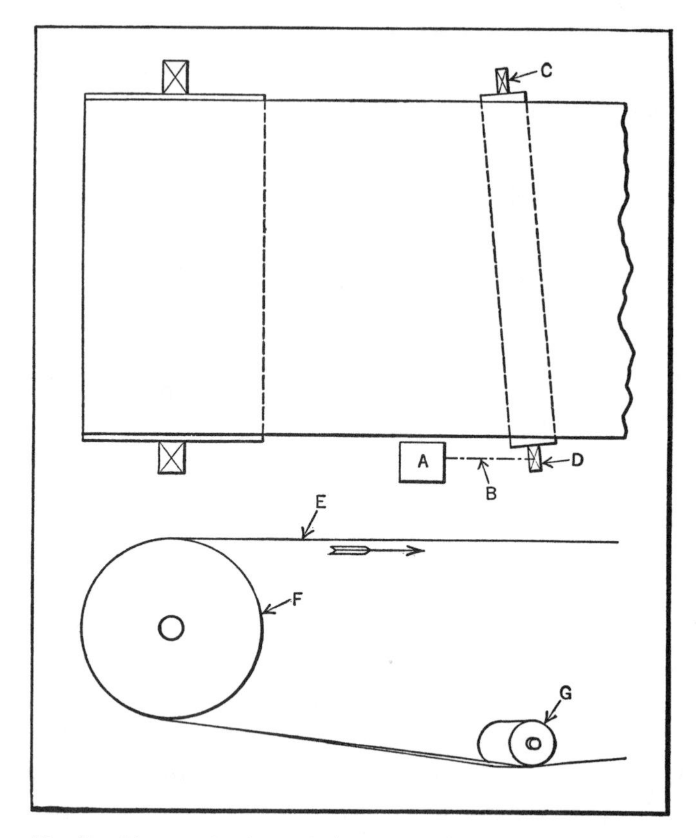

Fig. 18. Diagram Showing Swinging Idler Roll which is Actuated by Mechanism Shown in Fig. 19

and has closely spaced teeth cut on each side which are engaged by the pawl. Pawl *C* is pivoted on stud *F* in the vertical slide *G* and has an extension paddle *H* which is kept in contact with the edge of the belt by the counterweight *J*.

Slide *G* and pawl *C* are given a continuous vertical reciprocating movement by means of the pulley *K* through the crank *L* and the connecting-rod *M* which is pivoted at its upper end to the stud *F*. Pulley *K* receives its motion from another member of the conveyor (not shown). The screw engaging the nut on the swinging roll bearing is shown at N and is rotated by the ratchet wheel through the miter gears E.

With the pawl in the position shown, the belt is running in its central position on the driving roll; hence neither end of the pawl engages the ratchet wheel. However, if the belt were to creep, say, toward the left, the edge of the belt would force paddle H also toward the left, causing the pawl to swing on stud F. As a result, the hooked end of the pawl

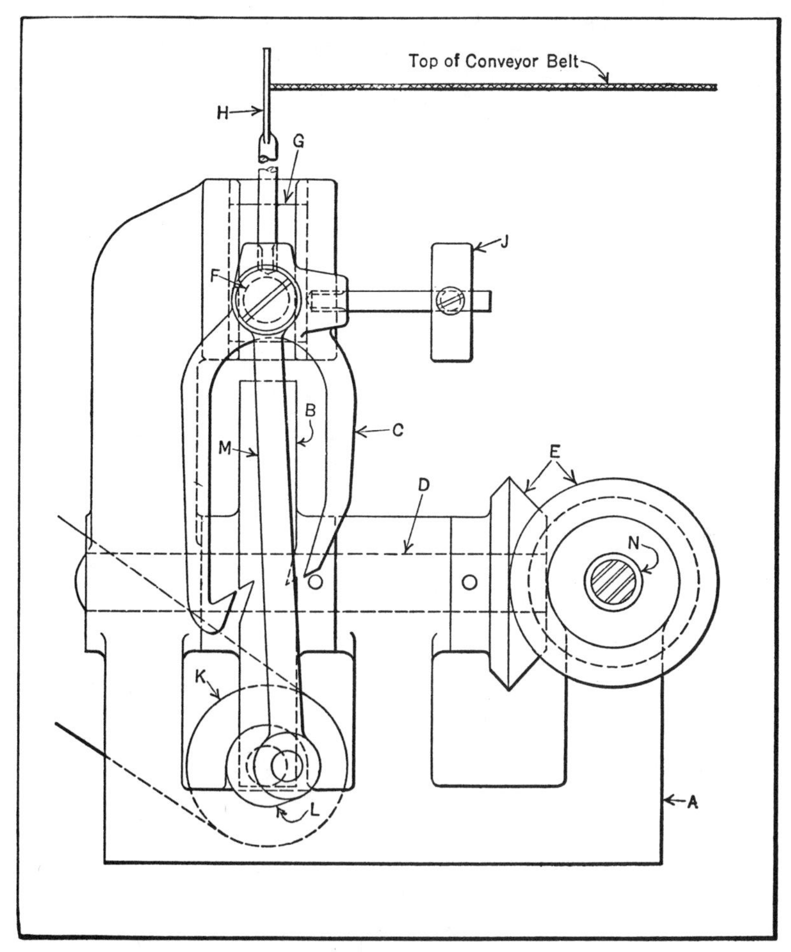

Fig. 19. Ratchet Mechanism for Automatically Swinging the Idler Roll into Position to Prevent Creep in a Wire-mesh Conveyor Belt

would engage the teeth in the ratchet wheel, rotating the ratchet wheel, miter gears, and screw *N*. As the screw rotates in the nut on the bearing roll, the roll is swung in the correct direction to guide the belt, through the angular position of the idler roll, back to the center of the driving roll. If the belt should creep toward the right, the movement of the paddle would cause the straight end of the pawl to engage the right-hand side of the ratchet wheel. As the teeth on this side of the wheel are cut opposite to those on the other side, the screw would rotate in the opposite direction and cause the idler roll, in turn, to swing in the opposite direction.

CHAPTER XVII

ENGINE VALVE DIAGRAMS AND THEIR APPLICATION IN STUDYING VALVE ACTION

The "valve-gear" or mechanism for operating the valve (or valves) which controls the admission and exhaust of steam in an engine cylinder is a comparatively simple mechanism especially if the engine has a single slide-valve; however, even the simplest form of valve-gear provides an interesting example of a mechanism requiring correct timing or relative action between certain main parts. This timing pertains to the action of the valve relative to the piston and steam ports, and it is secured first by proper design of the valve and its mechanism and finally by correct adjustment of the assembled parts. In connection with the design, so-called valve diagrams are used to determine in advance the action of a valve of given design. These valve diagrams and their application will be explained because of their relationship to the general subject of mechanism.

Diagrams which Show Valve Action.—When designing a slide-valve for a steam engine and the mechanism which operates the valve, it is desirable to be able to determine readily the position of the valve relative to the steam ports, for any given position of the crank or piston. Valve diagrams are commonly used for this purpose. These diagrams not only show graphically the relative positions of the valve and crank, but make it possible to design a valve with reference to a predetermined form of indicator card. Valve diagrams also indicate the effects of changes in the design of the valve on the steam distribution. In connection with steam engine work, certain problems or quantities relating to the point of cut-off, lead, etc., are assumed, and the remaining ones are required and may be determined by means of the valve diagrams. For instance, a designer might be given the point of cut-off, point of release, the lead, and the maximum port openings, the problem being to determine the valve travel, the outside and inside lap, and the angle of advance. By means of a suitable diagram, the valve travel, lap, etc., corresponding to these specified quantities may be readily determined. There are several different forms of valve diagrams, the *Zeuner* and the *Bilgram* diagrams being the ones most commonly used. The

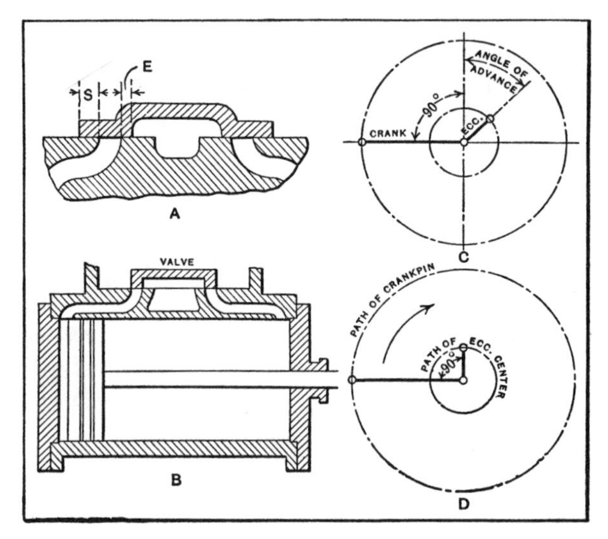

Fig. 1. Diagram Showing Slide Valves with and without Lap and Positions of Eccentric Relative to Crank

methods of laying out these diagrams and using them, in connection with steam engine work, will be described after considering some fundamental features of slide-valve design, so that the practical application of the diagrams may be more readily understood.

Position of Valve Relative to Ports.—A plain "D" slide-valve is represented at *A* in Fig. 1 on its seat and in mid-position. The *steam lap* or *outside lap S* is the amount by which the valve extends over the port on the admission side of the valve, when in mid-position on its seat. Similarly,

E is the *exhaust lap,* or *inside lap,* and is the amount by which the valve extends over the port on the exhaust side, when the valve is in mid-position on its seat. The necessity for having lap on a valve is shown by considering the lapless valve at *B*. Any movement of the valve to the right will admit steam behind the piston, and the other side of the piston will be open to the exhaust. The admission of steam on one side and the opening of the exhaust on the other will continue until the valve returns to its mid-position, which will occur when the piston is at the other end of its stroke. Such a valve arrangement as this would permit of no expansion of the steam in the cylinder; in other words, cut-off occurs at full stroke. This is an uneconomical type of valve.

For the lapless valve, the relative positions of the crank and eccentric are shown at *D*. It is necessary for the eccentric to be 90 degrees from the crank, in order that the valve shall be in mid-position when the piston is at the end of its stroke. Now, if there were any steam lap, it would be necessary to move the valve on its seat a distance to the right at least equal to the steam lap, in order that the port be open as soon as the piston starts on its stroke. By increasing the angle between the crank and the eccentric (for a direct motion), the valve can be moved along on its seat this amount. This angle is known as the *angle of advance,* and is indicated at *C*.

The Zeuner Valve Diagram.— The Zeuner valve diagram is illustrated in Fig. 2. As will be seen, there are two circles that are concentric and two smaller circles within the inner concentric circle. The larger circle need not be drawn to any particular scale, as it merely serves to represent the path of the crankpin. The smaller circle is either drawn to scale or full size, and has a diameter corresponding to the valve travel or twice the eccentricity. This is known as the *valve circle.* The two circles within the valve circle have a diameter equal to the radius of the valve circle, and are known as the *Zeuner circles.* The angle between the vertical center-line and the center-line upon which the two Zeuner circles are drawn is equal to the *angle of advance.* The cylinder is supposed to be on the left-hand side of the diagram, and the crank is turning in the direction indicated by the arrow.

For a head-end diagram or one for the head end of the valve and cylinder, the upper Zeuner circle is used for the admission or outward stroke, and the lower circle for the

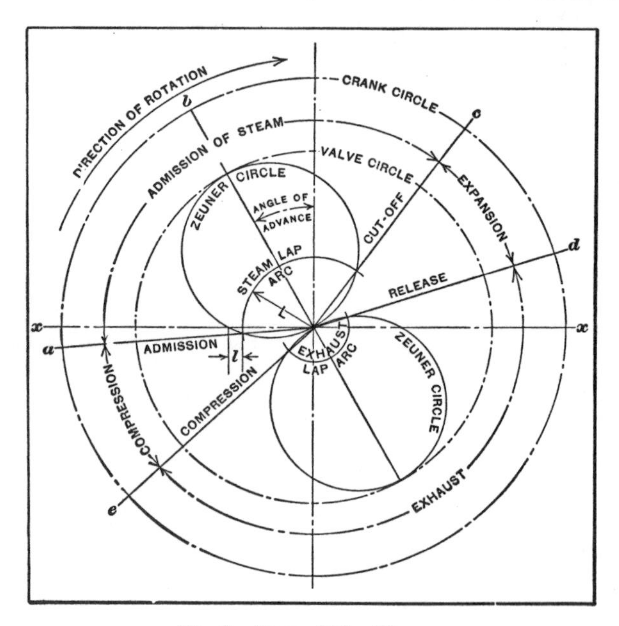

Fig. 2. Zeuner Valve Diagram

exhaust or return stroke. Fig. 2 is an example of a head-end diagram. With a center at the center of the valve circle, an arc with a radius equal to the steam lap is drawn, cutting the upper Zeuner circle. An arc with a radius equal to the exhaust lap is drawn on the lower circle in the same manner. These are called the *steam-lap* and the *exhaust-lap arcs.* The relation between the center-line of the crank for any given position, and these lap arcs and Zeuner circles, indicates the position of the valve. When the crank center-line crosses the steam-lap arc and the Zeuner circle, the port opening equals the distance between the steam-lap arc and the Zeuner circle measured along the crank center-line. For instance, when the crank center-line is at a, the valve is off center a distance L equal to the steam lap, and the port is about to be opened. When the crank has moved

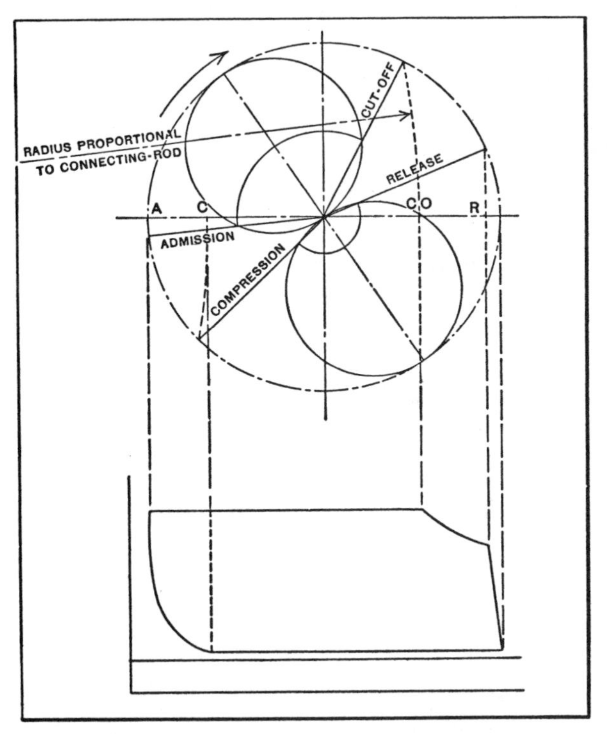

Fig. 3. Relation Between Zeuner Diagram and Indicator Card. Piston Positions are Marked A, for Admission; CO, for Cut-off; R, for Release; and C, for Compression

up to the horizontal center-line xx, representing the dead-center position, the port is opened an amount l equal to the lead, since l is the distance between the steam-lap arc and the Zeuner circle measured along the crank center-line. When the crank is at b, the maximum port opening occurs; at c, the steam is cut off; and at d, the exhaust port opens as shown by the intersection of the crank center-line with the lower Zeuner circle at the exhaust-lap arc. The exhaust port closes at *e* and compression begins. This cycle is then repeated.

From the foregoing, it will be seen that, as the crank center-line is rotated, it will intersect the valve circle, the two Zeuner circles, and the lap arcs. As the distance from the center of the valve circle to any point of the Zeuner circle, when measured along the crank center-line, shows the amount by which the valve has moved from its mid-position, the distance between the lap arc and the Zeuner circle shows the amount by which the port is uncovered.

Relation of Diagram to Indicator Card.—Since the valve diagram shows the relative positions of the crank and the valve for the various events of the steam engine cycle, it is possible to draw the valve diagram from the indicator card, or vice versa. Fig. 3 shows the relation between the two diagrams. The positions of the crank for admission, cut-off, release, and compression are indicated by the radial lines. To find the relative positions of the piston for these events, an arc is swung from the intersection of the crank center-line with the valve circle, to the horizontal center-line. The radius of this arc should be in the same proportion to the radius of the valve circle that the length of the connecting-rod of the engine is to the crank; that is, if the connecting-rod is five times as long as the crank, then the radius of the arc should be five times the radius of the valve circle. The center of the arc should be on the horizontal center-line extended to the left. The point which has been determined by this arc shows the relative distance which the piston has moved from the end of its stroke, for that particular position of the crank. These points can be projected down upon a diagram below to give the four principal points of the indicator card, the vertical distances on this card being determined by the relative steam pressures which exist at admission and at exhaust. The compression curve and the expansion line can be drawn as equilateral hyperbolas. Starting with the indicator diagram, the four piston positions on the valve circle diameter, where admission, cut-off, release, and compression occur, are found by projecting upward from these points on an indicator card to the center-line of the valve circle. Then, by swinging arcs from these points to the valve circle, the crank positions are determined, and the valve diagram is laid out accordingly.

Application of the Zeuner Diagram.— A construction which is often necessary in working the Zeuner diagram is shown at *A* in Fig. 4. A small circle is drawn around the intersection of the valve circle and the horizontal center-line. The radius of the circle is equal to the lead. A tangent *ab* to this circle, perpendicular to the center-line *oc* of the Zeuner circle, will cut the valve circle at the crank positions of admission and cut-off *oa* and *ob*. If the point of cut-off and the lead are known, as is generally the case when a steam engine is designed, this lead circle can be drawn, the cut-off point determined, and a line *ab* from the cut-off point be drawn tangent to the lead circle. A perpendicular to this line *oc*, through the center of the valve circle, gives the center-line of the Zeuner circles and determines the angle of advance. The lap arc is at once determined, as it will always cut the Zeuner circle at the intersection of the crank positions *d* and *e*.

The diagram for the head end of the cylinder has been referred to in the foregoing. The crank-end diagram is similarly constructed and used. As shown at *B*, Fig. 4, the Zeuner circle for determining the admission of steam and the cut-off is on the *lower* side of the horizontal center-line, and the circle for determining the release and compression is on the *upper* side; in other words, the positions of the two circles are reversed from the head-end positions. Lap arcs are drawn as before, but the lead circle is at the right end rather than the left. The positions of the piston, cor-

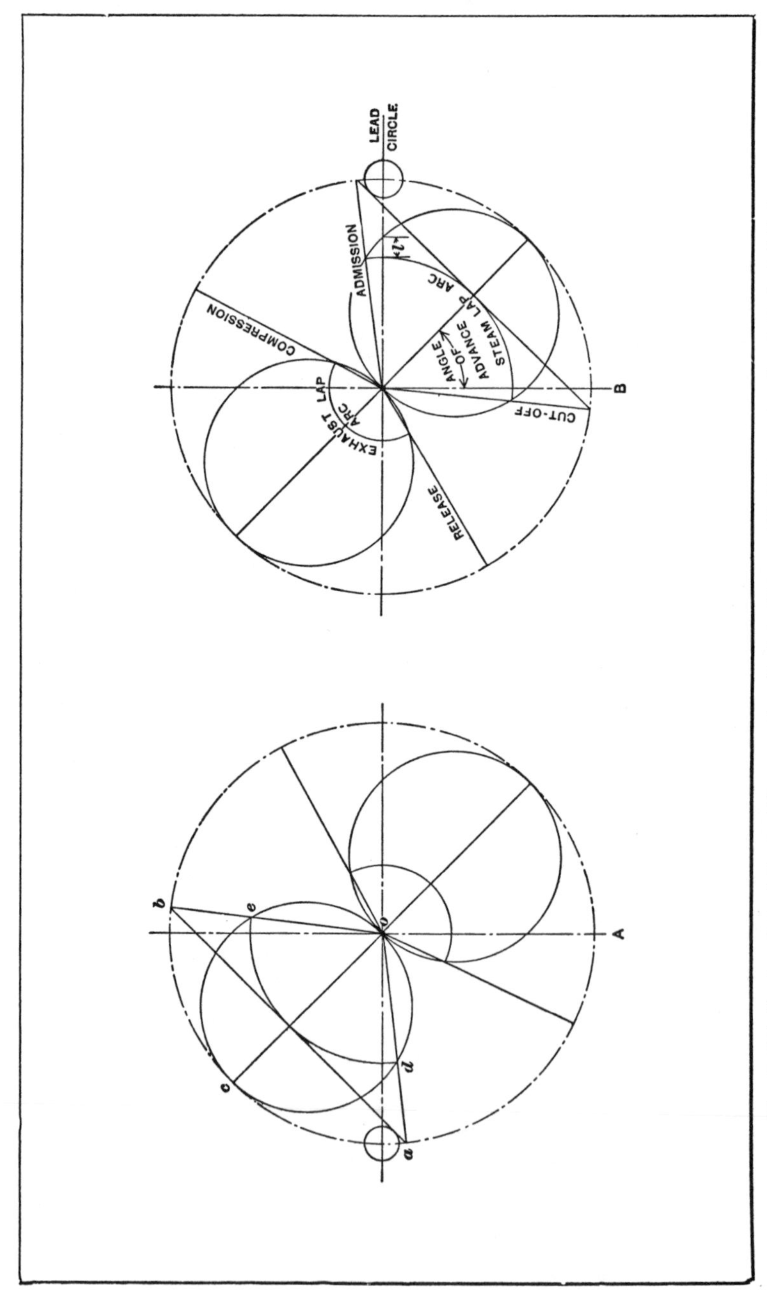

Fig. 4. (A) Construction of Zeuner Diagram when Lead is Known. (B) Diagram for Crank End of Cylinder

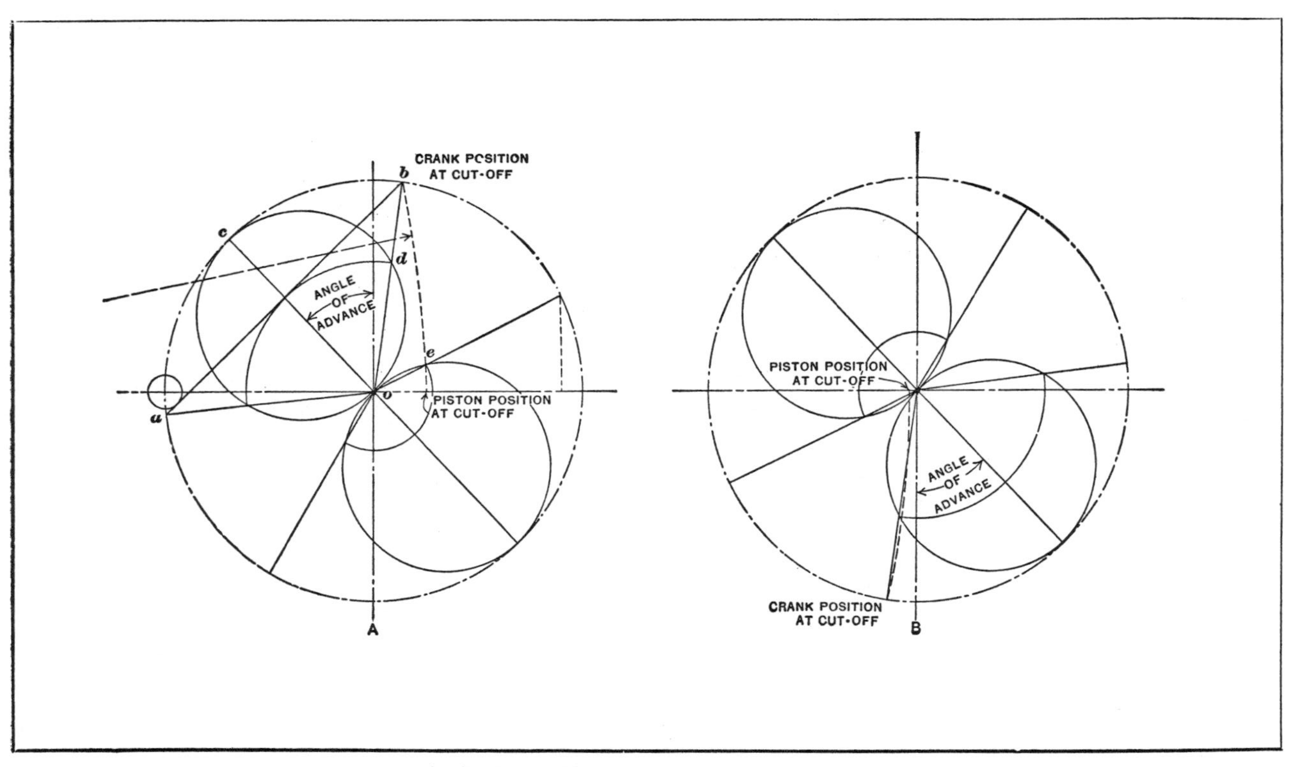

Fig. 5. Zeuner Diagrams Laid out from Given Data

responding to the crank positions at the four important points, are found by means of the arc proportional to the connecting-rod length, as before.

Construction of the Zeuner Diagram.—In order to illustrate the practical application of the Zeuner diagram, assume that a valve is to have a travel of 3 inches, a lead of 1/8 inch, and that the point of cut-off, under normal conditions, is to be at five-eighths stroke on the head end of the cylinder. The valve is to have an equal amount of outside lap on the two ends; hence, the cut-off will vary for the head and crank ends, owing to the angularity of the connecting-rod. As the result of this angularity, the amount of piston travel for the head- and crank-end strokes differs considerably for corresponding crank positions; consequently, a valve which has the same outside lap on both ends will give a shorter cut-off on the crank end than on the head end, and, in order to equalize the cut-off, it will be necessary to have a smaller lap on the crank end of the valve. If the amount of lap were varied, however, the lead would be unequal, and, as this is considered objectionable, it is common practice to make the lap equal and allow the variation in cut-off. Assume that the connecting-rod in this case is five times as long as the crank.

When constructing the Zeuner diagram, first draw a valve circle having a diameter equal to the valve travel, or 3 inches. Determine the piston position for five-eighths stroke, which represents the point of cut-off. (See Diagram *A*, Fig. 5.) Now, with a radius of 7 1/2 inches (or five times the radius of the valve circle) and with a center on the horizontal center-line of the valve circle extended, draw an arc upward to the valve circle, marking the position of the crank at the point of cut-off. Then, with a radius of 1/8 inch, draw the lead circle at the left of the diagram, as shown in the illustration. Draw a tangent *ab* to this circle which will terminate in the cut-off position. A perpendicular *oc* to this tangent and passing through the center of the valve circle will be the center-line of the two Zeuner circles. The angle between this perpendicular and the vertical is the angle of advance, and represents the angle between the eccentric and the crank. The steam-lap arc is drawn through the intersection *d* of the Zeuner circle with the crank position at cut-off, and will be found tangent to the line which was drawn tangent to the lead circle. The exhaust-lap arc is also drawn through the intersection *e* of the lower Zeuner circle with the crank position at release. The two laps may now be measured and the port openings for any position of the crank be determined. The maximum steam-port opening is measured along the center-line of the upper Zeuner circle between the valve circle and the lap arc.

To determine what will be the point of cut-off for the crank end of the cylinder, the process is as follows: Draw the valve circle as before, with a 3-inch diameter. Draw the lap arcs, reversing the positions as shown at *B*, Fig. 5, the steam-lap arc being in the lower right-hand quadrant and the exhaust-lap arc in the upper left-hand quadrant. The two Zeuner circles are next drawn, the center-line having the same angle of advance as in the diagram *A*. Now find the crank positions for the various events of the cycle by means of radial lines representing the different positions of the crank. Then, with a radius of 7 1/2 inches and with a center on the horizontal center-line extended to the left, as before, find the corresponding positions of the piston by swinging the arcs upward to the horizontal center-line.

Effect of Changing Eccentricity.—By means of the Zeuner diagram, it is possible to see clearly the effect of changing the eccentricity and the angle of advance in a slide-valve engine. If the eccentricity is increased without making other changes, the lead will be increased, cut-off occurs later, release sooner, and compression later. If the eccentricity is decreased, the lead is decreased, cut-off occurs earlier, the release later, and the compression earlier. Increasing the angle of advance (as the diagram, Fig. 2, shows) will cause the cut-off and all the other events to occur earlier, whereas decreasing it will have the opposite effect. These two methods of changing the operation of the valve are made use of in all types of shaft governors. A combination of the two, where the eccentricity is changed at the same time as the angle of advance, makes it possible to change the cut-off without changing the lead.

In the foregoing the piston has been assumed to be on the left of the diagram; if it were assumed to be on the right side, the Zeuner circles should be on the opposite sides of the vertical center-line and the other parts of the diagram similarly changed.

Summary of Principles of Zeuner Diagram.—When using the Zeuner diagram, it is well to have the following principles well in mind:

1. Any radial line on the valve circle represents a position of the crank.

2. The intercept on this radial line between the center and the Zeuner circle represents the movement of the valve from its mid-position.

3. The intercept on this radial line between the lap arc and the Zeuner circle represents the amount the port is open for that position of the crank.

4. The relative piston position can be determined, if the proportion of connecting-rod to crank is known.

5. The radial line representing the crank position for cut-off must pass through the intersection of the lap arc and the Zeuner circle. This also applies to release, compression, and admission.

6. A perpendicular to the crank position at the intersection with the lap arc and Zeuner circle will intersect the valve circle at a point coinciding with the center-line for the Zeuner circles; moreover, a perpendicular to any crank position at the intersection of the Zeuner circle will

intersect the valve circle at the intersection of the center-line for the Zeuner circles.

Bilgram Diagram.— The methods of laying out a Bilgram diagram for both the head and crank ends of a cylinder are shown at *A* and *B*, Fig. 6. The diameter ss_1 of the outer circle represents the stroke of the engine; this circle may be drawn to any convenient scale. A smaller circle is drawn from the center *a*, having a diameter equal to the travel of the valve. This valve-travel circle may also be drawn either to a reduced scale or to full size, if more convenient. Assuming that the inside and the outside lap and the lead are known, proceed as follows: Draw a line *ll* parallel to *ss*, and a distance above it equal to the required lead or amount of port opening at the beginning of the piston stroke. Next draw another circle from center *b* having a radius equal to the outside lap of the valve. As the diagram shows, this circle should have its center *b* on the valve-travel circle and should be tangent to line *ll*. About center *b*, draw a smaller circle having a radius equal to the inside lap of the valve. Now, to obtain the location of the center-line of the crank at the point of cut-off, draw a radial line *ac* tangent to the outside of the lap circle. A vertical line *cd* intersecting the stroke line *ss* shows approximately how far the piston travels before the steam is cut off. This vertical line will not locate the exact point of cut-off, owing to the angularity of the connecting-rod and the resulting variation between the movements of the crankpin and the piston. To obtain the exact position of the piston at the point of cut-off, line *cd* should be an arc having a radius proportional to the length of the connecting-rod and drawn from a center located at some point on an extension of center-line *ss*.

A line *ae* through center *b* indicates the angular position of the eccentric corresponding to a given amount of outside lap and lead, whereas, angle eas_1 equals the angle of advance. Line *af* tangent to the inside-lap circle (on the lower side for diagram *A*) locates the crank position when

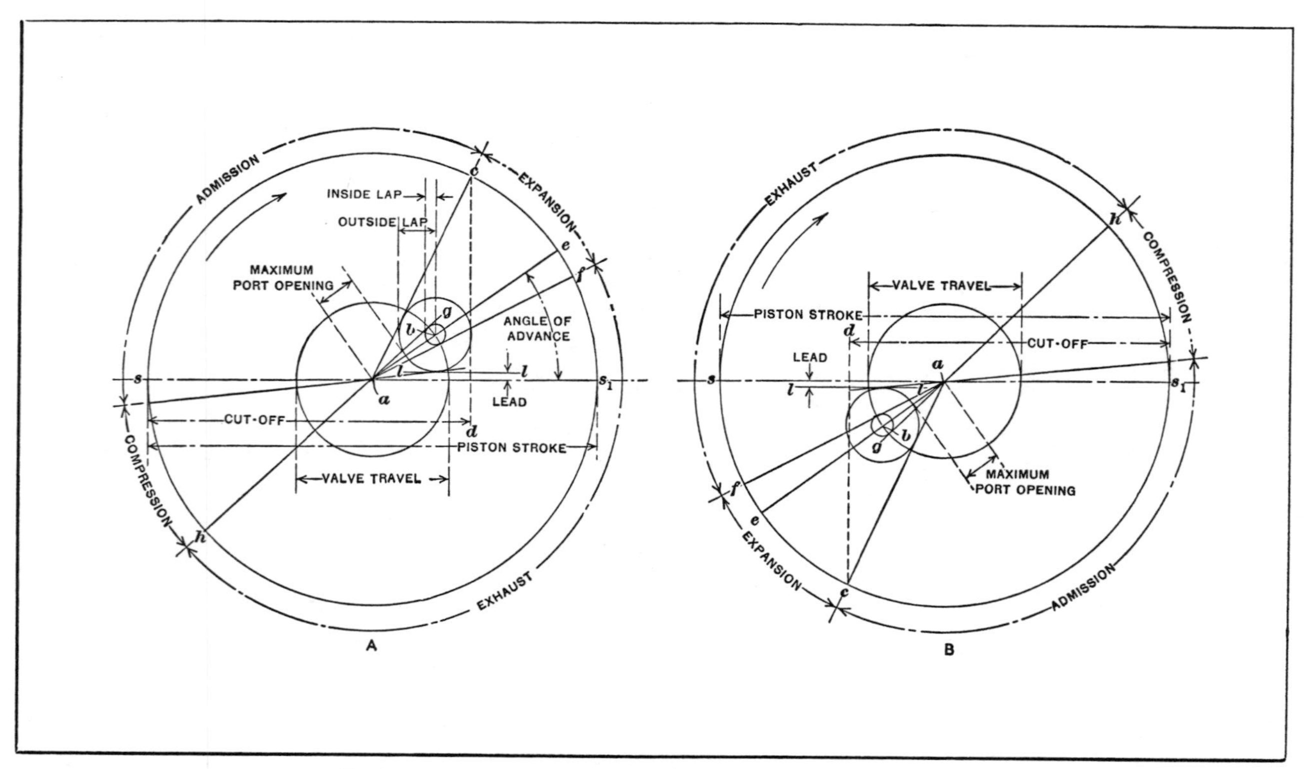

Fig. 6. Bilgram Valve Diagrams for Head and Crank Ends of Cylinder

the steam is released by the opening of the exhaust port, whereas line *gh*, tangent to the opposite side of the inside-lap circle and passing through center *a*, shows the position of the crank for the point of compression or the closing of the exhaust port. The maximum steam-port opening is represented by the distance from center *a* to the outside-lap circle, the measurement being taken along the center-line *ae*.

By studying these diagrams *A* and *B*, the effect of changes in the design of the valve may readily be determined. For instance, if the outside lap is increased (thus increasing the diameter of the lap circle), the point of cut-off will occur earlier in the stroke, giving a greater range of expansion; this change, however, will also cause an earlier compression, even if the inside lap is not increased, because the effect of enlarging the outside-lap circle is to change the position of the inside-lap circle and the angular position of line *gh*, so that the exhaust port is closed earlier. Increasing the inside lap also increases compression and delays the point of release or exhaust. By means of this diagram, if the point of cut-off, lead, maximum port opening, and point of compression were given, the necessary inside and outside lap and valve travel could be determined.

INDEX